Leukocyte Typing II

VOLUME 2

Human B Lymphocytes

Leukocyte Typing II

Volume 1 Human T Lymphocytes
Volume 2 Human B Lymphocytes
Volume 3 Human Myeloid and Hematopoietic Cells

Leukocyte Typing II

VOLUME 2
Human B Lymphocytes

Edited by
Ellis L. Reinherz   Barton F. Haynes
Lee M. Nadler   Irwin D. Bernstein

With 134 Illustrations

Springer-Verlag
New York Berlin Heidelberg Tokyo

Ellis L. Reinherz, M.D., Division of Tumor Immunology, Dana-Farber Cancer Institute, Boston, MA 02115 U.S.A.
Barton F. Haynes, M.D., Department of Medicine, Duke University School of Medicine, Durham, NC 27710 U.S.A.
Lee M. Nadler, M.D., Division of Tumor Immunology, Dana-Farber Cancer Institute, Boston, MA 02115 U.S.A.
Irwin D. Bernstein, M.D., Program in Pediatric Oncology, Fred Hutchinson Cancer Research Center, Seattle, WA 98104 U.S.A.

Library of Congress Cataloging in Publication Data
Main entry under title:
Leukocyte typing II.
Papers presented at the Second International Workshop on Human Leukocyte Differentiation Antigens, held in Boston, Sept. 17–20, 1984.
Includes bibliographies and indexes.
Contents: v. 1. Human T lymphocytes—v. 2. Human B lymphocytes—v. 3. Human myeloid and hematopoietic cells.
1. Leucocytes—Classification—Congresses.
2. Histocompatibility testing—Congresses. 3. Tissue specific antigens—Analysis—Congresses. I. Reinherz, Ellis L. II. International Workshop on Human Leukocyte Differentiation Antigens (2nd : 1984 : Boston, Mass.) III. Title: Leukocyte typing 2. IV. Title: Leukocyte typing two.
QR185.8.L48L48 1985 616.07'9 85-22229

Typeset by Bi-Comp Inc., York, Pennsylvania.

9 8 7 6 5 4 3 2 1

ISBN-13: 978-1-4612-9328-6 e-ISBN-13: 978-1-4612-4848-4
DOI:10.1007/978-1-4612-4848-4

Preface

The Second International Workshop on Human Leukocyte Differentiation Antigens was held in Boston, September 17–20, 1984. More than 350 people interested in leukocyte differentiation agreed to exchange reagents and participate in this joint venture. All in all, in excess of 400 antibodies directed against surface structures on T lymphocytes, B lymphocytes, and myeloid-hematopoietic stem cells were characterized. Because of the enormous quantity of serologic, biochemical, and functional data, *Leukocyte Typing II* has been divided into three volumes.

These books represent the written results of workshop participants. They should be helpful to both researchers and clinicians involved in scientific endeavors dealing with these broad fields of immunobiology. To those who delve into the various sections of the volumes, it will become evident that the work speaks for itself.

I am deeply indebted to the section editors, Barton F. Haynes, *Volume 1, Human T Lymphocytes,* Lee M. Nadler, *Volume 2, Human B Lymphocytes,* and Irwin D. Bernstein, *Volume 3, Human Myeloid and Hematopoietic Cells* for their major contributions in planning, executing, and summarizing the workshop, as well as council members John Hansen, Alain Bernard, Laurence Boumsell, Walter Knapp, Andrew McMichael, Cesar Milstein, and Stuart F. Schlossman. I would also like to thank the National Institutes of Health, World Health Organization, and International Union of Immunological Societies for making this meeting possible. Needless to say, I am most grateful to all of my colleagues who contributed to this effort and helped to accelerate the characterization of human immunobiology through their endeavors.

Ellis L. Reinherz, M.D.

Contents

Contributors

Kenneth C. Anderson Division of Tumor Immunology, Dana-Farber Cancer Institute, Department of Medicine, Harvard Medical School, Boston, Massachusetts 02115, U.S.A.

Ignacio Anegón Servicio de Inmunología, Hospital Clínic i Provincial, 08036 Barcelona, Spain

Gamil R. Antoun Department of Pediatrics and the Tom Baker Cancer Centre, University of Calgary, Calgary, Alberta, Canada T2N 4N1

Robert C. Atkins Nephrology Department, Prince Henry's Hospital, Melbourne, Australia

Emilio Berti First Department of Dermatology, University of Milan, Milan, Italy

Peter Beverley ICRF Human Tumor Immunology Group; School of Medicine, University College London, London WC1E 6JJ, U.K.

Luis Borche Servicio de Immunología, Hospital Clínic i Provincial, 08036 Barcelona, Spain

D. Bourel Groupe Universitaire de Recherche en Immunologie Fondamentale et Appliquée, Service de Médecine Interne, Département d'Hematologie, C.H.U., Rennes, France

Andrew W. Boyd Division of Tumor Immunology, Dana-Farber Cancer Institute, Department of Medicine, Harvard Medical School, Boston, Massachusetts 02115 U.S.A.

J. Garrett Bradley Department of Laboratory Medicine/Pathology, University of Minnesota Medical School, Minneapolis, Minnesota 55455, U.S.A.

K.M. Britten University Department of Pathology, General Hospital Southampton SO9 4XY, U.K.

Jo Ellen Brown University of Minnesota, Minneapolis, Minnesota 55455, U.S.A.

Pierre Carayon Immunotoxin Project, Research Center Clin-Midy, Sanofi Group, Montpellier, France

Giorgio Cattoretti National Cancer Institute of Milan, Milan, Italy

Edward A. Clark Department of Microbiology and Immunology, University of Washington, Seattle, Washington 98195, U.S.A.

Loran T. Clement Cellular Immunobiology Unit, Tumor Institute, University of Alabama in Birmingham, Birmingham, Alabama 35294, U.S.A.

B.B. Cohen Medical Research Council, Clinical and Population Cytogenetics Unit, Western General Hospital, Edinburgh, U.K.

Max D. Cooper Cellular Immunobiology Unit, Tumor Institute, University of Alabama in Birmingham, Birmingham, Alabama 35294, U.S.A.

Jeffrey Cossman Laboratory of Pathology, National Cancer Institute, National Institutes of Health, Bethesda, Maryland 20205, U.S.A.

Marilyn J. Crain Cellular Immunobiology Unit, Tumor Institute, University of Alabama in Birmingham, Birmingham, Alabama 35294, U.S.A.

Marco Cusini First Department of Dermatology, University of Milan, Milan, Italy

Cristina Cuturi Servicio de Inmunología, Hospital Clínic i Provincial, 08036 Barcelona, Spain

John F. Daley Division of Tumor Immunology, Dana-Farber Cancer Institute, Department of Medicine, Harvard Medical School, Boston, Massachusetts 02115, U.S.A.

Filippo de Braud National Cancer Institute of Milan, Milan, Italy

Domenico Delia National Cancer Institute of Milan, Milan, Italy

Bernd Dörken Medizinische Universitäts-Poliklinik, D-6900 Heidelberg, F.R.G.

Allen C. Eaves Terry Fox Laboratory, B.C. Cancer Research Centre, Vancouver, British Columbia, Canada

David Einfeld Immunobiology Group, Genetic Systems Corporation, Seattle, Washington 98121, U.S.A.

Patricia Elder Medical Research Council Clinical and Population Cytogenetics Unit, Western General Hospital, Edinburgh, U.K.

Ilse Marie Fastrup The University Department of Medicine and Hematology, Aarhus Amtssygehus, Denmark DK-8000

R. Fauchet Groupe Universitaire de Recherche en Immunologie Fondamentale et Appliquée, Service de Médecine Interne, Département d'Hématologie, C.H.U., Rennes, France

Douglas T. Fearon Department of Medicine, Harvard Medical School, Department of Rheumatology & Immunology, Brigham and Women's Hospital, Boston, Massachusetts 02115, U.S.A.

A. Feller Pathologisches Institute der Universität Kiel, Kiel, F.R.G.

David C. Fisher Division of Tumor Immunology, Dana-Farber Cancer Institute, Department of Medicine, Harvard Medical School, Boston, Massachusetts 02115, U.S.A.

Arnold S. Freedman Division of Tumor Immunology, Dana-Farber Cancer institute, Department of Medicine, Harvard Medical School, Boston, Massachusetts 02115, U.S.A.

Shu Man Fu Cancer Research Program, Oklahoma Medical Research Foundation, Oklahoma City, Oklahoma 73104, U.S.A.

Teresa Gallart Servicio de Inmunología, Hospital Clínic i Provincial, 08036 Barcelona, Spain

O. Margaret Garson Immunogenetics Research Unit, Cancer Institute, Melbourne, Australia 3000

K.C. Gatter Department of Haematology, John Radcliffe Hospital, Oxford OX3 9DU, U.K.

A. Gatzke Abt. Hämatologie, Klinikum Steglitz, Freie Universität Berlin, Berlin, F.R.G.

B. Genetet Groupe Universitaire de Recherche en Immunologie Fondamentale et Appliquee, Service de Medecine Interne, Departement d'Hematologie, C.H.U., Rennes, France

N. Genetet Groupe Universitaire de Recherche en Immunologie Fondamentale et Appliquée, Service de Médecine Interne, Département d'Hématologie, C.H.U., Rennes, France

Peter S. Giddy Nephrology Department, Prince Henry's Hospital, Melbourne, Australia

Josée Golay ICRF Human Tumor Immunology Group, School of Medicine, University College London, London WC1E 6JJ, U.K.

Shraga F. Goldmann Red Cross Blood Bank Ulm, Department of Transfusion Medicine, University of Ulm, D-7900 Ulm, F.R.G.

Walter Goldschmidts Pediatric Branch, Clinical Oncology Program, Division of Cancer Treatment, National Cancer Institute, National Institutes of Health, Bethesda, Maryland 20205, U.S.A.

B. Grosbois Groupe Universitaire de Recherche en Immunologie Fondamentale et Appliquée, Service de Médecine Interne, Département d'Hématologie, C.H.U., Rennes, France

Keith Guy Medical Research Council Clinical and Population Cytogenetics Unit, Western General Hospital, Edinburgh, U.K.

Günter J. Hämmerling Institut für Immunologie und Genetik, Deutsches Krebsforschungszentrum, D-6900 Heidelberg, F.R.G.

Kathleen R. Hagert University of Minnesota, Minneapolis, Minnesota 55455, U.S.A.

Wayne W. Hancock Nephrology Department, Prince Henry's Hospital, Melbourne, Australia

Toshiro Hara Cancer Research Program, Oklahoma Medical Research Foundation, Oklahoma City, Oklahoma 73104, U.S.A.

D.L. Hardie Department of Immunology, The Medical School, University of Birmingham, Birmingham B15 2TJ, U.K.

F. Herrmann Abt. Haematologie, Klinikum Steglitz, Freie Universitaet Berlin, Berlin, F.R.G.

R.D. Hockett, Jr. University of Minnesota, Minneapolis, Minnesota 55455, U.S.A.

Peter Hokland The University Department of Medicine and Hematology, Aarhus Amtssygehus, Denmark DK-8000

Keizo Horibe Human Immunogenetics Laboratory, Memorial Sloan-Kettering Cancer Center, New York, New York 10021, U.S.A.

J.C. Horowitz Division of Tumor Immunology, Dana-Farber Cancer Institute, Department of Medicine, Harvard Medical School, Boston, Massachusetts 02115, U.S.A.

Donald R. Howard Terry Fox Laboratory, B.C. Cancer Research Centre, Vancouver, British Columbia, Canada

W. Hunstein Medizinische Universitäts-Poliklinik, D-6900 Heidelberg, F.R.G.

Yoshifumi Ishii Department of Pathology, Sapporo Medical College, S1 W17 Sapporo 060, Japan

G.D. Johnson Department of Immunology, University of Birmingham, Medical School, Birmingham B15 2TJ, U.K.

D.B. Jones University Department of Pathology, General Hospital, Southampton SO9 4XY, U.K.

David G. Jose Immunogenetics Research Unit, Cancer Institute, Melbourne, Australia 3000

Lawrence K.L. Jung Cancer Research Program, Oklahoma Medical Research Foundation, Oklahoma City, Oklahoma 73104, U.S.A.

John H. Kersey University of Minnesota, Minneapolis, Minnesota 55455, U.S.A.

M. Khan Department of Immunology, University of Birmingham, Medical School, Birmingham B15 2TJ, United Kingdom

S. Kiesel Medizinische Universitäts-Poliklinik, D-6900 Heidelberg, F.R.G.

Kokichi Kikuchi Department of Pathology, Sapporo Medical College, S1 W17 Sapporo 060, Japan

James Kiwanuka Pediatric Branch, Clinical Oncology Program, Division of Cancer Treatment, National Cancer Institute, National Institutes of Health, Bethesda, Maryland 20205, U.S.A.

Jeroen Knops Institute für Immunologie und Genetik, Deutsches Krebsforschungszentrum, D-6900 Heidelberg, F.R.G.

Robert W. Knowles Human Immunogenetics Laboratory, Memorial Sloan-Kettering Cancer Center, New York, New York 10021, U.S.A.

Yasuo Kokai Department of Pathology, Sapporo Medical College, S1 W17 Sapporo 060, Japan

Norbert Kraft Nephrology Department, Prince Henry's Hospital, Melbourne, Australia

Bernhard Kubanek Red Cross Blood Bank Ulm, Department of Transfusion Medicine, University of Ulm, D-7900 Ulm, F.R.G.

H. Ladyman Department of Haematology, John Radcliffe Hospital, Oxford OX3 9DU, U.K.

F. Lancelin Groupe Universitaire de Recherche en Immunologie Fondamentale et Appliquée, Service de Médecine Interne, Département d'Hématologie, C.H.U., Rennes, France

Gilles J. Lauzon Department of Pediatrics and the Tom Baker Cancer Centre, University of Calgary, Calgary, Alberta, Canada T2N 4N1

Thierry Lavabre-Bertrand Service des Maladies du Sang, C.H.U., Montpellier, France

A.M. Lebacq-Verheyden Unité de Recherches sur les Maladies du Sang, Université Catholique de Louvain, Brussels, Belgium

Tucker W. LeBien Department of Laboratory Medicine and Pathology, University of Minnesota, Minneapolis, Minnesota 55455, U.S.A.

R. Leblay Groupe Universitaire de Recherche en Immunologie Fondamentale et Appliquée, Service de Médecine Interne, Département d'Hématologie, C.H.U. Rennes, France

Jeffrey A. Ledbetter Immunobiology Group, Genetic Systems Corporation, Seattle, Washington 98121, U.S.A.

Grace T.H. Lee Immunogenetics Research Unit, Cancer Institute, Melbourne, Australia 3000

N.R. Ling Department of Immunology, University of Birmingham, Medical School, Birmingham B15 2TJ, U.K.

B. Michael Longenecker Department of Immunology, University of Alberta, Edmonton, Alberta, Canada T6G 2H7

W.D. Ludwig Abt. Hämatologie, Klinikum Steglitz, Freie Universität Berlin, Berlin, F.R.G.

I.C.M. MacLennan Department of Immunology, University of Birmingham, Medical School, Birmingham B15 2TJ, U.K.

Ian T. Magrath Pediatric Branch, Clinical Oncology Program, Division of Cancer Treatment, National Cancer Institute, National Institutes of Health, Bethesda, Maryland 20205, U.S.A.

Abby L. Maizel Department of Pathology, Section of Pathobiology, The University of Texas Cancer Center, Houston, Texas 77030, U.S.A.

Gerald E. Marti Hematology Service, Clinical Pathology Department, National Institutes of Health, Bethesda, Maryland 20205, U.S.A.

M. Marty Groupe Universitaire de Recherche en Immunologie Fondamentale et Appliquée, Service de Médecine Interne, Département d'Hématologie, C.H.U., Rennes, France

D.Y. Mason Department of Haematology, John Radcliffe Hospital, Oxford OX3 9DU, U.K.

Robert T. McCormack Department of Laboratory Medicine/Pathology, University of Minnesota Medical School, Minneapolis, Minnesota 55455, U.S.A.

G. Merdrignac Groupe Universitaire de Recherche en Immunologie Fondamentale et Appliquée, Service de Médecine Interne, Département d'Hématologie, C.H.U., Rennes, France

Karin Meyer The University Department of Medicine and Hematology, Aarhus Amtssygehus, Denmark DK-8000

Patricia M. Michael Immunogenetics Research Unit, Cancer Institute, Melbourne, Australia 3000

Jordi Milà Servicio de Inmunología, Hospital Clínic i Provincial, 08036 Barcelona, Spain

Gerhard Moldenhauer Institut für Immunologie und Genetik, Deutsches Krebsforschungszentrum, D-6900 Heidelberg, F.R.G.

Frank Momburg Institut für Immunologie und Genetik, Deutsches Krebsforschungszentrum, D-6900 Heidelberg, F.R.G.

John Morgan Department of Pathology, Section of Pathobiology, The University of Texas Cancer Center, Houston, Texas 77030, U.S.A.

Marion Moxley Medical Research Council, Clinical and Population Cytogenetics Unit, Western General Hospital, Edinburgh, U.K.

Lee M. Nadler Division of Tumor Immunology, Dana-Farber Cancer Institute, Department of Medicine, Harvard Medical School, Boston, Massachusetts 02115, U.S.A.

P.D. Nathan Department of Immunology, University of Birmingham, Medical School, Birmingham B15 2TJ, U.K.

Leonard M. Neckers Laboratory of Pathology, National Cancer Institute, National Institutes of Health, Bethesda, Maryland 20205, U.S.A.

G. Pallesen University Institute of Pathology, Kommunehospitalet, DK-8000 Aarhus C, Denmark

Edward Park Division of Tumor Immunology, Dana-Farber Cancer Institute, Boston, Massachusetts 02115, U.S.A.

Carlo Parravicini Fifth Department of Pathology, University of Milan, Milan, Italy

Antonio Pezzutto Medizinische Universitäts-Poliklinik, D-6900 Heidelberg, F.R.G.

Glenn R. Pilkington Immunogenetics Research Unit, Cancer Institute, Melbourne, Australia 3000

Samuel J. Pirruccello Department of Laboratory Medicine/Pathology, University of Minnesota Medical School, Minneapolis, Minnesota 55455, U.S.A.

Stefania Pittaluga Laboratory of Pathology, National Cancer Institute, National Institutes of Health, Bethesda, Maryland 20205, U.S.A.

Jeffrey L. Platt Department of Laboratory Medicine/Pathology, University of Minnesota Medical School, Minneapolis, Minnesota 55455, U.S.A.

Philippe Poncelet Immunotoxin Project, Research Center Clin-Midy, Sanofi Group, Montpellier, France

Anand Raghavachar Red Cross Blood Bank Ulm, Department of Transfusion Medicine, University of Ulm, D-7900 Ulm, F.R.G.

A.M. Ravoet Unité de Recherches sur les Maladies du Sang, Université Catholique de Louvain, Brussels, Belgium

Frances Rawle ICRF Human Tumor Immunology Group, School of Medicine, University College London, London WC1E 6JJ, U.K.

H. Riehm Abt. Pädiatrie VI, Medizinische Hochschule Hannover, Hannover, F.R.G.

John T. Sandlund Pediatric Branch, Clinical Oncology Program, Division of Cancer Treatment, National Cancer Institute, National Institutes of Health, Bethesda, Maryland 20205, U.S.A.

Volker Schirrmacher Institute of Immunology and Genetics, German Cancer Research Center, D-6900 Heidelberg, F.R.G.

Stuart F. Schlossman Division of Tumor Immunology, Dana-Farber Cancer Institute, Department of Medicine, Harvard Medical School, Boston, Massachusetts 02115, U.S.A.

Reinhard Schwartz Institut für Immunologie und Genetik, Deutsches Krebsforschungszentrum, D-6900 Heidelberg, F.R.G.

Geraldine Shu Immunobiology Group, Genetic Systems Corporation, Seattle, Washington 98121, U.S.A.

Bruce Slaughenhoupt Division of Tumor Immunology, Dana-Farber Cancer Institute, Boston, Massachusetts 02115, U.S.A.

C.M. Steel Medical Research Council Clinical and Population Cytogenetics Unit, Western General Hospital, Edinburgh, U.K.

Tsuyoshi Takami Department of Pathology, Sapporo Medical College, S1 W17 Sapporo 060, Japan

Fumio Takei Terry Fox Laboratory, B.C. Cancer Research Centre, Vancouver, British Columbia, Canada

Takashi Takei Department of Pathology, Sapporo Medical College, S1 W17 Sapporo 060, Japan

Thomas F. Tedder Cellular Immunobiology Unit, Tumor Institute, University of Alabama in Birmingham, Birmingham, Alabama 35294, U.S.A.

E. Thiel Institute für Haematologie, München, F.R.G.

David A. Thorley-Lawson Departments of Pathology and Medicine, Tufts University School of Medicine, Boston, Massachusetts 02115, U.S.A.

Jane B. Trepel Laboratory of Pathology, National Cancer Institute, National Institutes of Health, Bethesda, Maryland 20205, U.S.A.

Ramón Vilella Servicio de Imnmunología, Hospital Clínic i Provincial, 08036 Barcelona, Spain

Jordi Vives Servicio de Inmunología, Hospital Clínic i Provincial, 08036 Barcelona, Spain

L. Walker Department of Immunology, University of Birmingham, Medical School, Birmingham B15 2TJ, U.K.

Janis J. Weis Department of Medicine, Harvard Medical School, Department of Rheumatology & Immunology, Brigham and Women's Hospital, Boston, Massachusetts 02115, U.S.A.

P. Wernet Medizinische Universitäts-Klinik, Tübingen, F.R.G.

James G. White University of Minnesota, Minneapolis, Minnesota 55455, U.S.A.

D.H. Wright University Department of Pathology, General Hospital, Southampton SO9 4XY, U.K.

Hiroo Yuasa Department of Pathology, Sapporo Medical College, S1 W17 Sapporo 060, Japan

Theodore F. Zipf Faculty of Medicine, University of Calgary, Calgary, Alberta, Canada T2N 4N1

Heddy Zola Flinders Medical Centre and Flinders University of South Australia, Bedford Park, South Australia 5042

Part I. Introduction

CHAPTER 1

B Cell/Leukemia Panel Workshop: Summary and Comments

Lee M. Nadler

Background

Since the First International Workshop two years ago, there has been extraordinary progress in the development and characterization of monoclonal antibodies directed against antigens expressed on normal and neoplastic B lymphocytes. Whereas the First International Workshop yielded eight CD clusters for T cells and five CD clusters for myeloid cells, no B cell clusters were identified. Two CD clusters (CD9 and CD10) which included the anti-p24 and anti-CALLA antibodies were established since they were considered to be useful in the subclassification of leukemias. Although there were no clusters of anti–B cell antibodies, a number of interesting antibodies were identified in the B cell panel. For example, several antibodies defining B cell-restricted antigens included: anti-B1, anti-Y29/55, anti-Tü1 (all directed against pan B antigens); anti-B2 and anti-FMC1 (directed against antigens expressed only on discrete stages of B cell differentiation, i.e., limited B cell antigens), and anti-BB1 (directed against a B cell activation antigen) (1–5). Similarly, several antibodies (e.g., L22, 3HBB2, and F8.11.13) defining B cell-associated antigens also appeared to be interesting (1). It was in this context and with the knowledge that many new antibodies directed against B cell/leukemia antigens had been recently developed that we undertook the planning of the B cell/ leukemia workshop.

B Cell/Leukemia Antigen Workshop: Conceptual and Technical Difficulties

The characterization of B cell-restricted, B cell-associated, and in some cases leukemia antigens is a much more difficult technical undertaking than is required for either T cell or myeloid antigens. This is due to the fact that B cells comprise only 5% of the peripheral blood mononuclear fraction. Techniques to enrich for B cells like E-rosette depletion increase the percentage of B cells to approximately 15–20% whereas subsequent adherence to remove macrophages generally only improves the purity to 50%. It is therefore very difficult to obtain a homogeneous population of normal human B cells from peripheral blood. Moreover, in addition to time required to purify B cells from a single individual, it is virtually impossible (short of leukophoresis) to obtain 10^8 B cells upon which to screen the large number of panel antibodies. Therefore, it was important to attempt to obtain B cells from human lymphoid tissues since cell suspensions from either T cell-depleted lymph node, tonsil, or spleen contain greater than 80% B cells. However, even with the use of lymphoid tissue cell suspensions and enrichment techniques, it is still very difficult to obtain a homogeneous population of human B cells. Considering this obstacle, it was therefore essential to use additional sources of normal and malignant B cells to demonstrate the specificity of the panel antibodies. These populations included B cell lymphoblastoid cell lines, B cell tumor lines, as well as tumor cells isolated from patients with non-T cell acute lymphoblastic leukemia (ALL), B cell chronic lymphocytic leukemia (CLL), and B cell non-Hodgkin's lymphoma (NHL). In addition to examining single-cell suspensions, we felt that it was crucial to confirm cellular specificity using *in situ* immunoperoxidase or alkaline phosphatase staining techniques to demonstrate antigens in lymphoid tissue sections. Using this multimodality approach, we believed that we could characterize the cellular expression of an antigen with confidence.

B Cell Leukemia/Panel Antibodies

Prior to selection of antibodies for the B cell/leukemia panel, investigators were required to provide preliminary data detailing the specificity of each antibody and to provide a titration curve demonstrating its reactivity on a known B cell population. The 52 antibodies selected to comprise the B cell panel (Table 1.1) were contributed by a total of 21 laboratories. The 21 leukemia panel antibodies (Table 1.2) were contributed by a total of 9 independent laboratories. These antibodies were then encoded with a Second International Workshop Code designation and this code was kept confidential until the time of the Workshop. In addition to specificity and titer, each investigator specified, if possible, the murine immunoglobulin subclass of each antibody and whether the antibody bound protein A and/or fixed rabbit or guinea pig complement (Table 1.3).

Target Cells for the B Cell/Leukemia Panel

To define the specificity of the B cell/leukemia panel antibodies, we developed a panel of normal and neoplastic cells and cell lines of multiple cellular lineages. This target cell panel is depicted in Table 1.4. As seen,

Table 1.1. B Cell workshop panel antibodies.

Workshop code	Antibody name	Laboratory origin	Investigators submitting
B1	I-2	73	Nadler
B2	BII6-2	94	Feller, Wacker
B3	29-132	60	Kraft
B4	HH1	20	Funderud
B5	B1	73	Nadler
B6	NUB1	71	Sagawa, Okubo, Matsuo, Yokoyama, Hagiwara, Shirnishi
B7	29-110	60	Kraft
B8	AB-1	20	Funderud
B9	B2	73	Nadler
B10	RW35-1C5	40	Habeshaw, Murray, Dhutt, Rainey
B11	MNM6	27	McMichael, Gotch
B12	UL-38	64	Raghavachar
B13	SJ12-3G2	31	Melvin
B14	B4	73	Nadler
B15	AL1a	67	LeBacq-Verheyden
B16	E5A7	21	Thompson
B17	HD28	102	Dörken, Moldenhauer, Schwartz
B18	AL1c	67	LeBacq-Verheyden
B19	PL-13	26	Horibe, Knowles
B20	PC-1	73	Anderson, Nadler
B21	SHCL-2	70	Schwarting
B22	2H7	66	Clark
B23	8B1	28	Naito, Flomenberg, Kernan, Dupont
B24	1F5	66	Clark
B25	HD6	102	Dörken, Moldenhauer, Schwartz
B26	PCA-1	73	Anderson, Nadler
B27	2-7	28	Naito, Flomenberg, Kernan, Dupont
B28	HD37	102	Dörken, Moldenhauer, Schwartz
B29	H616	66	Clark
B30	HH1 (purified)	20	Funderud
B31	HD39	102	Dörken, Moldenhauer, Schwartz
B32	9BA-5	66	Clark
B33	B2	73	Nadler
B34	B4	73	Nadler
B35	BL-13	7	Brochier
B36	BL-14	7	Brochier
B37	Blast-1	73	Thorley-Lawson, Nadler
B38	21D-10	47	Garrido
B39	Blast-2	73	Thorley-Lawson, Nadler
B40	SJ10-1H11	31	Melvin
B41	HB5	120	Tedder, Cooper
B42	B-7	113	Henke
B43	4G7	47	Levy
B44	HB6	120	Tedder, Cooper
B45	HB4	120	Tedder, Cooper
B46	UL-65	64	Raghavachar
B47	HB8	120	Tedder, Cooper
B48	HB9	120	Tedder, Cooper
B49	SHCL-1	47	Schwarting
B50	HB10	120	Tedder, Cooper
B51	HB11	120	Tedder, Cooper
B52	41H16	80	Mannoni

Table 1.2. B Cell workshop panel antibodies.

Workshop code	Antibody name	Laboratory origin	Investigators submitting
L1	SJ9-2E2	31	Melvin
L2	J5	101	Ritz
L3	L22	93	Royston, Dillman
L4	30	101	Ritz
L5	No sample		
L6	J13	101	Ritz
L7	3-3	28	Naito
L8	7-2	79	Martin
L9	6-4	28	Naito
L10	W8E7	47	Commercial
L11	J5	101	Ritz
L12	9-4	79	Martin
L13	L01-1	47	Commercial
L14	AL2	67	LeBacq-Verheyden
L15	AL3	67	LeBacq-Verheyden
L16	AL6	67	LeBacq-Verheyden
L17	SJ25-C1	31	Melvin
L18	J2	101	Ritz
L19	A2	113	Henke
L20	E20	113	Henke
L21	CLB-CALLA1	15	von den Borne
L22	CLB-Thromb2	15	von den Borne

antibodies were screened on fractionated peripheral blood cells to determine lineage specificity (targets 01–05). They were then tested on fractionated and unfractionated cells isolated from either lymph node, spleen, or tonsil in an attempt to confirm cellular specificity and also to determine the percentage of B cells reactive with each antibody (targets 06–08). Reactivity with bone marrow was undertaken to determine if an antigen was expressed on precursor cells of one or multiple lineages (target 09). To examine the expression of the antigen on activated B cells, B cells from peripheral blood or lymphoid tissues were stimulated with pokeweed mitogen (PWM) for either three or seven days (targets 10 and 11). To determine if an antigen was differentially expressed on subgroups of B cell tumors, the B cell/leukemia panel antibodies were screened on non-T ALLs (targets 12 and 14), B-CLLS (target 13), and B-NHLs (target 15). B cell lymphoblastoid, pre–B cell, and B cell NHL lines were also extensively tested (targets 16 and 17). In an attempt to determine if the antigens were expressed on hematopoietic precursors, antibodies were screened on fetal liver and bone marrow cells (targets 19 and 20). In addition to the formal target panel of B and non-B cells, we requested that, when possible, the antibodies be tested for reactivity with cell lines of myeloid and T cell origin as well as acute and chronic leukemias and lymphomas of T cell and myeloid origin.

Table 1.5 summarizes the actual number of cells tested by the cooperating laboratories examining the B cell panel antibodies. As seen 394 normal target cells, 87 cell lines, 200 leukemias, and 55 lymphomas were screened with the entire B cell antibody panel. This effort of screening 736 target cells resulted in the collection of 38,272 data points.

Table 1.6 summarizes the actual number of cells tested by laboratories examining the leukemia panel antibodies. As seen, 269 normal cells, 61 cell lines, 311 leukemic cells, and 73 lymphoma cells were evaluated with the entire leukemia antibody panel. This effort of screening 714 target cells resulted in the collection of 14,994 data points.

Laboratories Participating in the B Cell/Leukemia Panel Workshop

Table 1.7 summarizes by country the 44 laboratories participating in the Wet Workshop. The overwhelming majority of laboratories concentrated on phenotyping normal and leukemic cells. Six laboratories attempted to characterize the molecular nature of the antigens. Three laboratories undertook experiments to examine the functional relevance of the antigens. Finally, eight laboratories examined the cellular localization of panel antibodies *in situ* on normal and malignant tissues.

Analysis of Data

Following an initial examination of the data from the B cell panel, it was immediately evident that it would be very difficult to determine the precise percentage of B cells reactive with each antibody. This was due to the fact that most laboratories still only screened on peripheral blood E^- cells as a source of normal B cells (Table 1.5). Moreover, considering the diversity of techniques used to enumerate the number of positive cells, it appeared virtually impossible to compare the data from laboratory to laboratory on a statistical basis. It was decided that a different approach (i.e., nonstatistical) would be necessary to accurately compare the reactivity of antibodies. Therefore, the first step in data analysis was to attempt to determine the molecular weight of as many of the antigens as possible. This was greatly facilitated by the outstanding cooperation of the laboratories of LeBien; Clark; and Horibe and Knowles. By coordinating the efforts of these laboratories and even suggesting possible clusters based upon serologic data, it was possible to identify the molecular nature of many of these antigens. The results of these studies for the B cell/leukemia panel antibodies are summarized in Table 1.8. As seen in this table, the overwhelming majority of antigens could be precipitated by one or more laboratories.

The next stage of analysis was to determine the expression of these antigens on fractionated normal peripheral blood mononuclear cells and

Table 1.3. Properties of workshop antibodies.

Workshop number	Antibody name	Ig subclass	Protein A binding	Complement fixing
A. B cell panel				
B1	I-2	G2a	+	+
B2	BII6-2	?	?	?
B3	29-132	?	?	?
B4	HH1	G1	+	−
B5	B1	G2a	+	+
B6	NUB1	?	?	?
B7	29-110	G1	+	−
B8	AB-1	M	?	+
B9	B2	M	?	+
B10	RW35-1C5	M	?	?
B11	MNM6	G	?	?
B12	UL-38	?	?	?
B13	SJ12-3G2	G1	?	?
B14	B4	G1	−	−
B15	AL1a	G2c	+	−
B16	E5A7	M	?	+
B17	HD28	G2a	+	+
B18	AL1c	?	−	?
B19	PL-13	G1	?	−
B20	PC-1	M	?	+
B21	SHCL-2	G1	+/−	−
B22	2H7	G2b	+	?
B23	8B1	G2a	+	+
B24	1F5	G2a	+	+
B25	HD6	G1	?	−
B26	PCA-1	G1	+	−
B27	2-7	M	−	+
B28	HD37	G1	?	?
B29	H616	G2a	−	+
B30	HH1 (purified)	G1	?	?
B31	HD39	G1	?	?
B32	9BA-5	?	?	?
B33	B2	M	?	+
B34	B4	G1	−	−
B35	BL-13	G1	?	−
B36	BL-14	G1	?	−
B37	Blast-1	M	+	+
B38	21D-10	G	+	+
B39	Blast-2	G1	−	−
B40	SJ10-1H11	G1	?	?
B41	HB5	G2a	+	?
B42	B-7	G1	+	−
B43	4G7	G1	−	−
B44	HB6	M	?	?
B45	HB4	M	?	?
B46	UL-65	G	?	?
B47	HB8	M	?	?
B48	HB9	M	?	?
B49	SHCL-1	G2b	+	+
B50	HB10	G1	?	?
B51	HB11	M	?	?
B52	41H16	G2a	+	?

Table 1.3. (*Continued*)

Workshop number	Antibody name	Ig subclass	Protein A binding	Complement fixing
B. Leukemia panel				
L1	SJ9-2E2	G1	?	?
L2	J5	G2a	+	+
L3	L22	?	?	?
L4	J30	M	?	?
L5	No sample			
L6	J13	M	−	+
L7	3-3	G2b	+	+
L8	7-2	?	?	?
L9	6-4	G2a	+	+
L10	W8E7	G2a	+	+
L11	J5	G2a	+	+
L12	9-4	?	?	?
L13	L01-1	G2a	+	?
L14	AL2	G2b	−	+
L15	AL3	M	−	+
L16	AL6	G	−	+
L17	SJ25-C1	G1	?	?
L18	J2	M	−	+
L19	A2	G1	−	−
L20	E20	G1	−	−
L21	CLB-CALLA1	?	?	?
L22	CLB-Thromb2	?	?	?

Table 1.4. B Cell/leukemia target cell panel.

Target cell code	Cell type
01	Peripheral blood mononuclear cells
02	Peripheral blood T cells from normal donor (E^+)
03	Peripheral blood non-T cells from normal donor (E^-)
04	Adherent cells from normal donor (monocytes)
05	Polymorphonuclear leukocytes from normal donor (granulocytes)
06	Tonsil cells from normal donor
07	Lymph node cells from normal donor
08	Spleen cells from normal donor
09	Bone marrow cells from normal donor
10	PWM blasts, 3 days
11	PWM blasts, 5 days
12	Non-T, non-B, or pre-B ALL blasts
13	B-chronic lymphocytic leukemia cells
14	B-acute lymphoblastic leukemia blasts
15	B-lymphoma and cells from other B-malignancies
16	B-lymphoblastoid cell line
17	Non-T, non-B lymphoblastoid cell line
18	Chronic myeloid leukemia in blast crisis
19	Fetal liver
20	Fetal bone marrow
21	Other

Table 1.5. Panel of target cells for B cell panel.

Cell type	Total number of cell samples analyzed
Normal cells	
Peripheral blood mononuclear cells	122
Peripheral blood E-rosette-positive (T) cells	73
Peripheral blood E-rosette-negative (B) cells	47
Peripheral blood adherent cells	39
Polymorphonuclear leukocytes	58
Spleen cells	14
Lymph node cells	11
Tonsil cells	17
Bone marrow	13
Cell lines	
B cell lymphoblastoid	14
Pre-B leukemic	17
Burkitt	22
Myeloma	11
T cell	14
Myeloid	9
Leukemias	
Non-T ALL	61
T-ALL	17
B-CLL	80
T-CLL	7
CML stable phase	11
CML blast crisis	9
AML	15
Lymphomas	
WDLL	4
N/PDL	17
D/PDL (B)	5
DHL (B)	8
DM	2
NM	2
Myeloma	6
HCL	6
T-lymphoma	9

cell lines of varying lineages with the view to define the cellular expression of the antigen. For the B cell panel, two major classes of antigens were identified and are termed B cell restricted (only expressed on normal and neoplastic B cells) and B cell associated (expressed on B cells and cells of the other lineages). The reactivity of each subgroup of antibodies was then examined on neoplastic cells of several lineages to again confirm cellular specificity. The expression of these antigens on normal B cells as well as on leukemias and lymphomas of B cell origin suggested whether the antigen was expressed throughout ontogeny (pan B cell antigens) or was expressed only on limited stages of differentiation (limited B cell

Table 1.6. Panel of target cells for leukemia panel.

Cell type	Total number of cell samples analyzed
Normal cells	
Peripheral blood mononuclear cells	83
Peripheral blood E-rosette-positive (T) cells	40
Peripheral blood E-rosette-negative (B) cells	24
Peripheral blood adherent cells	26
Polymorphonuclear leukocytes	45
Spleen cells	9
Lymph node cells	8
Tonsil cells	7
Bone marrow	27
Cell lines	
B cell lymphoblastoid	12
Pre-B leukemic	8
Burkitt	19
Myeloma	7
T cell	8
Myeloid	7
Leukemias	
Non-T ALL	83
T-ALL	34
B-CLL	97
T-CLL	13
CML stable phase	9
CML blast crisis	16
AML	59
Lymphomas	
DWDLL	4
DPDL (B)	12
DH (B)	16
DM	5
NPDL	9
NM	3
Myeloma	9
HCL	6
T-lymphoma	9

antigens). This analysis allowed the phenotypic clustering of antigens as either B cell restricted or associated and into pan B vs. limited B cell antigens.

Using the clusters defined by molecular weight, specificity, and stage of differentiation, we then examined the clustered and non-clustered antigens with regard to their expression in tissue sections. Eight laboratories contributed data on the *in situ* expression of antigens obtained using either the immunoperoxidase or alkaline phosphatase techniques. Antigens could be divided into those which were restricted to B cell areas and those which were reactive with cells of other lineages (e.g., T cells or

Table 1.7. Laboratories participating in B cell/leukemia panel workshop.

Country	Lab #	Investigators	Workshop participation			
			Phenotyping	Biochem.	Function	Immuno hist.
Australia	39	Pilkington, Jose	x			
Australia	60	Kraft, Hancock, Giddy, Atkins				x
Belgium	67	Ravoet, Lebacq-Verheyden	x	x		
Canada	80	Mannoni	x			
Denmark	57	Pallesen				x
France	38	Poncelet, Lavabre-Bertrand	x			
France	54	Favrot	x			
France	85	Boucheix, Perrot	x			
France	96	Bourel, Genetet	x			
Israel	59	Gazit	x			
Italy	14	Berti, Parravicini, Cattoretti, Delia, de Braud				x
Japan	9	Morishima	x			
Netherlands	15	von dem Borne	x			
Netherlands	83	Lansdorp, van Mourik, Zeijlemaker	x			
Norway	20	Funderud, Godal	x			
Scotland	106	Steel, Elder, Cohen, Guy	x	x		
Spain	68	Gallart, Anegón, Curtain, Vives	x			
Sweden	107	Heldrup, Garwicz	x			
United Kingdom		Jones, Britten, Wright				x
United Kingdom	8	Horton	x			
United Kingdom	22	Golay, Beverley			x	
United Kingdom	27	McMichael, Gotch	x			
United Kingdom	40	Habeshaw, Murray, Rainey				x
United Kingdom	41	Mason				x
United Kingdom	110	MacLennan, Ling, Johnson, Khan, Nathan	x			x
United States	21	Thompson	x			
United States	26	Horibe, Knowles	x	x		
United States	31	Melvin, Peiper	x			
United States	34	LeBien		x		
United States	42	Winchester	x			
United States	52	Kersey, Gajl-Peczalska	x			
United States	66	Clark, Einfeld	x	x		
United States	73	Nadler, Boyd, Anderson, Freedman	x		x	
United States	86	Cossman, Neckers	x		x	
United States	104	Saunders, Cooper	x			
United States	101	Ritz	x			
United States	112	Paietta, Wiernik	x			

Table 1.7. (*Continued*)

Country	Lab #	Investigators	Workshop participation: Pheno-typing	Biochem.	Function	Immuno-hist.
United States	113	Henke	x			
West Germany	64	Raghavachar	x	x		x
West Germany	92	Wernet				x
West Germany	94	Feller	x			
West Germany	102	Dörken, Pezzutto, Moldenhauer, Schwartz, Hunstein	x			x
Yugoslavia	45	Janković, Popesković	x			

Table 1.8. Molecular characterization of B cell/leukemia panel antigens.

Workshop number	Antibody name	Molecular weight studies (Kd): LeBien	Clark	Horibe & Knowles	Ravoet & Lebacq-Verheyden
B1	I-2	34/29	33/28	35/32	31/37
B2	BII6-2	—	33/28	75	—
B3	29-132	—	33/28	—	29/34/72/89
B4	HH1	—	—	—	—
B5	B1	35	32	35	—
B6	NUB1	—	—	—	—
B7	29-110	—	135	125	—
B8	AB-1	75	95	—	—
B9	B2	140	145	140	—
B10	RW35-1C5	220	—	—	—
B11	MNM6	—	45	45(60–80/45/35)	45–50
B12	UL-38	—	—	—	—
B13	SJ12-3G2	—	—	—	—
B14	B4	90	95	87	—
B15	AL1a	—	—	—	110/180/220
B16	E5A7	—	—	—	—
B17	HD28	45/140	—	160/43	48
B18	AL1c	—	—	—	110/180/220
B19	PL-13	—	—	60–80/45/35	45–50/72–100
B20	PC-1	—	—	—	—
B21	SHCL-2	—	32	—	—
B22	2H7	35	32	—	—
B23	8B1	34/29	37	45/32	—
B24	1F5	35	32	35	—
B25	HD6	130	135	125	—
B26	PCA-1	—	—	26	—
B27	2-7	—	—	—	—
B28	HD37	90	95	87	—
B29	H616	90	85	85	—
B30	HH1 (Purified)	—	—	—	—
B31	HD39	130	135	125	—
B32	9BA-5	—	55	—	—

Table 1.8. (*Continued*)

Workshop number	Antibody name	Molecular weight studies (Kd)			
		LeBien	Clark	Horibe & Knowles	Ravoet & Lebacq-Verheyden
B33	B2	140	145	140	—
B34	B4	90	95	87	—
B35	BL-13	140	145	140	—
B36	BL-14	—	—	—	—
B37	Blast-1	45	45	43	—
B38	21D-10	—	33/24	—	24
B39	Blast-2	—	45	45(60–80/45/35)	45–50/72–100
B40	SJ10-1H11	—	135	125	—
B41	HB5	140	145	140	—
B42	B-7	66	85	75	30/35/72/90
B43	4G7	90	95	87	—
B44	HB6	—	—	—	—
B45	HB4	—	—	28	—
B46	UL-65	—	—	30	—
B47	HB8	45/55/65	—	—	—
B48	HB9	45/55/65	33/24	—	—
B49	SHCL-1	130	135	125	—
B50	HB10	220	—	45/32	—
B51	HB11	220	220	—	—
B52	41H16	45	45	40	—
L1	SJ9-2E2	—	—	—	—
L2	J5	100	100	100	103
L3	L22	85	90	90/70	—
L4	J30	24	33/24	—	23
L5	No sample				
L6	J13	100	100	220	—
L7	3-3	—	—	—	—
L8	7-2	34/29	33/28	35/32	31/37
L9	6-4	—	—	160	160
L10	W8E7	100	100	100	103
L11	J5	100	—	100	103
L12	9-4	—	220	220	—
L13	L01-1	90	90	90/70	97
L14	AL2	100	100	100	103
L15	AL3	—	—	—	—
L16	AL6	24	33/24	24	23
L17	SJ25-C1	90	95	87	—
L18	J2	24	33/24	24	23
L19	A2	90	90	90/70	97
L20	E20	90	90	90/70	—
L21	CLB-CALLA1	100	100	—	103
L22	CLB-Thromb2	24	33/24	24	—

dendritic reticulum cells). Antigens were then examined with regard to their localization within the secondary follicle. Antigens which were expressed on either the mantle zone or germinal center or both demonstrated unique patterns of antigen expression. Examination of these patterns frequently permitted clustering of antibodies. These clusters were generally consistent with those identified by phenotyping and molecular analysis.

This approach of determining clusters by molecular weight, cellular lineage reactivity, expression at distinct stages of normal and neoplastic B cell ontogeny, and *in situ* antigen localization permitted the identification of five new B cell-restricted CD clusters and one new B cell-associated CD cluster. Table 1.9 summarizes the division of the B cell panel into B cell-restricted and -associated antigens which are either clustered or non-clustered. These are then classified as either pan B or limited B cell antigens on the basis of their expression on normal and malignant B cells. Antibodies which are clustered by both cellular reactivity and molecular weight are grouped and those which do not cluster are individually listed. The B cell-associated antigens are also listed by their reactivity with fractionated peripheral blood cells. Table 1.10 summarizes a similar approach used to group the leukemia antigens. All but three antibodies clustered into groups by molecular weight and cellular reactivity. The majority of the antibodies were reactive with either CD9 or CD10 antigens.

In the subsequent sections, the data for individual antibodies will be reviewed in detail. It is clear that this approach has allowed for the identification of a number of B cell and leukemia antigen clusters. Moreover, as was true for the First International Workshop, a number of very interesting non-clustered antibodies have been identified which are destined to become clusters in future Workshops.

B Cell Panel Antibodies

B Cell-Restricted Antigens: Pan B/Clustered

Within the B cell antibody panel, two distinct clusters of antibodies are reactive with B cell-restricted antigens which appear to be expressed throughout B cell ontogeny. The first antibody cluster, which has been designated CD19, defines a glycoprotein of 95 Kd. The second antibody cluster, which has been designated CD20, defines a nonglycosylated phosphoprotein of 35 Kd. Each cluster will be examined in depth.

CD19 Cluster

The CD19 cluster contains four antibodies: B4 (B14,34) HD37 (B28), 4G7 (B43), and SJ25-C1 (L17). The prototype of this cluster is the B4 antigen

Table 1.9. Overview of B cell panel specificity.

Cellular specificity	B cell reactivity	B cell code designation (Ab name)	M.W. (Kd)	CD classification
1. *B cell restricted*				
Clustered	Pan B	B14,34 (B4), B28 (HD37), B43 (4G7), L17 (SJ25-C1)	95	CD19
Clustered	Pan B	B5 (B1), B22 (2H7), B24 (1F5)	35	CD20
Non-clustered	Pan B	B6 (NUB1)	?	—
		B8 (AB-1)	?	—
		B30 (HH1)	?	—
		B36 (BL-14)	?	—
Clustered	Limited B	B9,33 (B2), B35 (BL-13), B41 (HB5)	140	CD21
Clustered	Limited B	B7 (29-110), B25 (HD6), B31 (HD39), B40 (SJ10-1H11), B49 (SHCL-1)	135	CD22
Clustered	Limited B	B11 (MNM6), B19 (PL-13), B39 (Blast-2)	45	CD23
Non-clustered	Limited B	B21 (SHCL-2)	?32	—
		B46 (UL-65)	?30	—
		B2 (BII6-2)	?	—
		B3 (29-132)	?	—
		B42 (B-7)	?80	—
		B23 (8B1)	37	—
		B37 (Blast-1)	45	—
		B13 (SJ12-3G2)	?	—
		B20 (PC-1)	?	—
		B32 (9BA-5)	?55	—
2. *B cell associated*				
Clustered (B + M)	Pan B	B1 (I-2), L8 (7-2)	33/28	—
Clustered (B ± M + G)	Pan B	B47 (HB8), B48 (HB9)	45/55/65	24
Clustered (± B + G)	Limited B	B15 (AL1a), B18 (AL1c)	?	—
Clustered (B + T + G)	Pan B	B50 (HB10), B51 (HB11)	220	—

Non-clustered (B + T + M)	Pan B	B10 (RW35-1C5)	?	—
Non-clustered (B + T + M)	Pan B	B44 (HB6)	?	—
Non-clustered (B + T + M)	Pan B	B45 (HB4)	?	—
Non-clustered (B + G + M)	Pan B	B17 (HD28)	?45	—
Non-clustered (B + G + M)	Limited B	B26 (PCA-1)	26	—
Non-clustered (B + G + M)	Pan B	B52 (41H16)	45	—
Non-clustered (B + M)	Pan B	B4 (HH1)	—	—
Non-clustered (B + M)	Limited B	B12 (UL-38)	—	—
Non-clustered (B + M)	Limited B	B29 (H616)	?90	—

Table 1.10. Overview of leukemia panel specificity.

Cluster status	Leukemia panel designation (Ab name)	M.W. (Kd)	CD	Normal cellular reactivity				Leukemia reactivity (%)			
				B	T	M	G	Non-T ALL	T-ALL	B-CLL	AML
Clustered (CALLA)	L2 (J5), L6 (J13), L10 (W8E7), L11 (J5), L14 (AL2), L15 (AL3), L21 (CLB-CALLA1)	100	CD10	–	–	–	+	80	10	0	0
Clustered	L4 (J30), L16 (AL6), L18 (J2), L22 (CLB-Thromb2), B38 (21D-10)	24	CD9	+	–	+	+	60	30	20	60
Clustered (transferrin receptor)	L3 (L22), L13 (L01-1), L19 (A2), L20 (E20)	90	—	–	–	–	+	30	60	0	30
Non-clustered	L1 (SJ9-2E2)	?	—	–	–	–	–	20	20	0	0
	L7 (3-3)	?	—	–	–	–	–	10	40	20	0
	L9 (6-4)	160	—	+	+	+	+	40	70	20	70

initially reported by Nadler *et al.* (6,7). These antibodies define a B lineage-restricted antigen since their expression within the hematopoietic system is limited to normal and neoplastic B cells. The CD19 antigen is expressed on approximately 4–8% of peripheral blood mononuclear cells and on >90% of B cells isolated from either peripheral blood, spleen, lymph node, or tonsil (Table 1.11). In contrast, CD19 antigen is expressed on fewer than 5% of bone marrow mononuclear cells. The CD19 antigen was not detected on peripheral blood T cells, monocytes, or granulocytes. When compared to class II antigens or other previously reported B cell-restricted antigens [e.g., B1(8, 9)], the staining intensity of the CD19 antigen is relatively weak on both peripheral blood B cells and B cells isolated from lymphoid tissues (Table 1.11). The specificity and expression of the CD19 antigen in B cell ontogeny is further confirmed by its expression on cell lines. The CD19 antigen is expressed on all pre–B cell, B-lymphoblastoid, and Burkitt's lymphoma cell lines and on two of the four myeloma cell lines tested. It should be noted that unlike plasma cells isolated from myeloma patients, most myeloma cell lines express antigens which may be found on "earlier" B cells and therefore CD19 is probably not expressed on plasma cells. Confirming the lineage restriction of the CD19 antigen, it was not detected on T cell or myeloid lines.

The expression of the CD19 antigen on leukemia and lymphoma cells also confirmed its restricted lineage expression (Table 1.12). Virtually all non-T cell ALLs, B-CLLs, and B cell lymphomas express CD19 whereas it has not been detected on fresh or cryopreserved myeloma cells. CD19 is not expressed on tumor cells isolated from patients with acute and chronic myeloid leukemias nor is it expressed on T cell leukemias or lymphomas. The expression of the CD19 antigen on virtually all non-T cell ALLs, B-CLLs, and B cell lymphomas (regardless of histologic subtype) suggests that this antigen appears early in B cell ontogeny and is lost at the terminal stages of differentiation.

These results are consistent with the previous studies of Nadler *et al.* (6,7) and Anderson *et al.* (10) on the expression of anti-B4 on normal and malignant B cells and the studies of Pezzutto *et al.* (this volume, Chapter 33) and Hermann *et al.* (this volume, Chapter 31) examining the expression of HD37. In these studies, nearly 1000 lymphoid and myeloid tumors

Table 1.11. Clustered B cell restricted: Pan B. Reactivity with B cells.

CD	M.W.	PB %+ (*I*)[a]	Spleen %+ (*I*)	Pre-B lines $n = 4$ (*I*)	B-LBCL $n = 20$ (*I*)	Burkitt's lines $n = 5$ (*I*)	Myelom lines $n = 4$ (*I*
19	p95	90(+)	90(++)	4(+++)	20(+)	5(++)	2(+)
20	p35	90(+++)	90(+++)	2(++)	20(++)	5(+++)	3(+)

[a] Antigen intensity.

Table 1.12. Clustered B cell restricted: Pan B. Reactivity with leukemia and lymphoma cells.

CD	Cluster Pan B	Non-T ALL	B-CLL	B-lymphoma	Myeloma	AML	T-ALL
19	p95	90%	95%	95%	0	0	0
20	p35	50%	95%	95%	0	0	0

in total were examined for the expression of the CD19 antigen. The expression of CD19 on normal and neoplastic B cells was consistently B cell restricted and appeared to encompass all stages of normal and neoplastic cell ontogeny excluding the plasma cell. Moreover, in these studies it was clearly demonstrated that the CD19 antigen preceded all B cell-restricted antigens in early B cell ontogeny. In fact, the earliest identifiable and isolatable B cell expresses Ia and B4 (CD19) and lacks CALLA (CD10), B1 (CD20), cytoplasmic μ chains, and surface immunoglobulin (7,11,12). Moreover, all non-T cell ALLs thus far examined which express only the Ia and B4 antigens demonstrate rearrangements of the immunoglobulin μ-chain gene (7). The exquisite specificity of the CD19 antigen appears to be of diagnostic importance since μ-chain gene rearrangements have clearly been demonstrated in small numbers (5–10%) of T cell and myeloid leukemias and lymphomas whereas CD19 expression has not yet been demonstrated on these tumors (this report).

The expression of the CD19 antigen on *in situ* normal and neoplastic lymphoid and non-lymphoid tissue sections has been extensively examined in this Workshop (see multiple reports in Part III of this volume). The CD19 antigen appears to be limited in its expression to cells of B lineage. However, the CD19 antigen also appears to be expressed on dendritic reticulum cells although it is very difficult to resolve the question of whether these cells synthesize or simply absorb the antigen. The antigen is clearly expressed on the cell surface of the majority of lymphoid cells in the mantle zone and germinal center. As expected from single-cell suspension studies, CD19 is expressed on virtually all B cell lymphomas examined *in situ* using either the immunoperoxidase or alkaline phosphatase technique. That the expression of the CD19 antigen is limited to normal and neoplastic B cells in tissue sections confirms both its B cell specificity and pan B cell expression.

All antibodies in the CD19 cluster appear to define a single epitope by cross-blocking studies (see this volume, Chapter 12). To examine the expression of the CD19 antigen on activated and differentiating B cells, Freedman *et al.* (this volume, Chapter 37) examined the expression of the CD19 antigen following activation of normal resting B cells, employing the B cell mitogens: anti-Ig, protein A, and Epstein–Barr virus (EBV). The CD19 antigen did not significantly change its expression over six days in culture when assayed for expression of antigen using indirect immunofluorescence and flow cytometric analysis. At the present time, the function of the CD19 antigen is unknown.

CD20 Cluster

The CD20 cluster contains three antibodies: B1 (B5), 2H7 (B22), and 1F5 (B24). The prototype of this group is the B1 antigen which defines a 35-Kd nonglycosylated phosphoprotein (8,9,13). The CD20 cluster identifies a B cell-restricted antigen and demonstrates no detectable cross-reactivity with normal or malignant T or myeloid cells. Nearly identical in its expression to the CD19 cluster, the CD20 antigen is expressed on 4–8% of peripheral blood mononuclear cells and on >90% of B cells isolated from peripheral blood or lymphoid organs (Table 1.11). The staining intensity of the CD20 antigen is very strong and is only surpassed by that of class I and class II antigens. Approximately 5% of mononuclear cells isolated from normal bone marrow express the CD20 antigen. A major difference between the CD19 and CD20 clusters is the reactivity with pre-B cell lines and non-T cell ALL cells. The CD19 antigen is expressed on all pre-B cell lines and non-T cell ALLs tested whereas the CD20 antigen is expressed on approximately 50%. Otherwise the CD19 and CD20 antigens appear to have identical reactivity with virtually all target cell populations tested. It should again be emphasized that although the identical cellular populations express the CD20 antigen, the pattern and intensity of staining differ between these two clusters.

The expression of the CD20 antigen was also extensively examined in tissue sections. The CD20 antigen was expressed on primary follicular B cells, interfollicular B cells, and mantle zone and germinal center B cells. The expression of the antigen appears to be largely on the cell surface. In addition to its pan B cell expression there was some controversy as to whether it was also expressed on dendritic reticulum cells (see multiple reports in Part III of this volume).

Freedman *et al.* (this volume, Chapter 37) examined the expression of the CD20 antigen following activation of resting B cells *in vitro*. Virtually all resting B cells expressed the CD20 antigen. With activation with either protein A or EBV, the CD20 antigen was lost on approximately one-third of cells by six days in culture. Cells activated with anti-Ig alone did not appear to lose CD20 within six days. These studies confirm other reports (14; this volume, Chapters 35 and 36) demonstrating that the CD20 antigen is lost at the terminal stages of B cell differentiation. Of great interest are the studies of Clark *et al.* which suggest a functional role for the CD20 antigen (this volume, Chapter 38). Incubation of tonsillar B cells with antibodies in this cluster induce proliferation of resting B cells. Golay *et al.* in studies on the function of Workshop antibodies on normal and neoplastic B cell lines also demonstrate a regulatory role for the anti-

CD20 antigen (this volume, Chapter 39). Finally, Clark has demonstrated that all CD20 antibodies appear to identify a single epitope on the CD20 antigen (this volume, Chapter 12). In addition, he suggests that a fourth antibody (B21) may be part of this cluster since it cross-blocks the binding of the other three antibodies.

These studies on normal and neoplastic B cells suggest that the CD20 antigen is B lineage restricted and expressed throughout B cell ontogeny. Its expression on only 50% of non-T cell ALLs suggests that it appears later than the CD19 antigen. Previous studies by Nadler *et al.* (7,9) and Hokland *et al.* (11,12) demonstrated that only 50% of normal and neoplastic pre-B cells express the CD20 antigen. This antigen therefore appears to follow the CD19 antigen and the CD10 (CALLA) antigen in normal ontogeny. A functional role for this antigen at discrete stages of B cell activation is very interesting, considering that it is strongly expressed throughout most stages of ontogeny.

B Cell-Restricted Antigens: Limited B/Clustered

Within the B cell antibody panel, three distinct clusters of antibodies are reactive with B cell-restricted antigens which appear to be expressed at distinct stages of B cell ontogeny. The first antibody cluster, which has been designated CD21, defines a glycoprotein of 140 Kd. The second antibody cluster, designated CD22, defines a glycoprotein of 135 Kd. The final cluster, designated CD23, defines a B cell-restricted activation antigen.

CD21 Cluster

The CD21 cluster contains three antibodies: B2 (B9,B33), BL-13 (B35), and HB5 (B41). Two antibodies in this group, anti-B2 (15) and HB5 (16), have been previously extensively characterized. These antibodies define B lineage-restricted antigens since they react only with a subpopulation of normal and malignant B cells but are not reactive with normal or neoplastic T or myeloid cells. The CD21 antigen is expressed on approximately 5% of peripheral blood mononuclear cells and on greater than 90% of peripheral blood and lymphoid tissue B cells (Table 1.13). The expression

Table 1.13. Clustered B cell restricted: Limited B. Reactivity with B cells.

CD	Cluster Limited B	PB %+ (*I*)	Spleen %+ (*I*)	Pre-B lines *n* = 4 (*I*)	B-LBCL *n* = 20 (*I*)	Burkitt's lines *n* = 5 (*I*)	Myeloma lines *n* = 4 (*I*)
21	p140	90(+)	90(++)	0(−)	20(+/−)	2(+)	3(+)
22	p135	75(++)	75(+/++)	1(+)	10(+/−)	3(+)	1(+/−)
23	p45	0(−)	0(−)	0(−)	20(+++)	1(+)	4(++)

of the CD21 antigen is very weak on peripheral blood B cells and this antigen is more strongly expressed on B cells isolated from lymphoid tissues. The intensity of expression of the CD21 antigen on lymphoid tissue B cells is slightly weaker than the expression of the CD19 antigen and much weaker than the expression of CD20. Very few, if any, bone marrow mononuclear cells appear to express the CD21 antigen. Whereas the CD19 and CD20 antigens are expressed on virtually all B cell lines and tumors, the expression of the CD21 antigen is significantly more restricted. This antigen has not been detected on pre-B cell lines although it is expressed on a small subset of non-T cell ALLs (Tables 1.13 and 1.14). It is expressed on only some Burkitt lymphoma cell lines and, in fact, it is expressed only on those which are of African origin. This has been extensively studied by Magrath and his colleagues (this volume, Chapter 34). In contrast to the limited expression of CD21 on pre-B and Burkitt cell lines, all EBV-transformed lymphoblastoid B cell lines express the CD21 antigen although the intensity of its expression appears to be very weak (Table 1.13).

The expression of the CD21 antigen on leukemias and lymphomas of B cell origin is summarized in Table 1.14. It is expressed on very few non-T cell ALLs, on greater than 80% of B-CLLs, and on only 50% of B cell lymphomas. Further examination of the B cell lymphomas demonstrates that it is expressed on most poorly differentiated (i.e., centrocytic) lymphomas (18 of 20) and is expressed on very few large-cell lymphomas (centroblastic) (2 of 8). It is generally believed that the large-cell lymphomas correspond to more differentiated, activated B cells, suggesting that the CD21 antigen is lost with activation. In contrast, no myelomas tested expressed CD21 although most myeloma cell lines expressed the antigen. As stated above, this is probably due to the observation that most myeloma cell lines are either not true plasma cells or, alternatively, are derived from a very small subpopulation of plasma cells. Although not expressed on any normal T or myeloid cells, the CD21 antigen was expressed on several T cell leukemia lines (e.g., MOLT-4 and HPB-ALL). Moreover, T cell ALL cells isolated from patients demonstrated varying reactivities with each of the antibodies of the CD21 cluster. Of 17 T-ALLs tested, anti-B2 reacted with one, anti-BL-13 reacted with two, and anti-HB5 reacted with seven. The biological significance of the CD21 antigen on T cell ALLs is unknown at the present time. In contrast to the reactiv-

Table 1.14. Clustered B cell restricted: Limited B. Reactivity with tumor cells.

CD	Cluster Limited B	Non-T ALL	B-CLL	B-lymphoma	Myeloma	AML	T-ALL
21	p140	10%	80%	50%	0	0	0
22	p135	50%	25%	70%	0	0	0
23	p45	5%	60%	30%	0	0	0

ity with T cell leukemias, no myeloid leukemias or leukemic cell lines tested expressed the CD21 antigen.

The expression of the CD21 antigen on *in situ* lymphoid tissue sections was also extensively studied. The antigen was very weakly expressed on mantle zone B cells, more strongly expressed on marginal zone B cells, and it was difficult to determine whether the germinal center B cells expressed cell surface CD21 antigens. The antigen was very strongly expressed on the dendritic reticulum cell, suggesting that it might have been shed and passively absorbed rather than synthesized. These studies are extensively discussed by many authors in this report.

Freedman *et al.* examined the expression of the CD21 antigen following activation of resting B cells *in vitro* (this volume, Chapter 37). Regardless of the stimulus (anti-Ig, protein A, or EBV) the antigen was rapidly lost from the cell surface by three days *in vitro*. This is in distinct contrast to the persistent expression of the CD19 and CD20 antigens in the identical system. It is of interest to note that Clark (this volume, Chapter 12) demonstrated that at least two epitopes of the CD21 antigen have been defined: the anti-B2 and anti-BL-13 binding site and a second binding site for the anti-HB5 antibody. This is of interest considering that the HB5 antibody demonstrated slightly different patterns of expression with a number of target cells from those observed with the other two CD21 antibodies. This was true for some of the B cell leukemias and lymphomas; in addition, HB5 was expressed on some hairy cell leukemias whereas B2 and BL-13 were not. Moreover, several investigators suggested that a population of T cells might be reactive with anti-HB5 monoclonal antibody although this was not a consistent finding.

The function of the CD21 antigens has evoked recent intense interest. Iida Nadler, and Nussensweig initially demonstrated that the B2 antigen expressed the human C3d receptor (17) and this was subsequently confirmed by Fearon and his colleagues who showed that the HB5 antigen expressed the C3d receptor (18; this volume, Chapter 43). Neither the anti-HB5 nor the anti-B2 antibody blocked the binding of native C3d. Fingeroth, Tedder, and Fearon and their colleagues subsequently demonstrated that HB5 also expressed the EBV receptor (19; this volume, Chapter 43) and this was similarly demonstrated by Nadler *et al.* (this volume, Chapter 44). Whereas anti-HB5 did not block the binding of EBV, the anti-B2 monoclonal antibody blocked binding as well as the induction of proliferation and immunoglobulin synthesis by EBV.

In summary, the CD21 antigen, by its expression on normal and neoplastic B cells, appears to be B cell restricted. Its expression in B cell ontogeny appears to be limited and present data would suggest that it appears following the pre–B cell stage. This is concluded from its lack of expression on pre–B cell leukemias and the prior evidence of Hokland *et al.* that very few normal pre-B cells isolated from fetal liver and bone marrow or adult bone marrow express the B2 antigen (11,12). Most resting B cells appear to express the CD21 antigen, as has been demonstrated by its expression on peripheral blood and lymphoid tissue B cells as well as on most mantle zone B cells. However, with activation and CD21 antigen is lost from the cell surface and it is not expressed on activated normal or neoplastic B cells. The observation that the CD21 antigen expresses distinct binding sites for C3d and EBV is of great interest. It is now important to determine whether this antigen is important in the regulation of B cell function.

CD22 Cluster

The CD22 cluster contains five antibodies: HD6 (B25), HD39 (B31), 29-110 (B7), SJ10-1H11 (B40), and SHCL-1 (B49). This subgroup defines an antigen of 135 Kd which frequently demonstrates a second, slightly smaller band on SDS gels which may be due to differences in glycosylation. The identification and characterization of this cluster represents an important achievement of the Second International Workshop since the Workshop data represents the first detailed characterization of the specificity of this antigen.

The CD22 antigen is expressed on approximately 5% of peripheral blood mononuclear cells and on approximately 75% of B cells isolated from peripheral blood and lymphoid tissues (Table 1.13). The intensity of antigen expression on peripheral blood B cells is slightly less than the expression of the CD20 antigen but stronger than the expression of the CD19 or CD21 antigens (Table 1.13). The expression of the CD22 antigen on lymphoid tissue B cells is again stronger than that seen on peripheral blood and is similar to the CD19 intensity. Less than 1% of bone marrow mononuclear cells appeared to express the CD22 antigen. The CD22 antigen was not detected on T cells or myeloid cells, demonstrating its lineage restriction. In contrast to its expression on normal B cells, it was expressed on only 50% of EBV-transformed lymphoblastoid B cell lines and its intensity of expression was extremely weak. The CD22 antigen was expressed on one of four pre-B cell lines, on three of five Burkitt lymphoma lines, and on one of four myeloma cell lines (Table 1.13). It was not detected on any T cell or myeloid cell lines examined.

The expression of the CD22 antigen on B cell-derived leukemias and lymphomas is unique compared to all previously described B cell antigens (Table 1.14). For example, like the CD20 antigen, it is expressed on approximately 50% of non-T cell ALLs. The individual CD22 antibodies demonstrated distinct reactivities with non-T ALLs (of 56 non-T ALLs tested, SHCL-1 reacted with 29, 29-110 with 25, HD6 with 21, SJ10-1H11 with 14, and HD39 with only 11). In contrast to its expression on non-T cell ALL, fewer than 25% of B-CLLs expressed the CD22 antigen. In fact, the reactivity tended to be so weak that it was difficult to convincingly demonstrate that these tumors were positive. In contrast, the antigen is expressed on most (approximately 70%) B cell lymphomas. It is reactive with two-thirds of poorly differentiated/centrocytic lymphomas and two-thirds of large-cell/centroblastic lymphomas. Of note is its reactivity with hairy cell leukemia. Schwarting and his colleagues prepared SHCL-1 against hairy cells and clearly noted that this antibody reacted with hairy cell leukemia but not B-CLL. Finally, the CD22 antigen was not expressed on those myelomas tested. Although not expressed on any T cell leukemias or lymphomas, a small number of AMLs occasionally demonstrated reactivity with the CD22 antibodies. The reactivity did not appear to be consistently expressed on germinal center B cells whereas some felt that germinal center B cells were negative. The antigen did not react with dendritic reticulum cells and therefore was clearly B lineage restricted.

Activation of resting B cells with mitogens led to the rapid loss of the CD22 antigen (this volume, Chapter 37). By three days in culture, less than 25% of activated B cells still expressed this antigen. The loss of the antigen was very reminiscent of the CD21 antigen. In addition, Clark demonstrated at least two epitopes defined by the CD22 cluster antibodies (this volume, Chapter 12). HD6 appears to define one epitope and the other four antibodies defined another.

The expression of the CD22 antigen on normal and neoplastic B cells is very interesting. Its expression in B cell ontogeny and differentiation is clearly unique. It is expressed on some pre-B cells as demonstrated by its reactivity with non-T cell ALLs. It appears to be expressed on approximately 75% of resting B cells and is lost with activation. Consistent with this finding is the expression of CD22 on mantle zone B cells and its weak or absent expression on germinal center B cells. For most other antigens, this would be reflected in expression on B cell lymphomas. It is surprising to note that centrocytic and centroblastic lymphomas equally express the CD22 antigen. Its expression on HCL but not B-CLL is also of interest. Recent studies suggest that B-CLL is derived from a small subset of normal B cells that coexpress the B1 and T1 antigens. The lack of expression of the CD22 antigen on this subset is therefore not disturbing. At the present time the function of the CD22 antigen is unknown. In summary, the CD22 antigen probably appears during pre–B cell ontogeny, may be expressed on a subpopulation of resting B cells, and is lost from the cell surface when B cells are activated *in vitro* and *in vivo*.

CD23 Cluster

Three antibodies are contained in the CD23 cluster: MNM6 (B11), PL-13 (B19), and Blast-2 (B39). These antibodies precipitate a glycoprotein of 45 Kd.

The CD23 antigen is different from all B cell-restricted antigens thus far discussed since it is not expressed on resting B cells isolated from either peripheral blood or lymphoid tissues. It is also not expressed on T or myeloid cells. The three antibodies in the CD23 cluster were prepared by immunization with EBV-transformed B cell lymphoblastoid cell lines. It is therefore not surprising that all EBV-transformed lymphoblastoid lines strongly express the CD23 antigen (Table 1.13). The antigen was not expressed on pre-B cell lines, on one of five Burkitt lymphoma lines, and on all myeloma cell lines tested (Table 1.13). It was not expressed on any T cell or myeloid cell lines examined.

The expression of the CD23 antigen on leukemias and lymphomas was also examined. Very few, if any, non-T cell ALLs express CD23. Approximately two-thirds of B-CLLs express CD23 which is consistent with previous studies demonstrating that these tumors express other B cell activation antigens (20,21). In addition to B-CLL, approximately one-third of B cell NHLs express the CD23 antigen. These B cell NHLs appeared to correspond to activated B cells and therefore most nodular poorly differentiated lymphocytic lymphomas and some large-cell lymphomas expressed CD23. The CD23 antigen was not expressed on myeloma cells isolated from patients. The antigen was also not detected on any tumor cells isolated from patients with T cell or myeloid tumors.

In situ localization of CD23 demonstrated virtually no expression on resting mantle zone B cells in lymphoid tissue sections. In contrast, the CD23 antigen was clearly expressed on most germinal center B cells. It also appeared to be expressed on dendritic reticulum cells. The expression of an antigen on germinal center but not mantle zone B cells is clearly consistent with an antigen which is expressed exclusively on activated B cells.

In vitro activation of resting B cells clearly induced the expression of the CD23 antigen. Stimulation with anti-Ig, protein A, or EBV induced CD23 expression which peaked at day 3 and decreased significantly by day 6 (this volume, Chapter 37). All three antibodies in the CD23 cluster appeared to be directed against a single epitope on the CD23 molecule.

CD23 is the only cluster of antibodies defining a B cell-restricted activation antigen. The antigen is not expressed on pre-B or resting B cells. It appears with activation *in vitro* as well as with activation *in vivo*. The function of the CD23 antigen is presently unknown.

B Cell-Restricted Antigens: Non-Clustered

There are 16 B cell panel antibodies whose expression is restricted to B lymphocytes but which are not clustered. Of these 16 antibodies, four appear to be reactive with all B cells and therefore may be pan B cell antigens. The remaining 12 are expressed on some resting B cells or alternatively appear to be B cell activation antigens. To consider the expression of these antigens, they will be divided into three subgroups: 1) those which are expressed on all resting B cells, 2) those which are ex-

Table 1.15. Non-clustered B restricted: Expressed on all resting B cells. Reactivity with B cells.

Code	PB %+ (*I*)	Spleen %+ (*I*)	Pre-B lines *n* = 4 (*I*)	B-LBCL *n* = 20 (*I*)	Burkitt's lines *n* = 5 (*I*)	Myeloma lines *n* = 4 (*I*)
B6	90(+)	90(+/++)	2(+)	20(+/−)	1(+)	0
B8	90(+)	90(++)	4(++)	20(+/−)	5(+++)	1(+)
B30	90(+)	90(++)	0	10(+/−)	4(++)	0
B36	90(+)	90(+)	0	20(+)	5(++)	0

pressed on some resting B cells, and 3) those which are expressed on very few, if any, resting B cells. For the most part, the molecular nature of these antigens is still unknown (Table 1.8). Moreover, because of space limitations, description of the *in situ* tissue localization and function of these antigens will only be very briefly reviewed. This section will therefore represent a brief overview of some potentially very interesting, but as of yet, unclustered B cell antigens.

Antigens Expressed on All Resting B Cells

This group is composed of four antibodies: NUB1 (B6), AB-1 (B8), HH1 (B30), and BL-14 (B36). As seen in Table 1.15, most B cells isolated from peripheral blood and lymphoid tissues express these antigens. The intensity of expression for all four antigens appears to be equivalent to that seen for the CD19 and CD22 antigens. None of these antibodies are reactive with normal T or myeloid cells. The molecular weights for these four antigens are unknown. The major differences between these antigens become evident when their expression on B cell lines is examined. Each antibody appears to have a distinct fingerprint reactivity with B cell lines. Similar to the distinct patterns of reactivity with B cell lines, each antibody also appears to have a distinct pattern of reactivity with B cell tumors (Table 1.15 and 1.16). All four antibodies are reactive with most B-CLLs and B cell NHLs. In contrast, NUB1 (B6), HH1 (B30), and BL-14 (B36) are reactive with small numbers of non-T ALLs whereas AB-1 (B8) is reactive with most. Moreover, HH1 is reactive with a small number of T-ALLs and BL-36 is reactive with a small number of T-ALLs and AMLs. *In situ* antigen localization studies demonstrate that anti-HH1

Table 1.16. Non-clustered B restricted: Expressed on all resting B cells. Reactivity with tumor cells.

Code	Non-T ALL	B-CLL	B-lymphoma	Myeloma	AML	T-ALL
B6	5%	70%	70%	0	0	0
B8	70%	70%	70%	0	0	0
B30	10%	90%	80%	0	0	10%
B36	25%	95%	90%	0	20%	30%

antibody is reactive with mantle zone and germinal center B cells. Anti-BL-14 antibody stained mantle zone and germinal center B cells as well as dendritic reticulum cells. A more detailed description of the cellular expression of these antigens on tissue sections can be found in Part III of this volume. *In vitro* activation of resting B cells demonstrated only minimal, if any, change for each of these antigens by six days in culture. These studies, establishing the expression of these antigens on resting and activated normal B cells as well as on neoplastic B cells, suggest that these four B cell antigens are expressed on most stages of B cell ontogeny. The functional significance of any of these four antigens is presently unknown.

Antigens Expressed on Some Resting B Cells

This subgroup is made up of four antibodies which are not expressed on all B cells but demonstrate significant reactivity with some resting B cells. The first two antibodies SHCL-2 (B21) and UL-65 (B46) appear to be reactive with less than 50% of B cells isolated from either peripheral blood or lymphoid tissues (Table 1.17). The actual number of positive cells varies from individual to individual. The weak reactivity of both of these antigens makes it very difficult to accurately quantitate the number of positive cells. The second subgroup of antibodies, consisting of E5A7 (B16) and 2-7 (B27), appears to define a polymorphic B cell-restricted antigen. Both of these antibodies are either clearly unreactive with B cells isolated from peripheral blood and lymphoid tissues or are reactive with greater than 30% of B cells (Table 1.17). This pattern is very reminiscent of a polymorphic class II antigen. Of the four antibodies in this subgroup, none demonstrates reactivity with T or myeloid cells. The molecular nature of these antigens is unknown but at least one laboratory could precipitate the B21 antigen and another the B46 antigen (Table 1.8).

Examination of the reactivity of these antibodies with B cell lines and B cell tumors demonstrates several unique patterns of reactivity. The B21 antibody is reactive with pre–B cell and Burkitt cell lines but is not expressed on B-LBCL or myeloma lines (Table 1.17). It is expressed on some non-T cell ALLs, B-CLLs, and B-NHLs (Table 1.18). The expres-

Table 1.17. Non-clustered B restricted: Expressed on some resting B cells. Reactivity with B cells.

Code	PB %+ (*I*)	Spleen %+ (*I*)	Pre-B lines $n = 4$ (*I*)	B-LBCL $n = 20$ (*I*)	Burkitt's lines $n = 5$ (*I*)	Myeloma lines $n = 4$ (*I*)
? subpopulation						
B21	0–50(±/+)	0–30(+)	4(+)	0	3(+)	0
B46	0–20(+)	0–70(+)	1(+)	12(++)	1(+)	1(+)
Polymorphic						
B16	0/30–50(+)	0/30(+)	1(+)	2(+)	1(+)	0
B27	0/20(+)	0/30–50(+)	3(+/++)	10(++)	3(+/++)	2(+)

Table 1.18. Non-clustered B restricted: Expressed on some resting B cells. Reactivity with tumor cells.

Code	Non-T ALL	B-CLL	B-lymphoma	Myeloma	AML	T-ALL
? subpopulation						
B21	40%	50%	70%	0	0	0
B46	5%	20%	20%	0	0	0
Polymorphic						
B16	10%	30%	30%	0	20%	0
B27	30%	60%	70%	0	20%	0

sion on B-NHLs did not appear to localize to one or more histologically defined subgroups and appeared to be random. In contrast, the B46 antibody is not expressed on pre-B cell lines or Burkitt lines but is expressed on 12 of 20 B-LBCLs (Table 1.17). The B46 antibody is expressed on a small number of non-T cell ALLs, B-CLLs, and B-NHLs. Both antibodies appear to be B lineage restricted (Tables 1.17 and 1.18). The *in situ* expression and expression following activation differ from investigator to investigator and are therefore difficult to comment upon.

The B16 (E5A7) and B27 (2-7) antibodies represent a very different pattern of reactivity. As stated above, on an individual B cell population (either normal or neoplastic) the expression was either definitely positive or definitely negative. The intensity of antigen expression was moderate and similar to that of the CD19 antigen. The B16 antibody was unreactive with virtually every B cell line tested (Table 1.17) and similarly reacted with very few B cell tumors (Table 1.18). Moreover, the pattern of reactivity was again either definitely positive or negative, suggesting that this antigen is expressed on approximately 30% of individuals in a population. The B27 antigen, in contrast, was expressed on most B cell lines and tumor cells isolated from patients with B cell leukemias and lymphomas. Again the pattern of reactivity was moderately strong and clearly either positive or negative. This pattern of reactivity is reminiscent of an antigen expressed on approximately 50% of a population. Both antigens may therefore represent B cell-restricted, polymorphic antigens, a class of antigen not previously described.

Antigens Expressed on Very Few, If Any, Resting B Cells

This subclass has been divided into three "subgroups" on the basis of very preliminary data. The first subgroup appears to include one or more antibodies which may define or be related to the μ chain of IgM. By examination of Tables 1.8, 1.19, and 1.20, it is clear that these antibodies do not recognize a single antigen. Clark found that B2 (BI16-2) and B3 (29-132) precipitated an identical molecular weight structure but B42 (B-7) precipitated a different structure (Table 1.8). In contrast, in the studies of

Table 1.19. Non-clustered B restricted: Not expressed on resting B cells. Reactivity with B cells.

Code	PB %+ (*I*)	Spleen %+ (*I*)	Pre-B lines $n = 4$ (*I*)	B-LBCL $n = 20$ (*I*)	Burkitt's lines $n = 5$ (*I*)	Myeloma lines $n = 4$ (*I*)
(?) μ chain of IgM (p80)						
B2	10(+)	10(+)	0	0	3(+)	0
B3	0–50(+/++)	0–30(+)	2(++)	4(+)	2(+)	0
B42	30–40(+)	0–30(+)	1(+)	20(+)	3(+)	0
Activation antigens						
B23(?p37)	20(+)	20(+)	1(+)	20(++)	2(++)	1(+)
B37(p45)	20(+)	10(+)	0	20(++/+++)	2(+)	1(++)
Others						
B13	10(+)	10(+)	3(+/++)	10(+/−)	0	0
B20	10(+)	10(+)	0	0	0	0
B32	10(+)	10(+)	0	0	0	0

Horibe and Knowles B2 and B42 precipitated a similar molecular weight structure. Comparison of reactivity on normal B cells and B cell lines suggests that these antibodies identify three distinct antigens. Examination of their expression on B cell tumors demonstrates a similarity between B3 and B42. Again these antigens appear to be B lineage restricted. The data for expression on resting B cells and activated B cells, and *in situ* localization are as confused as the above phenotypic and molecular data. Most investigators suspect that the B42 antibody does in fact define the μ chain of IgM.

Two antibodies, 8B1 (B23) and Blast-1 (B37), define unique B cell activation antigens. The 8B1 antigen is similar to BB-1 previously described

Table 1.20. Non-clustered B restricted: Not expressed on resting B cells. Reactivity with tumor cells.

Code	Non-T ALL	B-CLL	B-lymphoma	Myeloma	AML	T-ALL
(?) μ chain of IgM (p80)						
B2	0	5%	40%	0	0	0
B3	10%	20%	30%	0	0	0
B42	10%	40%	30%	10%	0	20%
Activation antigens						
B23(?p37)	30%	40%	50%	0	20%	0
B37(p45)	0	40%	10%	0	0	0
Others						
B13	20%	10%	20%	0	10%	0
B20	0	10%	0	0	0	0
B32	10%	10%	10%	0	0	0

by Yokochi *et al.* (21) and the Blast-1 antigen has been previously characterized by Thorley-Lawson *et al.* (20). Very few, if any, resting B cells express either of these antigens. These antigens are strongly expressed on B-LBCL (Table 1.19) and on resting B cells activated *in vitro*. The B23 (8B1) antigen is expressed following stimulation with anti-Ig, protein A, or EBV whereas the B37 (Blast-1) antigen was expressed following anti-Ig activation by six days but expression under EBV induction required a longer interval (see this volume, Chapter 37).

The B23 antibody defines a glycoprotein of 37 Kd which is expressed on some non-T ALLs, B-CLLs, and B-NHLs (Table 1.20). The B37 antibody defines a 45-Kd antigen which is different from the CD22 antigen and is expressed on a subpopulation of B-CLLs and a very small number of B cell NHLs. These activation antigens appear to be very interesting and are very likely candidates for clusters in the next Workshop.

Finally, the three antibodies B13 (SJ12-3G2), B20 (PC-1), and B32 (9BA-5) also appear to be B lineage restricted. The molecular nature of these antigens is unknown. Very few resting B cells express these antigens and they are expressed on very few B cell lines or tumor cells (Tables 1.19 and 1.20). Very little can be said about these antigens from the studies undertaken in this Workshop. These antigens may be expressed on very discrete stages of B cell differentiation or alternatively on minor B cell populations.

B Cell Associated: Clustered

Pan B Cell

There are three clusters of antibodies which reactive with all B cells but also cross-react with cells of other lineages. In evaluating these B cell-associated antigens we will first identify the lineage expression (Table 1.21), then the expression on normal and B cell lines (Table 1.22), and finally examine the expression on leukemias and lymphomas (Table 1.23). These clusters will not be considered in great detail since the Second International Workshop has decided to stress lineage-restricted antigens.

The first cluster contains two antibodies I-2 (B1) and 7-2 (L8). These antibodies identify Ia-like class II nonpolymorphic antigens. The antigen

Table 1.21. Clustered B cell associated. Cellular reactivities.

Cluster	Resting B cells %+ (*I*)	T cells %+ (*I*)	Macrophage %+ (*I*)	Granulocyte %+ (*I*)	Myeloid lines $n = 5$	T lines $n = 4$
B1, L8	100	0	100(++)	0	3	2
B47, B48	90(+)	0	10(+/−)	100(+++)	0	2
B15, B18	0–40(+)	0	0	100(++)	0	0
B50, B51	100	100	100	0	4	1

Table 1.22. Clustered B cell associated. Reactivity with B cells.

Cluster	PB %+ (*I*)	Spleen %+ (*I*)	Pre-B lines *n* = 4 (*I*)	B-LBCL *n* = 20 (*I*)	Burkitt's lines *n* = 5 (*I*)	Myeloma line *n* = 4 (*I*)
B1, L8	100(+++)	100(+++)	4(+++)	20(+++)	4(+++)	4(++)
B47, B48	90(+)	100(+++)	4	8(+)	3(++)	2(+)
B15, B18	0–40(+)	10–40(+)	3	5(+)	1(++)	1(+)
B50, B51	90(++)	90(++)	2(+)	8(++)	3(+)	1(+)

is a glycoprotein of two chains of 29 Kd and 34 Kd. As has been previously extensively studied, the class II antigen is expressed on B cells and monocytes but is not expressed on resting T cells or granulocytes. The antigen is strongly expressed on virtually all B cells and B cell lines. In addition, nearly all B cell leukemias and lymphomas express this antigen except for myelomas. In addition, the antigen is expressed on virtually all AMLs and a subset of T cell ALLs.

CD24 Cluster

The second cluster also contains two antibodies, HB8 (B47) and HB9 (B48). Although not in the panel of this Workshop the prototype of this cluster is the BA-1 antibody initially reported by Abramson *et al.* (22). This subgroup is now designated the CD24 cluster. The CD24 antigen has been recently precipitated by LeBien (this volume, Chapter 18) and defines a three-chain structure of 45, 55, and 65 Kd. The expression of the CD24 antigen is restricted to B cells and granulocytes although a small number of monocytes also appear to be reactive. The CD24 antigen is expressed on virtually all B cells isolated from peripheral blood and lymphoid tissues. It is expressed on virtually all pre–B cell and Burkitt cell lines but is only expressed on less than 50% of B-LBCLs. This antigen is expressed throughout B cell ontogeny and is found on most non-T cell ALLs, and B cell NHLs. Of note is the observation that it is expressed on 50% of myelomas as well as a subpopulation of AMLs and T-ALLs. This antigen appears to be pan B on both normal and malignant cells. The function of the CD24 antigen is presently unknown.

The third pan B cell cluster includes two antibodies: HB10 (B50) and HB11 (B51). These antibodies define an antigen which appears to be related to the common leukocyte antigen and has a molecular weight of

Table 1.23. Clustered B cell associated. Reactivity with tumor cells.

Cluster	Non-T ALL	B-CLL	B-lymphoma	Myeloma	AML	T-ALL
B1, L8	100%	95%	95%	0	90%	30%
B47, B48	90%	90%	90%	50%	30%	10%
B15, B18	80%	60%	50%	20%	30%	15%
B50, B51	75%	90%	90%	0	70%	80%

220 Kd. This antigen is expressed on virtually all T cells, B cells, and monocytes but is unreactive with granulocytes. Similarly, it is reactive with all B cells isolated from peripheral blood and lymphoid tissues. In addition, 50–80% of mononuclear cells in bone marrow express this antigen. B cell lines generally express this antigen whereas pre–B cell and T cell lines do not (Table 1.22). Finally, virtually all non-T cell ALLs, B-CLLs, B-NHLs, AMLs, and T-ALLs express this antigen. Examination of the expression of this antigen *in situ* demonstrates that it is expressed on mantle zone and germinal center B cells. The biological significance of this common leukocyte antigen is unknown.

Limited B

The last B cell-associated cluster includes two antibodies: AL1a (B15) and AL1c (B18). The antigen defined by these antibodies is weakly expressed on a subpopulation of B cells but not on T cells or monocytes. In contrast, it is strongly expressed on virtually all granulocytes. The antigen is expressed on a subset of B cell lines and normal B cells isolated from peripheral blood and lymphoid tissues. In addition, approximately 70% of bone marrow mononuclear cells strongly express the antigen. The antigen is expressed on most non-T ALLs, 60% of B-CLLs, and 50% of B-NHLs (Table 1.23). It is also expressed on a subpopulation of AMLs and T-ALLs. The molecular structure and functional significance of this antigen are presently unknown.

B Cell Associated: Non-Clustered

There are ten antibodies which are expressed on B lymphocytes and do not appear to be clustered. This subgroup is very heterogeneous with varying expressions on normal and neoplastic hematopoietic cells. Because these antigens are not lineage restricted, they will only be mentioned very briefly. This is not to say that they may not be functionally important in B cell physiology, but the focus of this report is on clustered and lineage-restricted antigens.

Rather than dividing this subgroup into pan B and limited B, we will divide it according to expression on normal fractionated peripheral blood cells. As seen in Tables 1.24 and 1.25, the first subgroup demonstrates expression on B cells, T cells, and monocytes but is not expressed on granulocytes. The second subgroup is differentially expressed on B cells, not expressed on T cells, and is expressed on both monocytes and granulocytes. Finally, the third subgroup is expressed on B cells and monocytes but not on T cells or granulocytes. Within each subgroup, some antigens appear to be pan B cell and others appear to show more limited expression. We will briefly consider each subgroup.

Table 1.24. Non-clustered B-associated. Cellular activities.

Code	Resting B cells %+	T cells %+	Macro-phage %+	Granu-locyte %+	B-LBCL $n = 20$	Pre-B lines $n = 4$	Myelom lines $n = 4$
$T^+M^+G^-$							
B10	100	30–90	30	0	6	0	1
B44	100	30–90	40–60	0	20	1	4
B45	100	0–30	0–10	0	20	1	1
$M^+T^-G^+$							
B17(?p45)	100	0	30	30–80	18	0	2
B26(p26)	0	0	30–50	30–60	5	0	4
B38(p24)	0	0	50–90	20–50	0	2	0
B52(p45)	100	0	30–80	50–100	15	1	3
$M^+T^-G^-$							
B4	100	0	10–30	0	12	0	2
B12	20–30	0	20–50	0	10	3	4
B29(?p90)	20–30	0	30–50	0	20	4	4

$T^+M^+G^-$

Three antibodies made up this subgroup: RW35-1C5 (B10), HB6 (B44), and HB4 (B45). The molecular nature of these antigens is presently unknown. All three antigens appear to be pan B cell by their expression on resting and neoplastic B cells (Tables 1.24 and 1.25). B10 and B44 are clearly expressed on T cells and monocytes and similarly demonstrate reactivity with small numbers of myeloid and T cell tumors. The major difference between these two antigens is their reactivity with B-LBCL (Table 1.24). In contrast, the B45 antigen is expressed on much smaller numbers of T cells and monocytes and only reacts with a small percentage

Table 1.25. Non-clustered B-associated. Reactivity with tumor cells.

Code	Non-T ALL	B-CLL	B-lymphoma	Myeloid	AML	T-ALL
$T^+M^+G^-$						
B10	40%	95%	75%	10%	0	10%
B44	60%	95%	90%	10%	10%	40%
B45	30%	60%	50%	0	20%	0
$M^+T^-G^+$						
B17	50%	95%	75%	60%	25%	70%
B26	5%	0	10%	60%	20%	0
B38	60%	10%	10%	0	10%	10%
B52	20%	95%	60%	60%	20%	30%
$M^+T^-G^-$						
B4	10%	95%	30%	50%	10%	60%
B12	70%	50%	75%	10%	0	0
B29	30%	10–20%	50%	50%	50%	20%

of AML cells. This antigen must be more extensively studied to determine if the cross-reactivities are real before it can be established whether this should be considered a B cell-restricted antigen. The *in situ* expression and expression following activation can be noted in other reports in this volume.

M$^+$T$^-$G$^+$

The four antibodies in this group are HD28 (B17), PCA-1 (B26), 21D-10 (B38), and 41H16 (B52). This group of antibodies can be further divided into two subgroups on the basis of their expression on resting B cells. The B17 and B52 antibodies are expressed on most B cells and therefore are pan B cell antigens. In contrast, the B26 and B38 antigens are not expressed on resting B cells. As seen in Tables 1.24 and 1.25, the reactivities of B17 and B52 on normal and neoplastic B cells are very close and both may define the same p45 antigen. Although there were clearly differences in cellular expression, these two antigens may represent another cluster. The B26 antigen initially reported by Anderson *et al.* (23) is not expressed on resting B cells but is expressed on monocytes and granulocytes. It is expressed on a percentage of B-LBCLs and all myeloma lines tested. It is not expressed on non-T ALLs, B-CLLs, or B-NHLs but is expressed on all myelomas tested. It therefore appears to be a plasma cell-associated antigen. Although not expressed on T cell tumors, it clearly cross-reacts with myeloid malignancies. Finally, the B38 antigen appears to define a 24-Kd protein. This antigen is not expressed on resting B cells and therefore is clearly distinct from the CD9 cluster. The antigen is clearly expressed on normal myeloid cells but is not significantly expressed on myeloid tumors (Table 1.24 and 1.25). The antigen is reactive with non-T cell ALLs and with very few B-CLLs and B-NHLs. This pattern of reactivity suggests that this antigen has a very limited expression on pre–B cells and may be of considerable interest functionally.

M$^+$T$^-$G$^-$

This final subgroup contains three antibodies: HH1 (B4), UL-38 (B12), and H616 (B29). The B4 antigen appears to be a pan B cell antigen whereas the B12 and B29 antigens appear to have limited expression (Tables 1.24 and 1.25). The molecular nature of these structures is presently unknown. The B4 antigen is expressed on resting B cells and small numbers of monocytes. It is clearly expressed on myeloid and T cell tumors. It is expressed on very few non-T ALLs, most B-CLLs, and a small percentage of B-NHLs. In contrast to B4, the B12 and B29 antigens are only expressed on a subpopulation of B cells. Although only expressed on some resting B cells, they are definitely expressed on pre–B cell, B-LBCL, and myeloma lines. Similarly, there is definite expression of these antigens on non-T cell ALL and B-NHL cells. The major difference between these two antigens is the expression on B-LBCLs—B12, 10 of 20, and B29, 20 of 20. In addition, B12 shows minimal reactivity with T cell and myeloid tumors whereas B29 shows more significant activity.

Leukemia Panel Antibodies

The leukemia panel was composed of 21 antibodies (L1–L22; code designation L5 had no sample). These antibodies could be divided into five subgroups (Tables 1.2 and 1.10).

The first subgroup contained only one antibody, L17 (SJ25-C1), which turned out to be a CD19 antigen.

The next subgroup included all antibodies which reacted with the CALLA antigen (Table 1.10). This cluster (CD10) contained seven antibodies (Table 1.10) including the prototype J5 antibody described by Ritz and his colleagues (24). As seen in Table 1.10, the CALLA antigen is expressed on granulocytes, on 80% of non-T cell ALLs, and on a small number of T cell ALLs. Previous studies have demonstrated that CALLA is also expressed on a subpopulation of thymocytes (25), and cells in the fetal and adult kidney. In addition, the CALLA antigen is also expressed on a subpopulation of normal pre-B cells (7,11,12). To date, the function of the CALLA antigen is not known.

The third cluster defines the previously designated CD9 antigen. This antigen is a 24-Kd glycoprotein which is expressed on B cells, monocytes, and granulocytes. In the leukemia panel, five antibodies were contained in this cluster (Table 1.10). The prototypes of this cluster are the BA-2 (26) and J2 (27) antibodies. This antigen is expressed on leukemias of T, B, and myeloid origins (Table 1.10). The nature of this antigen has been extensively discussed in the First Workshop.

The fourth cluster includes four antibodies which are reactive with the transferrin receptor (gp 90 Kd). CD status has not been assigned to this molecule. The reactivity of these four antibodies is summarized in Table 1.10. This structure has been extensively studied in recent years.

Finally, three antibodies remain unclustered. Two of these antibodies 3-3 (L7) and 6-4 (L9) are very useful in identifying T cell leukemias and lymphomas (Table 1.10). L7 is not expressed on normal cells and appears to be a very interesting antigen. L1 (SJ9-2E2) is not expressed on normal cells but is expressed on a subgroup of non-T and T cell ALLs. The molecular nature of the L1 and L7 antigens is unknown. The L9 antigen appears to be a p160.

Summary and Future Directions

Figure 1.1 summarizes the expression of the clustered B cell and leukemia antigens of the Second International Workshop. Figure 1.2 summarizes the expression of these antigens in situ. This Workshop has been an

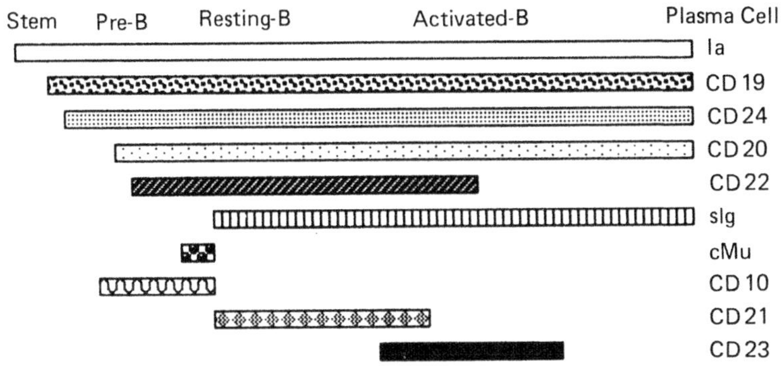

Fig. 1.1 Hypothetical model of B-cell antigen expression.

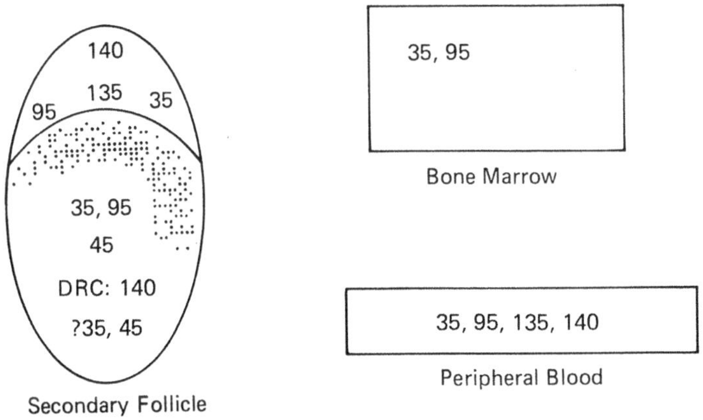

Fig. 1.2 Expression of antigens in tissue sections: Summary of B cell specific antigens.

enormous success and the participants should be congratulated. I predict that in the next Workshop, we will focus on molecules that regulate B-cell function.

Appendix

Addresses of Laboratories Submitting Antibodies or Participating in the Wet Workshop of the B Cell/Leukemia Section of the Second International Congress

Lab #7 J. Brochier, J.P. Magaud, G. Cordier, O. Gentilhomme
INSERM U80, Hopital Ed.-Herriot, Pav. P., 69374 Lyon Cedex, France

Lab #8 Michael A. Horton
Department of Haematology, St. Bartholomew's Hospital, London EC1A 7GE, United Kingdom
Lab #9 Yasuo Morishima, Saburo Monami, Yoshio Okumura
First Dept. of Internal Medicine, Nagoya University, 65 Tsurumai-cho, Showa-ku, Nagoya 466, Japan
Lab #14 Domenico Delia
Dept One. Sper "C", Instituto Nazionale Tumori, Via G. Venezia, 1, 20100 Milan, Italy
Giorgio Cattovetti
Cattedra di Puericultura, Universita de Milano, 20100 Milan, Italy
Lab #15 A.F.G. von den Borne, P.A.T. Tetteroo, M.B. Veer
Central Laboratory of the Netherlands Red Cross, Blood Transfusion Service, Plesmanlaan 125, 1066 CX Amsterdam, The Netherlands
Lab #20 Steinar Funderud, Heidi Kiil Blomhoff, Tore Godal
Laboratory for Immunology, The Norwegian Radium Hospital, Montebello, Oslo 3, Norway
Lab #21 John S. Thompson
University of Kentucky, Dept. of Medicine, RM. MN62, 800 Rose Street, Lexington, KY 40536
Lab #22 P.C. Beverley, J. Golay, S. Smith, G. Hariri
Imperial Cancer Research Fund, ICRF Human Tumour Immunology Group, University College of London, University Street, London WC1E 6JJ, United Kingdom
Lab #26 Robert W. Knowles, Keizo Horibe
Sloan-Kettering Institute, 1275 York Avenue, Box 41, New York, NY 10021
Lab #27 A.J. McMichael, F.M. Gotch
Nuffield Dept. of Medicine, John Radcliffe Hospital, Headington, Oxford OX3 9DU, United Kingdom
Lab #28 Kazuyuki Naito, Neal Flomenberg, Nancy Kernan
Human Immunogenetics Laboratory, Memorial Sloan-Kettering Cancer Center, 1250 First Avenue, New York, NY 10021
Lab #31 Susan L. Melvin
St. Jude Children's Research Hospital, 332 North Lauderdale, P.O. Box 318, Memphis, TN 38101
Lab #34 Tucker W. LeBien
University of Minnesota, Box 609 Mayo, Minneapolis, MN 55455
Lab #38 P. Poncelet
Center Research Clin-Midy, Ave Blayac, 34024 Montpellier Cedex, France
Lab #39 G.R. Pilkington, G.T.H. Lee, H. Thorne, D.G. Jose
Immunology Unit, Cancer Institute, 481 Lt. Lonsdale Street, Melbourne 3000, Australia

Lab #40 J.A. Habeshaw, Lesley J. Murray, Margaret Rainey
ICRF Medical Oncology Unit, St. Bartholomew's Hospital, West Smithfield, London EC1A 7BE, United Kingdom
Lab #41 D.Y. Mason
Dept. of Haematology, Nuffield Dept. of Pathology, John Radcliffe Hospital, Headington, Oxford OX3 9DU, United Kingdom
Lab #42 Robert Winchester, Jack Silver, Sanya Goyert
Mount Sinai School of Medicine, Hospital for Joint Diseases, Orthopaedic Institute, 301 East 17th Street, New York, NY 10003
Lab #45 Branislav D. Janković
Immunology Research Centre, Vojvode Stepe 458, 11221 Belgrade, Yugoslavia
Lab #47 Noel L. Warner
Becton Dickinson Monoclonal Center, 2375 Garcia Avenue, Mountain View, CA 94043
Lab #52 John H. Kersey
Univ. of Minnesota Hospitals and Clinics, 420 S.E. Delaware, Box 86, Minneapolis, MN 55455
Lab #54 Marie-Christine Favrot
Centre Leon Berard, Laboratoire d'Immunologie, INSERM U218, 28 rue Laennec, 69373 Lyon Cedex 2, France
Lab #57 Gorm Pallesen
University Institute of Pathology, Kommunehospitalet, DK-8000 Aarhus C, Denmark
Lab #59 Ephraim Gazit
Chaim Sheba Medical Center, Tel-Hashomer, Israel 52621
Lab #60 N. Kraft, W.W. Hancock, R.C. Atkins
Prince Henry's Hospital, St. Kilda Road, Melbourne 3004, Australia
Lab #63 W.P. Zeijlemaker
Central Lab. Netherlands Red Cross, Blood Transfusion Service, P.O. Box 9190, 1006 AD Amsterdam, The Netherlands
Lab #64 A Wolpl, A. Raghavachar, K. Koerner
Red Cross Ulm, Transfusion Medicine, Oberer Eselsberg 10, 79 Ulm 1, West Germany
Lab #66 Edward Clark
Immunobiology Group, Genetic Systems Corporation, 3005 First Avenue, Seattle, WA 98121
Lab #67 Anne-Marie Lebacq-Verheyden, Anne-Marie Ravoet
University of Louvain, Unité de Recherches sur les Maladies du Sang, UCL 30.52, Clos Chapelle-aux-Champs, 30, 1200 Brusells, Belgium
Lab #68 Teresa Gallart, Ignacio Anegón, Christina Cuturi, Jordi Vives
Servicio Immunogolía, Hospital Clínico y Provincial, Casanova 143, Barcelona 36, Spain
Lab #70 Roland Schwarting
Sloan-Kettering Institute, 1275 York Avenue, New York, NY 10021

Lab #71 Kimitaka Sagawa, Keiji Okubo, Yoshinobu Matsuo, M. Mitsuo Yokoyama
Dept. of Immunology, Kurume University School of Medicine, 67 Asahi-machi, Kurume 830, Japan
Shizuo Hagiwara, Masato Shiraishi
Nippon Reizo K.K. Research and Development Lab, 1-52-14 Kumegawa-cho, Hhigashimurayama 189, Japan
Lab #73 Lee M. Nadler
Dana-Farber Cancer Institute, 44 Binney Street, Boston, MA 02115
Lab #79 John A. Hansen, Paul J. Martin, Seymour J. Klebanoff, Patrick G. Beatty
Fred Hutchinson Cancer Research Center, 1124 Columbia Street, Seattle, WA 98104
Lab #80 Patrice Mannoni
Dept. of Immunology-Pathology, Medical Sciences Building, R 8-17, University of Alberta, Edmonton, Alberta, Canada
Lab #83 P.M. Lansdorp, P. van Mourik, W.P. Zeijlemaker
Dept. of Immunobiology, Central Lab. of Netherlands Red Cross, Blood Transfusión Service, Plesmanlaan 125, 1006 AD Amsterdam, The Netherlands
Lab #85 C. Boucheix, J.Y. Perrot, M. Mirshahi, C. Rosenfield, C. Soria, J. Soria
INSERM U253, Hôpital Paul Brousse, 16 bis, av. Paul Vaillant Couturier, 94804 Villejuif, France
Lab #86 Jeffrey Cossman, Leonard Neckers
Laboratory of Pathology, NCI, NIH, Building 10, Rm 2NID8, NIH, Bethesda, MD 20205
Lab #92 P. Wernet
Medizin Univ. Klinik, Alfried Miller Strasse 10, D7500 Tübingen, West Germany
Lab #93 I. Royston, R. Dillman
University of California, 3350 La Jolla Village Drive, VIIIE, San Diego, CA 92161
Lab #94 A.C. Feller, M.D.
Inst. of Pathology, University of Kiel, Hospitalstrasse 42, 2300 Kiel, West Germany
Lab #96 D. Bourel, N. Genetet et al.
Centre Régional de Transfusion et Groupe de Recherche en Immunologie, Fondamentale et Appliquée (GURIFA), Rue Pierre-Jean Gineste, 3500 Rennes, France
Lab #101 Jerome Ritz
Dana-Farber Cancer Institute, 44 Binney Street, Boston, MA 02115
Lab #102 B. Dörken, A. Pezzutto
Medizinische Universitäts-Poliklinik, Hospitalstrasse 3, D-6900 Heidelberg, West Germany

G. Moldenhauer, R. Schwartz, G.J. Hammerling
Institut für Immunologie und Genetik, Deutsches Krebsforschungszentrum, D-6900 Heidelberg, West Germany
Lab #104 Max D. Cooper, Sheila Saunders
Cellular Immunobiology Unit, University of Alabama, 224 Tumor Institute, University Station, Birmingham, AL 35294
Lab #106 C.M. Steel
MRC Clinical & Population Cytogenetics Unit, Western General Hospital, Crewe Road, Edinburgh EH4 2XU, United Kingdom
Lab #107 Jesper Heldrup, Stanislaw Garwicz
Pediatric Oncology, University Hospital, S221-85 Lund, Sweden
Lab #110 G. Johnson, N.R. Ling, I.C.M. MacLennan, P. Nathan
Dept. of Immunology, University of Birmingham, Vincent Drive, Edgbaston, Birmingham B15 2TJ, United Kingdom
Lab #112 Elisabeth Paietta, Peter H. Wiernik
Montefiore Medical Center, Albert Einstein College of Medicine, 111 East 210th Street, Bronx, NY 10467
Lab #113 Michael Henke
City of Hope National Medical Center, 1500 East Duarte Road, Duarte, CA 91010
Lab #120 Thomas F. Tedder, Max D. Cooper
Cellular Immunobiology Unit, University of Alabama, 224 Tumor Institute, University Station, Birmingham, AL 35294

References

1. Bernard, A., L. Boumsell, and C. Hill. 1984. Joint report on the First International Workshop on Human Leucocyte Differentiation Antigens: B2 protocol. In: *Leucocyte typing,* A. Bernard, L. Boumsell, J. Dausset, C. Milstein, and S.F. Schlossman, eds. Springer-Verlag, Berlin, Heidelberg, pp. 61–81.
2. Nadler, L.M., K.C. Anderson, M. Bates, E. Park, B. Slaughenhoupt, and S.F. Schlossman. 1984. Human B Cell Associated Antigens: Expression on Normal and Malignant B Lymphocytes. In: *Leucocyte typing,* A. Bernard, L. Boumsell, J. Dausset, C. Milstein, and S.F. Schlossman. Springer-Verlag, Berlin, Heidelberg, pp. 354–363.
3. Clark, E.A., and T. Kokochi. 1984. Human B cell and B cell blast-associated molecules defined with monoclonal antibodies. In: *Leucocyte typing,* A. Bernard, L. Boumsell, J. Dausset, C. Milstein, and S.F. Schlossman. Springer-Verlag, Berlin, Heidelberg, pp. 339–346.
4. LeBien, T.W., J.G. Bradley, D.R. Boue, J. Platt, A.F. Michael, and J.H. Kersey. 1984. B cells and kidneys: A "B + CALLA" workshop analysis. In: *Leucocyte typing,* A. Bernard, L. Boumsell, J. Dausset, C. Milstein, and S.F. Schlossman. Springer-Verlag, Berlin, Heidelberg, pp. 346–353.
5. Zola, H., J.G., Bradley, D.A. Brooks, P.J. Macardle, P.J. McNamara, H.A. Moore, and Nikoloutsopoulos. 1984. Human B cell lineage studied with monoclonal antibodies. In: *Leucocyte typing,* A. Bernard, L. Boumsell, J. Dausset,

C. Milstein, and S.F. Schlossman. Springer-Verlag, Berlin, Heidelberg, pp. 363–371.

6. Nadler, L.M., K.C. Anderson, G. Marti, M. Bates, E. Park, J.F. Daley, and S.F. Schlossman. 1983. B4, a human B cell associated antigen expressed on normal, mitogen activated, and malignant B lymphocytes. *J. Immunol.* **131:**244.
7. Nadler, L.M., S.J. Korsmeyer, K.C. Anderson, A.W. Boyd, B. Slaughenhoupt, E. Park, J. Jensen, F. Coral, R.J. Mayer, S.E. Sallan, J. Ritz, and S.F. Schlossman. 1984. The B cell origin of non-T cell acute lymphoblastic leukemia: A model for discrete stages of neoplastic and normal pre-B cell differentiation. *J. Clin. Invest.* **74:**332.
8. Stashenko, P., L.M. Nadler, R. Hardy, and S.F. Schlossman. 1980. Characterization of a new B lymphocyte specific antigen in man. *J. Immunol.* **125:**1678.
9. Nadler, L.M., P. Stashenko, J. Ritz, R. Hardy, J.M. Pesando, and S.F. Schlossman. 1981. A unique cell surface antigen identifying lymphoid malignancies of B cell origin. *J. Clin. Invest.* **67:**134.
10. Anderson, K.C., B. Slaughenhoupt, M.P. Bates, G. Pinkus, S.F. Schlossman, and L.M. Nadler. 1984. Expression of human B cell associated antigens on leukemias and lymphomas: A model of human B cell differentiation. *Blood* **63:**1424.
11. Hokland, P., P. Rosenthal, J.D. Griffin, L.M. Nadler, J. Daley, M. Hokland, S.F. Schlossman, and J. Ritz. 1983. Purification and characterization of fetal hematopoietic cells which express the common acute lymphoblastic leukemia antigen (CALLA). *J. Exp. Med.* **157:**114.
12. Hokland, P., L.M. Nadler, J.D. Griffin, S.F. Schlossman, and J. Ritz. 1984. Purification of the common acute lymphoblastic leukemia antigen (CALLA) positive cells from normal bone marrow. *Blood* **64:**662.
13. Oettgen, H.C., P.J. Bayard, W. vanEwijk, L.M. Nadler, and C. Terhorst. 1983. Further biochemical studies of the human B cell differentiation antigens B1 and B2. *Hybridoma* **2:**17.
14. Stashenko, P., L.M. Nadler, R. Hardy, and S.F. Schlossman. 1981. Expression of cell surface markers following human B lymphocyte activation. *Proc. Natl. Acad. Sci. U.S.A.* **78:**3848.
15. Nadler, L.M., P. Stashenko, R. Hardy, A. van Agthoven, C. Terhorst, S.F. Schlossman. 1981. Characterization of a human B cell specific antigen (B2) distinct from B1. *J. Immunol.* **126:**1941.
16. Tedder, T.F., L. Clement, and M. Cooper. 1984. Expression of C3d receptors during human B cell differentiation: Immunofluorescent analysis with the HB-5 monoclonal antibody. *J. Immunol.* **133:**678.
17. Iida, K., L.M. Nadler, and V. Nussenzweig. 1983. Identification of the membrane receptor for the complement fragment C3d by means of a monoclonal antibody. *J. Exp. Med.* **158:**1021.
18. Weis, J.J., T. Tedder, and D.T. Fearon. 1984. Identification of a 145,000 Mr membrane protein as the C3d receptor (CR2) of human B lymphocytes. *Proc. Natl. Acad. Sci. U.S.A.* **81:**881.
19. Fingeroth, J.D., J. Weis, T.F. Tedder, J.L. Strominger, P.A. Biro, and D.T. Fearon. 1984. Epstein–Barr virus receptor of human B lymphocytes is the C3d receptor CR2. *Proc. Natl. Acad. Sci. U.S.A.* **81:**4510.

20. Thorley-Lawson, D.A., R.T. Schooley, A.K. Bhan, and L.M. Nadler. 1982. Epstein–Barr virus superinduces a new human B cell differentiation antigen (B-last-1) expressed on transformed lymphoblasts. *Cell* **30:**415.
21. Yokochi, T., R.D. Holly, and E.A. Clark. 1982. B lymphoblast antigen (BB-1) expressed on Epstein–Barr virus-activated B cell blasts, B lymphoblastoid cell lines, and Burkitt's lymphomas. *J. Immunol.* **128:**823.
22. Abramson, C., J.H. Kersey, and T.W. LeBien. 1981. A monoclonal antibody (BA-1) primarily reactive with cells of human B lymphocyte lineage. *J. Immunol.* **126:**83.
23. Anderson, K.C., K. Park, M. Bates, R.C.F. Leonard, S.F. Schlossman, and L.M. Nadler. 1983. Antigens on human plasma cells identified by monoclonal antibodies. *J. Immunol.* **130:**1132.
24. Ritz, J., J.M. Pesando, J. Notis-McConarty, H. Lazarus, and S.F. Schlossman. 1980. A monoclonal antibody to human acute lymphoblastic leukemia antigen. *Nature* **283:**583
25. LeBien, T.W., personal communication.
26. Kersey, J.H., T.W. LeBien, C.S. Abramson, R. Newman, R. Sutherland, and M. Greaves. 1981. p24: A human hemopoietic progenitor and acute lymphoblastic leukemia-associated cell surface structure identified with a monoclonal antibody. *J. Exp. Med.* **153:**726
27. Hercend, T., L.M. Nadler, J.M. Pesando, S.F. Schlossman, E.L. Reinherz, and J. Ritz. 1981. Expression of a 26,000 dalton glycoprotein on activated human T cells. *Cell. Immunol.* **64:**192.

Part II. Serologic Specificity of B Cell/Leukemia Monoclonal Antibodies

CHAPTER 2

Analysis of the B Cell/Leukemia Workshop Monoclonal Antibodies Using an Immunoenzymatic Staining Assay and a Radioimmunoassay on Cells

Bernd Dörken, Gerhard Moldenhauer, Antonio Pezzutto, Reinhard Schwartz, Sophie Kiesel, and Werner Hunstein

Introduction

For the Second International Workshop on Human Leukocyte Differentiation Antigens we have tested the whole antibody panel of the B cell/leukemia Workshop with more than 50 different types of leukemias and lymphomas as well as a panel of normal cells. Included in the B panel are four antibodies produced by our group in Heidelberg (HD): B17 (HD28), B25 (HD6), B28 (HD37), and B31 (HD39). The two main reasons for testing such a large number of leukemias/lymphomas were:

1. Leukemias and lymphomas may be malignant counterparts of small subpopulations of normal cells or stages of differentiation which are underrepresented; therefore reactivity with such normal cell types can remain undetected. Thus leukemias and lymphomas present the opportunity of realizing cross-reactivity.
2. We are especially interested in antibodies as diagnostic reagents for leukemia typing. Therefore, antibodies with lineage specificity are of particular interest to us.

We have used two types of methods for testing:

1. Immunoenzymatic staining assay: this test is economical and allows testing of a large number of different cell types; in addition the detection of cytoplasmic antigens is possible.
2. RIA on cells: this test is especially useful when investigating weakly reacting monoclonal antibodies (mAbs).

Materials and Methods

Immunoenzymatic Staining Assay (IE) (1)

Preparation of Target Cells

Human peripheral blood mononuclear cells were isolated by Ficoll–Hypaque density gradient (FH) centrifugation. T cells were isolated from mononuclear cells by E-rosetting and fractionating the E^+ and the E^- cells on FH. The E^+ population was 95% $OKT3^+11^+$. The E^- preparation was enriched for B cells by removing the adherent monocytes (70% SIg^+). Monocytes were obtained by adherence to plastic dishes. Granulocytes were prepared from the cell pellet after FH centrifugation. Erythrocytes were removed by gravity sedimentation in the presence of 0.4% Dextran followed by lysis. Normal mononuclear bone marrow cells were recovered by FH centrifugation. Erythrocytes were prepared from the cell pellet after FH centrifugation. Tonsils were obtained at the time of routine tonsillectomy. Normal human thymocytes were obtained from patients during corrective cardiac surgery. Tissue specimens were finely minced and made into single-cell suspensions by extrusion through stainless steel mesh. Leukemic cells were obtained from peripheral blood and bone marrow of patients with leukemia or malignant lymphoma. Diagnosis was made using standard clinical, morphological, and cytochemical criteria. The histopathological diagnosis was determined according to the Kiel classification. In addition, leukemic cells were examined for the presence of various markers (HLA-DR, TdT, CALLA (J5), T antigens (OKT-series), OKM1, B1, BA1, surface and cytoplasmic Ig). 73 different types of leukemias and lymphomas were tested:

B leukemias/lymphomas (BL): B-type chronic lymphocytic leukemia (B-CLL) ($n = 12$); immunocytoma (IC) = lymphoplasmacytic lymphoma ($n = 7$); hairy cell leukemia (HCL) ($n = 4$); prolymphocytic leukemia (PLL) ($n = 3$); immunoblastic lymphoma (IB) ($n = 3$).
Plasma cell leukemia (PCL) ($n = 1$).
T leukemias/lymphomas (TL): T-CLL ($n = 3$); Sézary syndrome ($n = 4$); T type acute lymphoblastic leukemia (T-ALL) ($n = 4$).
Non-T acute lymphoblastic leukemia (ALL) ($n = 18$).
Acute myelo/monoblastic leukemia (ML) ($n = 14$).

Immunoenzymatic Staining

Target cells were plated in Terasaki plates (5×10^4/well) and fixed with 0.025% glutaraldehyde (RT, 10 min). Cells were incubated in sequence with mAb (10 μl, final dilution of 1:250 in Hank's containing 10% human serum, RT, 30 min, 3× wash), rabbit anti–mouse Ig (affinity-purified, 1 mg/ml, 10 μl, 1:50, RT, 30 min, 3× wash), and goat anti–rabbit Ig–

alkaline phosphate conjugate (Tago; 10 μl, 1 : 100, RT, 30 min, 3× wash). Staining was performed with a solution of fast-red-TR-salt (Sigma) and naphtol-AS-BI-phosphate (Sigma). For controls mAbs, B1, OKT3, HD11 (HLA-DR framework), and irrelevant mouse mAbs of different Ig isotypes were always included.

Radioimmunoassay on Cells (RIA) (2)

Cell Preparation

T cells were isolated from mononuclear cells by E-rosetting and fractionating the E^+ and E^- cells on FH (2×). The E^+ population was 95% $OKT3^+11^+$. Granulocytes were prepared from the cell pellet. Erythrocytes were removed by gravity sedimentation in the presence of 0.4% Dextran followed by lysis. The following cell lines were used:

T cell lines: Jurkat, JM-1, MOLT-4, CEM-C7.
Promyelocytic line HL-60.
B cell lines: Raji, Daudi, RAMOS, P3HR-1, BJAB, WI-L2 HF2, LICR-LON-HMY2.

Test Technique

Polyvinyl chloride microtiter plates (U-bottomed, Dynatech) were pretreated with BSA in order to saturate nonspecific sites. Viable cells (10^6/well) diluted in PBS plus 10% Gamma-Venin (Behringwerke) were added. mAb (50 μl, final dilution 1 : 250 in Hank's containing 10% human serum) were incubated in triplicates for 1 hr. The plates were washed 3 times and subsequently incubated with ^{125}I-labeled rabbit anti–mouse Ig (affinity-purified, chloramine-T iodination procedure, diluted in PBS-BSA-Gamma-Venin, 50 μl) for another hour. After washing, radioactivity in individual wells was measured. For controls mAbs B1, OKT3, HD11 (HLA-DR framework), and irrelevant mAbs of different Ig isotypes were always included.

Results

Immunoenzymatic Staining Assay (IE)

The mAb B31 (HD39) does not react with CLL/ALL cells in suspension. An unexpected result was the finding of a broad reactivity with tumors of these cell types in immunohistology (Table 2.1). Cytocentrifuge preparations revealed that the corresponding antigen is also expressed in cytoplasm. In contrast to studies of cells in suspension most cases of CLL and ALL were weakly positive in IE. Pretreatment of the cells with the deter-

Table 2.1. Reactivity of the monoclonal antibody B31 (HD39) with surface and cytoplasmic antigen.[a]

	Cells in suspension (IF)	Tissue sections (IE)	Cytocentrifuge preparations (IF)	Cells in Terasaki plates (IE) Glut.	Cells in Terasaki plates (IE) Glut. + BRIJ
CLL	0/26	8/8	6/6	4/6 (+)[b]	6/6 +
ALL/LB	0/20	9/10	7/8	6/6 (+)	6/6 +

[a] Abbreviations: IF = Indirect immunofluorescence; IE = immunoenzymatic staining; Glut. = fixation with 0.025% glutaraldehyde, 10 min RT; BRIJ = BRIJ 56 0.5%, 15 min RT; CLL = chronic lymphocytic leukemia; ALL = acute lymphoblastic leukemia; LB = lymphoblastic lymphomas (in tissue sections).
[b] (+) = Weak reaction; + = moderate reaction.

gent BRIJ intensified the staining. Therefore, it can be assumed that in this test during the fixation procedure, even without BRIJ treatment, cells become partially permeable so that cytoplasmic antigens become available for antibody binding. So it can be concluded that the spectrum of positive reactions with IE is broader as compared to the usual immunofluorescence staining of cells in suspension.

B Cell Panel

On the basis of the extensive number of tests performed with this method in combination with RIA data we were able to group the Workshop panel into four different categories:

Table 2.2. Antibodies with multiple cross-reactions (immunoenzymatic staining).[a]

	B	T	M	G	My		TL (n = 9)	ML (n = 9)	ALL (n = 13)	BL (n = 20)
B10	●[b]	●					2[c] ●	0	6 ●	19 ●
B41	○	○					2 ○	2 ○	3 ○	10 ○
B42				○			3 ○	2 ○	1 ●	7 ●
B44	●	●	○			E[d]	8 ●	3 ●	2 ●	19 ●
B45	●	●	○	○		E	6 ●	8 ○	11 ○	18 ●
B50	●	●	●				9 ●	9 ○	10 ●	19 ●
B51	●	●	●				5 ●	2 ○	8 ●	19 ●
B1	●	○	●		●		1 ●	8 ●	13 ●	19 ●
B7	●			○			0	2 ○	10 ●	19 ●
B12	●		●		○		0	5 ●	10 ●	15 ●
B23			○	○			3 ○	7 ○	3 ○	7 ○
B26					○		4 ○	6 ●	0	0
B29	○		●		○		1 ○	4 ●	5 ○	10 ○

[a] Abbreviations: B = Peripheral blood B cells (E^-); T = peripheral blood T cells (E^+); M = monocytes; G = granulocytes; My = bone marrow myeloid cells; TL = T cell leukemias/lymphomas; ML = myelo-monocytic leukemias; ALL = acute lymphoblastic leukemia; BL = B cell leukemias/lymphomas.
[b] ● = strong reaction; ○ = weak reaction.
[c] Number of positive cases.
[d] E = Reaction with erythrocytes.

Table 2.3. Antibodies which are not restricted to the B cell lineage but may be useful for leukemia typing (immunoenzymatic staining).[a]

	B	T	M	G	My		TL ($n = 11$)	ML ($n = 14$)	ALL ($n = 18$)	BL ($n = 29$)
B4	●						6[c] ○	4 ○	2 ●	25 ●
B17	●		○	○			9 ○	9 ○	5 ●	26 ●
B36	●			○			4 ○	6 ○	4 ●	26 ●
B52	●		○	●	●		2 ○	11 ●	3 ●	27 ●
B15	○			●	●		1 ○	3 ○	14 ●	20 ●
B18				●	●		0	0	12 ●	18 ●
B38					○	Th[b]	0	2 ●	7 ●	0
B47	●			●	●		2 ●	1	15 ●	28 ●
B48	●			●	●		2 ●	1	15 ●	26 ●

[a] For definitions of abbreviations and symbols, refer to Table 2.2
[b] Th = Reaction with thrombocytes.
[c] Number of positive cases.

1. Antibodies with multiple cross-reactions: 22 of the 52 B cell Workshop antibodies showed multiple cross-reactions with T cells and myelo-monocytic cells in IE (Tables 2.2 and 2.3). Apart from three cases (B7, B12, and B38), the results were confirmed by RIA (Tables 2.4 and 2.5). Considering the cross-reactivity pattern with myelo-monocytic cells the correlation between IE and RIA results was only partial. On the other hand RIA results were confirmed in every case by IE. Nine of the cross-reacting antibodies were put in a separate category because their reaction pattern was useful for the discrimination between different leukemias.

Table 2.4. Antibodies with multiple cross-reactions (RIA on cells).[a]

					Burkitt's					B-Lb	
	T	G	TL	HL-60	Raji	Daudi	RAMOS	P3HR1	BJAB	HF2	HMY2
B10	±[b]	−	−	−							
B41	+	−	±	−							
B42	++	++	++	++	++	++	++	++	++	++	++
B44	+	−	±	−							
B45	±	−	−	−							
B50	+	−	+	−							
B51	+	−	+	−							
B1	+	−	−	−							
B7	−	−	−	−	±	+	±	±	+	±	+
B12	−	−	−	−							
B23	−	±	−	−							
B26	−	+	+	++							
B29	−	+	+	−							

[a] Abbreviations: T = Peripheral blood T cells (E^+); G = granulocytes; TL = T cell lines (Jurkat, JM-1, MOLT-4, CEM-C 7); HL-60 = promyelocytic leukemia line; Burkitt's = Burkitt's lines; B-Lb = B-lymphoblastoid lines; HF2 = WI-L2 HF2; HMY2 = LICR-LON-HMY2.
[b] − = Negative (1–2 × background); ± = weakly positive (2–3 × background); + = positive (3–6 × background); ++ = strongly positive (>6× background).

Table 2.5. Antibodies which are not restricted to the B cell lineage but may be useful for leukemia typing (RIA on cells).[a]

					Burkitt's					B-Lb	
	T	G	TL	HL-60	Raji	Daudi	RAMOS	P3HR1	BJAB	HF2	HMY2
B4	±	+	±	−	+	++	++	++	+	+	+
B17	+	+	+	−	+	++	++	++	++	++	++
B36	−	±	±	−							
B52	±	++	−	+							
B15	−	+	−	−							
B18	−	±	−	−							
B38	−	−	−	−							
B47	−	++	−	−							
B48	−	++	+	+							

[a] For definitions of abbreviations and symbols, refer to Table 2.4.

Table 2.6. Reaction pattern of B cell workshop antibodies exhibiting cross-reactions but useful for leukemia typing.

Workshop antibody	Reaction pattern[a]
B4, B17 (HD28), B36, B52	BL +, ALL −
B15, B18	HCL −, PLL −, other BL +
B38	ALL +/−, BL −
B47, B48	ALL +, AML −

[a] BL = B leukemias/lymphomas; ALL = acute lymphoblastic leukemia; AML = acute myeloblastic leukemia; HCL = hairy cell leukemia

Table 2.7. Antibodies with undefined specificity (immunoenzymatic staining).[a]

	B	T	M	G	My	TL (n = 9)	ML (n = 9)	ALL (n = 13)	BL (n = 20)
B3								2[b] ○	4 ○
B6	○								1 ○
B9	○								1 ○
B11							1 ○		5 ○
B13							5 ○		
B16									
B19							1 ○		3 ○
B20									
B21	○							4 ○	5 ○
B32									
B33	○					1 ○			3 ○
B35	○							3 ○	7 ○
B37									1 ○
B39				○					3 ○
B46									2 ○

[a] For definitions of abbreviations and symbols, refer to Table 2.2.
[b] Number of positive cases.

Table 2.8. Antibodies with undefined specificity (RIA on cells).[a]

	T	G	TL	HL-60	Burkitt's					B-Lb	
					Raji	Daudi	RAMOS	P3HR1	BJAB	HF2	HMY2
B3	±	–	±	±							
B6	–	–	–	–	–	±	–	–	±	–	–
B9	–	–	–	–							
B11	–	–	–	–							
B13	–	–	–	–							
B16	–	–	–	–							
B19	–	–	–	–	–	+	±	–	–	++	+
B20	–	–	–	–							
B21	–	–	–	–	–	–	±	±	±	–	–
B22	–	–	–	–							
B33	–	–	–	–	–	–	–	–	–	±	±
B35	–	–	–	–	±	±	–	–	–	–	±
B37	–	–	–	–							
B39	–	–	–	–	–	++	–	–	++	+	+
B46	–	–	–	–							

[a] For definitions of abbreviations and symbols, refer to Table 2.4.

2. Antibodies which are not restricted to the B cell lineage but may be useful for leukemia typing (Tables 2.3, 2.5, and 2.6): B4, B17 (HD28), B36, and B52 reacted weakly with myelo-monocytic cells and T leukemias (Table 2.3). These antibodies strongly stained BL but not the majority of ALL (Table 2.6). B15 and B18 did not react with HCL and PLL, whereas other BL were positive in the majority of cases. B38, cross-reacting with thrombocytes, stained only ALL (7/18) but not any B-type leukemia. B47 and B48 are good reagents for the B lineage apart from their cross-reactivity with granulocytes; myelo-monocytic leukemias are negative.
3. Antibodies with undefined specificity: 15 antibodies are found in this

Table 2.9. B cell workshop antibodies reacting with cytoplasmic antigens (IE).[a]

Workshop antibody	CLL cells in Terasaki plates	
	Glut.	Glut. + BRIJ
B3, B11, B19, B35, B39, B40	–[b]	(+)/+
B25 (HD6), B31 (HD39), B49	(+)	+

[a] Abbreviations: IE = Immunoenzymatic staining; CLL = chronic lymphocytic leukemia; Glut. = fixation with 0.025% glutaraldehyde, 10 min RT; BRIJ = BRIJ 56 0.5%, 15 min RT.
[b] – = Negative reaction; (+) = weak reaction; + = moderate reaction.

category because of their weak reaction or their lack of reaction in IE (Table 2.7). We assume that the negative controls are included in this group. Out of the antibodies tested both with IE and RIA two reacted strongly with some B cell lines in RIA (B19 and B39) (Table 2.8). B3, B11, B19, B35, and B39 reacted with CLL cells only after pretreatment with detergent (Table 2.9). Therefore we assume that in these experiments the above mentioned mAbs react with cytoplasmic antigens.

4. B cell-specific antibodies: 13 antibodies were classified as specific for the B lineage. In addition, two further mAbs were included in this category because they are B cell specific apart from their cross-reactivity with single cases of T or myelo-monocytic leukemias in IE (Table 2.10). The specificity of the reaction was confirmed by RIA. However, two mAbs (B22 and B5) reacted weakly with T cells (Table 2.11). By testing the reactivities of a large panel of leukemias and lymphomas with the B cell-specific mAbs we could define three phenotype groups according to their reaction pattern (Table 2.12). Phenotype group A (B31, B40, B49, B25) reacted strongly with HCL and PLL; other BL and ALL were mostly weakly positive. A second characteristic of this phenotype group is that the positivity is intensified when CLL cells have been pretreated with the detergent BRIJ (Table 2.9). Therefore we can assume that these mAbs belong to the category reacting with cytoplasmic antigens. B22 and B24 (pheno-

Table 2.10. B cell-specific antibodies (immunoenzymatic staining).[a]

	B	T	M	G	My	TL (*n* = 11)	ML (*n* = 14)	ALL (*n* = 18)	BL (*n* = 29)
B30[b]	●					1[c] ○	2 ○	2 ●	25 ●
B27[b]	●					0	1 ●	7 ●	10 ●
B14	●					0	0	16 ●	28 ●
B28	●					0	0	16 ●	28 ●
B34	●					0	0	16 ●	29 ●
B43	●					0	0	16 ●	28 ●
B22	●					0	0	8 ●	23 ●
B24	●					0	0	6 ●	25 ●
B31	●					0	0	13 ○	23 ○
B40	●					0	0	9 ○	14 ○
B49	●					0	0	13 ●	24 ●
B25	●					0	0	13 ●	25 ○
B5	●					0	0	8 ●	28 ●
B8	●					0	0	13 ○	14 ○
B2						0	0	0	3 ○

[a] For definitions of abbreviations and symbols, refer to Table 2.2.
[b] B-cell specific except for rare cross-reactivity.
[c] Number of positive cases.

able 2.11. B cell-specific antibodies (RIA on cells).[a]

					Burkitt's					B-Lb	
	T	G	TL	HL-60	Raji	Daudi	RAMOS	P3HR1	BJAB	HF2	HMY2
27[b]	−	−	−	−	±	+	−	−	±	+	±
30[b]	−	−	−	−	−	±	±	±	+	±	−
14	−	−	−	−	±	+	+	+	+	±	+
28	−	−	−	−	+	+	±	±	+	±	±
34	−	−	−	−	±	+	+	+	+	±	+
43	−	−	−	−	+	+	+	+	+	+	+
22	±	−	−	−	++	++	++	++	++	++	++
24	−	−	−	−	+	++	++	++	++	++	++
31	−	−	−	−	±	±	±	−	±	+	+
40	−	−	−	−	±	±	±	±	±	±	++
49	−	−	−	−	+	++	++	++	++	++	++
25	−	−	−	−	±	+	±	±	±	±	+
5	±	−	−	−	+	++	++	++	+	++	++
8	−	−	−	−	−	−	−	−	±	−	−
2	−	−	−	−	+	−	−	−	−	−	−

For definitions of abbreviations and symbols, refer to Table 2.4.
B-cell specific except for rare cross-reactivity (cf. Table 2.10).

Table 2.12. Grouping of the B cell-specific antibodies.

Workshop antibody	Phenotype group	Reaction pattern with leukemia[b]	Specificity
B30[a]		BL +, ALL −	B cells
B27[a]		BL +^{s}/−, ALL −/+	B subset?
B31 (HD39), B40, B49, B25 (HD6)	A	HCL ++, PLL ++, other BL (+), ALL (+)/−	B cells (late B)[c]
B22, B24	B	BL +, ALL −/+	B cells
B14, B28 (HD37), B34, B43	C	BL +, ALL +	Pan B
B5[d]		BL +, ALL −/+	B cells
B8		BL (+), ALL (+)	B cells
B2		PCL +	plasma cells?

[a] B cell-specific except for rare cross-reactivity (cf. Table 2.10).
[b] BL = B leukemias/lymphomas; ALL = acute lymphoblastic leukemia; HCL = hairy cell leukemia; PLL = prolymphocytic leukemia; PCL = plasma cell leukemia; −/(+)/+/++ = negative/weak/positive/strong reaction, +^{s} = positive mainly with a distinct subpopulation.
[c] In studies of cells in suspension B31 (HD39) reacts only with HCL and PCL.
[d] Reaction pattern similar to phenotype group B except for one ALL.

type group B) do not react with the majority of ALL. Phenotype group C (B14, B28, B34, B43) has pan B character because of its broad spectrum of reactivity covering all BL and nearly ALL (16 positive cases out of 18 tested). B5 has a reaction pattern similar to the phenotype group B except for one ALL. B8 reacted weakly with BL and ALL. B2 was the only mAb reacting with the one tested case of plasma cell leukemia. B30, reacting strongly with BL, was negative in 16/18 ALL. Strikingly B27 reacted with a distinct subpopulation of BL.

Leukemia (L) Panel (Table 2.13)

Using both IE and RIA the L antibodies can be grouped into the following categories:

1. L12, L9, and L6 showed reactions with many cell types. It is assumed that this group contains the positive controls.
2. L8 reacted with all Ia-positive cell types.

Table 2.13. Reactivity pattern of the L panel.

	B	T	M	G	Th	Ly	My	Ery	Thy	TL	ML	ALL	BL
L12	●	■ ★	▲	♦ ★		●	♦		■	■ ★	♦ ★	●	●
L9	○	■ ★	▲				♦		■	■ ★	♦ ★	●	[illegible]
L6		□	△	♦		○	◇		□	□ ★	◇ ☆	●	○ [illegible]
L8	●	★				●	♦				♦	●	● [illegible]
L4				◇ ★	▲	○	♦		■	☆	♦	●	[illegible]
L16				◇	▲	○	♦		■	☆	♦	●	[illegible]
L18		☆		◇ ☆	▲	○	♦		■	★	♦	●	[illegible]
L22				◇ ☆	▲	○	♦		■	★	♦ ★	●	[illegible]
L19								▼	□	□ ★	★		
L20								▼		□ ★	★		[illegible]
L3								▼		★	☆		[illegible]
L13		☆						▼		★	★		[illegible]
L2				◇ ★		●			□	■		●	[illegible]
L10				◇ ★		●			■	■		●	[illegible]
L11		☆		◇		●			□	■ ☆		●	
L14				◇		○			□	■		●	
L15				◇		○			□	■		●	
L21						○			□	■ ★		●	[illegible]
L7					△	○			■	■ ★			
L17	●											●	● [illegible]
L1													

[a] Abbreviations and symbols: B = Peripheral blood B cells (E$^-$); T = peripheral blood T cells (E$^+$); M = monocyte G = granulocytes; Th = thrombocytes; Ly/My/Ery = lymphoid/myeloid/erythroid cells in bone marrow; Thy Thymocytes; TL = T lymphoblastic leukemia; ML = myelo/monocytic leukemia; ALL = acute lymphoblast leukemia; BL = B cell leukemia/lymphoma; ○ □ △ ◇ = weak reaction in IE; ● ■ ▲ ♦ ▼ = strong reaction in IE; ☆ weak reaction in RIA; ★ = strong reaction in RIA.

3. L4, L16, L18, and L22 reacted weakly with myeloid cells in bone marrow, moderately with thrombocytes, thymocytes, myeloid leukemias, and ALL in IE and with T and B lines in RIA.
4. L19, L20, L3, and L13 reacted with erythrocytes and erythroid cells in bone marrow in IE and with T and B lines and the HL-60 line in RIA.
5. L2, L10, L11, L14, L15, and L21 reacted weakly with granulocytes (except for L21) and strongly with lymphoblastic leukemias in IE and with Burkitt lines in RIA (except for L15).
6. L7 reacted weakly with thrombocytes and moderately with thymocytes and T lymphoblastic leukemias in IE and with T and B lines in RIA.
7. L17 reacted only with B-type leukemias/lymphomas and ALL.
8. L1 reacted only with two out of five Burkitt lines in RIA.

Discussion

The combined use of IE and RIA enabled us to group the mAbs of the B and L panel into distinct categories. The major advantages of the immunoenzymatic staining assay in Terasaki plates (1) are:

1. The test is sensitive because positive cells are easily detected with their strong red staining.
2. 400–500 cells of a monolayer can be evaluated all at once; therefore, small subpopulations of even less than 1% can be detected.
3. Preliminary information concerning specificity can be derived from morphology.
4. The test is economical (10 μl mAb per well) and rapid (2 hr); plates with fixed cells can be stored for months. Therefore we were able to test an extensive number of different cell types for the Workshop.

The major disadvantage is that cells are fixed with glutaraldehyde (however only at the low concentration of 0.025%). Therefore the possibility that surface antigens are destroyed or altered cannot be excluded. This is particularly relevant in the group of B cell antibodies with "undefined specificity." The major advantages of the RIA on cells (2) are:

1. Due to its high sensitivity this test is especially useful when investigating weakly reacting mAbs.
2. This test is performed without fixation of cells.

On this basis the combination of IE and RIA appears to be especially useful.

The nearly complete correlation between IE and RIA results concerning cross-reactivity made us confident about grouping together the 22 cross-reacting mAbs (Tables 2.2–2.5). Myelo-monocytic leukemias and T leukemias/lymphomas were found to be of major importance in excluding cross-reactivity. Nine out of the 22 mAbs were placed in a separate category because they appear to be useful reagents in discriminating between different types of leukemias. Four antibodies [B4, B17 (HD28), B36, B52] are characterized by their strong reaction with BL, lack of reaction with the majority of ALL, and their weak cross-reactivity with myelo-monocytic cells, a reactivity pattern similar to the antibody 41H.16 reported by Zipf *et al.* (3). B15 and B18 appear to be of interest because of their lack of reaction with mature B leukemias (HCL and PLL). B38, weakly cross-reacting with thrombocytes, is positive with ALL but not with BL.B47 and B48 belong to the same category as BA-1 (4), reacting with cells of the B lineage and granulocytes.

Among the antibodies "with undefined specificity" are two mAbs (B19 and B39) which are strongly positive with some B cell lines in RIA. Therefore these mAbs could be of interest although in IE they did not show any definite reaction with BL. An important finding is that B3, B11, B19, B35, and B39 stained CLL cells only after pretreatment with detergent (Table 2.9), suggesting that these mAbs react with cytoplasmic antigens.

Thirteen mAbs were found to be B cell specific; in addition, two other mAbs were placed in this category because they were specific except for cross-reactivity with single leukemias of non-B lineage. Three phenotypic groups could be defined with the above mAbs: The phenotype group A, which includes two of our antibodies [B31 (HD39), B40, B49, B25 (HD6)], is characterized by strong reaction with HCL and PLL. A similar reaction pattern was reported for the mAb FMC7 (5,6). The corresponding antigen for the group A antibodies has been characterized (see this volume, Chapter 7), whilst the FMC7 antigen has not been identified so far (7). In contrast to the data reported for FMC7 (7), the corresponding antigen for HD39 (B31 in group A) can be induced on CLL cells by TPA (see this volume, Chapter 46).

A characteristic finding for the group A antibodies is the intensification of the reaction after treatment of CLL cells with detergents. This suggests that these mAbs react with cytoplasmic antigens.

The interesting finding about HD39 (B31) is that it is differently expressed on cell surface and in cytoplasm during the process of B cell maturation (see this volume, Chapter 7). This finding stresses the importance of analyzing membrane and cytoplasmic expression of antigens in the characterization of B cell mAbs. This may also partly explain the differences between results of studies of cells in suspension and those obtained by immunohistology.

The reaction pattern BL^{+} $ALL^{-/+}$ of B22 and B24 (group B) and B5 can be compared to that of the established antibodies "B1" (8,9) and FMCl (10). The phenotype group C, which includes one HD antibody [B14, B28 (HD37), B34, B43], is characterized by a broad spectrum of reactivity covering all BL and 16 out of 18 cases of ALL. Therefore a pan B specificity can be postulated in accordance with the concept that the majority of ALL are early members of the B lineage. We were able to demonstrate that HD37 (B28 in group C) is a useful reagent for typing "unclassified" acute leukemias (see this volume, Chapter 33). The phenotype group C antibodies appear to recognize the same antigen as "B4" published by Nadler *et al.* (11,12) (see this volume, Chapter 3).

Most of the L panel mAbs reacted with many different cell types except for the mAbs L17 and L1; L17 is B cell specific and belongs to the phenotype group C. Nevertheless some of the L panel mAbs may be useful reagents for leukemia typing.

Summary

Using an immunoenzymatic staining assay and a radioimmunoassay on cells we were able to group the B antibodies (including four of our HD mAbs: B17, B25, B28, B31) according to their specificity:

1. B cell-specific antibodies: B14, B28, B34, B43; B22, B24; B31, B40, B49, B25; B5; B8; B2.
2. Antibodies which are B cell specific except for rare cross-reactivity with T-type or myelo-monocytic leukemias: B30; B27.
3. Antibodies which are not restricted to the B cell lineage but may be useful for leukemia typing: B4, B17, B36, B52; B15, B18; B38; B47, B48.
4. Antibodies with multiple cross-reactions: B10, B41, B42, B44, B45, B50, B51, B1, B7, B12, B23, B26, B29.
5. Antibodies with undefined specificity: B3, B6, B9, B11, B13, B16, B20, B21, B32, B33, B35, B37, B46 (weak or no reaction); B19, B39 (definite reaction only with a few B lines).

References

1. Dörken, B., A. Pezzutto, G. Moldenhauer, R. Schwartz, S. Kiesel, and W. Hunstein. 1984. An immunoenzymatic staining assay (ISA) for the rapid screening of monoclonal antibodies detecting membrane and cytoplasmic antigens. Submitted for publication (J. Immunol. Methods).
2. Hämmerling, G.J., U. Hämmerling, and J.F. Kearney. 1981. Production of antibody-producing hybridomas in rodent systems. In: *Monoclonal antibodies and T cell hybridomas*. Elsevier/North-Holland, Amsterdam, pp. 563–587.
3. Zipf, T.F., G.J. Lauzon, and B.M. Longenecker. 1983. A monoclonal antibody detecting a 39,000 M.W. molecule that is present on B lymphocytes and chronic lymphocytic leukemia cells but is rare on acute lymphocytic leukemia blasts. *J. Immunol.* **131**:3064.
4. Abramson, C.S., J.H. Kersey, and T.W. LeBien. 1981. A monoclonal anti-

body (BA-1) reactive with cells of human B lymphocyte lineage. *J. Immunol.* **126:**83.
5. Brooks, D.A., I.G.R. Beckman, J. Bradley, P.J. McNamara, M.E. Thomas, and H. Zola. 1981. Human lymphocyte markers defined by antibodies derived from somatic cell hybrids. *J. Immunol.* **126:**1373.
6. Catovsky, B.D., M. Cherchi, D. Brooks, and H. Zola. 1981. Heterogeneity of B-cell leukemia demonstrated by the monoclonal antibody FMC7. *Blood* **58:**406.
7. Zola, H., H.A. Moore, A. Hohmann, I.K. Hunter, A. Nikoloutsopoulos, and J. Bradley. 1984. The antigen of mature human B cells detected by the monoclonal antibody FMC7: studies on the nature of the antigen and modulation of its expression. *J. Immunol.* **133:**321.
8. Stashenko, P., L.M. Nadler, R. Hardy, and S.F. Schlossman. 1980. Characterization of a human B-lymphocyte-specific antigen. *J. Immunol.* **125:**1678.
9. Nadler, L.M., J. Ritz, R. Hardy, J.M. Pesando, and S.F. Schlossman. 1981. A unique cell surface antigen identifying lymphoid malignancies of B cell origin. *J. Clin. Invest.* **67:**134.
10. Brooks, A., I. Beckman, J. Bradley, P.J. McNamara, M.E. Thomas, and H. Zola. 1980. Human lymphocyte markers defined by antibodies derived from somatic cell hybrids. *Clin. Exp. Immunol.* **39:**477.
11. Nadler, L.M., K.C. Anderson, G. Marti, M. Bates, E. Park, J.F. Daley, and S.F. Schlossman. 1983. B4, a human B lymphocyte-associated antigen expressed on normal mitogen-activated, and malignant B lymphocytes. *J. Immunol.* **131:**244.
12. Nadler, L.M., S.J. Korsmeyer, K.C. Anderson, A.W. Boyd, B. Slaugenhaupt, E. Park, J. Jensen, F. Coral, R.J. Mayer, E. Sallan, J. Ritz, and S.F. Schlossman. 1984. B cell origin of non-T cell acute lymphoblastic leukemia. A model for discrete stages of neoplastic and normal pre-B cell differentiation. *J. Clin. Invest.* **74:**332.

CHAPTER 3

Analysis of Ten B Lymphocyte-Specific Workshop Monoclonal Antibodies

Gerhard Moldenhauer, Bernd Dörken, Reinhard Schwartz, Antonio Pezzutto, Jeroen Knops, and Günter J. Hämmerling

Introduction

The B cell/leukemia Workshop monoclonal antibodies (mAbs) have been extensively analyzed by our group employing indirect immunofluorescence (microscopy and cytofluorometry), immunoenzymatic staining, and radioimmunoassay on cells. Based on the results obtained, especially by immunoenzymatic staining and radioimmunoassay on cells, the B cell Workshop reagents can be divided into four different groups (see this volume, Chapter 2). Group 1 contains mAbs with multiple cross-reactions. In group 2 mAbs which show rare cross-reactions but may be useful for typing of leukemia and lymphoma are collected. mAbs with undefined specificity due to no or only weak reactivity are included in group 3. Finally, group 4 contains the B cell-specific mAbs.

The aim of this study was to further analyze the B cell-specific Workshop mAbs by means of binding inhibition studies and biochemical characterization of the recognized antigens.

Materials and Methods

Competitive Binding Inhibition Radioimmunoassay on Cells

This test was performed in flexible PVC plate as previously described (1). In general, 50 μl/well target cell suspension containing 10^6 viable cells were mixed with 50 μl Workshop mAb (final dilution 1 : 250) and were incubated for 1 hr. Then, 50 μl/well ^{125}I-labeled second mAb were added followed by another 1-hr incubation. The plate was washed three times with PBS plus 2% BSA by centrifugation and siphoning off supernatant. The dried plate was sliced with a hot wire and radioactivity in individual wells was measured in a gamma-counter.

Purification and Radiolabeling of mAbs

Workshop mAbs B28 (HD37) and B31 (HD39) produced by our group were precleared from ascitic fluid by two ammonium sulfate precipitations (50% saturation). Subsequently, monoclonal antibody was purified by DEAE Affi-gel blue (Bio-Rad) chromatography (2). Radioiodination of mAbs was performed by a slight modification of the chloramine T method (3) using 0.5 mCi [^{125}I]iodide (Amersham-Buchler) per 50 μg of mAbs.

Cell Surface Iodination and Immunoprecipitation

Cells from established human B cell lines LICR-LON-HMY2 and BJAB were radioiodinated by the lactoperoxidase method essentially as described by Goding (4). Usually, 10^7 cells (viability > 95%) were radiolabeled using 0.5 mCi [^{125}I]iodide and subsequently were solubilized in lysis buffer containing 1% NP-40 and proteinase inhibitors. Membrane glycoproteins were isolated from cell lysate by chromatography on lens culinaris lectin. For immunopreciptation a sandwich procedure was utilized consisting of mAb, purified goat anti–mouse Ig antibody, and protein A–Sepharose 4B (Pharmacia). After overnight incubation the adsorbent was washed and reduced in SDS sample buffer. Samples were subjected to polyacrylamide slab gel electrophoresis (SDS–PAGE) using a discontinuous buffer system.

Established mAbs

Several already established mAbs were included either for negative control or for comparison with Workshop mAbs. Antibodies HD11, HD40, and HD43 were generated by our group (unpublished results); mAb HD11 recognizes a framework determinant on HLC-class II antigen whereas HD43 as directed against a HLA-class II alloantigen. mAb HD40 reacts with an individual-specific antigen of a patient with centrocytic lymphoma. mAb W6/32 detects a monomorphic determinant on HLA-class I antigen (5). Antibodies B1 (6), B2 (7), and B4 (8) (Coulter) have been reported as being B cell specific whereas mAbs OKB2 (9) and OKB7 (9) (Ortho) obviously are not restricted to the B cell lineage. mAb OKT9 (Ortho) was described to be reactive with the transferrin receptor (10). The pan-B antibody (Dako) was described as a marker of the whole B cell lineage.

Results

We have concentrated our investigation on nine antibodies of the B cell Workshop (B14, B22, B24, B25, B28, B31, B40, B43, B49) and one antibody of the leukemia Workshop (L17) which were found to exhibit B cell-specific reactivity. mAbs B2 and B8 were excluded because they did not react with the target cells used in the study. Unfortunately, mAb B5 was also not included although it later turned out to be B cell specific. Based on their binding patterns these antibodies were subclassified into three different phenotypic subgroups (Table 3.1). mAb L17 was included due to its very similar reactivity to the antibodies in group C.

Three out of the ten selected mAbs were contributed by our laboratories, namely, B25 (HD6), B31 (HD39), and B28 (HD37), of which the first two were classified in group A whilst the latter belonged to group C. Thus, three probes were available for competitive binding inhibition assay of group A and C antibodies. This test is based on the fact that if two antibodies are directed against identical or proximate determinants of an antigen molecule, binding of labeled antibody should be inhibited by preincubation with unlabeled antibody.

Radioiodinated mAbs B31 and B28 were employed for binding inhibition radioimmunoassay on LICR-LON-HMY2 and BJAB target cells (Table 3.2). The binding of ^{125}I-labeled B31 antibody was strongly inhibited by mAbs B40 and B49 and, as expected, by B31 itself. These three antibodies belonged to group A. Surprisingly, mAb B25, although of group A, did not block binding of B31. When using labeled B28 antibody as probe all four mAbs of group C (B14, B28, B43, and L17) were inhibitory. With one exception (B25), the results clearly confirmed the grouping of the mAbs according to reactivity pattern. These data gave strong evidence that the four group C antibodies were directed against the same antigen which was distinct from a different antigen recognized by three out of four group A mAbs.

Next, it was attempted to biochemically characterize the corresponding antigens of the ten mAbs. Surface ^{125}I-labeled cell lysates of the human B cell lines LICR-LON-HMY2 and BJAB were used for immunoprecipitation. When employing crude cell lysate we failed to precipitate any antigen. Only when the glycoprotein fraction purified by lens culinaris lectin

Table 3.1. Grouping of the B cell workshop antibodies used for further analysis.

Workshop antibody	Phenotype group	Reaction pattern with leukemia[a]	Specificity
B31 (HD39), B40, B49, B25 (HD6)	A	HCL ++, PLL ++, other BL (+), ALL (+)/−	B cells (late B)[b]
B22, B24	B	BL +, ALL −/+	B cells
B14, B28 (HD37), B34, B43	C	BL +, ALL +	pan B

[a] BL = B leukemias/lymphomas; ALL = acute lymphoblastic leukemia; HCL = hairy cell leukemia; PLL = prolymphocytic leukemia; -/(+)/+/++ = negative/weak/positive/strong reaction.
[b] In studies of cells in suspension, B31 (HD39) reacts only with HCL and PLL.

Table 3.2. Competitive binding inhibition of B cell-specific workshop mAbs using CRIA.

Unlabeled mAb	Binding of ^{125}I-labeled second antibody			
	B31 (HD39)		B28 (HD37)	
	Cpm	%	Cpm	%
B14	27,086 ± 1022[a]	115	772 ± 164	7
B22	24,568 ± 2468	104	12,284 ± 673	106
B24	23,628 ± 996	100	12,612 ± 23	109
B25 (HD6)	21,948 ± 187	93	11,364 ± 922	98
B28 (HD37)	23,740 ± 2025	101	634 ± 59	5
B31 (HD39)	714 ± 116	3	11,372 ± 1182	98
B40	3158 ± 201	13	11,382 ± 30	98
B43	22,048 ± 277	94	818 ± 132	7
B49	432 ± 17	2	12,918 ± 1100	112
L17	25,252 ± 79	107	1322 ± 226	11
W6/32[b]	21,195 ± 1772	90	11,737 ± 527	101
HD11[c]	24,598 ± 2603	104	10,715 ± 1296	93

[a] Counts per minute ± SEM of three replicate assays.
[b] mAb reacting with a monomorphic determinant on HLA-class I.
[c] mAb recognizing a framework determinant on HLA-class II.

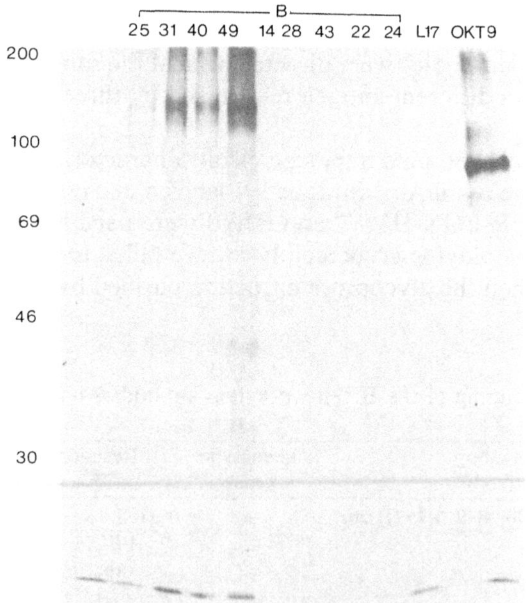

Fig. 3.1. SDS–PAGE analysis under reducing conditions of immunoprecipitates of indicated Workshop mAbs and OKT9 from lactoperoxidase ^{125}I-labeled cell lysates.

was utilized, the four mAbs of group A (B25, B31, B40, B49) specifically precipitated obviously the same protein. This glycoprotein antigen consisted of two polypeptide chains which corresponded in their electrophoretic mobility to M.W.'s of 130 and 140 Kd, both under reducing and nonreducing conditions (Fig. 3.1). Both human cell lines tested gave the same results. Considering the biochemical and the phenotypical analysis, and the binding inhibition data, there was strong evidence that the mAbs of group A were recognizing the identical target antigen. Since antibodies B25 (HD6) and B31 (HD39) did not block each other (see this volume, Chapter 7) but precipitated the same protein they were very likely directed against distinct epitopes of the same antigen.

In order to determine whether or not the group A and group C antibodies were related to already established B cell-reactive mAbs another series of binding competition experiments were performed. Radiolabeled antibodies B31 and B28 were used to compete with six commercially available mAbs listed in Table 3.3. The binding of mAb B31 was not influenced by any of these antibodies whilst binding of mAb B28 was strongly inhibited by the established antibody "B4." The degree of inhibition was as strong as obtained when using mAb B28 itself as homologous inhibitor. These results indicated that the group C antibodies which were claimed to detect the same antigen obviously are recognizing the "B4" antigen previously described by Nadler *et al.* (8). It remains unclear whether the four mAbs B14, B28, B43, and L17 are directed against the same or different epitopes on the "B4" antigen because it is possible that antibodies which bind to different epitopes that are in close vicinity can block each other by steric hinderance.

Table 3.3. Comparison of workshop mAbs B31 and B28 with established B cell antibodies (CRIA).

Unlabeled mAb	Binding of ^{125}I-labeled second antibody: B31 (HD39) Cpm	B31 (HD39) %	B28 (HD37) Cpm	B28 (HD37) %
B1 (Coulter)	43,142 ± 970[a]	102	31,122 ± 1592	98
B2 (Coulter)	46,678 ± 438	110	33,478 ± 3385	105
B4 (Coulter)	42,154 ± 1462	99	1254 ± 25	4
OKB2 (Ortho)	42,005 ± 210	99	30,566 ± 1575	96
OKB7 (Ortho)	40,810 ± 528	96	31,952 ± 1086	100
Pan-B (Dako)	40,098 ± 562	94	30,872 ± 2245	97
B31 (HD39)	1334 ± 110	3	33,021 ± 1964	104
B28 (HD37)	43,808 ± 2183	103	1584 ± 11	5
W6/32	41,310 ± 1411	97	32,230 ± 2876	101

[a] Cpm ± SEM of three replicate assays.

Table 3.4. Characterization of three distinct B cell-specific antigens defined by workshop mAbs.

Antigen (M.W.)	Workshop mAb	Reactivity	Epitopes
p 130/140 (130/140 Kd)	B25 (HD6), B31 (HD39), B40, B49	B cell subset (A)	(a) B25 (b) B31, B40, B49
?	B22, B24	B cell (B)	Not tested
"B4" (24/68 Kd)[a]	B14, B28 (HD37), B43, L17	Pan B cell (C)	One ?

[a] Ref. 8.

Conclusions

Our investigation of the ten Workshop mAbs included in this study is summarized in Table 3.4. These ten mAbs define three distinct antigens exclusively expressed on B lymphocytes. The group A mAbs B25, B31, B40, and B49 recognize a glycoprotein antigen composed of two polypeptide chains with apparent M.W.'s of 130 and 140 Kd. At least two different epitopes of the antigen can be distinguished by mAbs. Based on cellular distribution and biochemical characterization this antigen appears to be not previously discovered. mAbs B22 and B24 are grouped together because of their virtually identical patterns of reactivity on normal and neoplastic B cells. So far, the corresponding antigen could not be identified by immunoprecipitation. Finally, the group C mAbs B14, B28, B43, and L17 are directed against the "B4" antigen previously reported to be a protein antigen consisting of two chains with M.W.'s of 24 and 68 Kd. The question whether the four mAbs react with only one or more epitopes on the "B4" molecule is still open.

References

1. Hämmerling, G.J., U. Hämmerling, and J.F. Kearny (eds). 1981. Production of antibody-producing hybridomas in the rodent systems. In: *Monoclonal antibodies and T cell hybridomas*. Elsevier/North-Holland, Amsterdam, pp. 563–587.
2. Bruck, C., D. Portetelle, C. Glineur, and A. Bollen. 1982. One-step purification of mouse monoclonal antibodies from ascitic fluid by DEAE Affi-gel blue chromatography. *J. Immunol. Methods* **53**:313.
3. Greenwood, F.C., W.M. Hunter, and J.S. Glover. 1963. The preparation of ^{131}I-labeled human growth hormone of high specific radioactivity. *Biochem. J.* **89**:114.
4. Goding, J.W. 1980. Structural studies of murine lymphocyte surface IgD. *J. Immunol.* **124**:2082.
5. Barnstable, C.J., W.F. Bodmer, G. Brown, G. Galfre, C. Milstein, A.F. Williams, and A. Ziegler. 1978. Production of monoclonal antibodies to group

A erythrocytes, HLA and other human cell surface antigen—new tools for genetic analysis. *Cell* 14:9.
6. Stashenko, P., L.M. Nadler, R. Hardy, and S.F. Schlossman. 1980. Characterization of a human B lymphocyte-specific antigen. *J. Immunol.* **125:**1678.
7. Nadler, L.M., P. Stashenko, R. Hardy, C. van Agthoven, C. Terhorst, and S.F. Schlossman. 1981. Characterization of human B cell-specific antigen (B2) distinct from B1. *J. Immunol.* **126:**1941.
8. Nadler, L.M., K.C. Anderson, G. Marti, M. Bates, E. Park, J.F. Daley, and S.F. Schlossman. 1983. B4, A human B lymphocyte-associated antigen expressed on normal, mitogen-activated, and malignant B lymphocytes. *J. Immunol.* **131:**244.
9. Mittler, R.S., M.A. Talle, K. Carpenter, P. Rao, and G. Goldstein. 1983. Generation and characterization of monoclonal antibodies reactive with human B lymphocytes. *J. Immunol.* **131:**1754.
10. Goding, J.W., and G.F. Burns. 1981. Monoclonal antibody OKT 9 recognizes the receptor for transferrin on human acute lymphocytic leukemia cells. *J. Immunol.* **127:**1256.

CHAPTER 4

Screening of Workshop "B" Series Antibodies by Radioimmunobinding to Human Leukocyte Cell Lines and to Cells from Human Lymphoid Tumors

C. Michael Steel, Patricia Elder, and Keith Guy

Introduction

The fifty-two monoclonal antibodies of the Workshop B series were received in April, 1984. They were allowed to thaw and were held at 8°C for the period of study (April–August, 1984). Aliquots of the original samples were diluted 1:20 with RPMI 1640 culture medium containing 5% fetal calf serum (FCS), 5% horse serum (HS), and 0.02% sodium azide. The diluted antibodies were then tested for binding to a panel of cells from human lymphoid lines or from lymphoid tumors. The phenotypes of the cells forming the panel had already been determined but additional monoclonal antibodies of known specificities were included to act as positive or negative controls in the binding assays.

Materials and Methods

All test cells were fixed in 0.125% glutaraldehyde at 4°C for 5 min, followed by the addition of an equal volume of 0.15 *M* glycine buffer, pH 7.2, for a further 15 min. The fixed cells were washed twice in phosphate-buffered saline, pH 7.2, containing 1% w/v bovine serum albumin (PBS-BSA), then stored for up to two months at 4°C in the same solution with the addition of 0.02% sodium azide. Immediately before use they were resuspended and washed twice more in PBS-BSA.

The composition of the cell panel is shown in Table 4.1. All the cell lines were maintained routinely in the authors' laboratory in Ham's F10 medium with 5% FCS or in RPMI 1640 medium with 5% FCS and 5% HS. Fresh leukemic blood and lymphoma biopsy material were provided through the collaboration of colleagues in the Haematology Departments of the Western General Hospital and of Edinburgh Royal Infirmary and in the University Departments of Pathology and Clinical Oncology, to all of

Table 4.1. Panel of cells for radioimmunobinding assays.

Long-term cell lines

B cells

1. EB virus-transformed lines. Pools from large collection in authors' laboratory.[a]
2. Burkitt's lymphoma. Pools from 12 lines, including 2 (EB_4 and Ramos) which are EB virus negative.[a]
3. FAL_1 EB virus-transformed. From cord blood. Pre-B characteristics (neg. for surface and secreted lg, cytoplasmic μ chain +).
4. BLA_1 EB virus-transformed. From acute leukemia. Atypical growth pattern and grossly aneuploid.
5. Myeloma lines. Pool from RPMl 8226 (λ-secretor) and U266Bl (IgE secretor) (1,2).[a]

"Null" cells[b]

6. HL-60. From acute promyelocytic leukemia (3).
7. Reh. From acute lymphoblastic leukemia (4).[a]
8. K562. From chronic myeloid leukemia (5).[a]

"Early" T cells

9. MOLT-4 + CCRF.CEM (Pool). Both from acute lymphoblastic leukemia (6,7).[a]
10. HSB2. From acute lymphoblastic leukemia (8).[a]

Fresh leukemia/lymphoma cells

1. B follicular lymphoma. (Cells from excised node).
2. Hairy cell leukemia. (Cells from splenic deposit).
3. B chronic lymphatic leukemia (blood lymphocytes from two patients; <5% E-rosetting cells).
4. T cell chronic lymphatic leukemia (1 patient; leukocytosis +++, skin involvement. Cells were $OKT3^+$, $OKT4^+$, E-rosette +. No evidence of human T cell leukemia virus).
5. "Null" cell acute lymphoblastic leukemia. 1 adult patient. Cells were strongly positive for HLA-class II and for C-ALL antigen, negative for Ig or T cell markers.

[a] Phenotypes of lines indicated are detailed in Reference 9 though minor differences are found in the authors' laboratory, e.g., MOLT-4, CCRF.CEM, and HSB2 now show only low levels of terminal deoxynucleotidyl transferase and Reh is negative for MHC class II (Ia).
[b] "Null" in this context covers cells which are poorly differentiated and belong to neither B nor T lineage.

whom we are most grateful. Leukemia cells were separated from whole blood by flotation over Ficoll–Hypaque and single cell suspensions were produced from solid lymphoma biopsies by teasing out with sterile orange sticks. In all cases, viability of the test cells was checked, before fixation, by exclusion of 0.45% nigrosin (10). Only samples showing greater than 90% viability were included in the test panel.

B chronic lymphatic leukemia cells (B-CLL) from two patients were tested before and after culture in the presence of 100ng/ml 12-*O*-tetradecanoylphorbol 13-acetate (TPA) for 90 hr. TPA is known to produce morphological alteration in B-CLL cells and, in many cases, to induce the expression of new surface antigens (including HLA-DR and DQ determinants and those reactive with the monoclonal antibodies FMC7 and RFC4) (11–13).

The cells from patient 1 showed relatively high pre-TPA levels of HLA-class II antigens, including DQ, while those from patient 2 began with extremely low levels. Both were highly inducible as judged by the change in HLA-DQ expression (data not shown).

In order to detect cell cycle-dependent antigens, a mixed culture of human B lymphoblastoid cells representing a pool of 22 different lines was divided into two aliquots. One was maintained without refeeding for 9 days. There was no increase in cell numbers for the final 4 days and on harvesting at the end of this period, no dividing cells could be found in cytospin preparations. The parallel aliquot was fed every 24 hr by addition of fresh medium. Counting every 48 hr showed that the culture was in log phase growth when harvested and numerous mitotic figures were present in cytospin preparations. Fixed cells from both populations were included in the test panel.

Antibody Binding Assays

The Workshop antibodies (final dilution 1 : 200) were distributed in 50-μl aliquots in the wells of round-bottomed microtest II plates (Cooke microtiter). Additional antibodies in regular use in the authors' laboratory were added to further wells in dilutions known to saturate the binding sites on target cells under the assay conditions. These control antibodies are detailed in Table 4.2.

Test cells were adjusted to a concentration of 4×10^6/ml in PBS-BSA and distributed in 50-μl aliquots (2×10^5 cells) in the antibody-containing wells. After 40 min at room temperature, the cells were washed three times by centrifuging the whole plate, decanting the fluid, and resuspending the cells in PBS-BSA. To each well was then added 50μl of ^{125}I-conjugated sheep anti–mouse Ig (New England Nuclear) diluted to 1 μCi/ml in culture medium. Incubation continued for a further 40 min at room temperature followed by three more washes in PBS-BSA. The cell

Table 4.2. "Control" monoclonal antibodies used in radioimmunobinding assays.

Antibody	Specificity
CR 3.43[a]	Anti-MHC class II β chains, monomorphic.
H.92.1	Anti-MHC class I, monomorphic.
OKT1[b]	(Ortho) Anti-65,000 MW glycoprotein antigen of all T cells and some B-CLL cells.
J5[b]	(Coulter) Anti-"Common Acute Lymphoblastic Leukemia" Antigen—a marker for early hematopoietic cells.
PD7.26[a]	Anti–leukocyte common antigen.
DA6.127	Anti–human IgM.
BBMl	Anti-β_2-microglobulin.
MS120A	Anti–mouse sperm, unreactive with human leukocytes.
X63.NS1	Mouse plasmacytoma (non-secretor).

[a] Kindly provided by Dr DY Mason, Oxford.
[b] Used as 1 : 50 dilutions of stock Ig solution. All others used as undiluted hybridoma culture supernatant.

Table 4.3. Binding pattern of Workshop 'B' series antibodies to cell panel.[a]

	B cells										
								B-CLL			
								1		2	
	Pool B-LCL (range)	FAL 1	BL pool	BLA 1	8226/ U 266	B Lma	H.C.L.	No TPA	+ TPA	No TPA	+ TPA
B1	++/+++	+++	+++	+++	++	++	+	+++	+++	++	+++
B2	−/+	+	++	++	+	+++	−	−	−	−	−
B3	±/+	±	+	++	++	++	−	NT[b]	NT	−	−
B4	−/±	±	±	+	+	+	±	±	−	−	+
B5	−/±	±	±	+	+	+	±	−	−	−	±
B6	±/±	±	±	+	±	±	±	NT	NT	−	−
B7	+/++	++	+	+	++	+	++	NT	NT	+	+++
B8	±/+	±	±	±	+	±	+	NT	NT	++	±
B9	−/±	−	±	−	±	−	±	−	−	±	−
B10	±/++	+	+	±	+	+	−	NT	NT	−	−
B11	++/+++	++	±	++	±	±	−	±	++	−	±
B12	++/+++	++	+	++	++	+	+	NT	NT	−	++
B13	−/±	±	±	±	±	−	−	−	−	−	−
B14	−/±	−	±	±	±	±	+	−	±	−	±
B15	−/±	−	±	−	±	−	−	±	−	−	−
B16	−/±	−	±	±	±	−	−	−	−	−	±
B17	−/±	+	+	+	++	+	±	+	−	±	±
B18	−/±	−	+	+	±	−	−	±	−	−	±
B19	+/++	+	±	+	−	−	−	−	+	−	−
B20	±/+	+	++	+	+	+	−	NT	NT	−	−
B21	−/−	−	±	−	−	±	±	−	−	−	±
B22	−/+	±	±	+	±	++	++	±	±	−	−
B23	±/+	+	+	+	±	±	−	±	+	−	+
B24	−/±	−	±	+	−	+	+	±	−	−	±
B25	±/+	±	±	±	±	±	±	−	−	±	±
B26	−/±	±	±	±	±	−	±	−	−	−	−
B27	±/+	+	+	±	±	±	±	−	±	−	−
B28	−/+	+	+	±	±	±	+	±	±	−	+
B29	+/++	+	++	+	++	−	±	−	±	++	++
B30	−/±	−	±	+	±	±	+	−	−	−	±
B31	−/±	±	±	±	−	±	−	−	−	−	±
B32	−/+	−	±	±	±	−	−	−	−	−	±
B33	−/−	−	±	−	−	−	−	±	−	±	−
B34	−/±	−	±	−	−	±	±	±	−	−	±
B35	−/±	−	±	−	±	±	−	±	−	−	−
B36	−/⊥	±	±	+	−	++	+	+	−	−	±
B37	−/+	−	±	−	±	−	−	−	−	−	±
B38	−/−	−	−	−	±	−	−	−	−	−	−
B39	+/++	+	±	++	−	−	±	−	++	−	−
B40	−/±	−	±	±	−	−	±	−	−	−	−
B41	−/+	−	−	+	−	±	−	−	−	−	+
B42	−/+	−	+	+	±	++	±	+	+	−	−
B43	−/±	±	+	±	±	+	+	+	+	−	++
B44	−/+	−	+	−	+	+	+++	++	+	−	+
B45	+/++	+	++	+	++	+	++	NT	NT	−	+
B46	−/±	−	±	−	±	−	−	±	−	−	±
B47	−/+	±	±	±	±	+	±	++	+	−	++
B48	−/±	±	±	±	±	+	−	++	+	−	−
B49	−/±	±	±	±	±	−	−	NT	NT	−	±
B50	±/+	+	+	±	+	+	±	NT	NT	−	+
B51	−/±	±	−	±	−	±	+	±	−	±	−
B52	−/+	−	±	−	−	±	±	±	−	−	±
CR3.43	+++/+++	+++	+++	+++	+++	++	+++	+++	+++	++	+++
H.92.1	+++/+++	+++	+++	+++	+++	+++	+++	+++	+++	+++	+++
OKT1	−/−	−	−	−	NT	+	−	−	−	−	−
J5	±/±	±	+	±	−	+	±	+	+	+	−
PD7.26	+/++	++	++	±	NT	NT	NT	NT	NT	NT	NT
DA6.127	±/+	−	++	++	+	+++	−	NT	NT	−	+
BBM1	++/+++	NT	NT	˙NT	NT	+++	+++	NT	NT	+++	+++

[a] Antibodies are listed in left-hand column, cell panel as identified in Table 4.1 is across the top. Scoring − to +++ for each binding reaction is described in text.
[b] NT = not tested.

Null cells					T. cells		B-LCL (pool)		Interpretation	
									Strength	Selectivity
HL-60	Reh	K562	A.L.L.	MOLT-4/CCRF	HSB2	T-CLL	Log Ph.	Static		
	−	−	+	±	−	−	++	++	V. strong	B cell selective
	−	−	−	−	−	−	−	−	Strong	B cell selective (neg. on B-CLL)
	+	±	+	++	+	+++	−	±	Strong	Pan-reactive but neg. on B-CLL
	±	−	−	±	±	+	±	−	Weak	Pan-reactive
	−	−	−	−	±	±	+	−	Weak	B and T cell reactive
	+	±	−	+	+	+	+	±	Moderate	Pan-reactive
++	+++	+++	+	++	+	+++	++	−	V. strong	Pan-reactive
	+	+	−	+	±	+	−	−	Moderate	Pan-reactive
	±	±	−	−	−	±	−	−	V. weak	
	++	++	+++	+	±	++	±	−	Strong	Pan-reactive
	+	±	−	±	−	±	+++	++	V. strong	B cell selective
	++	+	−	±	±	+	+	+	Strong	Pan-reactive
	−	−	±	±	−	−	−	−	V. weak	
	−	−	−	−	−	−	±	−	Weak	B cell selective
	−	−	−	±	−	±	−	−	V. weak	
	−	−	−	±	±	±	−	−	V. weak/Neg.	
	±	−	±	+	±	+	−	−	Moderate	Pan-reactive
	−	−	±	±	−	±	−	−	Weak	? B cell selective
	−	±	−	−	−	±	++	++	Moderate	B cell selective
+	+++	+++	−	+	++	+	±	−	Strong	Pan-reactive
	−	±	+++	−	−	±	−	−	V. weak except Null A.L.L.	
	−	−	±	−	−	±	+	+	Moderate	B cell selective
	+	+	±	±	±	±	++	±	Moderate	Pan-reactive
	+	±	−	−	−	−	±	−	Wk/Mod	B and Null cells
	−	−	−	±	±	±	±	−	Weak	Pan-reactive
	−	−	−	±	−	±	±	−	Weak	Pan-reactive
	−	−	+++	±	−	±	+	−	Wk/Mod except Null A.L.L.	
	−	−	−	−	−	−	−	−	Moderate	B cell selective
	−	+	±	−	−	−	+	−	Mod/Str	B and Null cells
	−	−	−	±	−	−	±	−	Weak	? B cell selective
	±	−	−	±	−	−	−	−	V. weak	
	−	−	−	±	±	−	−	−	Weak	? Pan-reactive
	−	−	−	−	−	−	−	−	Negative	
	−	−	−	−	−	−	−	±	V. wk/Neg	
	±	−	−	−	−	±	±	−	V. wk/Neg	
	+	−	−	−	−	−	+	−	Moderate	B and Null cells
	−	−	−	+	−	±	−	−	Weak	? Pan-reactive
	−	−	±	−	−	−	±	−	Negative	
	−	−	−	±	−	±	++	+	Strong	? B cell selective
	−	−	−	−	−	−	−	−	Negative	
	−	−	±	±	−	±	−	±	Weak	? Pan-reactive
	±	−	±	±	±	±	±	±	Moderate	Pan-reactive
	−	−	++	−	−	−	±	±	Mod/Str	B and Null cells
	++	++	−	−	±	+	+	±	Strong	Pan-reactive
++	+++	+++	+	++	++	++	±	+	V. strong	Pan-reactive
	++	±	−	±	−	−	+	−	Mod/Wk	? B and Null cells
	±	+	++	±	−	+	+	±	Strong	Pan-reactive
	±	±	++	−	−	±	−	−	Variable	? B-CLL and A.L.L.
	−	−	−	+	±	+	−	−	Moderate	? T cell selective
+	++	+	−	+	+	++	+	−	Strong	Pan-reactive
	−	−	+	±	−	±	±	−	Weak	? Pan-reactive
	−	+++	+	±	−	±	+	−	Variable	+++ on K562 only
	−	−	+++	−	−	−	+++	+++		
++	++	+	+++	+	+++	+++	+++	+++		
	++	−	−	+	±	+++	−	−		
T	+	+++	++	+	NT	+	±	−		
T	NT	NT	NT	±	NT	++	+++	++		
	−	−	−	−	−	−	±	−		
++	++	−	+++	NT	+++	NT	+++	++		

pellets were finally taken up in 100 μl of PBS-BSA with 1% sodium dodecyl sulfate and transferred to LP3 tubes (Luckham) for gamma-counting.

Scoring

Since the counts obtained from iodinated anti–mouse Ig varied according to batch and age (storage) and in view of the variation in size and properties of the test cells, the following formula was adopted to facilitate comparisons between assays.

Background counts per minute (cpm) = mean of values obtained with mouse plasmacytoma culture supernatant (X63.NSl) and with an irrelevant mouse/mouse hybridoma culture supernatant (MSl.20A) as the "first antibody."

"100% positive" cpm = mean of values obtained from the two highest positive control antibodies (one of these was always anti-MHC class I, the other usually anti-MHC class II but in the case of T or Null cells it could be anti-β_2-microglobulin, J5, or OKT_1) plus the four highest values obtained from among the B series antibodies.

A score for the binding of each individual antibody was then calculated as:

$$\frac{\text{Individual cpm} - \text{background cpm}}{\text{``100\% + ve'' cpm} - \text{background cpm}} \times 100$$

A score of less than 10 was recorded as −, 11–25 as ±, 26–50 as +, 51–75 as ++, and >75 as +++.

Results

The scores for binding of each antibody to each of the test cell populations are recorded in Table 4.3. Five separate aliquots of pooled B lymphoblas-

Table 4.4. Classification of workshop B series antibodies.[a]

Strength of binding	Selectivity			
	Pan-reactive	B cell selective	B+ null cells	Other
Strong	3, 7, 10, 12, 20, 44, 45, 47, 50	1, 2, 11, 39	43	48, 52
Moderate	6, 8, 17, 23, 42	19, 22, 28	24, 29, 36, 46	21, 27, 49
Weak	4, 25, 26, 32, 37, 41, 51	14, 18, 30		5
V. weak/ neg.	9, 13, 15, 16, 31, 33, 34, 35, 38, 40			

[a] Antibodies, identified by Workshop B series number, have been grouped according to the data set out in Table 4.3.

toid cell lines (each containing contributions from eight to fifteen different lines) were tested and the scores obtained on the separate occasions showed a good measure of consistency. The range is shown in Table 4.3.

Fresh (unfixed) and glutaraldehyde-fixed cells from the same pool of B lymphoblastoid lines were compared directly for ability to bind the Workshop antibodies. There was excellent agreement between the two sets of scores (data not shown), indicating that fixation by the method described had not introduced any serious artifact into the assay.

The interpretation of the pattern of reactivity of each antibody with the test panel is included in Table 4.3 and a rough grouping of the antibodies, based on these patterns, is given in Table 4.4.

Discussion

The indirect radioimmunobinding assay used in the present study gives some information on the distribution of the antigens detected by the Workshop B series of antibodies and, provided the combining sites are saturated, it indicates their relative abundance.

Among the most interesting of the antibodies are those which give very strong reactions with only one or two of the cell types tested; for example, B21 and B27 appear to be highly selective for the single sample of Null cell ALL in the panel while B52 binds strongly to K562 which can be induced to differentiate in the direction of erythroleukemia (14) but under standard culture conditions must be classified as a poorly differentiated cell line.

The two examples of B-CLL were selected for investigation of the difference in HLA-class II expression before and after TPA treatment and, though the comparison is incomplete, they show a number of divergent reactions. B29 binds to cells from patient 2 only. The determinants recognized by B11 and B39 appear to be TPA-inducible in cells from patient 1 but not in those from patient 2 while the reverse applies to antibodies B43 and B47 (and possibly also B41). Two other antibodies, B7 and B12, show enhancement of binding to the cells of patient 2 after TPA induction while B8 shows the opposite trend. Unfortunately, these three were not tested on the cells from patient 1.

The comparison of cells from a static (nondividing) culture with those from the same pool of B lymphoblastoid lines in log phase growth appears to identify at least two antigens expressed only in the latter state, namely those recognized by antibodies B7 and B23. In the former case, the corresponding antigen is inducible on B-CLL cells by TPA and there is at least a suggestion that the same is true for the determinant recognized by B23.

It would obviously be unwise to make any definitive claims about the properties of the B series antibodies on the basis of the present findings since no kinetic studies have been undertaken and since the cell panel is rather small. However the data, when combined with those from other contributors to the Workshop, should contribute to a more complete classification.

Summary

Fifty-two Workshop B series antibodies have been tested in a radioimmunobinding assay against a panel of well-characterized cells from established human leukocyte lines and from individual cases of leukemia or lymphoma. The antibodies were classified according to their patterns of reactivity as being either pan-reactive, B cell selective, selective for B and "Null" cells, selective for other populations, or unreactive. A small number of antibodies appear to detect antigens of special interest, either because of their highly restricted distribution or because their expression is related to the state of activation of the cells which bear them.

References

1. Matsuoka, Y., G.E. Moore, Y. Yagi, and D. Pressman. 1967. Production of free light chains by a hemopoietic cell line derived from a patient with multiple myeloma. *Proc. Soc. Exp. Biol. Med.* **125:**1246.
2. Nilsson, K. 1971. Characteristics of established myeloma and lymphoblastoid cell lines derived from an E myeloma patient. A comparative study. *Int. J. Cancer* **7:**380.
3. Collins, S.J., R.C. Gallo, and R.E. Gallacher. 1977. Continuous growth and differentiation of human myeloid leukemia cells in suspension culture. *Nature* **270:**347.
4. Rosenfeld, C., A. Goutner, C. Choquet, A.M. Venuat, B. Kayibanda, J.L. Pico, and M.F. Greaves. 1977. Phenotypic characteristics of a unique non-T, non-B acute lymphoblastic leukemia cell line. *Nature* **267:**841.
5. Lozzio, C.B., and B.B. Lozzio. 1975. Human chronic myelogenous leukemia cell line with positive Philadelphia chromosome. *Blood* **45:**321.
6. Minowada, J., T. Ohnuma, and G.E. Moore. 1973. Rosette-forming human lymphoid cell lines. 1. Establishment and evidence of origin from thymus-derived lymphocytes. *J. Natl. Cancer Inst.* **49:**891.
7. Foley, G.E., H. Lazarus, S. Farber, B.G. Uzman, B.A. Boone, and R.E. McCarthy. 1965. Continuous culture of human lymphocytes from peripheral blood of a child with acute leukemia. *Cancer* **18:**522.
8. Adams, R.A., A. Flowers, and B.J. Davis. 1968. Direct implantation and serial transplantation of a human acute lymphoblastic leukemia in hamsters. *Cancer Res.* **28:**1121.
9. Minowada, J. 1978. Markers of human leukemia–lymphoma cell lines reflect hematopoietic cell differentiation. In: *Human lymphocyte differentiation: Its application to cancer,* INSERM symposium No 8, B. Serrou and C. Rosenfeld, eds. Elsevier/North Holland, Amsterdam, pp. 337–344.

10. Kaltenbach, J.P., M.H. Kaltenbach, and W.B. Lyons. 1958. Nigrosin as a dye for differentiating live and dead ascites cells. *Exp. Cell Res.* **15:**112.
11. Totterman, T.H., K. Nilsson, and C. Sundstrom. 1980. Phorbol ester-induced differentiation of chronic lymphocytic leukemia cells. *Nature* **288:**176.
12. Guy, K., V. van Heyningen, E. Dewar, and C.M. Steel. 1983. Enhanced expression of human la antigens by chronic lymphocytic leukemia cells following treatment with 12-*O*-tetradecanoylphorbol-13-acetate. *Eur. J. Immunol.* **13:**156.
13. Caligaris-Cappio, F., G. Janossy, D. Campana, M. Chilosi, L. Bergin, R. Foa, D. Delia, M.C. Guibellino, P. Preda, and M. Gobbi. 1984. Lineage relationship of chronic lymphocytic leukemia and hairy cell leukemia: studies with TPA. *Leuk. Res* **8:**567.
14. Andersson, L.C., M. Jokinen, and C.G. Gahmberg. 1979. Induction of erythroid differentiation in the human leukemia cell line K562. *Nature* **278:**364.

CHAPTER 5

Expression of Lymphocyte Differentiation Antigens in Immunodeficiency Diseases

Thomas F. Tedder, Loran T. Clement, Marilyn J. Crain, and Max D. Cooper

Immunodeficiency diseases are frequently caused by defects in lymphocyte differentiation resulting in inadequate synthesis of immunoglobulin (Ig) or the lack of antigen-specific antibodies (1). The failure of B cells to differentiate into plasma cells results from either inherent B cell defects, the selective absence of helper T cell activity, excessive suppressor T cell activity, or Epstein–Barr virus infection (2,3). The most common deficiency is the failure of IgA production. Patients with selective IgA deficiency (IgA^-) have B cells of normal phenotypes except their IgA B cells are immature and coexpress surface IgM (4). In addition, these patients may be concomitantly deficient in other IgG isotypes (5). Patients with common variable immunodeficiency (CVI) have normal numbers of B and T cells of normal phenotypes, but few if any plasma cells (6). Similarly, X-linked immunodeficiency (XLA; X-linked agammaglobulinemia) results from abortive B cell differentiation. Males with this disease form normal numbers of bone marrow pre–B cells, but are deficient in B cells and plasma cells (7). The frequency of circulating B cells is 100-fold lower than that in normals, but the T cell lineage appears to be unaffected since cell-mediated immunity is intact and circulating T cells and T cell subsets are normal in number, proportion, and function. Patients with acquired immunodeficiency disease syndrome (AIDS) have a preferential loss of helper T cells that results in suppression of T cell function (8). In addition, B cells from these patients appear to be polyclonally activated with elevated numbers of cells spontaneously secreting Ig (9), yet these patients are unable to produce specific antibodies.

The analysis of cell surface antigens present on resting lymphocytes has proven useful in the study of immunodeficiency diseases by demonstrating selective loss of particular lymphocyte subpopulations. However, most markers have proven to be poor indicators of lymphocyte maturation and functional ability since these antigens are present during many stages of differentiation. For example, patients with CVI possess normal numbers of lymphocytes bearing surface Ig, despite their inability to produce antibodies. In this report, we characterize four new antibodies which identify B cells in particular stages of maturation and use these to assess the developmental state of T and B cells from patients with IgA^-, CVI, XLA, and AIDS. The HB-4 monoclonal antibody identifies an antigen exclusively present on a subpopulation of B cells and a subpopulation of large granular lymphocytes with NK cell activity (10). The HB-5 antibody identifies a 145,000-M.W. antigen that is present on mature B cells (11). This molecule is the C3d receptor (CR2) and a receptor for Epstein–Barr virus (12). Another antibody, HB-7, uniquely reacts with immature lymphocytes, does not react with mature B cells, but identifies plasma cells (13). HB-7 identifies a 45,000-M.W. molecule that is closely related if not identical to the T10 antigen. T cell maturity was assessed with the HB-10 antibody that identifies a subpopulation of immature T cells (14). The HB-10 antigen is also present on all B cells.

Materials and Methods

Patients and Cells

Mononuclear cells were isolated from heparinized blood samples obtained from healthy donors or from patients with immunodeficiency diseases. The frequencies of B cells, T cells, and monocytes were determined by cell marker analysis as described previously (11). Patients with IgA^- had normal serum Ig levels but were markedly deficient in serum IgA, and had normal numbers of B cells, T cells, and monocytes. Patients with CVI had low levels of all serum Ig isotypes but possessed nearly normal numbers of B cells, T cells, and monocytes. Patients with XLA were markedly deficient in serum Ig and B cells but possessed nearly normal frequencies of T cells and monocytes. Patients with AIDS had elevated levels of serum IgG and IgA. MNC isolated from patients with AIDS had normal frequencies of B cells but were deficient in helper T cells expressing the Leu 3 antigen.

Antibodies and Immunofluorescence Analysis

The HB-4 monoclonal antibody was produced by fusion of the Ag8.653 myeloma with spleen cells from a mouse immunized with the human B cell line BJAB (10). HB-5, HB-7, and HB-10 antibodies were produced as described (11,13,14). Suspensions of viable cells were analyzed for surface antigen expression by incubation with the HB-4, HB-5, HB-7, or HB-10 antibodies (50 μg/ml), followed by tetramethylrhodamine isothiocyanate-conjugated goat anti–mouse Ig isotype-specific antibodies as described (11). After washing, the cells were co-stained with fluorescein isothiocyanate (FITC)-conjugated, affinity-purified goat antibodies to human μ heavy chains or with Leu 4 antibodies (Becton Dickinson) followed by the appropriate FITC-conjugated mouse heavy-chain isotype specific reagent. Two-color fluorescence staining of fixed cytocentrifuge cell preparations was visualized using a Leitz Orthoplan fluorescence microscope.

Results

Normal Lymphocyte Expression of HB-4, -5, -7, and -10 Antigens

The HB-4 antibody identified a subpopulation of mature, resting B cells as determined by the examination of B cells during development and by functional analysis of HB-4$^+$ B cells (Fig. 5.1) (10). HB-4 reacted with fewer than 5% of IgM$^+$ cells in fetal liver and bone marrow, but with 25% in fetal spleen and approximately 40% in newborn blood. HB-4 reacted on average with 66% of IgM$^+$ B cells from adult blood. In contrast, only 13% of IgM$^+$ B cells from adult spleen, tonsil, and lymph node were HB-4$^+$. HB-4$^+$ B cells from blood could be induced to proliferate by cross-linkage of their surface Ig but not by T cell-derived growth factors. The subpopulation of activated B cells from adult blood which responds to T cell-derived differentiation factors was HB-4$^-$, as were plasma cells. Therefore, HB-4 reacts with a subpopulation of mature B cells but not with activated B cells from blood.

The HB-5 antibody identifies the CR2 molecule and is described in detail in Chapter 43 of this book. Detectable levels of this receptor are expressed during early B cell development (Fig. 5.1).

The HB-7 monoclonal antibody reacted with 95% of B cells from fetal tissues, 50% of B cells from newborn blood, and 60% of B cells from adult marrow, but fewer than 10% of B cells from adult blood (13). The HB-7 antigen was not expressed by the subpopulation of activated B cells found

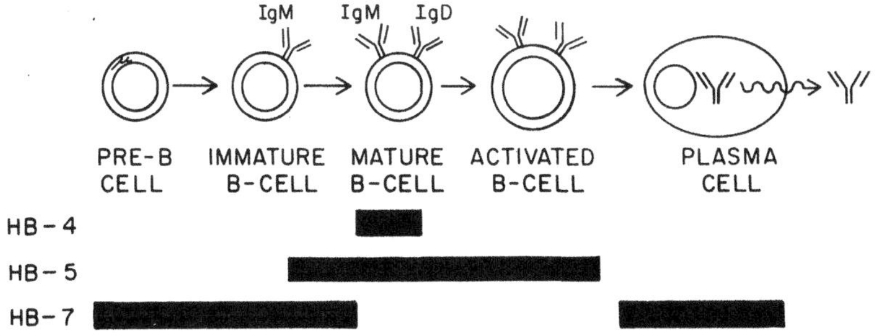

Fig. 5.1. Expression of the HB-4, 5, and 7 antigens during different stages of B cell development.

in adult blood but was present on plasma cells. Therefore, this antigen is unique in that it is expressed early in B cell differentiation, is not detectable on mature B cells, but appears again with terminal differentiation (Fig. 5.1).

The HB-10 antibody reacts with all B cells and a subpopulation of T cells (14) and is described in Volume 1 (Chapter 16) of this series. The HB-10 antigen was expressed by a subset of T3$^+$ mature thymocytes and by most T cells in fetal bone marrow and newborn blood. In adult blood, HB-10 reacted with 65% of the Leu 4$^+$ T cells, but with fewer T cells in adult spleen and tonsil (27%). Among helper T cells from blood the HB-10 antigen marks a subpopulation of immature T cells that can produce B cell growth factors but is deficient in the ability to produce B cell differentiation factors.

B Cell Expression of HB-4 Antigen in Immunodeficiencies

The results described above indicate that HB-4 reacts with a subpopulation of mature B cells but not with activated B cells. Patients with IgA$^-$ possessed frequencies of HB-4$^+$ B cells similar to those of normals (Table 5.1). Patients with CVI were heterogeneous with respect to B cell expression of the HB-4 antigen. B cells from five of the 16 patients examined expressed HB-4 antigen at a frequency that was significantly lower than normal while all of one patient's B cells were HB-4$^-$. All five patients with XLA showed a significant decrease in HB-4$^+$ B cells. HB-4 antigen expression by B cells from patients with AIDS varied. More than 90% of two patients' B cells were HB-4$^+$ while only 25% of one patients' B cells expressed this antigen. The other two patients were like normals in the frequency of HB-4$^+$ B cells.

B Cell Expression of HB-5 Antigen in Immunodeficiencies

In contrast to HB-4, the HB-5 antibody reacts with both mature and activated B lymphocytes in blood. Most B cells from IgA$^-$ patients and

Table 5.1. B and T cell expression of differentiation antigens in various immunodeficiency diseases.

Patients (number)	Mean % of cells which expressed HB antigen (range)[a]			
	IgM$^+$ B cells		Leu 4$^+$ T cells	
	HB-4$^+$	HB-5$^+$	HB-7$^+$	HB-10$^+$
Normals (12)	66 (37–93)	95 (83–99)	10 (1–32)	65 (46–75)
IgA$^-$ (4)	50 (30–75)	79 (42–93)	5 (0–16)	58 (28–76)
CVI (16)	46 (0–94)	88 (63–100)	16 (1–55)	50 (22–85)
XLA (5)	25 (3–35)	15 (9–24)	70 (23–90)	87 (70–99)
AIDS (5)	72 (25–99)	39 (0–76)	20 (0–50)	28 (12–40)

[a] Reactivity was determined by two-color indirect immunofluorescence microscopy.

patients with CVI expressed the HB-5 antigen at normal frequencies. In contrast, B cells from XLA patients were rarely HB-5^+. Patients with AIDS also had significantly lower frequencies of HB-5^+ B cells.

B Cell Expression of HB-7 Antigen in Immunodeficiencies

The HB-7 antibody is unique in reacting with immature B cells and plasma cells but not with mature and activated B cells. B cells from patients with IgA^- and CVI rarely expressed the HB-7 antigen. The B cells from one patient with CVI did express significantly higher frequencies of HB-7 antigen than normals. This patient's B cells also rarely expressed the HB-4 antigen suggesting most of the B cells were immature. Most B cells from patients with XLA expressed the HB-7 antigen. B cells from one of four AIDS patients expressed HB-7 at a higher frequency than normal.

T Cell Expression of HB-10 Antigen in Immunodeficiencies

The phenotype of T cells was also analyzed in patients with immunodeficiencies. Leu 4^+ T cells from patients with IgA^- were like normals in regards to HB-10 antigen expression as were most T cells from patients with CVI. However, T cells from six patients with CVI expressed significantly lower frequencies of HB-10 antigen than normals. Their T cells were similar to those found in the AIDS patients where only one fourth of T cells were HB-10^+. In contrast, virtually all of the T cells from patients with XLA were HB-10^+.

Discussion

Multiple changes in cell surface antigens occur during lymphocyte development. In some cases, phenotypic changes also correlate with changes in functional ability. However, until we understand the role these cell surface molecules play in lymphocyte function, we will not know if these phenotypic changes are strictly associated with the functional abilities of the cells which bear them. Nonetheless, these antigens provide useful markers for examining the lymphocytes of patients with immunodeficiency diseases in the context of normal B and T cell maturation.

Most patients with IgA^- or CVI appeared to have normal numbers and phenotypes of lymphocytes as defined by HB-4, -5, and -7 antigen expression by B cells and HB-10 expression by T cells. However, the B cells from four patients with CVI appeared to be activated since their B cells rarely expressed the HB-4 antigen but were HB-$5^+,7^-$. Thus, these B cells must be blocked from further maturation since these patients are deficient in serum Ig. In contrast, the B cells from one patient with CVI had a phenotype most often found in newborn blood (HB-$4^-,5^+,7^+$) suggesting that the B cells from this patient were unable to mature beyond an early stage of differentiation. T cells from six of the CVI patients were also significantly different from normals in that the majority of them had a mature or activated HB-10^- phenotype. Three of these patients had normal B cells, two had activated B cells, and one had immature B cells. Therefore, CVI represents a heterogeneous disease with some patients possessing either B and T cells of normal phenotypes, immature B cells, activated B cells, or activated T cells.

The majority of the few B cells present in patients with XLA were of the immature phenotype, HB-4^-,5^-,7^+. Although the basis for the apparent excessive wastage of immature B cells in these patients remains unknown, it is apparent that the majority of the B cells which progress beyond the pre–B cell stage of development do not mature. Therefore, this lesion in cell development affects B cells at a very early stage of development. The frequency of T cells in these patients was equivalent to that of normals but the majority were of the immature HB-10^+ phenotype. These results suggest either that B cells may play an important role in activating T cells *in vivo* or that the XLA gene defect may result in a block of B and T cell maturation.

B cells from most AIDS patients were odd in having phenotypes inconsistent with either normal mature B cells or activated B cells. The B cells of these patients were mostly HB-7^- with variable yet generally high expression of the HB-4 antigen and significantly decreased expression of CR2 identified by HB-5 antibodies. Therefore, their B cells may be activated by stimuli which do not represent physiological activation. The reason for reduced CR2 expression is unknown but these patients have high titers of antibodies against Epstein–Barr virus. Therefore, reduced CR2 expression may be related to virus infection, to circulating immune complexes, or to abnormal B cell activation. The T cells from these patients also had an activated phenotype.

The importance of using new monoclonal antibodies to assess lymphocyte maturation in patients with immunodeficiencies should be emphasized. Through the simultaneous use of antibodies which react with cells during discrete yet different stages of maturation, such as HB-4, HB-5, and HB-7, we are better able to identify the blocks which occur at different stages of lymphocyte differentiation in immunodeficient patients. Phenotypic characterization such as this should aid in the further identification of the different defects which cause these diseases.

Summary

T and B cell differentiation was examined in patients with common variable immunodeficiency (CVI), selective IgA deficiency (IgA$^-$), X-linked agammaglobulinemia (XLA), and acquired immunodeficiency syndrome

(AIDS) using monoclonal antibodies to four lymphocyte differentiation antigens. These antibodies are: 1) HB-4, which identifies a subpopulation of resting B cells; 2) HB-5, which reacts with the C3d receptor present on mature B cells; 3) HB-7, which identifies immature B lymphocytes; and 4) HB-10, which reacts with virgin, but not activated or memory T cells. B cells from all four IgA^- patients had normal phenotypic profiles. B cell maturation in CVI patients was heterogeneous: cells from 11 of 16 patients had normal antigenic phenotypes, while B cells from four CVI patients had normal HB-5 and HB-7 antigen expression but rarely expressed the HB-4 antigen, suggesting they were preactivated. One CVI patient's B cells were of an immature phenotype. The vast majority of the limited numbers of IgM B cells from five XLA patients expressed the immature $HB\text{-}4^-,5^-,7^+$ phenotype. B cells from AIDS patients were mostly $HB\text{-}7^-$ with variable expression of the HB-4 antigen and significantly decreased expression of the HB-5 antigen. Although T cells from IgA^- and most CVI patients had normal HB-10 antigen expression, T cells from five CVI patients and most AIDS patients appeared to be activated. In contrast, the circulating T cell population in XLA patients was phenotypically similar to that of normal newborns.

Acknowledgments. This work was supported by grants CA 16673 and CA 13148 from the National Cancer Institute; 1-608, March of Dimes Birth Defects Foundation; and 5M01-RR-32 DRR/NIH.

References

1. Rosen, F.S., M.D. Cooper, and R.J.P. Wedgwood. 1984. The primary immunodeficiencies. *New England J. Med.* **311**:235.
2. Siegal, F.P., M. Siegal and R.A. Good. 1978. Role of helper, suppressor and B-cell defects in the pathogenesis of the hypogammaglobulinemias. *New England J. Med.* **299**:172.
3. Reinherz, E.L., M.D. Cooper, S.F. Schlossman, and F.S. Rosen. 1981. Abnormalities of T cell maturation and regulation in human beings with immunodeficiency disorders. *J. Clin. Invest.* **68**:699.
4. Conley, M.E., and M.D. Cooper. 1981. Immature IgA cells in IgA deficient patients. *New England J. Med.* **305**:495.
5. Oxelius, V-A., A-B. Laurell, B. Lindquist, H. Golebiowska, V. Axelsson, J. Bjorkander, and L.A. Hanson. 1981. IgG subclasses in selective IgA deficiency: importance of IgG2-IgA deficiency. *New England J. Med.* **304**:1476.
6. Preud'homme, J.L., C. Friscelli, and M. Seligmann. 1973. Immunoglobulins on the surface of lymphocytes in fifty patients with primary immunodeficiency disease. *Clin. Immunol. Immunopathol.* **1**:241.
7. Pearl, E.R., L.B. Vogler, A.J. Okos, W.M. Crist, A.R. Lawton, and M.D. Cooper. 1978. B lymphocyte precursors in human bone marrow: an analysis of normal individuals and patients with antibody-deficiency states. *J. Immunol.* **120**:1169.

8. Gottlieb, M.S., R. Schroff, H.M. Schanker, J.D. Weisman, P.T. Fan, R.A. Wolf, and A. Saxon. 1981. *Pneumocystis carinii* pneumonia and mucosal candidiasis in previously healthy homosexual men: evidence of a new acquired cellular immunodeficiency. *New England J. Med.* **305:**1425.
9. Lane, H.C., H. Masur, L.C. Edgar, G. Whalen, A.H. Rook, and A.S. Fauci. 1983. Abnormalities of B-cell activation and immunoregulation in patients with the acquired immunodeficiency syndrome. *New England J. Med.* **309:**453.
10. Tedder, T.F., L.T. Clement, and M.D. Cooper. 1983. Use of monoclonal antibodies to examine differentiation antigens on human B cells. *Fed. Proc.* **42:**415A.
11. Tedder, T.F., L.T. Clement, and M.D. Cooper. 1984. Expression of C3d receptors during human B cell differentiation: Immunofluorescence analysis with the HB-5 monoclonal antibody. *J. Immunol.* **133:**678.
12. Fingeroth, J.D., J.J. Weis, T.F. Tedder, J.L. Strominger, J.A. Barbosa, and D.T. Fearon. 1984. Epstein–Barr virus receptor of human B lymphocytes is the C3d receptor CR2. *Proc. Natl. Acad. Sci. U.S.A.* **81:**4510.
13. Tedder, T.F., L.T. Clement, and M.D. Cooper. 1984. Discontinuous expression of a membrane antigen (HB-7) during B lymphocyte differentiation. *Tissue Antigens* **24:**140.
14. Tedder, T.F., M.D. Cooper, and L.T. Clement. 1984. Differential production of B cell growth and differentiation factors by phenotypically distinct subpopulations of human helper T cells. *Fed. Proc.* **43:**1826A.

CHAPTER 6

Use of Two Monoclonal Anti–Human B Cell Antibodies in the Study of Early B Cell Differentiation

Anand Raghavachar, Shraga F. Goldmann,
and Bernhard Kubanek

Materials and Methods

Monoclonal Antibodies

UL-38 and UL-90 were raised against fresh CALLA$^+$, cμ^- leukemia cells. Details of the fusion protocol and properties of the antibodies are described elsewhere (1). Briefly somatic cell hybridization and growth of hybridomas UL-38 and UL-90 was performed essentially as described by Köhler and Milstein (2). As fusion partners PAI-O BALB/c myeloma cells (3), a nonproducing subclone of the mouse myeloma cell P3x63Ag8, were used (obtained from Dr. J.W. Stocker, Basel). mAbs BA-1 (4), BA-2 (5), and BA-3 (5) were obtained from Hybritech Inc., San Diego, CA, B1 (6) and J5 (7) from Coulter, Hialeah, FL, and VIL-A1 (8) was a kind gift of Dr. W. Knapp, Vienna. All antibodies were used at a concentration of 1 μg/ml.

Leukemia Cells/Cell Lines

Mononuclear cells were isolated from heparinized peripheral blood and/or bone marrow samples of leukemia patients using Ficoll–Hypaque density gradient centrifugation (9). The diagnosis of leukemia was made by standard clinical, morphological, and cytochemical criteria (10); by these criteria more than 90% of the cells in the samples were leukemic cells. The hematopoietic cell lines used in this study and their properties are listed in Table 6.1 [for references see (11), (12)].

Table 6.1. Comparison of expression of UL-38, BA-1, BA-2, BA-3, B1, and UL-90 antigens on hematopoietic cell lines.

Cell line	Origin	Reactivity[a] with					
		UL-38	BA-1	BA-2	BA-3	B-1	UL-90
Reh-6	cALL	+	+	+	+	−	−
NALL-1	cALL	+	+	+	+	−	−
KM-3	cALL	+	+	+	+	−	−
Nalm-6	pre–B-ALL	+	+	+	+	+	−
BALL-1	B-ALL	−	+	−	−	+	−
Daudi	BL	+	+	−	+	−	−
BJAB	BL	+	−	+	+	+	−
Raji	BL	+	−	−	+	+	−
X308	BL	+	+	−	−	+	−
RAMOS	BL	+	−	−	+	+	−
KE-37	T-ALL	−	−	−	−	−	−
HSB-2	T-ALL	−	−	−	−	−	−
MOLT-4	T-ALL	−	−	−	−	−	−
CEM-ATCC	T-ALL	+	−	−	+	−	−
KG-1	AML	−	−	−	−	−	−
ML-1	AML	−	−	−	−	−	−
ML-2	AML	−	−	−	−	−	−
ML-3	AML	−	−	−	−	−	−
K562	CML-BC	−	+	−	−	−	−
HL-60	APL	−	−	−	−	−	−

[a] Results from at least 5 experiments. + indicates >20% reactive cells − indicates <20% reactive cells.

Detection of Antibody Reactivity

The binding of mAbs was assessed by indirect immunofluorescence with saturating concentrations of mAb culture supernatant or ascites and fluoresceinated rabbit anti–mouse IgG $(Fab')_2$ or IgM $(Fab')_2$ second antibodies (Cappel Labs, West Chester, PA). Appropriate controls of mouse IgG or mouse IgM were performed. Fluorescence of cells was evaluated using a Zeiss microscope with incident light illumination and equipped for the dual-wavelength method. The TdT-assay was performed by immunofluorescence according to Stass *et al.* (13).

Differentiation Induction *in vitro*

Differentiation induction experiments were done as described by Nadler *et al.* (14). Cells were resuspended in RPMI 1640 medium containing 20% FCS (1×10^6/ml) and cultures were maintained at 37°C in a humidified atmosphere in the presence of 1.5 n*M* TPA (Sigma, Munich, FRG). Samples collected before culture and 4 hr and 24 hr after exposure to TPA were washed and labeled for immunofluorescence as described above.

Results

Preliminary Characterization of Established Hybridomas UL-38 and UL-90

Two weeks after fusion the culture supernatants were tested for reactivity with surface antigens of the immunizing cell, peripheral blood cells, and hematopoietic cell lines. From this, clone UL-38 was selected because of its unique reactivity with cells and cell lines of various B cell differentiation stages. The pattern of reactivity with cell lines can be seen from Table 6.1, and that with leukemia cells of different subtypes from Table 6.2. Although the UL-38 antibody is reactive with all stages, from earliest to latest, of B cell differentiation, this report will focus on earliest B cell differentiation stages. A clone from another fusion experiment, UL-90, showed selective reactivity with CALLA-positive leukemia cells from patients (Table 6.3), but did not react with any of the cell lines shown in Table 6.1 nor with normal peripheral blood cells.

AUL Cells

The distribution of phenotypes defined by the expression of OKIa1, UL-38, BA-1, CALLA, TdT, and myeloperoxidase is listed in Table 6.4. In all

Table 6.2. Reactivity of UL-38 with cells from leukemia/lymphoma patients.

Classification	No. positive/no. tested
1. AUL	
(E^-, sIg^-, $CALLA^-$, Cytochem.$^-$, DR^+)	7/9
2. cALL	
(E^-, sIg^-, $CALLA^+$, $c\mu^-$)	24/24
3. pre–B-ALL	
(E^-, sIg^-, $CALLA^+$, $c\mu^+$)	3/3
4. B-ALL	
(E^-, sIg^+, $CALLA^-$)	2/2
5. T-ALL	
(E^+, sIg^-, $CALLA^-$, T-Ag$^+$)	0/4
6. cT-ALL	
($E^{\pm}$, sIg^-, $CALLA^+$, T-Ag$^+$, DR^+)	0/2
7. CLL	
(E^-, sIg^+, $CALLA^-$, DR^+)	3/5
8. CML-BC"M"	
(My-Ag$^+$, E^-, sIg^-, $CALLA^-$)	0/4
9. CML-BC"L"	
(My-Ag$^-$, E^-, sIg^-, $CALLA^+$)	3/3
10. AML and AMML	
(My-Ag$^+$, E^-, sIg^-, $CALLA^-$)	0/15
11. B-NHL	11/14

Table 6.3. Reactivity of UL-90 with cells from leukemia/lymphoma patients.

Classification	No. positive/no. tested
1. AUL (E$^-$, sIg$^-$, CALLA$^-$, Cytochem.$^-$, DR$^+$)	0/5
2. cALL (E$^-$, sIg$^-$, CALLA$^+$, cμ^-)	16/16
3. pre–B-ALL (E$^-$, sIg$^-$, CALLA$^+$, cμ^+)	0/4
4. B-ALL (E$^-$, sIg$^+$, CALLA$^-$)	0/1
5. T-ALL (E$^+$, sIg$^-$, CALLA$^-$, T-Ag$^+$)	0/2
6. cT-ALL (E$^\pm$, sIg$^-$, CALLA$^+$, T-Ag$^+$, DR$^+$)	0/1
7. CLL (E$^-$, sIg$^+$, CALLA$^-$, DR$^+$)	0/4
8. CML-BC"M" (My-Ag$^+$, E$^-$, sIg$^-$, CALLA$^-$)	0/5
9. CML-BC"L" (My-Ag$^-$, E$^-$, sIg$^-$, CALLA$^+$, cμ^-)	2/2
10. AML and AMML (My-Ag$^+$, E$^-$, sIg$^-$, CALLA$^-$)	0/6
11. B-NHL	0/3

of the patients examined there was a strong expression of nuclear TdT and a lack of reactivity for CALLA and myeloperoxidase. The investigation of nine patients revealed three major subgroups: Ia$^+$, UL-38$^-$, BA-1$^-$ (2/9); Ia$^+$, UL-38$^+$, BA-1$^-$ (3/9); and Ia$^+$, UL-38$^+$, BA-1$^+$ (4/9). No myeloid surface antigens were unequivocally demonstrated (data not shown). To exclude the possibility that UL-38 identified either a constant or variable portion of an Ia-like HLA-D/DR related antigen, studies of Ia-antigen-positive cells were undertaken. Screening studies outlined above demonstrated that Ia-antigen-positive myeloid cell lines were totally unre-

Table 6.4. Phenotypes of AUL-blast cells.[a]

Patient	OKIa	UL-38	BA-1	J5	TdT	Peroxidase
1	++	+	–	–	++	–
2	++	–	–	–	++	–
3	++	–	–	–	++	–
4	++	+	–	–	++	–
5	++	++	++	–	++	–
6	++	+	–	–	++	–
7	++	++	++	–	++	–
8	++	++	++	–	++	–
9	++	++	++	–	++	–

[a] +, Weak reactivity; ++, strong reactivity; –, no reactivity.

active as were monocytes and activated T cells (data not shown). A large number of malignancies expressing the Ia-antigen were also completely negative with anti-UL-38 (Table 6.2).

Differentiation Induction of AUL Cells

The cellular origin of two Ia^+, UL-38^+, BA-1^- cases of AUL has been further investigated by induction of *in vitro* differentiation with TPA. In patient number 1 (Table 6.4) no phenotypic changes could be detected up to 72 hr after exposure of the cells to TPA. In patient number 4, however, after a 24-hr culture in the presence of TPA, 68% of the cells reacted with J5 and there was also a significant increase in the intensity of fluorescence with UL-38. There were no other changes in reactivity detectable with respect to BA-1, TdT, myeloperoxidase, or cytoplasmic staining for IgM heavy chains.

CALLA-Positive Leukemic Cells Including Pre-B-ALL

The cALL patient group showed a great heterogeneity of surface marker pattern and antigen density distribution; nine distinct phenotypes were identified in twenty patients evaluated. This report will focus on the reactivity with mAb UL-90. As can be seen from Table 6.5, 20% of the CALLA-positive adult non-T, non-B ALL cases were pre–B-ALL as

Table 6.5. Phenotype of CALLA-positive adult non-T, non-B ALL.[a]

Patient	J5	BA-3	ViL-A1	BA-1	BA-2	B1	UL-90	Cμ	sIg
1	++	++	++	++	++	++	+	–	–
2	++	+	+	++	–	++	++	–	–
3	++	++	++	++	–	–	++	–	–
4	+	ND	ND	++	++	–	+	–	–
5	ND	++	++	++	++	++	–	+	–
6	++	++	++	++	++	++	++	–	–
7	++	++	ND	++	ND	–	++	–	–
8	++	++	ND	++	–	+	++	–	–
9	+	+	+	++	+	+	+	–	–
10	++	++	++	++	++	–	+	–	–
11	++	++	++	++	++	++	–	+	–
12	++	++	ND	++	+	++	++	–	–
13	++	++	++	+	–	–	++	–	–
14	++	++	ND	++	+	–	++	–	–
15	+	++	+	++	–	++	++	–	–
16	++	+	+	++	+	++	–	+	–
17	++	++	++	++	+	+	+	–	–
18	++	ND	ND	++	++	–	++	–	–
19	++	++	++	++	–	++	–	+	–
20	+	+	ND	++	+	++	+	–	–

[a] +, Weak reactivity; ++, strong reactivity; –, no reactivity; ND, not done.

assessed by demonstration of cytoplasmic IgM heavy chains (cμ) and of these all showed the UL-90 negative marker profile. On the other hand, among the cμ^-, CALLA$^+$ group, there was no patient who displayed the UL-90 negative phenotype. Evaluation of the UL-90 negative classes did not give any characteristic pattern of B marker expression despite the lack of cμ.

Discussion

Several immunological and enzymatic markers have provided a tool for the subdivision of ALL into four major subgroups (15), namely, cALL, T-ALL, unclassified ALL or AUL, and B-ALL. This paper reports on the use of two new monoclonal antibodies in the study of AUL and cALL. Both mAbs, UL-38 and UL-90, were raised against CALLA$^+$, cμ^- leukemia cells and are B lineage specific (1). The present study confirms and extends previous investigations utilizing mAbs. In the AUL group three subgroups were identified: a) Ia$^+$, UL-38$^-$, BA-1$^-$; b) Ia$^+$, UL-38$^+$, BA-1$^-$; c) Ia$^+$, UL-38$^+$, Ba-1$^+$. The majority of the cases examined showed the UL-38$^+$ phenotype, which might indicate that most cases of AUL are of B cell origin. Supporting evidence for this hypothesis was obtained by *in vitro* induced differentiation of blast cells in one case of Ia$^+$, UL-38$^+$, BA-1$^-$ AUL. In contrast to Nadler *et al.* (14) we were able to induce CALLA expression, confirming the theory that CALLA is an early B cell-associated antigen. The phenotype of the induced AUL cells was incomplete, showing no B marker with the exception of CALLA. Evidence presented by Korsmeyer *et al.* (16) indicates that CALLA is a marker of early B cell precursors, since rearrangement of immunoglobulin heavy-chain genes occurs in these cells. On the other hand, recent data suggest that heavy-chain gene rearrangement is not restricted to the cells of the B cell lineage (17) and that such rearrangement may also be present in cells from phenotypically defined acute nonlymphocytic leukemia. It is clear from this study that no single cell surface antigen determinant examined strictly predicts the configuration of immunoglobulin genes and vice versa. Yet the combination of UL-38 antigen expression and immunoglobulin gene configuration may provide insight into the cellular origin in a given case of undifferentiated leukemia.

Another part of the present study deals with the application of UL-90 and additional mAbs to enable further characterization of the CALLA-positive subgroup of non-T ALL. The UL-90 antibody defines a subpopulation of cALL and does not react with cells of earlier or later stages of B cell differentiation. Cells of T lineage and myeloid lineage are completely negative as are all the leukemia cell lines tested in this study including the CALLA-positive cell lines. However, we have no explanation for the discrepancy between UL-90 antigen expression on established CALLA-positive cell lines and on CALLA-positive patient cells. The cell lines might represent an alternative line of differentiation or anomalous expression of antigens on a malignant cell. Our multimarker analysis confirms the heterogeneity of cALL as described by Nadler *et al.* (14) but in addition a new subgroup could be defined. Twenty percent of the cALL cases examined showed the pre–B cell phenotype, i.e., $CALLA^+$, $c\mu^+$. All of these showed no reactivity with our UL-90 antibody, whereas the cases of cALL in the strict sense, i.e., $CALLA^+$, $c\mu^-$, expressed the UL-90-defined antigen. There was no correlation detectable between UL-90 reactivity and other B marker expression in the few cases studied, except the lack of cμ in the UL-90-positive cases and the expression of cμ in the UL-90-negative cases. Thus the UL-90 antibody seems to define the early stages of B cell differentiation, exemplified by pre–B-ALL or cμ-negative cALL. Based on the phenotypes identified in this investigation and the considerations given above, a hypothetical model of earliest B cell development is presented in Fig. 6.1. We are aware of the small number of cases studied and of the preliminary character of our observations. Thus, our antibodies UL-38 and UL-90 will need further investigation with a larger number of malignant cells. The application of these mAbs in combi-

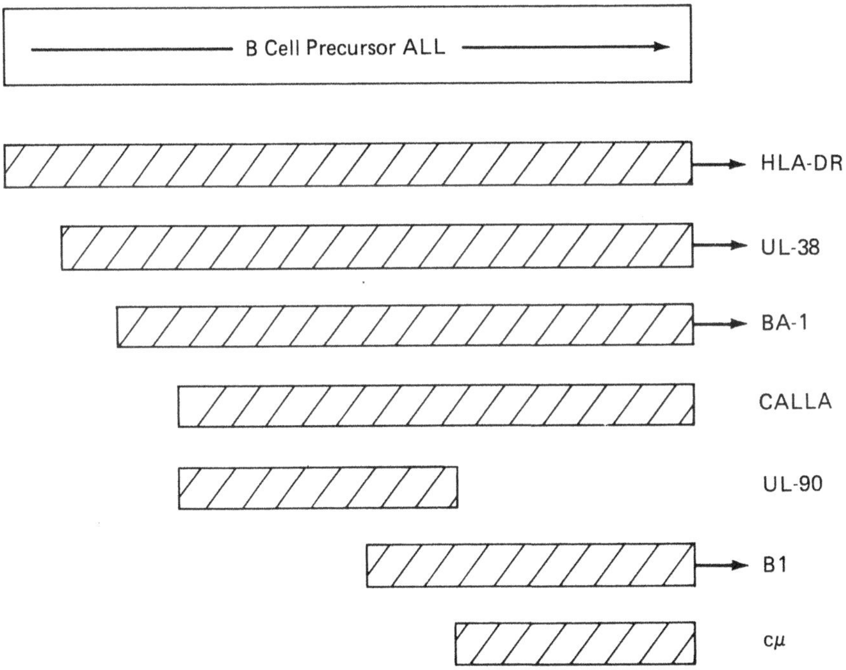

Fig. 6.1. Hypothetical scheme of early B cell differentiation based upon MoAb phenotypes in acute leukemia.

nation with examination of the immunoglobulin gene rearrangements might further our understanding of normal B cell differentiation.

Summary

The development of hybridoma techniques for the production of monoclonal antibodies (mAbs) has enabled the identification of cell surface determinants selectively expressed by B cells at varying stages of differentiation (18). We have generated a series of cytotoxic mAbs raised against CALLA-positive lymphocytic leukemia cells that define B cell lineage-specific antigens. Of these, UL-38 seems to identify most of the patients with the Ia^+, $CALLA^-$, $BA\text{-}1^-$ form of non-T ALL. The cellular origin of the Ia^+, $UL\text{-}38^+$, $CALLA^-$ non-T cell ALL has been further investigated by *in vitro* studies with TPA, an agent known to promote cellular differentiation. TPA was capable of inducing the expression of CALLA in one of two patients with the Ia^+, $UL\text{-}38^+$, $BA\text{-}1^-$ phenotype of AUL. Another mAb, termed UL-90, defines a subpopulation of the CALLA-positive non-T cell ALL and does not react with AUL or the non-T cell ALL of later B cell differentiation stages, which are CALLA negative. By serological analysis these antibodies have been shown to be different from B cell-associated mAbs known so far. These studies suggest that AUL and cALL are heterogeneous and represent a spectrum of early B cell differentiation. A scheme of early B cell differentiation based upon the phenotypes defined by mAb analysis is postulated.

Acknowledgment. This work was supported by the Deutsche Forschungsgemeinschaft, SFB112/B13.

References

1. Raghavachar, A., K. Koerner, G. Sawastzki, A. Ganser, and B. Kubanek. A new monoclonal antibody UL-38, reactive with all stages of B cell differentiation. Manuscript in preparation.
2. Köhler, G. and C. Milstein. 1975. Continuous cultures of fused cells secreting antibody of predefined specificity. *Nature* **256**:495.
3. Stocker, J.W., H.K. Forster, V. Miggiano, C. Stähli, G. Staiger, B. Takacs, and T.H. Staehelin. 1982. Generation of two new mouse myeloma cell lines "PAI" and "PAI-O" for hybridoma production. *Res. Discl.* **217**:155.
4. Abramson, C.S., J.H. Kersey, and T.M. Le Bien. 1981. A monoclonal antibody BA-1 reactive with cells of human B lymphocyte lineage. *J. Immunol.* **126**:83.
5. Le Bien, T.M., J. Kersey, S. Nakazawa, K. Minato, and J. Minowada. 1982. Analysis of human leukemia/lymphoma cell lines with monoclonal antibodies BA-1, BA-2 and BA-3. *Leuk. Res.* **6**:299.
6. Nadler, L.M., P. Stashenko, R. Hardy, A. van Agthoven, C. Terhorst, and

S.F. Schlossman. 1981. Characterization of a human B cell specific antigen (B2) distinct from B1.
7. Ritz, J., J.M. Pesando, J. Notis-McContary, H. Lazarus, and S.F. Schlossman. 1980. A monoclonal antibody to human acute lymphoblastic leukemia antigen. *Nature* **283:**583.
8. Knapp, W., O. Majdic, P. Bettelheim, and K. Liszka. 1982. VIL-A1, a monoclonal antibody reactive with common acute lymphoblastic leukemia cells.
9. Böym, A. 1968. Separation of leukocytes from blood and bone marrow. *Scand. J. Clin. Lab. Invest.* **97:**21s.
10. Wintrobe, M.W. 1981. *Clinical hematology,* 8th ed. Lea & Febiger, Philadelphia.
11. Tobinai, K., M. Hirose, K. Minato, and M. Schimoyama. 1981. The reaction specificity of various monoclonal antibodies and cellular origin and differentiation of cultured cell lines derived from human leukemias and lymphomas. *Japan J. Clin. Oncol.* **11:**469.
12. Minowada, J., E. Tatsumi, K. Sagawa, M.S. Lok, T. Sugimoto, K. Minato, L. Zgoda, L. Prestine, L. Kover, and G. Gould. 1984. A scheme of human hematopoietic differentiation based on the marker profiles of the cultured and fresh leukemia–lymphomas: the result of workshop study. In: *Leucocyte Typing,* A. Bernard, L. Boumsell, J. Dausset, C. Milstein, and S.F. Schlossman, eds. Springer-Verlag, Berlin, Heidelberg, p. 519–527.
13. Stass, S.A., H.R. Schumacher, T.P. Keneklis, and F.J. Bollum. 1979. Terminal deoxynucleotidyl transferase immunofluorescence of bone marrow smears: experience in 156 cases. *Am. J. Clin. Pathol.* **72:**898.
14. Nadler, L.M., J. Ritz, M.P. Bates, E.K. Park, K.C. Anderson, S.E. Sallan, and S.F. Schlossman. 1982. Induction of human B cell antigens in non-T cell acute lymphoblastic leukemia. *J. Clin. Invest.* **70:**433.
15. Foon, K.A., R.W. Schroff, and R.P. Gale. 1982. Surface markers on leukemia and lymphoma cells: recent advances. *Blood* **60:**1.
16. Korsmeyer, S.J., A. Arnold, A. Bakshi, J.V. Ravetch, U. Siebenlist, P.A. Hieter, S.O. Sharrow, T. Le Bien, J.H. Kersey, D.G. Poplack, P. Leder, and T.A. Waldmann. 1983. Immunoglobulin gene rearrangement and cell surface antigen expression in acute lymphocytic leukemias of T cell and B cell precursor origins. *J. Clin. Invest.* **71:**301.
17. Rovigattis, U., J. Mirro, G. Kitchingman, G. Dahl, J. Ochs, S. Murphy, and S. Stass. 1984. Heavy chain immunoglobulin gene rearrangement in acute non lymphocytic leukemia. *Blood* **63:**1023.
18. McKenzie, I.F.C., and H. Zola. 1983. Monoclonal antibodies to B cells. *Immunol. Today* **4:**10.

CHAPTER 7

Characterization of a Human B Lymphocyte-Specific Antigen Defined by Monoclonal Antibodies HD6 and HD39

Gerhard Moldenhauer, Bernd Dörken, Reinhard Schwartz, Antonio Pezzutto, and Günter J. Hämmerling

Introduction

Studies of human B lymphocyte differentiation have been based traditionally on cell morphology in connection with the detection of cytoplasmic and membrane immunoglobulin (1). Together with some non-lineage restricted cell surface markers including HLA-class II antigen, the receptor of the Fc portion of IgG, the complement component C3 receptor, and others, several distinct stages of differentiation from the pluripotent stem cell via the mature B lymphocyte to the antibody-secreting plasma cell have been defined. The understanding of B cell maturation and differentiation was further improved by studies of lymphatic leukemias and malignant lymphomas which have been shown to display characteristics similar to normal cells at equivalent stages of differentiation (2).

During the past few years, several laboratories have reported the development of monoclonal antibodies (mAbs) which define B cell-associated or even B cell-specific antigens (3–12). These antibodies have evolved as valuable tools for the analysis of discrete stages of B cell differentiation, for definition of B cell subpopulations, and for a better classification of B cell-derived leukemias and lymphomas (13). In future a more complete spectrum of B cell-specific mAbs will not only facilitate diagnosis, monitoring, and follow-up of B cell neoplasms but may also be used for therapy, e.g., with respect to autologous bone marrow transplantation.

In this communication we describe the generation and characterization of two mAbs—designated HD(Heidelberg)6 (B25 Workshop nomenclature) and HD39(B31)—which define a B lymphocyte-specific antigen expressed preferentially on late stages of B cell maturation.

Materials and Methods

Hybridoma Production and Screening

Spleen cells from BALB/c mice previously immunized with fresh cells obtained from a patient with hairy cell leukemia were fused with NS1-Ag4/1 myeloma cells (14). Hybridoma supernatants were screened employing an immunoenzymatic staining assay in Terasaki plates and standard immunofluorescence methods.

Preparation of Target Cells

Human peripheral blood mononuclear cells (PBMC) were isolated by Ficoll–Hypaque density gradient centrifugation. Granulocytes were prepared from the cell pellet; contaminating cells were removed by gravity sedimentation in the presence of Dextran followed by lysis. Monocytes were obtained by adherence to plastic dishes. B and T cells were isolated from PBMC by repeated E-rosetting. The E^- preparation was enriched for B cells by removing the monocytes utilizing a Sephadex G 10 column. Normal mononuclear bone marrow cells were recovered by Ficoll–Hypaque centrifugation. Leukemia cells were prepared from peripheral blood and bone marrow of patients with leukemia or malignant lymphoma. The histopathological diagnosis was determined according to the Kiel classification.

Antibody Binding Assays

Indirect immunofluorescence was performed on cells in suspension using FITC-conjugated goat anti–mouse IgG/IgM (Tago) and fluorescence microscopy or flow cytometry (Ortho Diagnostic System). Double-marker analysis of B lymphocytes was carried out by subsequent staining with TRITC–goat anti–human polyvalent Ig (Nordic).

The immunoenzymatic staining assay in Terasaki plate is described elsewhere (see this volume, Chapter 2).

Antibody binding to cytoplasmatic antigens was detected on cytocentrifuge preparations of cells which were either fixed with acetone or treated with detergent (BRIJ 56, Sigma) after glutaraldehyde fixation.

A cellular radioimmunoassay (CRIA) on viable cells (10^6/well) was done in microtiter plate. Reactivity of mAbs was measured by binding of ^{125}I—labeled rabbit anti–mouse Ig. For competitive binding inhibition studies DEAE Affigel blue-purified mAbs HD6 and HD39 were directly radiolabeled with [^{125}I]iodide. Target cells were first incubated with cold mAb (inhibitor) and subsequently with radioiodinated second mAb (competitor).

Cryostat sections of normal human lymph node were stained according to the PAP technique employing aminoethylcarbazole (AEC) as chromogenic substrate.

Radiolabeling and Immunoprecipitation

LICR-LON-HMY2, BJAB, and P3HR-1 cells were surface radioiodinated by a modification of the lactoperoxidase method (15). Cell lysates were purified by affinity chromatography on lens culinaris lectin. For immunoprecipitation a sandwich procedure was used consisting of mAb, purified goat anti–mouse Ig antibody, and protein A–Sepharose 4B. The adsorbent was reduced in SDS sample buffer and then subjected to electrophoresis on discontinuous polyacrylamide slab gels (SDS–PAGE).

Results

The mouse monoclonal antibodies HD6 and HD39 were raised against hairy cell leukemia. Both were found to belong to the immunoglobulin isotype IgG1 K. They did not fix complement nor did they bind to protein A.

Reactivity with Cells from Normal Individuals

mAbs HD6 and HD39 reacted with 3% of peripheral blood mononuclear cells from the blood of 10 individuals tested by indirect immunofluorescence (Table 7.1). After separation of PBMC into E^- and E^+ cells both antibodies bound to less than 1% of the E^+ fraction containing most of the T cells. No reactivity with T cells, monocytes, granulocytes, erythrocytes, and platelets was observed. About 50% of the monocyte-depleted E^- cells, consisting mainly of B lymphocytes, were stained with HD6 and HD39. Double-marker analysis of sIg^+ cells revealed that both antibodies were reactive with a subpopulation representing 60–70% of immunoglobulin-bearing lymphocytes. In bone marrow only a small number of lymphoid cells, approximately 2–3%, were positive with the two antibodies whereas half of the tonsil cells and one third of the spleen cells were recognized. These results indicated that HD6 and HD39 were exclusively expressed on a major subpopulation of B cells. Both mAbs exhibited very similar binding patterns on normal cells.

Reactivity with Leukemic Cells

HD6 and HD39 were found to be directed also against antigens expressed on certain types of B cell neoplasms. All ten cases of hairy cell leukemia (HCL) reacted strongly with both antibodies (Table 7.2). In addition, most

Table 7.1. Expression of HD6 and HD39 antigens on normal cells using indirect immunofluorescence.

Cell type	Number of tests	% of cells stained with mAb	
		HD6	HD39
Peripheral blood			
PBMC	10	3 (2–5)[a]	3 (1–4)
B (E^-)[b]	2	48 (37–59)	50 (41–60)
B (sIg^+)[c]	5	63 (36–71)	73 (45–86)
T (E^+)	5	1 (0–2)	1 (0–2)
Monocyte	5	3 (2–4)	2 (1–3)
Granulocyte	5	0	0
Erythrocyte	3	0	0
Platelets	3	0	0
Bone marrow	14		
Lymphoid		3 (1–6)	2 (1–5)
Myeloid		0	0
Erythroid		0	0
Thymus	3	1 (0–3)	1 (0–1)
Tonsil	3	55 (45–59)	51 (41–59)
Spleen	1	35	37

[a] Percent positive cells (range).
[b] E^- fraction depleted of monocytes by passage through Sephadex G 10.
[c] Double-marker staining.

Table 7.2. Reactivity of HD6 and HD39 with leukemic cells using indirect immunofluorescence.

Classification	HD6	HD39	HD37[a]
T-CLL, Sézary[b]	0/10[c]	0/10	0/10
T-ALL	0/6	0/6	0/6
Non-T ALL	11/29 (+)[d]	2/25 (+)	27/30
B-CLL	6/32	0/26	25/25
B-PLL	5/5	3/4	4/4
HCL	10/10	10/10	10/10
IC	1/21	0/18	19/20
MM	0/6	0/6	0/6
AML/AMML	0/15	0/15	0/15

[a] mAb HD37 was included for comparison (pan B reactivity).
[b] CLL = Chronic lymphocytic leukemia, Sézary- = Sézary syndrome, ALL = acute lymphoblastic leukemia, PLL = prolymphocytic leukemia, HCL = hairy cell leukemia, IC = immunocytoma (lymphoplasmocytic lymphoma), MM = multiple myeloma, AML = acute myeloblastic leukemia, AMML = acute myelomonoblastic leukemia.
[c] Expressed as number of positive cases / total number tested.
[d] Faint staining.

cases of prolymphocytic leukemia (B-PLL) were also positive. Six out of 32 B cell chronic lymphatic leukemias (B-CLL) were reactive with HD6; in contrast, HD39 was always negative using this target cell. Other types of B cell malignancies and tumors of T cell and myeloid origin did not carry the corresponding antigen. Some cases of non-T type acute lymphoblastic leukemia (non-T ALL) were faintly stained. The expression of HD6 antigen on leukemic cells always appeared to be broader in terms of differentiation stages when compared to HD39. Studies of HD39 binding to leukemic cells using membrane staining of cells in suspension and cytoplasmic staining of cytocentrifuge preparations fixed with acetone or treated with glutaraldehyde and detergent revealed that the HD39 antigen also occurred in the cytoplasm of B cells which were obviously negative for surface expression (Table 7.3).

Reactivity with Human Cell Lines

By using a radioimmunoassay on cells, HD6 and HD39 were tested on established human cell lines (Table 7.4). All the Burkitt's lymphoma lines and the B lymphoblastoid lines showed positive reactions. Only the Burkitt's lymphoma line RAMOS was weakly positive. Compared to the binding of mAb HD11 recognizing a monomorphic determinant on HLA-class II antigen, the binding of HD6 and HD39 was substantially weaker, indicating that the total amount of antigen expressed is relatively low. There was no reactivity with four T cell lines and the prolymphocytic leukemia line HL-60.

Immunohistological Staining

Cryostat sections of human lymph node were stained with HD6 and HD39. The two antibodies predominantly stained the lymphoid follicles

Table 7.3. Reactivity of HD39 with leukemic cells: comparison of cell surface staining and cytoplasmic staining by indirect immunofluorescence.

Classification	Number tested	Number of leukemias stained with HD39	
		Surface[a]	Cytoplasm[b]
T-CLL	2	0	0
T-ALL	2	0	0
Non-T ALL	4	0	4
B-CLL	4	0	4
HCL	3	3+[c]	3++

[a] Membrane staining of cells in suspension.
[b] Cytoplasmic staining of cytocentrifuge preparations.
[c] + = Moderate reaction, ++ = strong reaction.

Table 7.4. Reactivity of HD6 and HD39 with human cell lines measured by CRIA.

Cell type	HD6	HD39	HD11[a] (HLA-class II)
Burkitt's lymphoma			
Raji	++[b]	++	+++
Daudi	++	++	+++
P3HR-1	++	+	–
RAMOS	+	+	+++
BJAB	++	++	+++
B lymphoblastoid			
WI-L2-HF2	++	++	+++
LICR-LON-HMY2	++	++	+++
T cell			
Jurkat	–	–	–
JM-1	–	–	–
MOLT-4	–	–	–
CEM-C7	–	–	–
Promyelocytic leukemia			
HL-60	–	–	–

[a] mAb HD 11 directed against a framework determinant of HLA-class II antigen was included for comparison.
[b] – = negative, + = weakly positive (3–5-fold higher than background), ++ = positive (5–10-fold higher than background), +++ = strongly positive (>10-fold higher than background).

(Fig. 7.1). Most of the cells in the mantle zone were strongly positive whereas only a subpopulation of cells in the germinal center was stained. Some few positive lymphoid cells were seen in the interfollicular region.

Biochemical Characterization of Target Antigen

Surface-radioiodinated cell lysates of cell lines LICR-LON-HMY2, BJAB, and P3HR-1 were utilized to isolate the corresponding antigen by immunoprecipitation. Only when the glycoprotein fraction isolated by lens culinaris lectin was used, mAb HD6 as well as HD39 specifically precipitated a protein consisting of two polypeptide chains with apparent M.W.'s of 130 and 140 Kd (Fig. 7.2). As shown by identical electrophoretic mobility on reduced and nonreduced SDS–PAGE slab gels the antigen was not linked by disulfide bonds. These findings indicated that HD6 and HD39 were directed against identical antigens.

Competitive Binding Inhibition Studies

In order to determine whether or not the same antigen was recognized by the two antibodies, cross-inhibition radioimmunoassay was performed. For this, binding of radiolabeled HD6 was competed by preincubation with increasing amounts of unlabeled HD39 and vice versa (Table 7.5). If

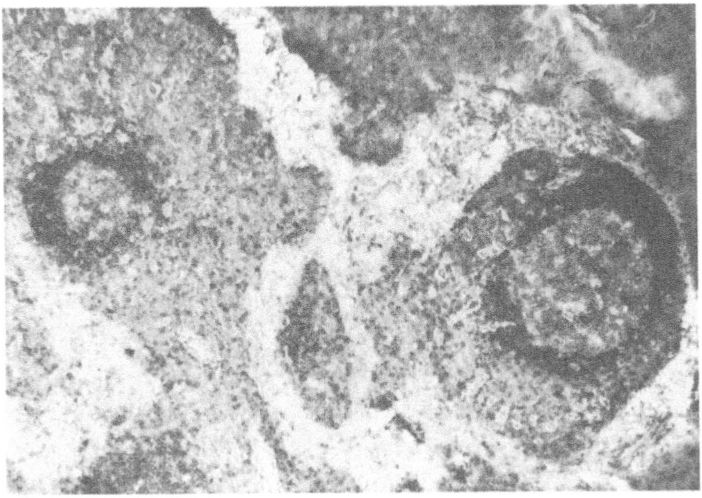

Fig. 7.1. Immunohistological staining of a human lymph node with HD39.

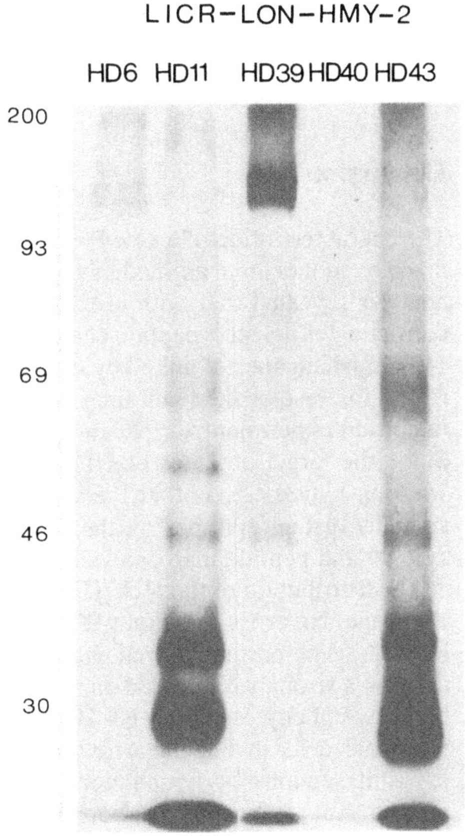

Fig. 7.2. SDS–PAGE analysis under reducing conditions of immunoprecipitates of indicated mAb from ^{125}I-labeled cell lysates. HD11 and HD43 are directed against HLA-class II antigens. HD40 is an irrelevant monoclonal antibody applied for control.

Table 7.5. Competitive binding inhibition of HD6 and HD39 using CRIA.

Unlabeled mAb	Binding of ^{125}I-labeled mAb: HD6 Cpm	HD6 %	HD39 Cpm	HD39 %
HD6 10 μg[a]	736 ± 45[b]	2	52,721 ± 1548	103
1 μg	3602 ± 386	10	49,146 ± 229	96
0.1 μg	19,216 ± 516	54	50,604 ± 944	99
HD39 10 μg	36,238 ± 2266	102	874 ± 53	2
1 μg	32,292 ± 3863	91	1966 ± 394	4
0.1 μg	38,782 ± 880	109	18,793 ± 1231	37
HD37 10 μg[c]	34,946 ± 4094	98	51,530 ± 1617	101

[a] Counts per minute ± SEM of three replicate assays.
[b] Amount purified mAb per well of microtiter plate.
[c] mAb HD37 (pan B reactivity) was included for control.

the two antibodies were directed against identical or proximate determinants of the antigen, binding of the labeled antibody should have been inhibited. Surprisingly, HD6 did not block the binding of HD39 nor did HD39 inhibit the binding of HD6. Although each antibody could completely inhibit its own binding. Thus, it became evident that HD6 and HD39 were recognizing distinct epitopes.

Discussion

The characterization of a new B cell-specific differentiation antigen recognized by monoclonal antibodies HD6 and HD39 is described. Biochemical analysis revealed that both mAbs precipitate a cell surface glycoprotein consisting of two polypeptide chains with apparent M.W.'s of 130 Kd and 140 Kd which are not linked by disulfide bonds. The antigen was isolated from three established cell lines of different origin. As shown by cross-inhibition experiments, HD6 and HD39 recognize two distinct epitopes. Since the target antigens of HD6 and HD39 appear to be identical our working hypothesis is that the two antibodies are directed against two spatially distant epitopes on the same molecule. This can be proven by 2-D gel and peptide map analysis which are currently in progress.

The distribution of the HD6/HD39 antigen are strictly limited to cells of B lineage. No cross-reaction with T lymphocytes, monocytes, and granulocytes, using normal as well as malignant cells, was ever observed. The antigen is strongly expressed on a major subpopulation of SIg^+ peripheral blood lymphocytes (about 60–70%), on tonsil cells, spleen cells, and B lymphoid cells in bone marrow whereas plasma cells are negative. In cryostat sections of lymph node only the B cell-dependent areas are stained. Taken together, the corresponding antigens of HD6 and HD39

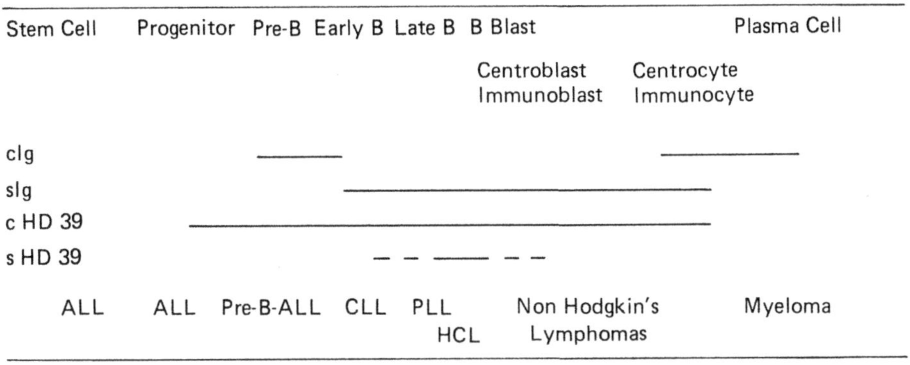

c = cytoplasmic; s = surface; ALL = Acute Lymphoblastic Leukemia; CLL = Chronic Lymphocytic Leukemia; PLL = Prolymphocytic Leukemia; HCL = Hairy Cell Leukemia

Fig. 7.3. Hypothetic B cell maturation scheme.

antibodies are phenotypically and biochemically different from other already described B cell antigens.

When testing leukemia cells some differences in surface expression of HD6 and HD39 became evident. HD39 reacted with less cases of certain B cell leukemias and was more restricted to mature B cell malignancies than HD6. This might be explained by the finding that the antibodies recognize distinct epitopes. The HD39 determinant may be lost or may not be accessible for binding of antibody due to conformational changes of the molecule. Surprisingly, the HD6/HD39 antigen exhibited a remarkably different distribution pattern when antigen expression on the cell surface and in the cytoplasm are compared. Most B cell tumors with HD6/HD39-negative cell surface show antibody binding to the cytoplasmic antigen. Even most cases of ALL are weakly positive in cytoplasm. These findings suggest that the HD6/HD39 antigen expressed on the cell surface is restricted to the late stages of B cell differentiation; in contrast, the same antigen in the cytoplasm seems to be a marker for the entire B cell lineage (Fig. 7.3). This observation stresses the importance of analyzing both membrane and cytoplasmic antigen expression and may also explain discrepancies between studies using intact cells in suspension and immunohistological staining.

Summary

A novel B lymphocyte-specific differentiation antigen is defined by mAbs HD6 and HD39 which recognize two distinct epitopes of a glycoprotein with M.W.'s of 130 and 140 Kd. A major subpopulation of B cells carries the antigen on the cell surface. In terms of differentiation, antigen expression on the membrane is restricted to late stages of B cell maturation whereas in the cytoplasm the antigen is present along the whole B cell lineage. Due to its distribution in normal and malignant cells and its biochemical properties the HD6/HD39 antigen is unique and is not related to any previously described B cell-associated antigen.

References

1. Fu, S.M., R.J. Winchester, and H.G. Kunkel. 1974. Occurrence of surface IgM, IgD, and free light chains on human lymphocytes. *J. Exp. Med.* **139:**451.
2. Stein, H., and G. Tolksdorf. 1980. Development and differentiation of the T-cell and B-cell systems: a perspective. In: *Malignant lymphoproliferative diseases.* J.G. van der Tweel, C.R. Taylor, and F.T. Bosman, eds. Leiden University Press, The Hague, pp. 13–30.
3. Stashenko, P., L.M. Nadler, R. Hardy, and S.F. Schlossman. 1980. Characterization of a human B lymphocyte-specific antigen. *J. Immunol.* **125:**1678.
4. Nadler, L.M., P. Stashenko, R. Hardy, C. van Agthoven, C. Terhorst, and S.F. Schlossman. 1981. Characterization of human B cell-specific antigen (B2) distinct from B1. *J. Immunol.* **126:**1941.
5. Nadler, L.M., K.C. Anderson, G. Marti, M. Bates, E. Park, J.F. Daley, and S.F. Schlossman. 1983. B4, a human B lymphocyte-associated antigen expressed on normal, mitogen-activated, and malignant B lymphocytes. *J. Immunol.* **131:**244.
6. Brooks, D.A., I. Beckman, J. Bradley, P.J. McNamara, M.E. Thomas, and H. Zola. 1980. Human lymphocyte markers defined by antibodies derived from somatic cell hybrids. I. A hybridoma secreting antibody against a marker specific for human B lymphocytes. *Clin. Exp. Immunol.* **39:**477.
7. Brooks, D.A., I.G.R. Beckman, J. Bradley, P.J. McNamara, M.E. Thomas, and H. Zola. 1981. Human lymphocyte markers defined by antibodies derived from somatic cell hybrids. IV. A. monoclonal antibody reacting specifically with a subpopulation of human B lymphocytes. *J. Immunol.* **126:**1373.
8. Abramson, C.S., J.H. Kersey, and T.W. LeBien. 1981. A monoclonal antibody (BA-1) reactive with cells of human B lymphocyte lineage. *J. Immunol.* **126:**83.
9. Kersey, J.H., T.W. LeBien, C.S. Abramson, R. Newman, R. Sutherland, and J. Greaves. 1981. p. 24: A human hemopoietic progenitor and acute lymphoblastic leukemia associated cell surface structure identified with monoclonal antibody. *J. Exp. Med.* **153:**726.
10. Mittler, R.S., M.A. Talle, K. Carpenter, P. Rao, and G. Goldstein. 1983. Generation and characterization of monoclonal antibodies reactive with human B lymphocytes. *J. Immunol.* **131:**1754.
11. Zipf, T.F., G.J. Lanzon, and B.M. Longenecker. 1983. A monoclonal antibody detecting a 39,000 m.w. molecule that is present on B lymphocytes and chronic lymphocytic leukemia cells but is rare on acute lymphocytic leukemia blast. *J. Immunol.* **131:**3064.
12. Wang, C.Y., W. Azzo, A. Al-Katib, N. Chiorazzi, and D.M. Knowles. 1984. Preparation and characterization of monoclonal antibodies recognizing three distinct differentiation antigens (BL1, BL2, BL3) on human B lymphocytes. *J. Immunol.* **133:**684.

13. Anderson, K.C., M.P. Bates, B.L. Slanghenhoupt, G.S. Pinkus, S.F. Schlossman, and L.M. Nadler. 1984. Expression of human B cell-associated antigens on leukemias and lymphomas: a model of human B cell differentiation. *Blood* **63:**1424.
14. Köhler, G., and C. Milstein. 1975. Continuous cultures of fused cells secreting antibody of predefined specificity. *Nature* **256:**495.
15. Goding, J.W. 1980. Structural studies of murine lymphocyte surface IgD. *J. Immunol.* **124:**2082.

CHAPTER 8

Six Distinct Antigen Systems of Human B Cells as Defined by Monoclonal Antibodies

Yoshifumi Ishii, Tsuyoshi Takami, Hiroo Yuasa, Takashi Takei, Yasuo Kokai, and Kokichi Kikuchi

Introduction

Human B cells have been phenotypically characterized by the presence of various cell surface markers including surface membrane immunoglobulins (SmIg), Ia-like antigens, and receptors for immunoglobulin Fc (FcR) and complement C3 (CR). In recent years, a number of cell surface markers, either specific for or associated with human B cells, have also been identified using monoclonal antibodies (mAbs) to delineate human B cell development (1).

In the present study, we have generated six distinct mAbs (L22–L27), which detect six distinct antigen systems expressed either in the cytoplasm of human B cells (L26) or on their cell surfaces (L22, L23, L24, L25, and L27). We wish to report herein on the tissue distribution and chemical characteristics of these human B cell-specific antigens as well as on their functional role in the induction of antibody synthesis by human B cells.

Materials and Methods

Cells and Tissues

Human lymphoid tissues including thymuses, tonsils, lymph nodes, and bone marrow were used. Peripheral blood mononuclear cells (PBM) were separated from defibrinized blood by Ficoll–Conray gradient centrifugation. A variety of human hematopoietic cell lines were maintained in cultures with RPMI 1640 medium plus 10% fetal calf serum.

Monoclonal Antibodies

Six distinct monoclonal antibodies (L22–L27) produced in our laboratories were extensively studied in this work. All of these six clones were derived from hybrids of NS-1 mouse myeloma cells with splenocytes from BALB/c mice immunized with human tonsil B cells, and produced monoclonal antibodies of subclass IgG1 (L22, L25, L27), IgG2a (L24, L26), or IgG2b (L23).

Immunofluorescence and Cytofluorography

Viable lymphoid cells were stained by indirect membrane immunofluorescence using mouse mAb and FITC-conjugated goat antibody to mouse Ig. Cytofluorographic analysis of the stained cells was performed on a FACS analyzer (Becton Dickinson, Sunnyvale, CA). For the detection of cytoplasmic (L26) antigen, the cytosmears of the lymphoid cells were fixed with acetone and stained by indirect immunofluorescence.

Immunoperoxidase Staining

Human lymphoid tissue sections were stained by an avidin–biotin–peroxidase (ABC) method using mouse mAb, biotinated goat anti-mouse Ig, and avidin–biotin–peroxidase complexes (2).

Immunoprecipitation and SDS–Polyacrylamide Gel Electrophoresis (SDS–PAGE)

Radioiodinated cell surface components solubilized from tonsil lymphocytes or the glycoprotein fraction isolated from the labeled cell lysate by lentil lectin-coupled Sepharose 4B column were used for immunoprecipitation with mAbs (3). The precipitates were dissociated with SDS and 2-mercaptoethanol and were subjected to SDS–PAGE. After electrophoresis, the gel slab was dehydrated and radioactivity in the gel slab was visualized by autoradiography.

Enzyme-Linked Immunosorbent Assay (ELISA)

PBM (10^6/ml) were cultured in the presence of pokeweed mitogen (5 μg/ml) and/or MAb (1 μg/ml) for 7 days. The culture supernatants were assayed for IgG and IgM content by the ELISA technique adapted for Sandwich enzyme immunoassay (4).

Results

Reactivity of Anti–B Cell mAbs with Normal Lymphoid Cells

All of six mAbs reactive with human B cells labeled neither human thymus nor peripheral T cells. As indicated in Table 8.1, L25, L26, and L27 reacted with the majority of B cells present in the blood and lymphoid tissues. L22, L23, and L24, on the other hand, had selective reactivity with human B cell subpopulations, and it was found that L22 labeled 20–30% of B cells in lymphoid tissues whereas L23 and L24 stained 60–80% of these B cells present in the blood and lymphoid tissues. All but one mAb did not label any cell types including granulocytes, monocytes, platelets, and erythrocytes, though L23 appeared to faintly cross-react with granulocytes and monocytes.

Lymphoid Tissue Localization of Cells Identified by Anti–B Cell mAbs

All of our mAbs (L22–L27) had selective reactivity with lymphoid follicles of lymph nodes and tonsils, where B cells preferentially localize (Fig. 8.1). With pan B cell reagents including L25, L26, and L27, both small and large lymphocytes located in the mantle zone and in the germinal center, respectively, were equally stained. In addition, L25 labeled large dendritic or interdigitating cells located in the thymus-dependent areas of lymphoid tissues and in the thymic medulla, which might be identical to lymphoid dendritic cells (LDC) associated with accessory cell function (5). L22, on the other hand, labeled small lymphocytes in the mantle zone but not large lymphocytes in the germinal center. L23 and L24 gave

Table 8.1. Reactivity of L22–L27 mAbs with human hematopoietic cells.[a]

Cells	Number of samples	L22	L23	L24	L25	L26	L27	HLA-DR
Thymus	4	0	0	0	1	0	0	1
Tonsil	10	23	51	53	59	63	62	68
T cell	5	2	4	4	4	5	5	5
B cell	5	29	62	65	85	93	93	98
Blood								
Lymphocyte	20	1	6	7	9	10	10	14
Granulocyte	5	0	14	0	0	0	0	0
Monocyte	5	0	10	0	0	0	0	92
Platelets	5	0	0	0	0	0	0	0
Red cell	5	0	0	0	0	0	0	0
Bone marrow	3	1	9	3	3	3	3	16

[a] Data are expressed as mean percentage of positive cells stained by respective mAb.

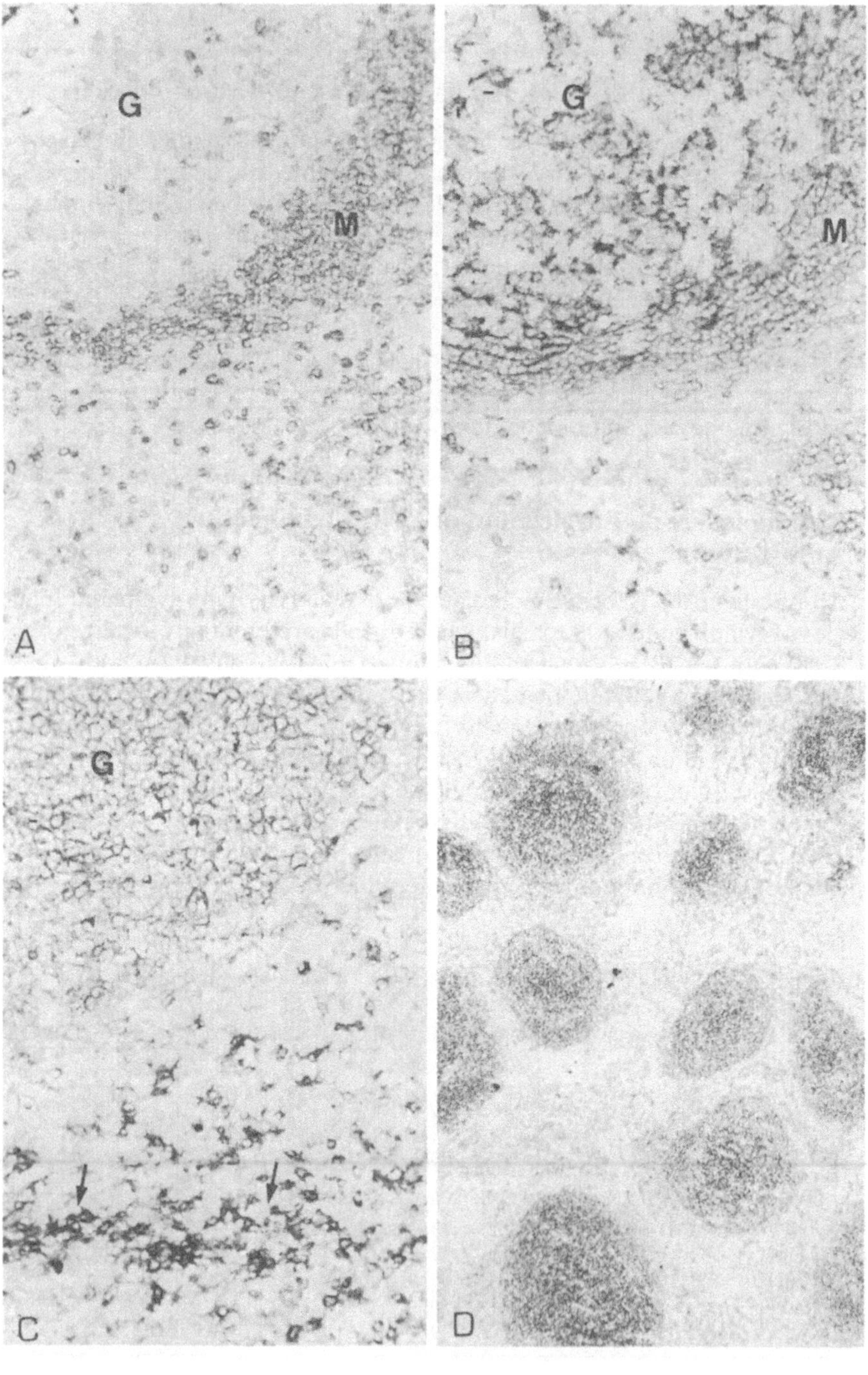

Fig. 8.1. Immunoperoxidase staining of lymph node (A,B) and tonsil tissue sections (C,D) stained with L22 (A), L24 (B), L25 (C), and L26 (D) mAbs. L22 labels

similar staining patterns and labeled small lymphocytes in the mantle zone stronger than large lymphocytes in the germinal center. These L23 and L24 antibodies also cross-reacted with follicular dendritic cells (FDC) that distribute in the superficial pole (light zone) of the germinal center (6). All six mAbs tested did not stain plasma cells in these lymphoid tissues.

Cell Line Distribution of Antigens Identified by Anti–B Cell mAbs

A panel of human hematopoietic cell lines was used to assess the specificity of L22–L27 binding (Table 8.2). As expected, L25 antigen had the broadest distribution on these cell lines among six mAbs, and was expressed on all of the B cell-derived cell lines including Epstein–Barr virus (EBV)-transformed B cell lines. B cell-type acute lymphatic leukemia (B-ALL), and Burkitt's lymphoma cell lines, as well as on pre-B and common ALL (cALL) cell lines, the latter of which have been thought to be derived from B cell progenitors representing the early stage of B cell ontogeny (7). L26 and L27 co-existed in the cytoplasm and plasma membrane of the same B cell lines, respectively, but pre-B and cALL cell lines were entirely negative for L26 and L27 antigens tested so far. Whereas L22 reacted with none of the cell lines tested, L23 and L24 antigens were expressed on B-ALL and Burkitt's lymphoma cell lines but little on EBV-transformed B cell lines. None of the six mAbs tested were reactive with T cell, myeloid, monocytic, and erythroid cell lines.

Expression of Antigens Defined by Anti–B Cell mAbs in Human B Cell Malignancies

We tested L22–L27 antigen expression in tumor cells derived from various types of B cell malignancies (Table 8.3). All these six antigens were not expressed in either myelogenous (n = 6) or T cell-type leukemias (n = 13). As indicated in Table 8.3, cALL consistently expressed L25 but not others. B cell-type chronic lymphocytic leukemia (B-CLL), on the other hand, expressed all six distinct B cell antigens (L22–L27). When malignant B cell lymphoma cases were tested for their antigen expression, it was found that most of the cases lacked L22, whereas L25 and L26 were expressed in the majority of these cases tested. L23, L24, and L27 were

---

small lymphocytes in the mantle zone (M) of lymphoid follicle. L24 stains mantle zone lymphocytes stronger than germinal center (G) cells. It also cross-reacts with follicular dendritic cells present in the germinal center. L25 labels B cells in lymphoid follicle and dendritic or interdigitating cells (arrows) located in the thymus-dependent area of lymphoid organs. L26 intensely stains all lymphoid follicles present in the tonsil.

Table 8.2. Cell line distribution of human B cell antigens defined by mAbs.

Cell lines	L22	L23	L24	L25	L26	L27	HLA-DR
T cell lines	−[a]	−	−	−	−	−	−
CCRF-CEM	−	−	−	−	−	−	−
MOLT-4F	−	−	−	−	−	−	−
Peer	−	−	−	−	−	−	−
SKW-3	−	−	−	−	−	−	−
TALL-1							
EBV-B cell lines	−	−	−	+	+	+	+
Adult blood CESS	−	−	+/−	+	+	+	+
TAKE	−	−	−	+	+	+	+
MATS	−	−	−	+	+	+	+
KOI	−	−	−	+	+	+	+
OKA	−	+/−	−	+	+	+	+
KON	−	−	−	+	+	+	+
Cord blood U125	−	−	+/−	+	+	+	+
U203	−	+/−	+/−	+	+	+	+
U1129	−	−	+/−	+	+	+	+
U0110							
Burkitt's lymphoma cell lines	−	+/−	+/−	+	+	+	+
Raji	−	+	+	+	+	+	+
Daudi	−	+	+	+	+	+	+
Eb-3	−	−	−	−	+	+	+
P3-HR-1	−	−	−	−	−	−	+/−
NAMALVA	−	+	+	−	+	+	+
BJAB	−	+	+	+/−	+	+	+
RAMOS	−	+	+	+/−	+	+	+
RAMOS B-7[b]							
B-ALL cell line	−	+	+	+	+	+	+
BALL-1							
Myeloma cell lines	−	−	−	+	+	+	+
RPMI-8226	−	−	−	−	−	−	+
SKO-007							
Pre-B cell line	−	−	−	+	−	−	+
NALM-1							
cALL cell lines	−	−	−	+	−	−	+
Reh	−	−	−	+	−	−	+
KM-3							
Non-lymphoid cell lines	−	−	−	−	−	−	−
KG-1	−	−	−	−	−	−	−
ML-2	−	−	−	−	−	−	−
K562	−	−	−	−	−	−	−
U 937							

[a] More than 50% fluorescent cells are positive and less than 2% fluorescent cells are negative. When 5–15% of weakly stained cells are found, they are judged as +/−.
[b] RAMOS cell line infected with EBV.

able 8.3. Distribution of antigens in various types of B cell malignancies as detected by nmunofluorescence

Tumors	No. of patients	Number reactive with antibodies									
		L22	L23	L24	L25	L26	L27	Leu 1	T10	DR	SmIg
ALL[a]	9	0	0	1	9	2	2	0	9	9	0
-CLL	9	9	9	9	9	9	9	9	4	9	9
CL	2	1	1	1	2	2	2	0	1	2	2
on-Hodgkin's lymphoma[b]											
Follicular mixed	1	1	1	1	1	1	1	0	1	1	1
Follicular large cell	2	0	1	2	1	2	2	0	2	2	2
Diffuse small cleaved	2	1	2	2	2	2	2	0	2	2	2
Diffuse large cell	6	0	3	3	5	6	6	0	6	6	6
Small non-cleaved cell	3	0	2	3	3	3	3	0	3	3	3
Immuno-blastic	2	0	0	2	2	2	0	0	3	3	3
lasma cell leukemia	2	0	0	0	0	0	0	0	2	1	1
Ayeloma	1	0	0	0	0	0	0	0	7	0	0

This type was identified by the presence of cALL antigen on the cell membrane.
Diagnosis was made on the basis of the NCI Working Formulation.

not expressed on malignant B cells derived from some lymphoma cases, most of which were histologically diagnosed as either the large cell or immunoblastic type of B cell lymphomata (8). Plasma cell myeloma and leukemia cells were invariably devoid of all these six antigens.

Chemical Characterization of Antigens Defined by Anti–B Cell mAbs

SDS–PAGE profiles of immunoprecipitates formed between ^{125}I-labeled tonsil cell lysate and anti–B cell mAbs (L22–L27) can be seen in Fig. 8.2. L23 and L24 precipitated glycoproteins with molecular weights (M.W.) of 205Kd and 145Kd, respectively, both of which could interact with lentil lectin. L26 antigen consisted of at least two noncovalently associated subunits with 33Kd and 30Kd M.W., which had no affinity to lentil lectin. L22, L25, and L27 failed to precipitate any specific components from the ^{125}I-labeled tonsil cell lysate.

Functional Studies

Evolution of antigen expression during the activation of B cells with PWM was first analyzed using our anti–B cell mAbs. L22, L23, and L24

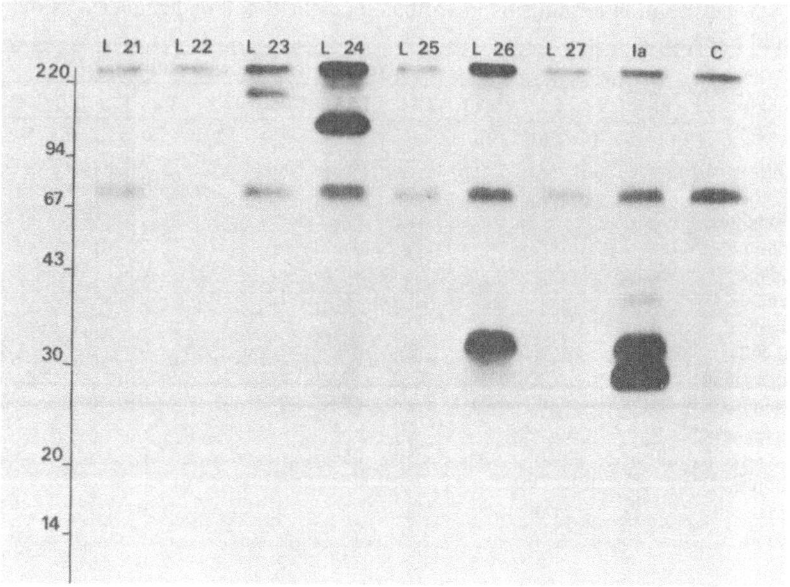

Fig. 8.2. SDS–PAGE profiles of antigens precipitated by L22–L27 mAbs from ^{125}I-labeled tonsil cell lysate. Whereas L22, L25, and L27 failed to precipitate any specific components, L23 and L24 could precipitate 205 Kd and 145 Kd M.W. glycoproteins, respectively. L26 detected a major 33 Kd M.W. component, which was associated with a minor 30 Kd M.W. band. Anti-HLA-DR mAb demonstrated the subunit structure of an Ia molecule composed of α and β chains. C (control).

antigens disappeared from blood B cells by 4 days, and L27 by 5 days after PWM stimulation, whereas L25 and L26 was sustained in these cells until 7 days after the activation.

In the following experiments, we studied the effect of L22–L27 mAbs on Ig synthesis of B cells induced by PWM (Fig. 8.3). Whereas L22, L26, and L27 had no effect on Ig production by blood B cells, L25 moderately suppressed both IgM and IgG synthesis by these PWM-stimulated B cells. L25 showed the strongest inhibition when added to the cultures at day 0, and lost its inhibitory effect if it was added 2 or more days after the stimulation. L23 and L24 showed inhibitory effect on IgM but not IgG synthesis, which was more profound with L24 than with L23.

Discussion

The relationship of L22–L27 antigen expression to human B cell development can be hypothetically summarized as follows. During the early

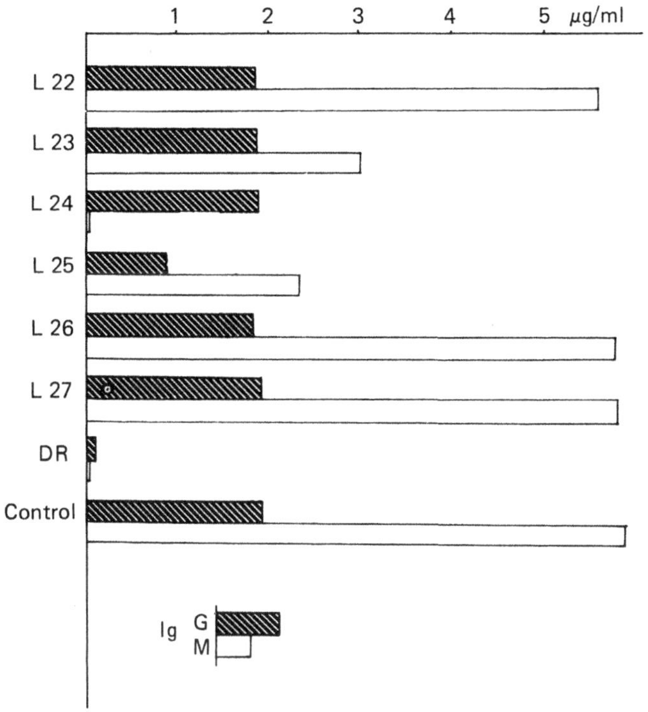

Fig. 8.3. Effect of various anti–B cell mAbs (L22–L27) on IgM and IgG synthesis by B cells stimulated with PWM.

stages of B cell ontogeny, L25 first appears on B cell progenitors as represented by cALL and their cell lines, most of which have been demonstrated to have rearranged Ig H-chain genes without expressing any classes of Ig in their cytoplasm and plasma membrane (9,10). Following L25, L26 and L27 might be expressed on B cells, as reflected by their occasional expression in cALL cases. All six B cell antigens (L22–L27) are positive in B-CLL associated with SmIg and Leu 1, and if B-CLL, as suggested, corresponds to the leukemic counterpart of early or young B cells capable of differentiating into more mature B cells (11), L22–L27 antigens all become fully expressed at this particular stage of B cell ontogeny.

Immunohistological studies have shown that resting small B cells located in the mantle zone of lymphoid follicles express all six antigen systems (L22–L27), whereas activated large B cells in the lymphoid germinal centers lack L22 and express a reduced amount of L23 and L24. These two cell types equally expressed L25, L26, and L27. All these six antigens, however, could not be detected in plasma cells, suggesting that they must be lost from B cells during their differentiation into antibody-secreting cells. Sequential changes in antigen expression of blood B cells induced by PWM suggest that B cells lose L22, L23, L24, L27, L26, and L25 in turn during their activation and differentiation into Ig-secreting cells. This is also supported by the data obtained with human B cell tumors corresponding to the later stages of B cell differentiation, indicating that these B cell tumors mostly lack L22, and a reduced number of these cases also lack L23, L24, and L27. Like normal plasma cells, plasma cell leukemia and myeloma cells expressed none of the antigens defined by our six mAbs.

We next sought to determine whether any of our six mAbs detected molecules that regulate B cell function. Using PWM-driven polyclonal activation of B cells, it was shown that some of these antibodies had an inhibitory effect on Ig production and secretion by blood B cells, when the antibody was added to PWM-stimulated cultures. Thus, whereas L22, L26, and L27 had no effect on Ig synthesis, L25 moderately suppressed both IgM and IgG synthesis. More interestingly, L23 and L24 showed inhibitory effect on IgM but not IgG synthesis by PWM-stimulated B cells, which was more clearly observed with L24 than with L23. It seems possible that L23 and L24 are preferentially expressed on a particular subset of human B cells, from which IgM- rather than IgG-producing cells are primarily derived (12).

In conclusion, mAbs may thus be utilized as probes to determine the functional role of the molecules with which they react, and further to analyze the mechanism involved in the development of lymphocyte function.

Summary

Six monoclonal antibodies (mAbs) that recognize six distinct antigens of human B cells, termed L22, L23, L24, L25, L26, and L27, respectively, were generated. Among these antigens, L26 was not or little expressed on B cell surfaces, and was identified to be the cytoplasmic antigen of human B cells. The other five B cell antigens were clearly expressed on the cell membrane. Immunochemical studies showed that L23 and L24 were glycoproteins with 205 Kd and 145 Kd molecular weights, respectively, both of which interacted with lentil lectin. L26 consisted of noncovalently associated 33 Kd and 30 Kd M.W. components, which could not bind to lentil lectin and were distinct from Ia-like antigens including DP, DQ, and DR. As judged from tissue distribution studies using normal, cultured, and neoplastic human hematopoietic cells, L22, L23, and L24 were expressed on human B cell subpopulations, whereas L25 was found on most B-lineage cells. Thus, L22 was selectively expressed on resting small B cells in the mantle zone of lymphoid follicles as well as B cell-type CLL (B-CLL) cells. L23 and L24 were expressed on approximately two-thirds of B cells in blood and lymphoid tissues, where small B cells in the mantle zone were labeled with L23 and L24 stronger than were large B cells in the germinal center. L25, L26, and L27 were found in most B cells present in both blood and lymphoid tissues, and L25, but not L26 and L27, was consistently expressed on fresh and cultured common ALL (cALL) cells. In addition, L25 was shared with lymphoid dendritic cells located in the T-zone of lymphoid organs, whereas L23 and L24 existed in follicular dendritic cells present in the germinal center. Functional studies employing pokeweed mitogen (PWM)-induced B cell differentiation demonstrated that, while L25 mAb suppressed both IgM and IgG synthesis by PWM-stimulated B cells, L23 and L24 showed inhibitory effect on IgM but not IgG synthesis. L22, L26, and L27 mAbs had no inhibitory or augmenting effect on this system.

References

1. McKenzie, I.F.C., and H. Zola. 1983. *Immunol. Today* **4:**10.
2. Hsu, S.M., L. Raine, and H. Fanger. 1981. *J. Histochem. Cytochem.* **29:**577.
3. Ishii, Y., T. Kon, T. Takei, J. Fujimoto, and K. Kikuchi. 1983. *Clin. Exp. Immunol.* **53:**31.
4. Fujimoto, J., S. Levy, and R. Levy. 1983. *J. Exp. Med.* **159:**752.
5. Van Voorhis, W.C., J. Valinsky, E. Hoffman, J. Liban, L.S. Hair, and R.M. Steinman. 1983. *J. Exp. Med.* **158:**174.
6. Mandel, T.E., R.P. Phipps, A. Abbot, and J.G. Tew. 1980. *Immunol. Rev.* **53:**29.
7. Greaves, M.F. 1981. *Cancer Res.* **41:**4752.
8. The Non-Hodgkin's Lymphoma Pathologic Classification Project. 1982. *Cancer* **49:**2112.
9. Korsmeyer, S.J., P.A. Hieter, J.V. Ravetch, D.G. Poplack, P. Leder, and T.A. Waldmann. 1981. In: *Leukemia markers,* W. Knapp, ed. Academic Press, New York, p. 85.
10. Sacchi, N., T.V. LeBien, S. Trost, D. Breviario, and F.J. Bollum. 1984. *Cell. Immunol.* **84:**65.
11. Johnstone, A.P. 1982. *Immunol. Today* **3:**343.
12. Kuritani, T., and M.D. Cooper. 1982. *J. Exp. Med.* **153:**839.

CHAPTER 9

B-C1, B-C2, B-C3: Monoclonal Antibodies against B Cell Differentiation Antigens

Ignacio Anegón, Ramón Vilella, Teresa Gallart,
Cristina Cuturi, Luis Borche, Jordi Milà, and Jordi Vives

Introduction

B lymphocyte ontogeny and differentiation has been mainly studied in the framework of the Ig system and with other phenotypic markers such as Ia-like antigens (1,2), receptors for the Fc portion of IgG (3,4), for complement components (5,6), and mouse erythrocytes (7,8). Many of these cell markers, with the exception of surface immunoglobulins, lack cell specificity.

Monoclonal antibodies (mAbs) have been extremely useful in the definition of T cell subsets and T cell differentiation pathways. In the last years, several laboratories have developed mAbs reacting with B cell-specific or associated antigens (9–18) in order to obtain similar kinds of information on B cell lineage. These mAbs have resolved many of the problems of B cell markers listed above such as the lack of cell specificity of many of them, but more mAbs are needed in order to define with detail the sequential differentiative steps, the subpopulations, and their contributions to the functional properties of human B cells.

In this communication we describe the initial characterization of three monoclonal antibodies—B-C1, B-C2, and B-C3—recognizing B cell antigens and capable of defining, in conjunction with other known B cell markers, stages of differentiation along the B cell axis.

Material and Methods

Production of Monoclonal Antibodies

B-C monoclonal antibodies were produced according to the standard procedure described by Kohler and Milstein (19) with slight modifications, as previously described (20). P3/NS1/1-Ag4-1 myeloma cells (kindly provided by Dr. M. Spitz) were fused with BALB/c splenocytes hyperimmunized with cells from a B-prolymphocytic leukemia (B-PLL) for the production of B-C1 and B-C2 and with cells from a common acute lymphoblastic leukemia (cALL) for the production of B-C3.

Screening of hybridoma supernatants was done by an ELISA technique similar to that described by R.H. Kennett (21) using paraformaldehyde-fixed immunizing cells and rabbit anti–mouse immunoglobulin peroxidase-conjugated antiserum (Nordic) as developing reagent. Positive hybridomas were cloned several times and subsequently maintained by i.p. injection in BALB/c mice. Monoclonal antibodies containing ascites were used in all subsequent experiments. CRIS-1 and EDU-1 are two monoclonal antibodies produced in our laboratory; CRIS-1 has been included in the CD5 (T p67) cluster of differentiation (22) and EDU-1 recognizes monomorphic determinants of human Ia-like antigens (23).

Isolation of Cell Populations

Peripheral blood mononuclear cells (PBMC) were obtained by Ficoll–Hypaque (F–H) density gradient centrifugation following standard procedures. Granulocytes were isolated from the sediment formed during F–H gradient centrifugation by erythrocyte lysis with Tris-buffered NH_4Cl. E positive and E negative cells were separated on a F–H gradient according to their capacity to form E-rosettes with *Vibrio cholerase* neuraminidase-(Bohrinwerke) treated sheep erythrocytes. T cells were recovered by treating E-positive pellets with Tris-buffered NH_4Cl to lyse erythrocytes. Monocytes were obtained by adherence to plastic culture dishes. Cell suspensions from lymphoid tissues were obtained by teasing apart the tissue until the cells were separated from the connective stroma.

Immunofluorescent Assays

Monoclonal antibody reactivity with single cell suspensions or frozen tissue sections was determined using indirect immunofluorescence with monoclonal antibodies at a saturating dilution. As the second antibody, goat anti–mouse IgG antiserum, FITC- or TRITC-conjugated (Meloy and Cappel), was used. All techniques with cell suspensions were done in the presence of 10% human AB serum, 0.02% sodium azide. Incubation and washing steps were done at 4°C.

F(ab')2 fragments of goat anti–human Ig antiserum (FITC- or TRITC-conjugated, TAGO or Kallestad) were used to detect membrane (m) and cytoplasmic (c) immunoglobulins (Ig). Reactivity of the monoclonal antibodies with plasma cells was assessed by two-color fluorescence, in which monoclonal antibody staining was carried out in suspension (using TRITC-conjugated second antibody), then air-dried on microscope slides, fixed, and stained for cIgs using FITC-conjugated anti–human Igs. Indirect immunofluorescence of mAbs was assessed by immunofluorescence microscope or by flow cytometry (Becton Dickinson FACS Analyzer) using log amplification for volume and fluorescence analysis. Dead cells and erythrocytes were gated out by their electronic volume.

Leukemia/Lymphoma Cells

Diagnosis was based on standard clinical, cytomorphological, cytochemical, and immunological criteria. All cell markers were evaluated either simultaneously or later on cryopreserved cells for each leukemia/lymphoma cell sample, with the exception of B-C3 which was not studied on all the leukemias or lymphomas included in this study. Populations were scored as positive when 20% of malignant cells bound mAb.

Cell Cultures

Plasma cells were generated in a seven-day culture of PBMC (1×10^5 cells/ml) with pokeweed mitogen (PWM, Sigma) at 1 μg/ml final concentration. PB E-negative adherent-depleted cells (1×10^5) were co-cultured with 1×10^5 irradiated (2,000 r) autologous PB E-positive cells and stimulated with PWM (1 μg/ml) in the presence of several dilutions of B-C mAbs and the proliferative response evaluated after 4 days of culture by ^{3}H-thymidine incorporation. Also splenic mononuclear cells were stimulated with PWM in the presence of B-C mAbs at several dilutions. In other experiments splenic non-T cells (2×10^5) as well as splenic mononuclear cells (2×10^5) and PBMNC (2×10^5) were cultured with several dilutions of B-C mAbs without mitogens and the proliferative response evaluated after 3 days of culture. All these cultures were done in round-bottomed 96-well microplates (Linbro) and the final volume in each well was 0.2 ml. RPMI 1640 culture medium (Gibco) was used in all these experiments, supplemented with 1% glutamine (Gibco) and containing 10% FCS (Gibco).

Modulation Experiments

Modulation experiments were carried out with B-C1 and B-C2 mAbs on spleen non-T cells as described by Reinherz *et al.* (24). Unreactive ascites obtained from a mouse i.p.-injected with NS1 cells served as negative control. Following modulation at 37°C for various time periods (1 hr and 20 hr), the reactivity of modulated cells with B-C1 and B-C2 mAbs was determined by indirect immunofluorescence as indicated above. Cells were examined in a Leitz immunofluorescence microscope.

Blocking Experiments

B-CLL and E⁻ cells reactive with B-C1, B-C2, and/or B-C3 were incubated with these antibodies at various dilutions, washed, and then rosette formation with mouse erythrocytes (Mr) or with EAC (EACr) was evaluated and compared with that with untreated cells. B-CLL cells positive for B-C monoclonal antibodies were preincubated with heat-aggregated human Ig (25) and then evaluated for their reactivity with B-C mAbs.

Western Blot Analysis

Western blot analysis was done on samples of solubilized cells subjected to SDS–PAGE in nonreducing conditions according to Laemmli (26), then transferred electrophoretically to a nitrocellulose sheet which was washed, blocked, and probed with mAbs, washed and then incubated with rabbit anti–mouse immunoglobulin peroxidase-conjugated antiserum (DAKO). The enzymatic activity was developed with DAB/H202, as described by Tombin *et al.* (27).

Results

Reactivity of B-C mAbs with Normal Lymphoid Cells

Table 9.1 shows the reactivities of B-C1, B-C2, B-C3, and other mAbs tested on fractionated peripheral blood nucleated cells. The cells stained by B-C1 and B-C2 were confined to the E^- fraction of PBMC and the results essentially paralleled those obtained with B1 and/or BA-1. Both B-C1 and B-C2 were unreactive with E-positive cells and monocytes. B-C2 was also positive with granulocytes. B-C3 reacted weakly with 30% of E^- cells, on some E^+ cells, and with granulocytes.

Table 9.1. Reactivity patterns of B-C mAbs on peripheral blood nucleated cells.[a]

mAb	PBMC[b] (n = 6)	E^- (n = 7)	E^+ (n = 5)	Monocytes (n = 3)	Granulocytes (n = 4)
B-C1	8 ± 1	30 ± 14	1 ± 0	2 ± 1	0
B-C2	7 ± 2	29 ± 16	1 ± 0	3 ± 1	61 ± 21
B-C3	26 ± 10	14 ± 4	24 ± 5	1 ± 1	52 (n = 1)
B1/BA-1	7 ± 2	35 ± 16	1 ± 0	3 ± 1	1 ± 1[c]
CRIS-1 (p67)	74 ± 7	7 ± 5	79 ± 10	ND[d]	ND
Mo2/OKM1	13 ± 4	30 ± 11	2 ± 2	66 ± 6	85 ± 12[e]

[a] Numbers indicate mean percentage of positive cells ± SD.
[b] Peripheral blood mononuclear cells.
[c] B1 reactivity.
[d] Not determined.
[e] OKM1 reactivity.

Table 9.2. Double-staining for mIgs and mAbs on E$^-$ peripheral blood mononuclear cells.[a]

mAb	lx$^+$ mAb^{+b}	lx$^-$ mIg$^+$ mAb$^+$	lx$^-$ mIg$^+$ mAb$^-$	lx$^-$ mIg$^-$ mAb$^+$	lx$^-$ mIg$^-$ mAb$^-$
B-C1	0	30 ± 2.4	1.5 ± 0.5	0	67.5 ± 2.3
B-C2	0	27 ± 3	1 ± 0.5	0	71 ± 3
B-C3	0	14 ± 5	26 ± 0.5	4.5 ± 0.5	55 ± 4
B1	0	27 ± 2.5	1 ± 0.5	0	71 ± 3
BA-1	0	31 ± 4	2 ± 1	0	66 ± 4.5
CRIS-1 (p67)	0	0.5 ± 0.4	36 ± 4	19 ± 4	44 ± 6

[a] Each entry represents the mean percentage ± SD for three normal donors.
[b] lx = latex (latex positive cells were 27 ± 5%).

None of these BC mAbs reacted with red blood cells or platelets. To further characterize the relative population of E-negative cells with B-C mAbs, two-color experiments were performed in order to identify reactivities with anti-Ig and mAb simultaneously. For this purpose, monocytes present in PB E-negative cell fractions were labeled by latex particle ingestion and they were excluded in counting mIg-positive cells. E-negative cells were incubated at 37°C and washed with warm PBS. As shown in Table 9.2, B-C1 and B-C2 were only expressed on the mIg-positive cells. There were no cells negative for mIg and positive for B-C1 or B-C2 mAb. B-C1 and B-C2 appeared to be expressed on more than 90% of mIg-positive cells but not on all them since there was a small percentage of mIg-positive cells without ingested latex particles and negative for B-C1 or B-C2. Small percentages of such cells were also present with B1 or BA-1 combinations. These cells may be B lymphocytes but it is more probable that they are cells with bound cytophilic IgG, either monocytes that did not ingest latex particles, or the non-T, non-B subset of lymphocytes (latex$^-$, mIg$^-$, CRIS$^-$ = 44 ± 6%; see Table 9.2) with Fc receptors.

Tables 9.3 and 9.4 show B-C reactivities on unseparated and fractionated cell suspensions from lymphoid organs. The pattern of reactivity is

Table 9.3. Reactivity of B-C mAbs antibodies on unseparated cell suspensions of lymphoid organs.[a]

mAb	Spleen (n = 1)	Tonsil (n = 2)	Lymph node (n = 5)	Bone marrow (n = 2)	Thymus (n = 2)
B-C1	22	19 ± 8	18 ± 9	2 ± 1	0
B-C2	32	16 ± 5	26 ± 5	24 ± 9	0
B-C3	ND[b]	11 (n = 1)	10 ± 5 (n = 3)	6 (n = 1)	63 ± 6
B1/BA-1	20	23 ± 7	17 ± 6	2 ± 1	0
CRIS-1 (p67)	50	51 ± 14	60 ± 10	8 ± 4	70 ± 10
EDU-1 (DR)	20	31 (n = 1)	15 ± 4	10 ± 2	1 ± 1

[a] Values are mean percentage of positive cells ± SD.
[b] Not determined.

Table 9.4. Reactivity of B-C monoclonal antibodies on fractionated cell suspensions of lymphoid organs.[a]

	Lymph node		Spleen		Tonsil
	E^-	E^+	E^-	E^+	E^-
	(n = 1)	(n = 1)	n = 3	n = 1	n = 1
B-C1	52	0	42 ± 3	1	77
B-C2	48	2	49 ± 8	1	80
B-C3	10	2	12 ± 1	ND[b]	12
B1/BA-1	50	0	42 ± 6	2	78
CRIS-1 (p67)	1	90	6 ± 2	85	7
EDU-1 (DR)	54	1	48 ± 5	1	ND

[a] Values are mean percentage of positive cells ± SD.
[b] Not determined.

essentially the same as for PB nucleated cells. B-C1 and B-C2 were expressed only on E-negative cell fractions and the percentages of reactive cells on each sample were very close to those of B1 and/or BA-1. The exception was bone marrow cells, with B-C2 reacting on a greater percentage of cells than BA-1. B-C3 reacted mainly with a fraction of E-negative cells and with a large proportion of thymocytes. Table 9.5 shows the results of double-labeling experiments on unfractionated cell suspensions from a lymph node. The reactivities were very similar to those found with E-negative PBMC. (see Table 9.2).

In order to study the expression of B-C mAbs during the terminal stages of B cell differentiation, we analyzed the cIg-containing cells obtained by stimulation of PBMC with PWM. There were 10% of cIg-positive cells on day 7 of culture, the majority of them plasma cells but also some lymphoblastoid cells. Table 9.6 shows that B-C mAbs were unreactive with the cIg-positive cells while 90% of them reacted with EDU-1 mAb (anti-Ia). Although multiple myeloma (MM) plasma cells do not express Ia antigens, it has been reported that high percentages of PWM-stimulated plasma cells are Ia positive (28,29).

Table 9.5. Double-staining for mIgs and mAbs on unfractionated lymph node cell suspension.[a]

	mIg^+mAb^+	mIg^+mAb^-	mIg^-mAb^+	mIg^-mAb^-
B-C1	39	3.5	0	57
B-C2	31	1.5	0	61
B-C3	12	25	4	41
BA-1	26	2	0	72
CRIS-1 (p67)	2	33	62	2

[a] Values are mean percentage of positive cells.

Table 9.6. Reactivity of B-C mAbs with cytoplasmic Ig containing (cIg+) cells obtained by stimulation of PBMC with PWM in a 7-day culture.

mAb	% cIg+ cells reacting with mAb
B-C1	0
B-C2	0
B-C3	0
EDU-1 (DR)	90
CRIS-1 (p67)	0
BA-1	0

Reactivity of B-C mAbs with Cells from Lymphoproliferative Disorders (LPD) (Table 9.7)

Non-T Acute Lymphoblastic Leukemia (Non-T ALL)

The neoplastic cells from all these cases expressed TdT and Ia antigens (EDU-1). CALLA antigen (J5) was present on 8 cases and 2 cases stained faintly for cIg; these were B1 positive. All non-T ALL cases were unreactive with a CD5 reagent (CRIS-1). B-C2 was moderately to strongly reactive with 9 of 10 cases (1 null-ALL unreactive). B-C1 was moderately reactive with relatively low percentages of cells on the two pre-B ALL, and unreactive with all other more immature ALL. B-C3 was strongly reactive with 6 of 9 cases of ALL.

B Cell Chronic Lymphocytic Leukemia (B-CLL)

All cases were positive for Ia antigen and mouse rosettes (Mr) while B1, mIg, and CRIS-1 were reactive in most but not all B-CLL. FMC7 and CALLA were expressed in a low proportion (~20%) of these B-CLL. One patient had an associated serum monoclonal paraprotein. B-C2 was moderately to strongly reactive on a high percentage of cells on 19 of 19 B-CLL. B-C1 reactivity with 10 of 19 B-CLL showed considerable variations, both in terms of intensity and percentage of positive cells. B-C3 reacted weakly with 7 of 15 B-CLL tested.

B Cell Prolymphocytic Leukemia (B-PLL)

The cells of all cases tested expressed Ia antigens, B1, mIg, and FMC7 strongly. Mr and CRIS-1 were positive on two different cases. B-C1 and B-C2 were strongly positive in all B-PLL tested. B-C3 reacted weakly and in low percentages in one case.

Table 9.7. Expression of B-C mAbs on leukemias and lymphomas.[a]

Cell type[b]	Reactivity with mAb: no. positive/no. tested (mean ± SD)[c]		
	B-C1	B-C2	B-C3
Null-ALL (EDU-1$^+$, J5$^-$, cIg$^-$, CRIS-1$^-$, B1$^-$)[d]	0/2	1/2 (58)	2/2 (52 ± 26)
CommonALL (EDU-1$^+$, J5$^+$, cIg$^-$, B1$^-$, mIg$^-$, CRIS-1$^-$)	0/6	6/6 (51 ± 12)	3/5 (55 ± 20)
Pre B-ALL (EDU-1$^+$, J5$^+$, cIg$^+$, B1$^+$, mIg$^-$, CRIS-1$^-$)	2/2 (22 ± 1)	2/2 (51 ± 10)	1/2 (35)
B-CLL (EDU-1$^+$, J5$^{-/+}$, B1$^{+/-}$, mIg$^+$, CRIS-1$^{+/-}$, Mr$^+$, FMC7$^{-/+}$)	10/19 (48 ± 24)	19/19[e] (61 ± 18)	7/15 (56 ± 26)
B-PLL (EDU-1$^+$, J5$^-$, B1$^+$, mIg^{++}, CRIS-1$^{+/-}$, Mr$^{-/+}$, FMC7^{++})	3/3 (63 ± 21)	3/3 (51 ± 17)	1/3 (22)
B-NHL (EDU-1$^+$, J5$^{+/-}$, B1$^+$, mIg^{++}, CRIS-1$^{-/+}$, Mr$^{-/+}$, FMC7$^{+/-}$)	5/6[f] (42 ± 11)	5/5[f] (62 ± 20)	1/3 (37)
MM[g] (EDU-1$^-$, J5$^-$, B1$^-$, mIg$^{+/-}$, cIg$^+$, CRIS-1$^-$, Mr$^-$, FMC7$^-$)	0/2	0/2[h]	0/2
T-ALL and T-LL (EDU-1$^-$, B1$^-$, mIg$^-$, CRIS-1$^+$, TdT$^{+/-}$, other anti-T mAbs)	0/5	0/5	3/5 (41 ± 29)
T-CLL (EDU-1$^-$, B1$^-$, mIg$^-$, CRIS-1$^{+/-}$, other anti-T mAbs)	0/4	0/4	0/4
AML (EDU-1$^+$, J5$^-$, B1$^-$, CRIS-1$^-$, TdT$^-$, other anti-myeloid mAbs)	0/9	0/9	9/9 (52 ± 23)

[a] Defined by standard clinical and hematological criteria. Phenotypic analysis done in parallel with B-C mAbs on the same cells.
[b] ALL, Acute lymphoblastic leukemia; CLL, chronic lymphocytic leukemia; PLL, prolymphocytic leukemia; NHL, Non-Hodgkin's lymphoma; MM, multiple myeloma; T-LL, T-lymphoblastic lymphoma; AML, acute myeloblastic leukemia.
[c] Positive cases when >20% cell samples reacted with mAbs. Mean ± SD of reactive cells in positive cases.
[d] CRIS-1 is a mAb produced locally and included in the CD5 (T + B-CLL, p67) group; EDU-1 is a mAb produced locally that reacts with a monomorphic DR antigen.
[e] One case with monoclonal serum paraprotein.
[f] Two cases with monoclonal serum paraprotein.
[g] Bone marrow plasma cells.
[h] Double-labeling with B-cytoplasmic Ig.

B Cell Non-Hodgkin's Lymphomas (B-NHL)

These cases included two with serum monoclonal paraprotein. This small group of lymphomas was heterogeneous in phenotype, but all cases were positive for Ia antigen B1, and mIg. Some were CALLA and FMC7 positive and a few showed a weak to moderate reactivity with CRIS-1 and small percentages of Mr. B-C1 reacted moderately to strongly in 5 of 6 cases, including one case with serum monoclonal paraprotein. B-C2 reacted strongly in all cases B-C3 stained faintly 1 of 3 cases tested.

Interestingly the cytofluorographic analysis of B-C2 reactivity using a two-parameter "dot plot" display (log. amplified fluorescence intensity vs. electronic volume) showed that in most B-LPD the brightest cells were the larger ones. The existence on the same cell population of large cells that were weakly stained suggests that this higher intensity of fluorescence on large cells may reflect higher antigen density rather than an increased number of fluorochromes per cell due to a larger size. This phenomenon was not observed with B-C1. However, B-C1 reactivity, both in terms of fluorescence intensity and percentage of reactive cells, increased on the B-LPD corresponding to the more differentiated stages of B cell lineage e.g., B-PLL, B-NHL (see Table 9.7 and Fig. 9.1). The above comments on the dependence of fluorescence intensity on both cell size and antigen density are also relevant to B-C1 reactivity, but in both cases direct antigen density measurements are needed to clarify this point.

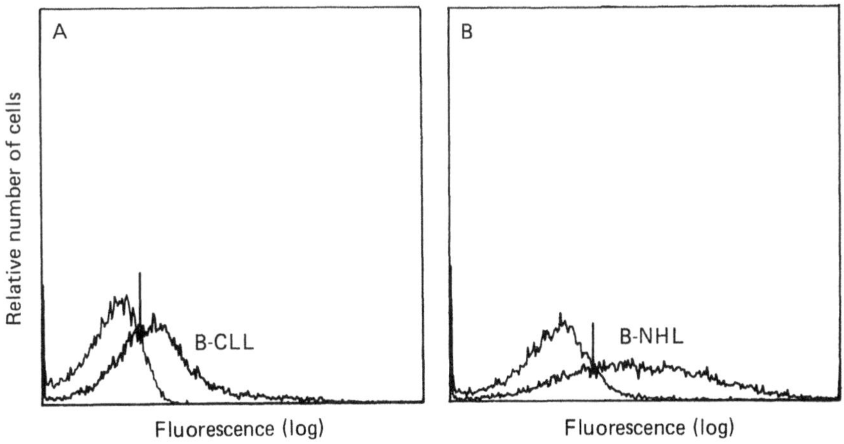

Fig. 9.1. FACS analyzer histograms of reactivities of B-C1 mAb with B-CLL cells (a) and B-NHL cells (b). Background fluorescence staining (vertical line) was obtained by incubating cells with unreactive control ascites and developing with FITC-conjugated goat anti–mouse IgG.

Multiple Myeloma (MM)

Bone marrow samples of multiple myeloma showed strong cIg staining and were negative for all other cell markers tested.

Since cellular composition in this bone marrow was heterogeneous, and B-C2 reacted with a significant percentage of cells of these samples, double labeling for cIg and B-C2 was performed to exclude its reactivity with plasma cells. Very low percentages of positive cells were observed with B-C1 and B-C3 on these bone marrow samples.

T Cell Lymphoproliferative Disorders (T-LPD)

The phenotype of 9 cases was very heterogeneous when analyzed with several anti-T mAbs. B-C1 and B-C2 were unreactive with all cases. B-C3 reacted moderately to strongly with 3 of 5 immature T-LPD and was negative with the 3 T-CLL.

Acute Myeloid Leukemias (AML)

All these cases were TdT negative and expressed Ia antigens. These leukemias were divided on the basis of the FAB classification supplemented with cytochemical reactions: 1, M5, 2 M4, 1 M2, and 5 M1. B-C3 reacted strongly with all these AML.

None of these cases were reactive with B-C1 or B-C2 mAbs.

B-C Reactivity with Cell Lines

Table 9.8 shows B-C reactivities on cell lines. B-C1 reacted with Burkitt's lymphoma cell lines. It was negative on KM-3, the other B cell lineage line tested, and also negative on T cell line MOLT-4 and on myeloid line K562. B-C2 was unreactive with all B and T cell lines and stained a small fraction of the myeloid K562 line. B-C3 was reactive with KM-3 and MOLT-4 cell lines.

Table 9.8. Reactivity of B-C mAbs on cell lines.

Cell line	B-C1 %	B-C2 %	B-C3 %
Raji	95	0	0
Raji mutant	95	0	0
Daudi	95	0	0
KM-3	0	3	90
MOLT-4	0	0	95
K562	0	35	1

Blocking Experiments

Studies were conducted in order to demonstrate possible mouse erythrocyte, C3b, or Fc receptor reactivity with B-C mAbs.

Pretreatment of a B-CLL, reactive with the three mAbs, with heat-aggregated human IgG, failed to diminish the percentage of reacting cells when compared with B-CLL cells not exposed to aggregated IgG. Similarly, pretreatment of B-CLL cells and E-negative PBMC with B-C mAbs at various dilutions failed to inhibit Mr and EACr formation when compared with cells pretreated with unreactive ascites (data not shown).

Modulation Experiments

No significant modulation was observed within 1 hr after incubation of spleen non-T cells with B-C1 and B-C2 mAbs (data not shown). However, after incubation for 20 hr at 37°C, significant loss of surface antigen was observed in comparison with spleen non-T cells incubated with unreactive ascites control (Table 9.9).

Proliferative Assays

The proliferative response of E^- adherent depleted cells to PWM in the presence of B-C mAbs and E^+ irradiated cells was evaluated in order to explore the possibility that these mAbs could have some effect on B cell proliferation. Figure 9.2 shows preliminary results, indicating that B-C mAbs do not seem to influence significantly, under these experimental conditions, the proliferative response of B cells. In addition, when B-C1 and B-C2 mAbs were added at various dilutions at the beginning of the culture of total spleen mononuclear cells stimulated with PWM no specific effect was observed on the proliferative response, as shown in Fig. 9.3. B-C mAbs did not induce by themselves a proliferative response in spleen non-T cells, spleen mononuclear cells, or PBMNC (date not shown).

Table 9.9. Antigenic modulation experiments on spleen non-T cells.

Modulating mAb	Reactivity of modulated cells with mAb B-C1	B-C2
B-C1	25%[a]	63% (++)
B-C2	74% (++)	28%[a]
Unreactive ascites control	77% (++)	66% (++)

[a] Cells exhibiting few weak immunofluorescent patches; (+) fluorescence intensity.

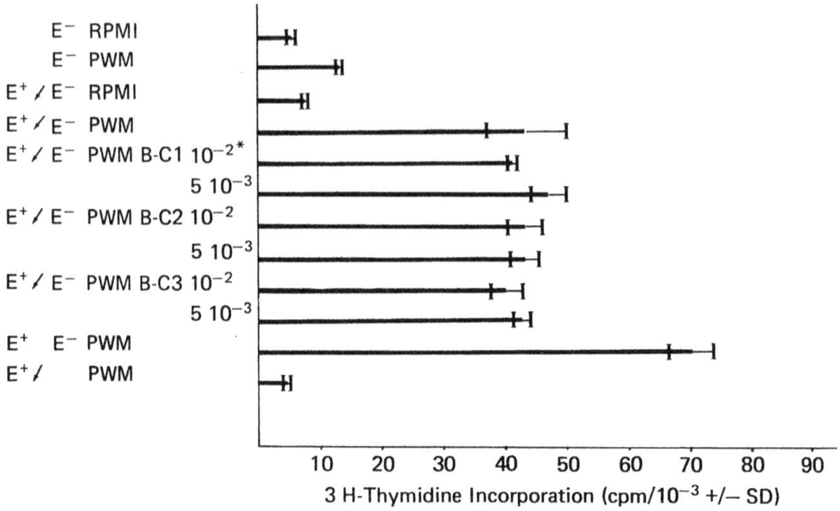

Fig. 9.2. Co-cultures of peripheral blood E-positive and autologous E-negative cells stimulated with PWM in presence of B-C mAbs at several dilutions. No effects on the proliferative response of E-negative cells to PWM were observed. Arrows indicate irradiation (2000 r).

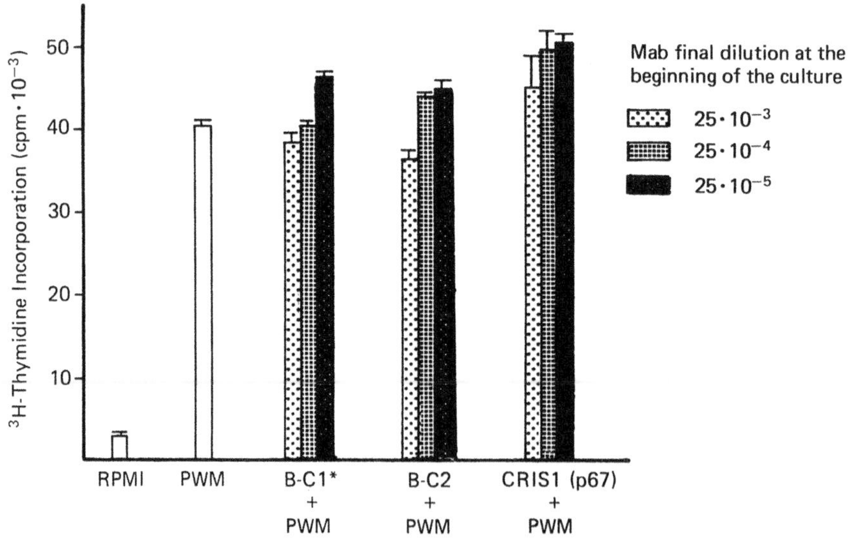

Fig. 9.3. Cultures of spleen mononuclear cells stimulated with PWM in presence of B-C1, B-C2, and CRIS-1 (CD5, gp67) mAbs at several dilutions. No specific effects on the proliferative response were detected.

Reactivity of B-C1, B-C2, and B-C3 on Adult Kidney

Preliminary data for B-C reactivities on cryostat sections of two adult kidneys indicated that B-C1 reacted weakly with glomerular visceral epithelium, and on some glomerulis stained weakly parietal epithelium. B-C2 reacted with glomerular parietal epithelium and with the epithelium of selected cortical tubules. B-C3 showed no reactivity with the kidney sections tested.

Molecular Weight (M.W.) Determination of the Antigens Defined by B-C mAbs

Western blot analysis under nonreducing conditions of the antigens detected by B-C mAbs is shown in Fig. 9.4. Lane 1 shows the results of B-C2 tested with the lysates of B-CLL cells. Attempts to define the B-C2 antigen by Western blot analysis have not met with success, despite the strong expression of the antigen on the B-LPD cells tested. Lane 2 shows the result for B-C3 mAbs with lysates of MOLT-4 cells. It reacted with a single band of 26 Kd. Lanes 3 and 4 show the results for B-C1 with lysates

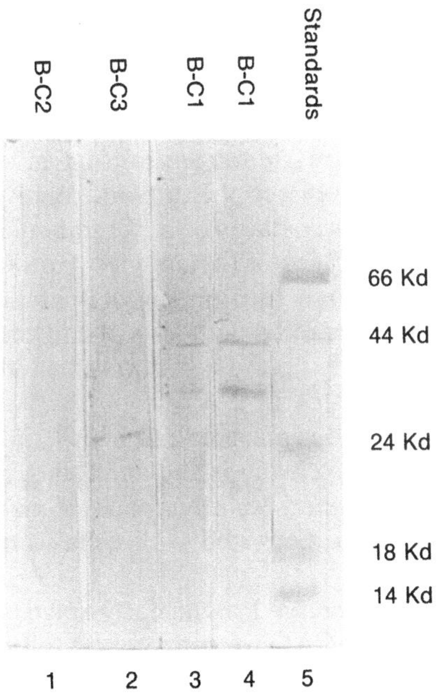

Fig. 9.4. Western blot analysis of B-C mAbs under nonreducing conditions. Lane 1, B-C2 tested with B-CLL lysates. Lane 2: B-C3 tested with MOLT-4 lysates. Lanes 3 and 4, B-C1 tested with Raji and Daudi cells, respectively. As can be seen, B-C2 failed to detect any polypeptide chain although its antigen was strongly expressed on cells tested. B-C1 detected a major band of 33–35 Kd and another minor one of 40 Kd, and B-C3 detected a single band of 26 Kd.

of Raji and Daudi cells, respectively. In both cases, B-C1 detected a major band with an apparent M.W. of 33–35 Kd and another minor one of 40 Kd. B-C3 and B-C2 gave the same results when tested with several leukemia cells (data not shown). Radioimmunoprecipitation techniques are now being applied to confirm these results.

Discussion

The results with normal cells suggest that B-C1 reacts, within the hematopoietic system, with a B cell-specific antigen, while B-C2 detects a B cell-associated antigen present on B cells and granulocytes. Both mAbs were, in our hands, useful pan-B mAbs reacting with more than 90% of B cells present in peripheral blood and lymphoid organs. These results closely paralleled those obtained with B1 and BA-1 mAbs. Both B-C1 and B-C2 mAbs were unreactive with peripheral T lymphocytes, thymocytes, T cell lines, and T-LPD cells. B-C2 mAb reacted with granulocytes and the erythromyeloid cell line K562 as well as with a high percentage of bone marrow nucleated cells. These findings raise the possibility that B-C2 may be reactive with both B cell and granulocyte precursors present in the bone marrow. However B-C2 was negative with M1 and M2 AML cases, suggesting that its antigen is not expressed on early stages of myeloid cells.

B-C3 seems to recognize an antigen present in subsets of B cells, T cells, and granulocytes from peripheral blood and lymphoid organs. It is strongly expressed one early stages of B, T, and myeloid lineages since it reacted with high percentages of thymocytes and of KM-3 and MOLT-4 cell lines, representing immature phenotypes of B and T cell lineage, respectively (30). It also reacted with all AML tested, in 3 of 5 immature T-LPD, and predominantly on the less differentiated stages of the B cell axis (see below).

In Fig. 9.5, a schematic representation of B-C mAb expression along the B cell differentiation pathway is shown. This scheme is based on the reactivities of B-C mAbs, and other lymphocyte markers tested in parallel, with normal lymphoid cells and with cells from 40 cases of B cell malignancies.

B-C2 appears to be expressed at the early stages of B cell differentiation, probably preceding CALLA antigen and following Ia antigens, as suggested by the fact that it was reactive with 9 of 10 non-T ALL, including a CALLA-negative DR-positive null-ALL, and reacted with a proportion of bone marrow cells which, although not directly proved, may include B cell precursors. Its expression probably spans all stages of B cell differentiation since it had a constant and high reactivity with the B-LPD tested and with virtually all normal B cells from peripheral blood and lymphoid organ cell suspensions B-C3 was expressed strongly on a high

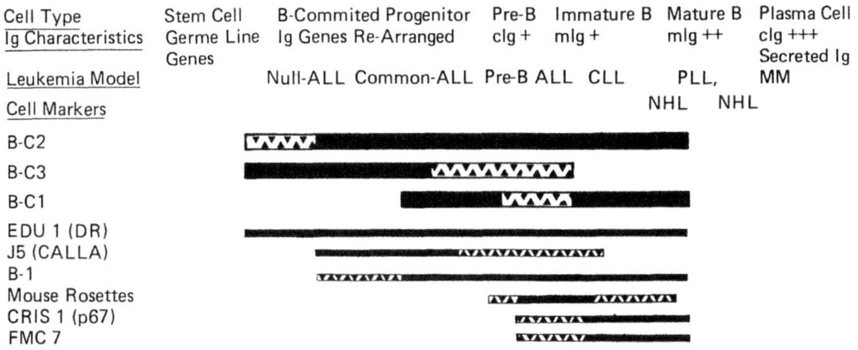

Fig. 9.5. Schematic representation of B-C mAb expression along the B cell differentiation pathway based on the reactivities of B-C mAbs and other lymphocyte markers tested in parallel on the same normal lymphoid cells and on cells of 40 cases of B-LPD. ALL, Acute lymphoblastic leukemia; CLL, B cell chronic lymphocytic leukemia; PLL, B cell prolymphocytic leukemia; NHL, B cell non-Hodgkin's lymphoma; MM, multiple myeloma. Continuous and discontinuous lines: positive with more than or less than 70% of B cell malignancies, respectively.

percentage of null and common ALL and it may precede B-C2 expression since there was a B-C3-positive B-C2-negative null-ALL. As expected from studies on normal B cells (30% faint reactivity of mIg-positive cells) it was weakly positive on only a fraction of B-CLL, NHL, and B-PLL cases.

B-C1 appears later than B-C2 and B-C3 (unreactive with null and common ALL) and seems to be expressed from the pre–B cell stage, since it was positive with the two pre-B ALL tested. Its expression appears to include most but not all the stages of B cell differentiation. It is present on virtually all B cells present on normal peripheral blood or lymphoid organs but its heterogeneous expression on B-CLL may reflect its presence on only a fraction of the small normal cell population, proposed as the normal counterpart of B-CLL (31), or expressed in lower antigen density as compared with the remaining B cells. Its expression seems to increase on more differentiated B-LPD (B-PLL,NHL). Normal plasma cells derived from a PWM-driven system and multiple myeloma plasma cells were unreactive with all B-C mAbs.

The pattern of reactivity of these mAbs with normal cells, cell lines, malignant cells, and kidney structures, as well as the immunochemical characterization of their antigens, seem to exclude the possibility that they recognize immunoglobulins, Ia antigens, mouse erythrocyte, C3b, and Fc receptors. In addition, blocking experiments with B-C mAbs did not induce changes in EACr or Mr formation, and preincubation of B cells with aggregated IgG did not influence B-C reactivity.

These mAbs seem to be distinct biochemically and by their pattern of reactivities from others previously published anti-B mAbs, such as B2 (p140) (9), BA2 (11), FMC1 (12), FMC7 (13), BB-1 (gp37) (14), CD10 (non-T, non-B, p100) (15), B4 (p95) (16), BL2 (p68) (17), OKB1 (gp168), OKB4 (gp87), OKB7 (gp175) (18). B-C1 and B1 (p35) (8) seem to be distinct based on B-C1 reactivity with adult kidney structures that have been described as unreactive with B1 (32). B-C2 seems to be different from BA-1 (p30) (10) since, unlike B-C2, this mAb was reactive with a small percentage of bone marrow cells, and was reactive with the KM-3 cell line. B-C2 has similarities with the recently described OKB2 (18) but the relationship between these mAbs remains to be clarified in simultaneous testing. Although B-C3 and CD9 (non-T, non-B, p24) reagents share some characteristics, they seem to be different, since B-C3 did not react with platelets, monocytes, and adult kidney structures (15).

B-C1, B-C2, and B-C3 are valuable mAbs to define normal B cell differentiation stages, and, in conjunction with other known lymphocyte markers, they are useful tools in the immunological characterization of lymphoproliferative disorders.

Summary

In this report three mAbs are described—B-C1 and B-C2, raised against a B-prolymphocytic leukemia, and B-C3 against a common acute lymphoblastic leukemia. Cytofluororographic analysis and double-labeling experiments with normal mononuclear cells, E^+, E^-, from peripheral blood (PB), secondary lymphoid organs (lymph nodes, tonsil, spleen), and thymocytes showed that B-C1 and B-C2 reacted only with B cells. No reactivity existed with PWM-induced plasma cells. The results obtained with cells from 32 acute and chronic B cell lymphoproliferative disorders (B-LPD) disclosed that B-C1 recognizes an antigen present from the pre-B stage and its reactivity/intensity seems to increase in the more differentiated B-LPD. The antigen detected by B-C2 appears to be present from very early stages along the B cell axis and its expression spans all stages of B cell differentiation with a constant number (31/32) of B-LPD positive and a high percentage of reactivity. B-C1 and B-C2 were negative with all the acute and chronic T cell LPD and acute myeloblastic leukemias tested, although B-C2 reacted with granulocytes. B-C3 reacted mainly with the less differentiated stages of B, myeloid, and T cell lineages, although it also detected small percentages of PB mononuclear cells and granulocytes. B-C1 and B-C2 had no mitogenic properties on spleen mononuclear cells, and they did not influence in a specific way their proliferative response to PWM. Modulation experiments on non-T spleen cells and immunochemical characterization by Western blot confirmed that the B-C1 (M.W.: 33–35 Kd), B-C2, and B-C3 (M.W.: 36 Kd) mAbs recognize different antigens as suggested by the above pattern of reactivity. The results obtained with B-C1, B-C2 and B-C3 are discussed in relation to other known B cell surface markers.

Acknowledgment. This work was supported by grants 0653 from CAICYT and 82/63, 82/64, and 82/66 from FIS.

References

1. Winchester, R.J., S.M. Fu, P. Vernet, H.G. Kunkel, B. Dupont, and C. Jerslid. 1975. Recognition by pregnancy serum of non-HLA alloantigen selectively expressed on B lymphocytes. *J. Exp. Med.* **141:**924.
2. Humphreys, R.E., J.M. McCune, L. Chess, H.C. Herrman, D.J. Malenks, D.L. Mann, P. Parham, S.F. Schlossman, and J.L. Strominger. 1976. Isolation and immunologic characterization of a human B lymphocyte specific cell-surface antigen. *J. Exp. Med.* **144:**98.
3. Dickler, H.B., and H.G. Kunkel. 1972. Interaction of the aggregated γ-globulin with B lymphocytes. *J. Exp. Med.* **136:**191.
4. Winchester, R.J., S.M. Fu, and H.G. Kunkel. 1975. IgG on lymphocyte surfaces: technical problems and the significance of a third cell population. *J. Immunol.* **114:**1210.
5. Bianco, C., P. Patrick, and V. Nussenzweig. 1970. A populations of lymphocytes bearing a membrane receptor for antigen–antibody complement complexes. *J. Exp. Med.* **132:**702.
6. Gupta, S.R., A. Good, and F.P. Siegal. Rosette formation with mouse erythrocytes. II. A marker for human B and non-T lymphocytes. *Clin. Exp. Immunol.* **25:**319.
7. Gupta, S.R., A. Good, and F.P. Siegal. 1976. Rosette formation with mouse erythrocytes. III. Studies in patients with primary immunodeficiency and lymphoproliferative disorders. *Clin. Exp. Immunol.* **26:**204.
8. Shashenko, P., L.M. Nadler, R. Hardy, and S.F. Schlossman. 1980. Characterization of a human B lymphocyte-specific antigen. *J. Immunol.* **125:**1678.
9. Nadler, L., P. Shashenko, R. Hardy, A. van Agthoven, C. Terhorst, and S.F. Schlossman. 1981. Characterization of human B cell specific antigen (B2) distinct from B1. *J. Immunol.* **126:**1941.
10. Abramson, C.S., J.H. Kersey, and T.W. LeBien. 1981. A monoclonal antibody (BA-1) reactive with cells of human B lymphocyte lineage. *J. Immunol.* **126:**83.
11. Kersey, J.H., T.N. LeBien, C.S. Abramson, R. Newman, R. Sutherland, and M. Greaves. 1981. p24: A human leukemia-associated and lymphohemapoietic progenitor cell surface structure identified with monoclonal antibody. *J. Exp. Med.* **153:**726.
12. Brooks, D.A., I. Beckman, J. Bradley, P.J. McNamara, M.E. Thomas, and H. Zola. 1980. Human lymphocyte markers defined by antibodies derived from somatic cell hybrids. I. A hybridoma secreting antibody against a marker specific for human B lymphocytes. *Clin. Exp. Immunol.* **39:**477.
13. Brooks, D.A., I.G.R. Beckman, J. Bradley, P.J. McNamara, M.E. Thomas, and H. Zola. 1981. Human lymphocyte markers defined by antibodies derived

from somatic cells hybrids. IV. A monoclonal antibody reacting specifically with a subpopulation of human B lymphocytes. *J. Immunol.* **126:**1373.
14. Yokochi, T., R. Holly, and E. Clarck. 1981. B lymphoblast antigen (BB-1) expressed on Epstein–Barr virus activated B cell blasts, B lymphoblastoid cell lines, and Burkitt's lymphomas. *J. Immunol.* **128:**823.
15. Bernard, A., L. Boumsell, C. Hill. 1984. Joint report of the First International Workshop on Human Leukocyte Differentiation Antigens by the investigators of the participating laboratories: B2 protocol. In: *Leucocyte typing,* A. Bernard, L. Boumsell, J. Dausset, C. Milstein, and S. F. Schlossman, eds. Springer-Verlag, Berlin, Heidelberg, pp. 61–82.
16. Nadler, L.M., Morti G. Anderson, M. Bates, E. Park, J.F. Daley, and S.F. Schlossman. 1983. B4, a human B lymphocyte associated antigen expressed on normal, mitogen activated and malignant B lymphocytes. *J. Immunol.* **131:**244.
17. I.D.M., Knowles, B. Tolidjan, C.C. Marboe, J.P. Halper, W. Azzo, and C.Y. Wang. 1983. A new human B-lymphocyte surface antigen (BL2) detectable by a hybridoma monoclonal antibody: distribution on benign and malignant lymphoid cells. *Blood* **62:**191.
18. Mitler, R.S., M.N. Table, K. Carpenter, P.E. Rao, and G. Goldstein. 1983. Generation and characterization of monoclonal antibodies reactive with human B lymphocytes. *J. Immunol.* **131:**1754.
19. Kohler, G., and C. Milstein. 1975. Continuous cultures of fused cells secreting antibody of predefined specificity. *Nature* **256:**495.
20. Vilella, R., J. Yague, and J. Vives. 1983. Monoclonal antibody against HLA-A Aw32+A25. Is HLA-Aw32 an allele with no unique antigenic determinant? *Hum. Immunol.* **6:**53.
21. Kennett, R.H. 1981. Enzyme linked antibody assay with cells attached to polyvinyl chloride plates. In: *Monoclonal antibody hybridomas: a new dimension in biological analyses,* R.H. Kennett, T.J. McKearn, and K.B. Bechtol, eds. Plenum Press, New York, London, pp. 376–377.
22. Bernard, A., L. Boumsell, C. Hill. 1984. Joint report of the First International Workshop on Human Leukocyte Differentiation Antigens by the investigators of the participating laboratories: T2 protocol. In: *Leucocyte typing,* A. Bernard, L. Boumsell, J. Dausset, C. Milstein, and S. F. Schlossman, eds. Springer-Verlag, Berlin, Heidelberg, pp. 61–82.
23. Colombany, J., V. Lepage, and J. Kalil. 1983. HLA monoconal antibodies registry: second listing. *Tissue Antigens* **22:**97.
24. Reinherz, E.L., S.C. Mener, K.A. Fitzgerald, R.E. Hussey, H. Levine, and S.F. Schlossman. 1982. Antigen recognition by human T lymphocytes is linked to surface expression of the T3 molecular complex. *Cell* **30:**735.
25. Dickler, H.B. 1974. Studies of the human lymphocyte receptor from heat-aggregated or antigen-complexed immunoglobulin. *J. Exp. Med.* **140:**508.
26. Laemmli, U.K. 1970. Cleavage of structural proteins during the assembly of the head of bacteriophage T4. *Nature* **227:**680.
27. Tombin, H., T. Staelelin, and J. Gordon. 1979. Electrophoretic transfer of proteins from polyacrylamide gels to nitrocellulose sheets: procedure and some applications. *Proc. Natl. Acad. Sci. U.S.A.* **76:**4350.
28. Halper, J., S.M. Fu, C.Y. Wang, R. Winchester, and H.G. Kunkel. 1978.

Patterns of expression of human ''Ia-like'' antigens during the terminal stages of B cell development. *J. Immunol.* **120:**1480.

29. Zola, H., P.J. McMamara, H.A. Moore, I.J. Smart, D.A. Brooks, I.G.R. Beckman, and J. Bradley. 1983. Maturation of human B lymphocytes—Studies with a panel of monoclonal antibodies against membrane antigens. *Clin. Exp. Immunol.* **52:**655.
30. Minowada, J., E. Tatsumi, K. Sagawa, M.S. Lok, T. Sugimoto, K. Minato, L. Zgoda, L. Prestine, L. Kover, and D. Gould. 1984. A scheme of human hematopoietic differentiation based on the markers profiles of the cultured and fresh leukemia lymphomas. The result of Workshop study. In: *Leucocyte typing,* A. Bernard, L. Boumsell, J. Dausset, C. Milstein, and S.F. Schlossman, eds. Springer-Verlag, Berlin, Heidelberg, pp. 519–527.
31. Caligaris-Cappio, F., M. Gobbi, M. Bofill, and G. Janossy. 1982. Infrequent normal B lymphocytes express features of B chronic lymphocytic leukemia. *J. Exp. Med.* **155:**623.
32. Platt, J.L., T.W. LeBien, and A.F. Michel. 1983. Stages of renal ontogenesis identified by monoclonal antibodies reactive with lymphohemopoietic differentiation antigens. *J. Exp. Med.* **157:**155.

CHAPTER 10

Human B Cells: Is FMC7 a Marker for Relatively Mature B Cells or Does It Define a Population Equivalent to the LyB5-Negative Mouse B Cells?

Heddy Zola

T lymphocytes have been shown to undergo functional differentiation as they mature, such that two or more subsets can be found in mature T cell populations. These distinct cells derive from common progenitors, but cannot be converted from one subset to the other. Thus the differentiation pathway may be said to fork into parallel paths. Does the B cell lineage also fork into parallel pathways, or are all the different phenotypes which can be observed merely stages along a single maturation pathway? The strongest evidence for distinct, parallel B cell subsets comes from studies on the murine LyB5 antigen. $LyB5^+$ cells differ from $LyB5^-$ cells in phenotype and function, and one subset is absent from neonatal blood and from adult blood in congenitally immune-deficient CBA/N mice [reviewed by Scher (1); Huber (2)]. Furthermore no report has appeared suggesting that $LyB5^+$ cells can be induced to lose expression, or mature LyB5 negative B cells induced to express the antigen. The results are generally interpreted in terms of two parallel B cell pathways, although the possibility that $LyB5^-$ are less mature cells, which eventually mature into $LyB5^+$ cells, has not been formally excluded.

We have described a monoclonal antibody, designated FMC7, which reacts with approximately half the surface immunoglobulin (SmIg) positive B cells in human blood (3). We have previously interpreted our results to suggest that FMC7 reacts with the relatively mature B cells in blood (4, 5), but a reexamination of the information available suggests that FMC7 may be a marker for the human equivalent of the murine $LyB5^-$ subset. In this paper the relative merits of the two hypotheses are discussed in relation to all available data.

Materials and Methods

FMC7 is a mouse IgM monoclonal antibody produced by fusion of myeloma P3-NS1-AG4-1 (6) and spleen cells from a BALB/c mouse immunized with human B lymphoblastoid cells of the HRIK cell line. The general properties of the antibody have been described (3). This paper discusses results drawn from a number of papers, referred to below, and the methodology is described in detail in those papers.

Results and Discussion

FMC7 reacts with a cell membrane antigen which either is protein or depends on membrane protein for its expression, and protein synthesis for its reexpression after removal from the cell membrane (7). The antigen is an integral membrane component, readily capped, and interacts with the cytoskeleton (7). These properties are consistent with a function as receptor for an extracellular signal and transmission of the signal to cytoplasmic components.

FMC7 reacts with 3–6% of mononuclear cells in normal blood. Reactive cells are B lymphocytes, but not all B cells react. The proportion of B cells reacting with FMC7 depends on the method used, but in our hands has ranged from 33% in dual-fluorescence testing with antibody against SMIg(G + A + M) to 64% calculated from the percentages of cells reacting, in separate tests, with anti-SMIg(G + A + M) and with FMC7 (3). In routine use the number of cells in peripheral blood mononuclear preparations scored positive with FMC7 is usually about 50% of the number scored positive for SMIg(G + A + M). The number of cells scored positive usually approximates SMIgM-positive cells, but double-marker tests show that IgM-positive and FMC7-positive populations are not entirely coincident. Staining of blood B cells is generally weak, and consequently not all workers who have used the antibody would agree with the findings described. Staining of positive cell lines is generally very bright, indicating that staining on blood cells is limited by the amount of antigen in the cell membrane, rather than the affinity of the antibody.

FMC7 reacts with only a few B cell lines. Of greater significance is the finding (8) that most chronic lymphocytic leukemia (CLL) cells are negative, while prolymphocytic leukemia (PLL) cells are generally positive. Hairy cell leukemia cells are also reactive (9). Acute lymphoblastic leukemia and multiple myeloma are negative, as are normal B cell precursors and plasma cells, and B cells activated by pokeweed mitogen (4). On the basis of these results we have concluded that FMC7 reacts with an antigen which is absent from B cell precursors, is expressed on a proportion of SMIg-positive B cells, and is lost as B cells are activated (4, 7). Based largely on the reactivity with PLL rather than CLL we have suggested that the proportion of B cells which are reactive are more mature than

Table 10.1. Comparison of expression of FMC7 in human B cells with LyB5-negative mouse B cells.

Mouse: LyB5-negative cells	Human: FMC7-positive cells
About 50% of B cells in normal adult blood	About 50% of B cells in normal adult blood
100% of B cells in neonate	100% of B cells in cord blood
100% of B cells in XID mice	100% of B cells in some immune deficiencies

those which do not express the antigen (4, 7). In agreement with this, Robert *et al.* (10) reported that the FMC7⁻ cells of a patient with CLL expressed the antigen during activation by lipopolysaccharide.

However, other attempts to induce $FMC7^-$ cells to express the antigen by culturing with "maturation-inducing agents" have been unsuccessful (7). In cord blood, moreover, all B cells react with FMC7 (11), a finding difficult to reconcile with the idea of a relatively mature subpopulation. Furthermore, FMC7 accounts for all the B cells in some immune deficiency states (12). These findings suggest an analogy between FMC7-positive cells in man and LyB5-negative cells in the mouse. Table 10.1 shows that the similarity in terms of the phenotypes of normal adult, normal neonate, and immune-deficient adult is strong, at least superficially. However, the immune deficiencies involved are different—an X-linked defect in the CBA/N mouse as compared with selective IgA deficiency and a proportion of common variable immune deficiency in man. Neither of these conditions shows any strong genetic component, and neither is X-linked.

LyB5-positive and negative populations in mice differ in important functional properties: $LyB5^-$ cells respond to antigen only if it is presented by an MHC-compatible cell, while $LyB5^+$ cells respond in a genetically unrestricted manner [reviewed by Singer *et al.* (13)]; the two subsets differ in the types of antigen to which they respond (14); $LyB5^+$ cells proliferate in response to anti-immunoglobulin whereas $LyB5^-$ cells need a second signal (15).

Whether or not the $FMC7^+$ population is directly analogous to the $LyB5^-$ population, it will be important to compare $FMC7^+$ and $FMC7^-$ fractions in functional assays in order to determine whether FMC7 serves as a marker for a functional subset of B lymphocytes. Such studies are under way in our laboratory and in other laboratories.

Summary

FMC7 is a monoclonal antibody which reacts with a membrane protein expressed on about 50% of B cells in human blood. B cell precursors and plasma cells do not express the antigen. It has been proposed that FMC7 is a marker for relatively mature B cells, largely on the basis of its reactivity with prolymphocytic and hairy cell but not chronic lymphocytic leukaemias. However, in cord blood and certain immune-deficient patients FMC7 reacts with all circulating B cells. This suggests a homology between $FMC7^{+}$ human B cells and $LyB5^{-}$ mouse B cells, which comprise 50% of normal adult mouse B cells but 100% of B cells in neonatal normal mice and the immune-deficient CBA/N mutant line. LyB5 provides the first clear evidence of functionally disparate sub-lineages within the mouse B cell lineage, and by analogy FMC7 may allow the identification of the two functionally distinct subsets of human B cells.

Acknowledgments. The possibility that FMC7 delineates a functional B cell subset, and the analogy with $LyB5^{-}$ cells, has occurred independently to other workers. I acknowledge stimulating discussions and correspondence with Drs. Robin Callard and Pat Mongini, and correspondence on the immune deficiency aspect with Dr. S. Gupta. Work in the author's laboratory is supported by grants from the Anti Cancer Foundation of the Universities of South Australia and the Australian National Health and Medical Research Council.

References

1. Scher, I. 1982. CBA/N immune defective mice; Evidence for the failure of a B cell subpopulation to be expressed. *Immunol. Rev.* **64:**117.
2. Huber, B.T. 1982. B cell differentiation antigens as probes for functional B cell subsets. *Immunol. Rev.* **64:**57.
3. Brooks, D.A., I.G.R. Beckman, J. Bradley, P.J. McNamara, M.E. Thomas, and H. Zola. 1981. Human lymphocyte markers defined by antibodies derived from somatic cell hybrids. IV. A monoclonal antibody reacting specifically with a subpopulation of human B lymphocytes. *J. Immunol.* **126:**1373.
4. Zola, H., P.J. McNamara, H.A. Moore, I.J. Smart, D.A. Brooks, I.G.R. Beckman, and J. Bradley. 1983. Maturation of human B lymphocytes—Studies with a panel of monoclonal antibodies against membrane antigens. *Clin. Exp. Immunol.* **52:**655.
5. Zola, H., J. Bradley, D.A. Brooks, P. Macardle, P.J. McNamara, H. Moore, and A. Nikoloutsopoulos. 1984. The human B cell lineage studies with monoclonal antibodies. In: *Leucocyte typing,* A. Bernard, L. Boumsell, J. Dausset, C. Milstein, and S.F. Schlossman, eds. Springer-Verlag, Berlin, Heidelberg, p. 363.
6. Kohler, G., C.S. Howe, and C. Milstein. 1976. Fusion between immunoglobulin secreting and non-secreting lines. *Eur. J. Immunol.* **6:**292.
7. Zola, H., H. Moore, A. Hohmann, I.K. Hunter, A. Nikoloutsopoulos, and J. Bradley. 1984. The antigen of mature human B cells detected by the monoclonal antibody FMC7: studies on the nature of the antigen and modulation of its expression. *J. Immunol.* **133:**321.
8. Catovsky, D., M. Cherchi, D. Brooks, J. Bradley, and H. Zola. 1981. Heterogeneity of B cell leukemias demonstrated by the monoclonal antibody FMC7. *Blood* **58:**406.

9. Worman, D.P., D.A. Brooks, N. Hogg, H. Zola, P.C.L. Beverley, and J.C. Cawley. 1983. The nature of hairy cells; A study with a panel of monoclonal antibodies. *Scand. J. Haematol.* **30**:223.
10. Robert, K.H., G. Juliusson, and P. Biberfeld. 1983. Chronic lymphocytic leukemia cells activated in vitro reveal cellular changes that characterize B-prolymphocytic leukaemia and immunocytoma. *Scand. J. Immunol.* **17**:397.
11. Zola, H., H.A. Moore, J. Bradley, J.A. Need, and P.C.L. Beverley. 1983. Lymphocyte sub-populations in human cord blood: analysis with monoclonal antibodies. *J. Reproductive Immunol.* **5**:311.
12. Gupta, S., D.A. Brooks, J. Bradley, and H. Zola. 1985. Monoclonal antibody-defined B lymphocyte subpopulations in primary immune deficiency disorders. *J. Clin . Lab. Immunol.* **16**:59.
13. Singer, A., A. Yoshihiro, M. Shigeta, K.S. Hathcock, A. Ahmed, C.G. Fathman, and R.J. Hodes. 1982. Distinct B cell subpopulations differ in their genetic requirements for activation by T helper cells. *Immunol. Rev.* **64**:137.
14. Mond, J.J. 1982. Use of the T lymphocyte regulated type 2 antigens for the analysis of responsiveness of $Lyb5^+$ and $Lyb5^-$ B lymphocytes to T lymphocyte derived factors. *Immunol. Rev.* **64**:99.
15. DeFranco, A.L., J.T. Kung, and W.E. Paul. 1982. Regulation of growth and proliferation in B cell subpopulations. *Immunol. Rev.* **64**:161.

CHAPTER 11

Spontaneous Mouse Erythrocyte Rosette Formation (M-RFC) with Human B Lymphocytes: Diagnostic Value in B-Lymphoproliferative Diseases and Lack of Relationship to B Cell Protocol Monoclonal Antibodies

Peter Hokland, Karin Meyer, and Inge Marie Fastrup

After the original description of rosette formation between mouse erythrocytes and chronic lymphocytic leukemia B lymphocytes (1), mouse rosette formation (M-RFC) has been shown to identify a subset of normal peripheral blood B lymphocytes (2) that may functionally be less mature than non-M-RFC B lymphocytes (3). Due to the fragility of the rosette formation, separation studies and functional characterization of these B cell subsets have not been possible. On the other hand, M-RFC has been extremely useful in leukemia phenotyping, and it is now well-documented that this marker is present mainly on B-CLL cells, and to a much lesser extent on other B-lymphoproliferative diseases. Bearing in mind that blood and tissue samples from patients with these diseases often contain significant numbers of T cells, which may influence the interpretation, we decided to reevaluate M-RFC by the simultaneous use of an extensively characterized B cell marker, the B1 monoclonal antibody (mAb), that reacts with all mature B-lymphoproliferative diseases (4). Finally, in search of reagents that might define the mouse erythrocyte receptor on human B lymphocytes, we have examined the B cell protocol mAbs for inhibitory activity in the mouse rosette assay.

Materials and Methods

Patients

Peripheral blood was obtained from 10 CALLA$^+$ B1$^+$ ALL patients, 10 AML patients, and 33 B-CLL patients. In addition, lymph node tissue containing more than 30% monoclonal B cells from 21 patients with

B-lymphoma and spleen tissue from 5 patients with B hairy cell leukemia were examined. Finally, as controls, peripheral blood from 10 normal donors was analyzed.

Determination of M-RFC/B1 Ratio

Mouse rosette-forming cells were enumerated using unsensitized mouse erythrocytes from BALB/c mice (MRBC) as previously described (5). The number of B lymphocytes in the various cell suspensions was determined as the number of cells reactive with mAb B1 in standard indirect immunofluorescence assays.

The affinity of the various malignant B cell populations for mouse erythrocytes was determined by the ratio between M-RFC and the number of cells reactive with B1.

Mouse Rosette Inhibition Tests

Lymphocytes from five patients with B-CLL were labeled with B cell Workshop mAbs in dilutions ranging from 1 : 50 to 1 : 250 under standard conditions. After two washes each sample, containing 2×10^6 cells, was divided into two, the first half being processed for indirect immunofluorescence using a rabbit anti-mouse FITC-conjugated antibody (Dakopatts, Copenhagen, Denmark), while the other half was rosetted with MRBC. Apart from the above mentioned Workshop antibodies the following controls were included: T3 (UCHT-1, kindly provided by Dr. P.C. Beverly), B1 (kindly provided by Dr. L.M. Nadler), and MY4 (kindly provided by Dr. J.D. Griffin). As a positive control for rosette inhibition, E-rosetting was performed using normal peripheral blood lymphocytes preincubated with or without various dilutions of the anti-T11 B67.1 mAb (kindly provided by Dr. B. Perussia), which reacts with the sheep erythrocyte receptor.

Results

The mouse/B1 ratio was evaluated in a number of patients with B-lymphoproliferative diseases that express the B1 pan B antigen as well as in AML patients and normal donors. As can be seen from Fig. 11.1 this ratio is similar in normal donors and acute myeloid leukemia patients at presentation. In contrast, as can be seen from Fig. 11.2, when this ratio was evaluated in 33 B-CLL patients the mean ratio was 0.75 with only three patients exhibiting ratios less than 0.5. This is highly significant when compared to the ratio seen in normal donors ($p < 0.01$, student's t test). Finally, regarding other B-lymphoproliferative diseases, mean ratios less

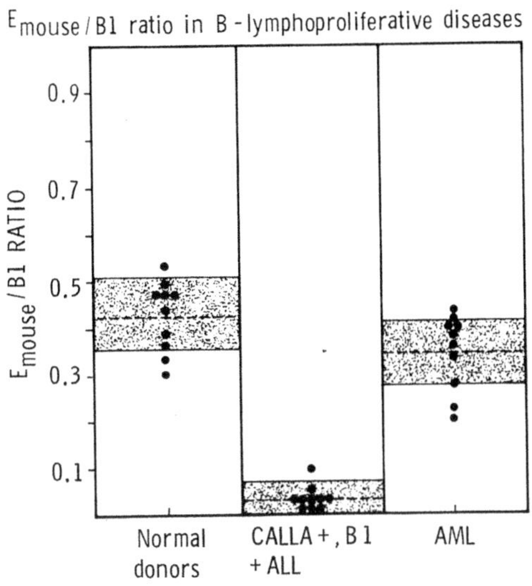

Fig. 11.1. E_{mouse}/B1 ratios in peripheral blood from 10 normal donors, 10 patients with newly diagnosed CALLA$^+$, B1$^+$ acute lymphoblastic leukemia (ALL), and 10 patients with newly diagnosed acute myeloid leukemia (AML).

than 0.1 were observed when testing pre-B ALL, B-lymphoma, and B hairy cell leukemia patients (Figs. 11.1 and 11.2).

The effect of the B cell protocol mAbs on mouse rosette formation was tested on all antibodies, but in Table 11.1 only B cell-specific antibodies have been shown, since none of the other antibodies showed inhibitory effect. Whereas most of the antibody groups defined by biochemical methods showed high degrees of activity with B-CLL cells in indirect immunofluorescence, none of them inhibited mouse rosette formation to significant extents, even though the gp35 group and especially the B36

Table 11.1. Effect of B cell protocol mAbs on spontaneous mouse rosette formation (M-RFC) with B-CLL cells.[a]

mAb no.	Reactivity in IIF	M-RFC
14, 28, 34, 43 (gp95)	90–100 (+ → ++)[b]	+++[c]
5, 22, 24, 36 (gp35)	90–100 (+(+) → +++)	++ → +++
9, 33, 35, 41 (gp140)	90–100 (+ → +++)	+++
7, 25, 31, 40, 49 (gp135)	90–100 ((+) → +(+))	+++
11, 19, 39 (gp45)	90–100 ((+) → +(+))	+++
2, 3, 42 (gp80)	5–45 ((+) → +)	+++
6, 8, 30 (unclustered)	90–100 (++ → +++)	+++

[a] Five patients; B1 > 90%.
[b] Percentage positive cells (antigen density).
[c] M-RFC: +++, no inhibition; ++, 20–40% inhibition; +, >40% inhibition.

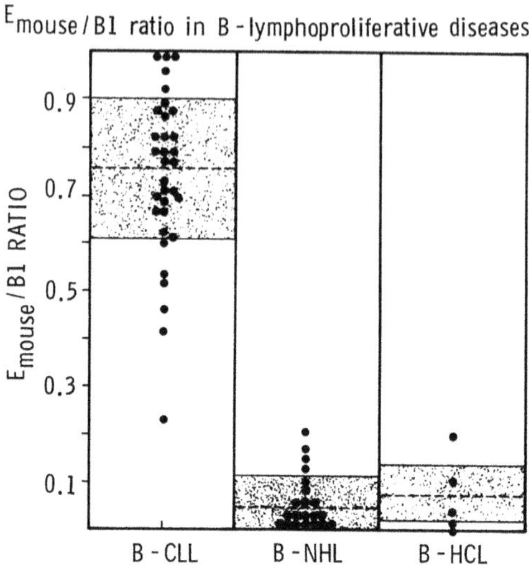

Fig. 11.2. E_{mouse}/B1 ratios in peripheral blood in newly diagnosed B cell-derived neoplasias: 33 patients with chronic lymphocytic leukemia (B-CLL), 10 malignant lymphoma patients, and 5 hairy cell leukemia (HCL) patients.

antibody showed some inhibition. Under similar conditions rosette formation between normal peripheral blood T lymphocytes and sheep erythrocytes was completely abolished by T11 antibodies, indicating that the experimental conditions for demonstrating rosette inhibition were suitable.

Discussion

The implications of these experiments are twofold: Firstly, we have shown that by expressing the number of M-RFC as a fraction of the total number of B cells in a given population, as identified by the pan B cell antibody B1, a very reliable parameter can be defined that can be used even in situations where the number of monoclonal B cells is low. Expression of M-RFC data in this way should thus lead to a much needed standardization. Needless to say, these results also clearly demonstrate the value of M-RFC, especially in distinguishing between B-CLL and malignant B-lymphoma. This is of value not only when phenotyping such patients initially, but also for demonstrating the transition from M-RFC$^+$ B-CLL to M-RFC$^-$ B-lymphosarcoma.

Given the fact that M-RFC requires stringent optimization for obtaining consistent results, it was of value to investigate whether any of the B cell antibodies submitted to the Workshop might define the mouse erythrocyte receptor on human B lymphocytes, since such a reagent could obviate the tedious use of MRBC. It was therefore disappointing to note that none of the antibodies showed the kind of inhibition that could be easily demonstrated with T11 antibodies in the standard E-rosette formation assay (Table 11.1).

Recent attempts to characterize the mouse erythrocyte receptor have shown that it contains phosphatidylethanolamine in a complex containing glycoprotein and a subclass of albumin (6). It is at present unclear why none of the monoclonal anti–B cell antibodies raised thus far have identified this receptor, but it might well be that the erythrocyte binding structure is not immunogenic to the same species that is able to form the rosettes. Alternatively, steric hindrance by the complexed glycoproteins may be involved. Fusions performed in rat systems or the use of enzyme-treated B lymphocytes as immunogens in mice systems might resolve this.

Summary

The affinity of different malignant B lymphocytes for mouse erythrocytes has been evaluated by estimating the fraction of B lymphocytes (identified with mAb B1) that formed mouse rosettes. This mouse/B1 ratio gave meaningful results even in situations where the monoclonal B cell compartment constituted only a small fraction of the total number of leukocytes. Whereas the ratio was increased in B-CLL, other $B1^+$ B cell diseases including pre-B-ALL, B-lymphoma, and hairy cell leukemia showed mean ratios of less than 0.1. Thus, this parameter is excellent in the differential diagnosis of B-lymphoproliferative diseases. When examining the B cell Workshop panel of mAbs for inhibition in the mouse rosette assay, none had significant activity. The implication of these findings have been discussed.

Acknowledgment. This work was supported by grants from the Danish MRC and the Danish Cancer Society.

References

1. Stathopoulos, G., and E.V. Elliott. 1974. Formation of mouse and sheep-red-blood-cell rosettes by lymphocytes from normal and leukaemic individuals. *Lancet* **I:**600.
2. Gupta, S., and M.H. Grieco. 1975. Rosette formation with mouse erythrocytes: Probable marker for human B-lymphocytes. *Int. Arch. Allergy Appl. Immunol.* **49:**734.
3. Lucivero, G., A.R. Lawton, and M.D. Cooper. 1981. Rosette formation with mouse erythrocytes defines a population of human B-lymphocytes unresponsive to pokeweed mitogen. *Clin. Exp. Immunol.* **45:**185.

4. Nadler, L.M., J. Ritz, R. Hardy, J.M. Pesando, and S.F. Schlossman. 1981. A unique cell surface antigen identifying lymphoid malignancies of B cell origin. *J. Clin. Invest.* **67**:137.
5. Hokland, P., and J. Ellegaard. 1981. Immunological studies in chronic lymphocytic leukemia—I. Elucidation of subset heterogeneity. *Leuk. Res.* **5**:341.
6. Zalewski, P.D., L. Valente, and I.J. Forbes. 1984. A phosphatidylethanolamine-containing complex on human B-cells that mediates rosette formation with mouse erythrocytes. *J. Immunol.* **132**:2491.

Part III. Biochemical Analysis of Antigens Defined by the Workshop B Cell/Leukemia Panel

CHAPTER 12

Human B Cell Surface Molecules Defined by an International Workshop Panel of Monoclonal Antibodies

Edward A. Clark and David Einfeld

Introduction

A primary goal of the B cell section of the First and Second International Workshops on Leukocyte Differentiation Antigens has been to identify groups of monoclonal antibodies (mAbs) which have a similar specificity and react with the same biochemically defined antigen. The antigens detected by the First Workshop panel were identified by immunoprecipitation and PAGE gel analysis (1,2); assignment of antibodies to specific groups was also possible based on serologic cluster analysis (3) and cross-blocking studies (1). Two clusters of differentiation (CD) groups were identified: CD9-specific mAbs react with a 24-Kd polypeptide p24 present on non-T, non-B acute lymphocytic leukemias; CD10-specific antibodies recognize the well-known p100 CALLA antigen. At least six other antigens were identified, but since only one or two antibodies recognized these antigens, they were not given an International CD designation.

In this study, we have attempted to biochemically define the antigens detected by the more than 70 mAbs in the B cell/leukemia Workshop panel. Over a dozen distinct polypeptides were precipitated by mAbs in the panel. Select prototype mAbs to different polypeptides were then purified and conjugated with fluorescein so that other appropriate mAbs in the Workshop panel could be tested for their ability to block the binding of the group prototype. In this way we have been able to determine which mAbs react to the same molecules and to similar or distinct epitopes on these molecules. These results help provide a foundation for detailed studies of the structure and function of B cell antigens.

Materials and Methods

Monoclonal Antibodies

The Second International Workshop panel of mAbs to B cell/leukemia-associated antigens comprised 51 mAb reagents in the B cell section (B1–B52, no B12) and 21 mAbs in the leukemia section (L1–L22, no L5). Ascites of these antibodies were used at 1:250 dilution for indirect immunofluorescence (IF) studies with an affinity-isolated FITC-goat $F(ab')_2$ anti–mouse Ig reagent (Tago, Burlingame, CA) and at a 1:40 dilution for radioimmunoprecipitation experiments. For direct IF studies, mAbs were purified from ascites fluid and conjugated with fluorescein as follows: 100 μl of freshly filtered saturated sodium sulfate (pH 7.4) were added to 250 μl of ascites fluid at room temperature in a microfuge tube; after incubation for 2 hr at RT, tubes were microfuged, the sodium sulfate aspirated, and pellets resuspended in 100 μl of 0.29 *M* bicarbonate buffer, pH 9.3. A 1-mg/ml solution of FITC in dimethyl sulfoxide was prepared and 15 μl added to the antibody solution with vortexing. After 1 hr at RT, 885 μl of PBS with 0.1% BSA and 0.1% azide were added to each tube. Using this microconjugation method, which is Jeff Ledbetter's modification to the method of Goding (4), 80% of the more than 20 mAbs treated in this way were active at doses of 0.1–2 $\mu l/10^6$ target cells.

Radioimmunoprecipitation

Cells radioiodinated with ^{125}I using lactoperoxidase were lysed in Tris buffer containing 0.5% NP-40 and 0.1% SDS. Antigens were precipitated by incubation of lysate with 10 μl of diluted ascites (1/10) for 16 hr at 4°C followed by a 2-hr incubation with Sepharose-conjugated affinity-isolated goat anti–mouse IgG and IgM (12.5 λ of beads at 8 mg/ml). Following the method of K. Shriver (unpublished) (5), the beads were washed twice with 50 m*M* Tris containing 0.45 *M* NaCl, 0.5% NP-40, and 0.1% SDS, pH 8.3; twice with 100 m*M* Tris, pH 8.0, containing 0.25 *M* lithium chloride and 1% freshly added 2-mercaptoethanol; and once with 20 m*M* Tris, containing 50 m*M* NaCl and 0.5% NP-40, pH 7.5. After elution into SDS–PAGE sample buffer, the bound material was electrophoresed under reducing conditions in 7% or 10% discontinuous polyacrylamide slab gels. Bands were visualized by exposure of dried gels to Kodak X-Omat film using enhancing screens.

Two-Color Flow Cytometry

Lymphoid cells were stained with fluorescein-conjugated (green) or phycoerythrin (PE)-conjugated (red) mAb as previously described (6). After 30 min at 4°C, cells were washed twice and analyzed on a FACS IV cell sorter with the appropriate 560-nm dichroic mirror beam splitter and 540-nm short-pass (green) and 580-nm long-pass (red) filters in front of the photomultiplier tubes. Forty thousand cells/test were analyzed. The data were gathered and presented on three-dimensional plots of cell number vs. log of green fluorescence vs. log of red fluorescence in a 64 × 64 grid. Approximately every 4–5 dots represents a doubling of fluorescence.

Cross-Blocking Analysis

FITC-conjugated mAbs were titered against antigen-positive target cells such as Daudi, Raji, or Nalm-6 cell lines. An antibody concentration just above saturation of 2×10^5 cells was selected; 50 μl of ascites fluid (1 : 250 dilution) were added to the cell suspension 15 min prior to adding the FITC-conjugated antibody for a 30-min incubation on ice (1).

Results and Discussion

In order to identify antigen-positive target cells for biochemical studies, using indirect IF and flow cytometry, we first screened the B cell/leukemia panel of mAbs for reactivity against tonsillar B lymphoctyes and a set of B cell lines (Daudi, Raji, Nalm-6, Corinna II, SB-1, 8226). In the B cell panel 47 of 51 mAbs were reactive against one or more of these targets (B16, B18, B32, and B46 were not), and 20 of 21 mAbs in the leukemia panel were reactive (L7 was not). The reactive mAbs were then tested for their ability to immunoprecipitate ^{125}I-labeled antigens from cell membrane lysates (e.g., Figs. 12.1–12.3). Based on preliminary biochemical results, mAbs reacting with distinct polypeptides were selected for purification and conjugated with fluorescein. The additional immunoprecipitations and blocking studies were performed in order to cluster and to subdivide these groups by blocking of prototype antibodies. Where two or more antibodies reacted with the same epitope, the antibodies have been

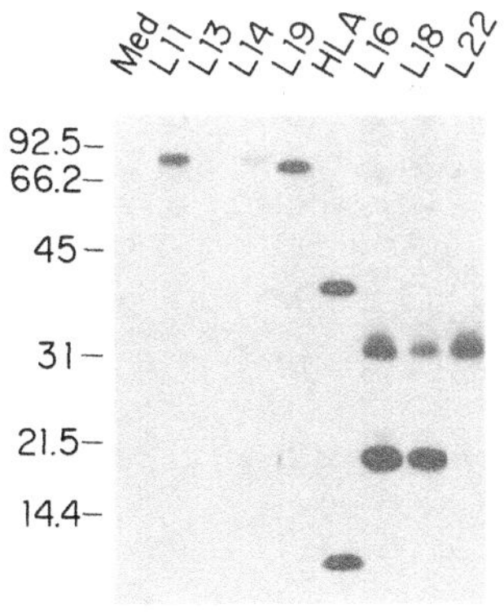

Fig. 12.1. Radioimmunoprecipitation slab gel electrophoresis of leukemia-associated antigens using ^{125}I-labeled Nalm-6 cells under reducing conditions. The p100 CALLA antigen (L11, L14), p90 transferrin receptor (L13, L19), and the p33/24 antigens (L16, L18, L22) are evident.

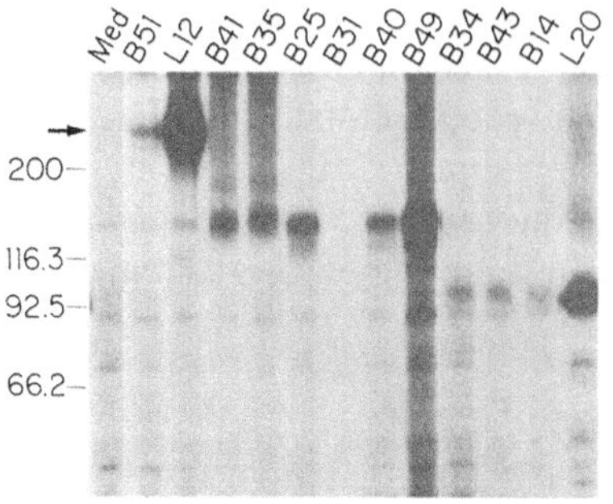

Fig. 12.2. Radioimmunoprecipitation of B cell-associated antigens. ^{125}I-labeled Daudi cells with anti-p220 antibodies (B51, L12), anti-Bp145 antibodies (B41, B35), anti-Bp135 antibodies (B25, B31, B40, B49), anti-Bp95 antibodies (B34, B43, B14), and anti-pTF$_r$ antibody (L20).

listed as a cluster group (Table 12.1) according to previous practice (3). Individual antibodies reactive with distinct epitopes but similar-sized antigens which may or may not be the same antigen are listed in Table 12.2. In this paper, space does not permit detailing the cross-blocking data. Key results are summarized in Tables 12.3 and 12.4. In every case, except the partial blocking examples noted, at least two blocking tests were performed and the results were clear and unequivocal. Of course, blocking of the binding of one monoclonal antibody by another does not necessarily mean they recognize the same epitope; phylogenetic or functional studies can further distinguish closely associated epitopes (7).

Leukemia-Associated Panel

The CD9 Group

In the previous workshop, a group of mAbs reacted with a 24-Kd polypeptide, p24, first described by Kersey *et al.* (8). In this study, a group of six antibodies reacted with the p24 antigen (L4, L16, L18, L22, B38, and B48), but also precipitated a second 33-Kd band (Fig. 12.1). One antibody, B38, did not precipitate clearly defined antigen but was placed in the group based on its ability to block the binding of fluorescein-conjugated L18 antibody. As in the previous workshop (1) three epitope groups

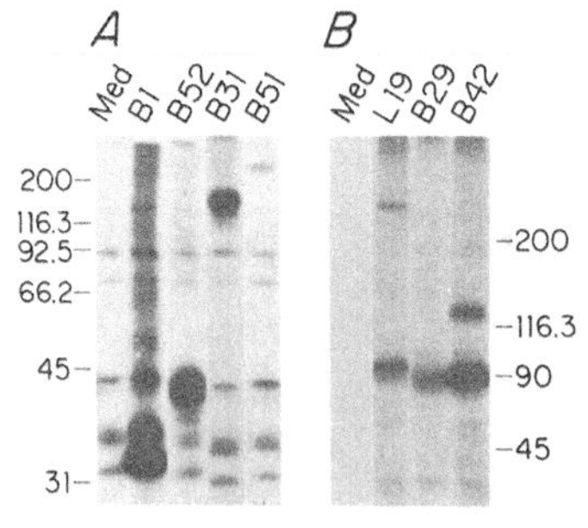

Fig. 12.3. Radioimmunoprecipitation of miscellaneous B cell antigens not forming clear cluster groups. (A) Anti-DR (B1), anti-Bp45 (B52), anti-Bp135 (B31), and p220 (B51) on Daudi cells; (B) Anti-transferrin receptor (L19), anti-p85 (B29), and Bp85/μ chain (B42) antigens on Daudi cells.

Table 12.1. Classification of major B cell/leukemia workshop antibody groups based on serologic, biochemical, and cross-blocking analyses.

Group	Antigen (M.W.)	Workshop antibodies	Cross-blocking subgroups	Comments
1. CD9	p33/24 (33,000/24,000)	L4(J30), L16(AL6), L18(J2), L22(CLB-THROMB-2), B38(21D10), B48(HB-9)	(1) L4, L18, B38 (2) L16, L22 (3) B48	B cell *associated*
2. CD10	p100 (100,000)	L2/L11(J5), L6(J13), L10(W8E7), L14(AL2), L21(CLB-cALLA-1)	(1) L10, L14, 24.1 (2) L2, L6, L11, L21	CALLA, B cell *associated*
3. TF_r	p90 (90,000)	L3(L22), L13(L01.1), L19(A-2), L20(E-20)	None	Transferrin receptor, B cell *associated*
4. CD21 ($C3d_r$)	Bp145 (145,000)	B9/B33(B2), B35(BL13), B41(HB-5)	(1) B9, B33, B35 (2) B41	C3d receptor/EBV receptor
5. CD22	Bp135 (135,000)	B7(29-110), B31(HH1), B25(HD6), B40(SJ10), B49 (Leu 14)	(1) B7, B31, B40, B49 (2) B25	Pan B cell antigen
6. CD19	Bp95 (95,000)	B8(AB1), B14/B34(B4), B28(HD37), B43(Leu 12), L17(SJ25C.1)	None	Pan B/pre–B cell *specific* antigen
7. CD20	Bp35 (32,000)	B5(B1), B21(5-HLL-2), B22(2H7), B24(1F5)	None	Pan B cell *specific* antigen
8. ____	p____?	B4(HH1), B17(HD28), B36(BL14)	None	Pan B cell

Table 12.2. Other B cell-associated antigens detected by B cell/leukemia workshop panel.

Antigen	Workshop antibodies	Blocking mAb	Blocking	Comments
p33/38	B1(Ia), B2(BII-62), B3(29-132), L8(7.2)	B1	None	DR-like
Bp37	B23(BB-1)	B23	None	B activation antigen
Bp45 (CD23)	B52(41-H16), B11(MHM6), B37(BLAST-1), B39(BLAST-2)	B52	None	B activation antigens, more than one?
p55	B32(9BA5)	B32	B30?	B and hematopoietic cells
p85	B29(H616)	B29	None	B cells/PMN
Bp85	B42(B7)	B42	None	B cells, IgM?
p220	B51(HB-11), L12(9.4), T104(3AC5)	B51, T104	None	Common leukocyte
Plasma	B20(PC-1), B26(PCA-1)	—	—	Plasma cell
Unknown	L1(SJ9-2E2), L9(6-4), L15(AL3)	—	—	Leukemia
Unknown	B6(NU-B1), B10(RW35), B13(5J12), B15(ALIA), B19(TL-13), B27(2-7), B36(HHI), B44(HB-6), B47(HB-8), B50(HB-10)			B cell

were evident based on cross-blocking studies: L16 and L22, unlike L4 and B38, only partially blocked fluorescein-L18 antibody; B48 precipitated identical polypeptides but did not block L18.

The CD10/CALLA Group

The p100 CALLA antigen was precipitated by five different antibodies in the Workshop (L2/L11, L6, L10, L14, and L21). Examples in Fig. 12.1 show p100 migrating slightly higher than the p90 transferrin receptor. Unlike the previous workshop where only one epitope on CALLA was detected by five antibodies (1), the current panel apparently detects two different epitopes. The somewhat surprising results are outlined in Table 12.3. Whereas the anti-CALLA mAb 24.1 blocked all antibodies in the

Table 12.3. Cross-blocking of Anti-CD10 (CALLA) specific monoclonal antibodies.

	Blocking of anti-CALLA antibody			
Test mAbs	24.1	L2	L6	L10
24.1, L10, L14	+	+	+	+
L2, L6, L11, L21	−	+	+	+
Control L17	−	−	−	−

able 12.4. Cross-blocking of B cell-associated mAbs.

	Blocking of fluorescein-conjugated mAbs (antigen group)							
Antibody groups	B35 (Bp145)	B49 (Bp135)	B14 (Bp95)	B24 (Bp32)	B36 (p?)	B52 (Bp45)	B42 (p85)	B29 (p85)
9, B33, B35	+	−	−					−
41	−	−	−					−
7, B31, B40, B49	−	+	−					−
25	−	−	−					−
8, B14, B28, B34, B43, L17		−	+					−
5, B21, B22, B24			−	+	−	−		−
4, B17, B36			−	−	+	−		−
52			−	−	−	+	−	−
37–B39, B44–B47, B50			−			−	−	−
42			−			−	+	−
29			−	−		−	−	+

panel, not all of the other mAbs blocked 24.1 even though they did block L2 and L21. The cause for this "one-way" blocking is not clear, but this phenomenon has been observed for T cell antigens (9). A point of emphasis—as the reader may appreciate, the CALLA antigen is not B cell specific or even restricted to hematopoietic cells.

The p90 Transferrin Receptor (TF_r)

Four distinct antibodies react with the p90 transferrin receptor (L3, L13, L19, and L20) which in its reduced form migrates on gels as a characteristic 180-Kd dimer. All four antibodies blocked the binding of fluorescein-conjugated L20.

B Cell-Associated Panel

The remaining major antigen cluster groups appeared to be B cell specific based on our limited serologic studies of peripheral blood and tonsillar leukocytes.

The Bp145, C3d, or CR2 Receptor (CD21)

Three antibodies (B9/B33, B35, and B41) in the Workshop panel reacted with a 145-Kd polypeptide distinct from a slower migrating 135-Kd molecule, Bp135 (Fig. 12.2). To confirm that Bp145 and Bp135 were distinct, cross-blocking studies were performed: none of the Bp135 specific antibodies blocked fluorescein-conjugated B35 and reciprocally none of the anti-Bp145 specific antibodies blocked fluorescein-conjugated B49 (Table 12.4). The B41 antibody (HB-5) did not block the binding of B35 or B9. This is of particular interest since Iida and coworkers (10) have found that

Bp145 is the receptor for the C3d complement component. Recently, using the HB-5 mAb, Fingeroth *et al.* (11) reported that Bp145 is also the receptor for Epstein–Barr virus. Similar results have been reported concurrently by Nadler *et al.* (this volume, Chapter 44) using their mAb, B2. However, unlike HB-5, the B2 antibody alone blocks EBV binding, suggesting that the epitopes detected by B2 and HB-5 may be functionally distinct.

The Bp135 Pan B Cell Antigen (CD22)

Five mAbs in the Workshop reacted with a 135-Kd B cell antigen (B7, B25, B31, B40, and B49). Four antibodies blocked the binding of B49, while one (B25) did not. Once again, two distinct epitopes apparently are being detected on this new interesting molecule (Fig. 12.2).

The Bp95 Pan B Cell Antigen (CD19)

Five different mAbs reacted with a structure of 95 Kd. Antigens precipitated by three of these antibodies are shown together in Fig. 12.2. This antigen was first described by Nadler *et al.* using antibody B4 (12). All of the antibodies blocked the binding of the Workshop antibody B14 (B4) (Table 12.4).

The Bp35 Pan B Cell Antigen (CD20)

Three antibodies in the Workshop precipitated a 35-Kd polypeptide (B5, B22, and B24) and a fourth antibody (B21) blocked fluorescein-conjugates of the other three (Table 12.4). Fluorescein-conjugates of B5, B22, and B24 all are blocked by unconjugated antibodies in the group. This antigen too was first described by Nadler and coworkers using the antibody B1 (13). Recently, we have discovered that Bp35 may play an important role in B cell activation (14): monoclonal anti-Bp35 antibody 1F5 stimulates proliferation of resting tonsillar B cells. The characteristics of this induction are described in Chapter 38 of this volume. Bp35 was previously called "Bp32" prior to this workshop.

A Biochemically Undefined Cluster Group

Three monoclonal antibodies blocked the binding of fluorescein-conjugated B36 (B4, B17, and B36) and thus recognize the same molecule (Table 12.4). Tissue and cell line distribution were similar to that of the Bp35 antigen, but two-color analyses clearly indicated the "B36" antigen is not Bp35. Our biochemical results were equivocal.

Other B Cell-Associated Antigens Not Forming Clear Cluster Groups

As usual several anti-DR-like antibodies made their way into the panel and will not be discussed further [Fig. 12.3(A), Table 12.2). Several antibodies have been listed, based on prior publications as blast-associated antigens (15–17). One antibody, B52, precipitated a 45-Kd polypeptide [Fig. 12.3(A)] which is similar to that reported for the B11 (MHM6) and the B37 (B-LAST1) and B39 (BLAST-2) antigens (16,17). However, B11, B37, and B39 did not precipitate detectable polypeptides; nor did they block the binding of fluorescein-conjugates of B52 or B23. The B23 (BB-1) and B32 (9BA5) are antibodies we developed and react with 37-Kd and 55-Kd polypeptides, respectively (5). However, with the Workshop reagents, we could not detect these polypeptides.

Two antibodies reacted with polypeptides of approximately 85 Kd [Fig. 12.3(B)]. However, two-color analyses of these antibodies gave distinct results: B42 reacted strongly with Bp35^{dull} resting B cells while B29 (H616) reacted strongly with Bp35^{bright} B cell blasts [(5), and data not shown]. The B42 stains in a similar fashion to our mAb to μ chain.

Three antibodies reacted with a 220-Kd member of the T200 family on B cell lines [Fig. 12.3(B)]: B51, L12, and our antibody T104 (3AC5) (5). The epitopes detected by these three antibodies are different and the tissue distribution of the antigens are distinct. For example, 9.4 reacts with all B cells and T cells while 3AC5 reacts with all B cells and only some T cells (5).

The remaining antibodies in the Workshop neither precipitated defined antigens in our hands nor blocked any of our available fluorescein-conjugated antibodies. Some antibodies, e.g., B20, B26, only reacted with myeloma lines, and immunochemical studies were not pursued.

Two-Color Flow Cytometric Analyses

Using fluorescein-conjugated B cell-associated antibodies for green staining and phycoerythrin (PE)-conjugated anti-Bp35 or anti-HLA-DR antibodies for red staining, we have defined two distinct B cell subpopulations in tonsils (6). The two-color data are presented on a three-dimensional plot of cell number versus log green fluorescence (left) versus log red fluorescence (right) (Fig. 12.4, top left). About every 4–5 dots represents a doubling of fluorescence. Staining of tonsillar lymphocytes with PE-conjugated (red) anti-Bp35 reveals three classes of cells (Fig. 12.4, top right): Bp35^{-}, Bp35^{dull}, and Bp35^{bright}. The Bp35^{dull} cells are IgM^{bright} (Fig. 12.4, middle left) dense resting B cells (6), while the Bp35^{bright} cells are IgM$^{dull/neg}$ and are more buoyant activated cells. Because direct conjugates are used, the relative fluorescence is a good indicator of antigen density.

Using the fluorescein-conjugates of the Workshop mAbs, we examined the expression of B cell antigens on the Bp35^{dull} dense (mantle zone) tonsillar cells and Bp35^{bright} preactivated (germinal center) cells (6). Several distinct patterns were evident. The first "IgM-like" pattern (Fig. 12.4, middle), where the antigen is strongly expressed on Bp35^{dull} cells and weakly or not at all on Bp35^{bright} cells, was seen for B52 (Bp45), B23

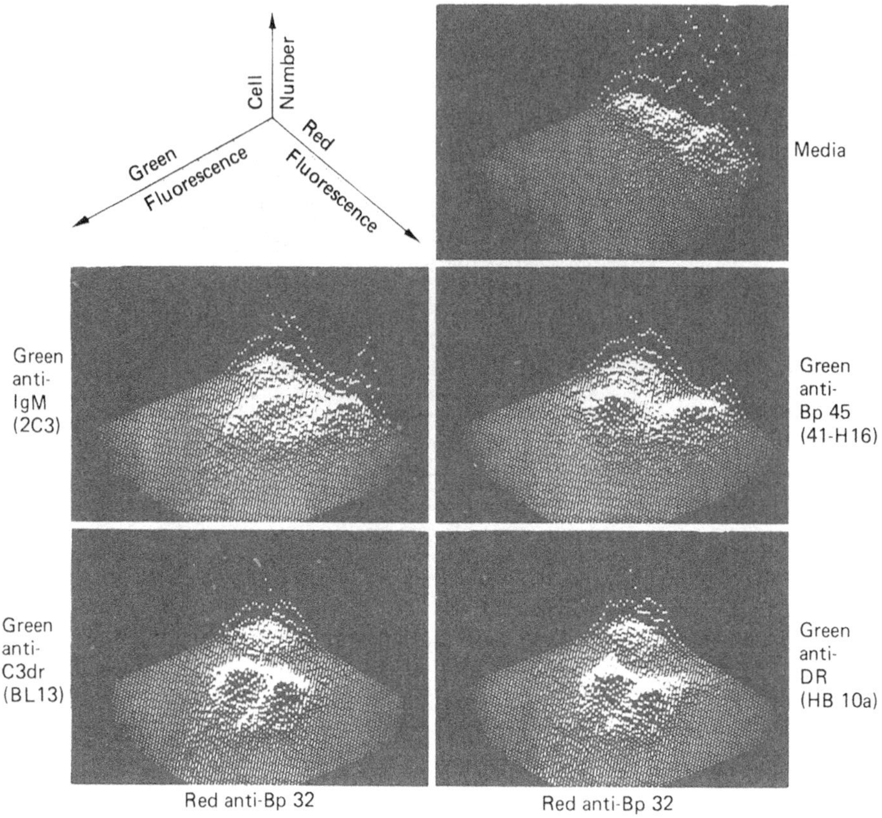

Fig. 12.4. Expression of B cell Workshop antigens on tonsillar B cell subsets using two-color immunofluorescence analysis. Upper left panel shows how data are plotted: cell number (vertical) versus log green fluorescence (left) versus log red fluorescence (right). All samples were stained with PE-conjugated (red) anti-Bp32. IgM (2C3) and Bp45 (41-H16) were at higher density on $Bp32^{dull}$ cells but not $Bp32^{bright}$ cells (middle). The $C3d_r$ (BL13) and HLA-DR (HB10a) antigens generally were expressed at higher levels on $Bp32^{dull}$ cells than on $Bp32^{bright}$ cells (bottom).

(Bp37), and B42 (p85). This was surprising since two of the antigens, Bp45 and Bp37, are thought to be B cell activation antigens.

A second "DR-like" pattern, where the antigen is expressed at high density but at somewhat lower level on $Bp35^{bright}$ cells (Fig. 12.4, bottom), was detected for Bp145 $C3d_r$ specific antibodies and DR/DQ specific antibodies. Thus, dense tonsillar B cells in secondary follicles express high levels of DR antigens, suggesting they may be distinct from primary resting or circulating B cells (18). The presence of Bp45 and Bp37 on these dense cells also suggests they are not typical resting cells. Could they be memory B cells?

A third pattern, where the antigen was expressed equally on both populations, was seen for Bp135 and Bp95 antigens (Fig. 12.5, top). Finally, the fourth pattern, where the Bp35^{bright} population expressed more of the antigen than Bp35^{dull}, was evident for B29 (H616) and B36 (Fig. 12.5, middle). The CD9 and TF_r specific antibodies showed some increased binding on the Bp35^{bright} cells but the results were not very dramatic (Fig.

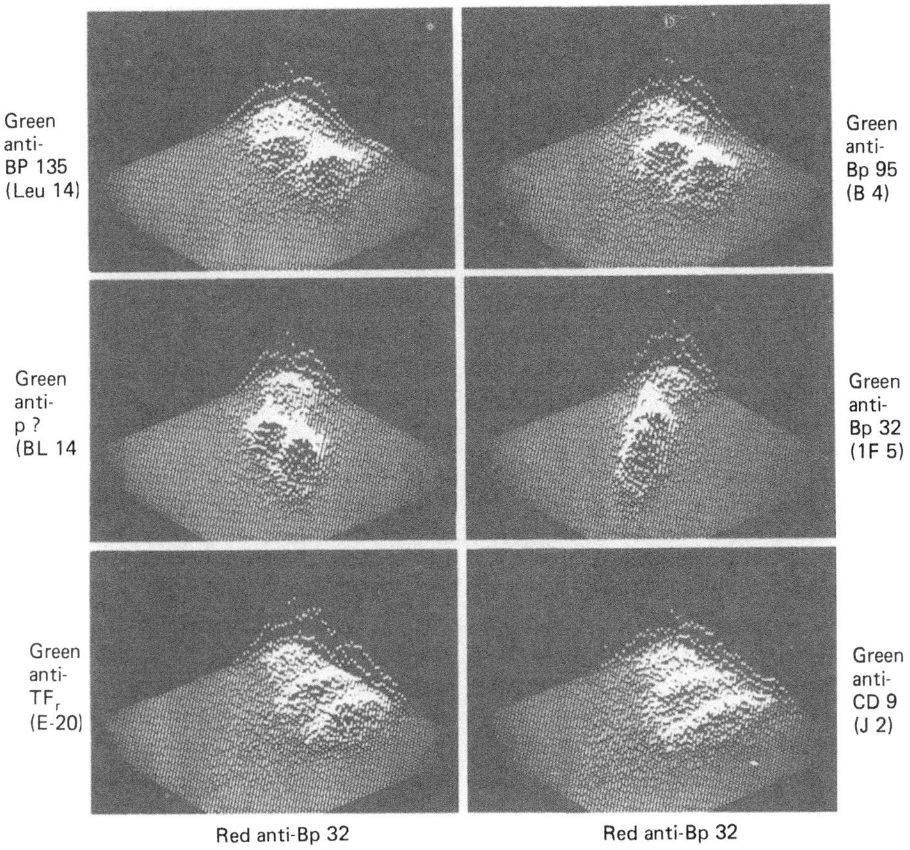

Fig. 12.5. Expression of B cell and leukemia Workshop antigens on tonsillar B cell subsets using two-color immunofluorescence analysis; see Fig. 12.4 for details. The Bp135 (Leu-14) and Bp95 (B4) antigens are expressed equally on dense and buoyant tonsillar B cells (top). The antigen detected by BL13 is expressed at slightly higher levels on Bp35^{bright} cells (middle left), while anti-Bp35 versus a second anti-Bp35 antibody (1F5) shows identical expression. The TF receptor (E-20) and CD9 p33/24 antigen (J2) are expressed at somewhat higher levels on Bp35^{bright} cells.

12.5, bottom). These results may reflect *in vivo* changes in membrane antigen phenotype occurring during B cell activation in secondary follicles and it is important to compare them to changes seen with *in vitro* activation (see, e.g., this volume, Chapter 36).

Also shown in these figures is red-anti-Bp35 (2H7) versus green-anti-Bp35 (1F5) as an example of a pattern of identical or coordinate expression (Fig. 12.4). None of the antigens tested showed coordinate expression with either Bp35 or HLA-DR (data not shown).

In summary, now that the structure of a family B cell-specific and B cell-associated antigens have been defined, distinct epitopes have been identified, and expression of the antigens on B cell subsets is known, the stage is set for investigating the function and coordinate interactions of these surface molecules. The function of the Bp35 and Bp145 antigens are now being actively studied, and we can look forward to the next workshop where much will have been learned about the function and molecular biology of these molecules.

Acknowledgment. This work was supported in part by Genetic Systems Corporation and by NIH grant CA34199 (CA39935).

References

1. Clark, E.A., and T. Yokochi. 1984. Human B cell and B cell blast-associated surface molecules defined with monoclonal antibodies. In: *Leucocyte typing,* A. Bernard, L. Boumsell, J., Dausset, C. Milstein, and S.F. Schlossman, eds. Springer-Verlag, Berlin, Heidelberg, p. 339–346.
2. LeBien, T.W., J.G. Bradley, D.R. Bone, J.L. Platt, A.F. Michael, and J.H. Kersey. 1984. B cells and kidneys: A "B + CALLA" workshop analysis. In: *Leucocyte typing,* A. Bernard, L. Boumsell, J. Dausset, C. Milstein, and S.F. Schlossman, eds. Springer-Verlag, Berlin, Heidelberg, p. 346–353.
3. Bernard, A., L. Boumsell. 1984. B2 protocol summary. In: *Leucocyte typing,* A. Bernard, L. Boumsell, J. Dausset, C. Milstein, and S.F. Schlossman, eds. Springer-Verlag, Berlin, Heidelberg, p. 61–81.
4. Goding, J.W. 1976. Conjugation of antibodies with fluorochromes: Modification to the standard methods. *J. Immunol. Methods* **13:**215.
5. Clark, E.A., J.A. Ledbetter, P.A. Dindorf, R.C. Holly, and G. Shu. 1985. Polypeptides on human B cells associated with cell activation. *Human Immunol.,* in press.
6. Ledbetter, J.A., and E.A. Clark. 1985. Surface phenotype and function of tonsillar germinal center and mantle zone B cell subsets. *Human Immunol.,* in press.
7. Clark, E.A., Martin, P.J., Hansen, J.A., and Ledbetter, J.A. 1983. Evolution of epitopes on human and nonhuman primate lymphocyte cell surface antigens. *Immunogen.* **18:**599.
8. Kersey, J.H., T.W. LeBien, C.S. Abramson, R. Newman, R. Sutherland, and M. Greaves. 1981. p24: A human leukemia-associated and lymphohemopoietic progenitor cell surface structure identified with monoclonal antibody. *J. Exp. Med.* **153:**726.

9. Martin, P.J., J.A. Ledbetter, E.A. Clark, P.G. Beatty, and J.A. Hansen. 1984. Epitope mapping of the human surface suppressor/cytotoxic T cell molecule Tp32. *J. Immunol.* **132:**759.
10. Iida, K., L. Nadler, and V. Nussenzweig. 1983. Identification of the membrane receptor for the complement fragment C3d by means of a monoclonal antibody. *J. Exp. Med.* **158:**1021.
11. Fingeroth, J.D., J.J. Weis, T.F. Tedder, J.L. Strominger, P.A. Biro, and D.T. Fearon. 1984. Epstein–Barr virus receptor of human B lymphocytes in the C3d receptor CR2. *Proc. Natl. Acad. Sci. U.S.A.* **81:**4510.
12. Nadler, L.M., K.C. Anderson, G. Marti, M. Bates, E. Park, J.F. Daley, and S.F. Schlossman. 1983. B4, a human B lymphocyte-associated antigen expressed on normal, mitogen-activated, and malignant B lymphocytes. *J. Immunol.* **131:**244.
13. Stashenko, P., L.M. Nadler, R. Hardy, and S.F. Schlossman. 1980. Characterization of a human B cell-specific antigen. *J. Immunol.* **125:**1678.
14. Clark, E.A., G. Shu, and J.A. Ledbetter. 1985. Role of Bp35 cell surface polypeptide in human B cell activation. *Proc. Nat. Acad. Sci. USA* **82:**1766.
15. Yokochi, T., R.D. Holly, and E.A. Clark, 1982. B lymphoblast antigen (BB-1) expressed on Epstein–Barr virus-activated B cell blasts, B lymphoblastoid cell lines, and Burkitt's lymphomas. *J. Immunol.* **128:**823.
16. Rowe, M., E.K. Hildreth, A.B. Rickinson, and M.A. Epstein. 1982. Monoclonal antibodies to Epstein–Barr virus-induced transformation-associated cell surface antigens: Binding patterns and effect upon virus specific T-cell cytotoxicity. *Int. J. Cancer* **29:**373.
17. Thorley-Lawsen, D.A., R.T. Schooley, A.K. Bhan, and L.M. Nadler. 1982. Epstein–Barr virus superinduces a new human B cell differentiation antigen (B-LAST1) expressed on transformed lymphoblasts. *Cell* **30:**415.
18. Ledbetter, J.A., P.J. Martin, and E.A. Clark. 1985. Mantle zone and germinal center B cells respond to different activation signals. *Proc. Sixteenth Leukocyte Culture Conf.* in press.

CHAPTER 13

Structural Analysis of Cell Surface Molecules Recognized by Leukemic Cell/B Cell Panel Antibodies

Tucker W. LeBien, J. Garrett Bradley, Jeffrey L. Platt, and Samuel J. Pirruccello

Introduction

During our participation in the First International Workshop on Human Leucocyte Differentiation Antigens we focused our efforts on a serological and immunochemical analysis of the "B + CALLA" Workshop antibodies (1). We herein describe similar studies conducted under the aegis of the Second International Workshop.

Materials and Methods

All the methods employed in this study have been previously described in reports from this laboratory. These include analysis of monoclonal antibody binding to leukemic cells (2), analysis of monoclonal antibody binding to renal parenchyma (3), and immunochemical characterization of cell surface molecules by radioimmunoprecipitation (RIP) and sodium dodecyl sulfate–polyacrylamide gel electrophoresis (SDS–PAGE) (4).

Results

Leukemia Cell Panel of Monoclonal Antibodies

We screened these antibodies for reactivity with the pre-B acute lymphoblastic leukemia (ALL) cell line NALM-6 by indirect immunofluorescence, and found 16/21 to be reactive. The antibodies that did not react included L1, L7, L9, L12, and L15. The 16 reactive antibodies were then used to immunoprecipitate ^{125}I-labeled NALM-6 NP-40 lysates, and Fig. 13.1 shows the results obtained with 13 of them. Five antibodies (L2, L10, L11, L14, L21) recognized a 100-kilodalton (Kd) protein, four antibodies

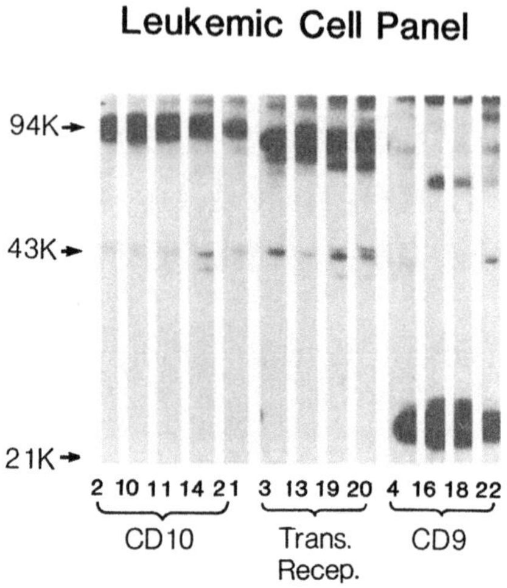

Fig. 13.1. SDS–PAGE and autoradiograph of ^{125}I-labeled NALM-6 cell lysates immunoprecipitated with leukemic cell panel antibodies.

(L3, L13, L19, L20) recognized a 90-Kd protein, and four antibodies (L4, L16, L18, L22) recognized a 24-Kd protein with a minor component of 26 Kd. The 26-Kd component was not an artifact in this particular gel and its structural relationship to p24 has been reported elsewhere (5). Coupled with serologic data, it was apparent that the 100-Kd protein is the common acute lymphoblastic leukemia antigen (CALLA) designated in the First Workshop as CD10 (nT-nB, p100), and the p24 molecule is CD9 (nT-nB, p24). Analysis of the immunoprecipitates obtained with antibodies L3, L13, L19, and L20 revealed the presence of a 180-Kd protein under nonreducing conditions, and a 90-Kd protein under reducing conditions (Fig. 13.2). This migration profile is consistent with the reported structure of the transferrin receptor (6,7). Of the other three antibodies that bound to NALM-6, L8 precipitated a 29/34-Kd complex consistent with HLA-DR, but no precipitates were obtained with L6 or L17.

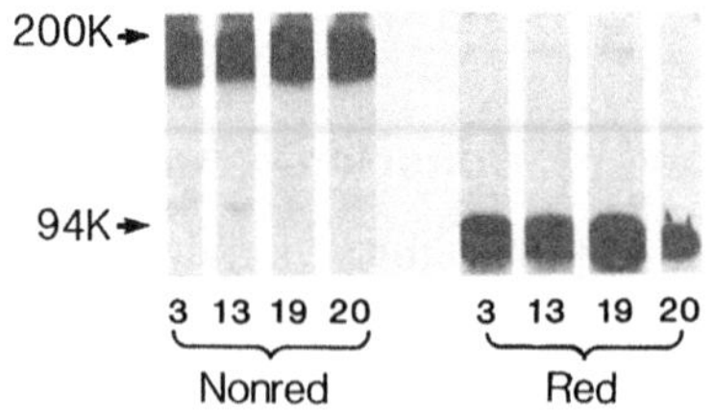

Fig. 13.2. SDS–PAGE and autoradiograph of ^{125}I-labeled NALM-6 cell lysates immunoprecipitated with leukemic cell panel antibodies L3, L13, L19, and L20. Nonred = no 2-ME, red = with 2-ME.

Although not conforming to our serologic and structural criteria, both antibodies L6 (J13, Ref. 8) and L15 (AL3, Ref. 9) recognize CALLA. J13 and AL3 are mouse and rat IgM monoclonals, respectively, and may not have been recognized by our goat anti-mouse Ig reagents.

B Cell Panel of Monoclonal Antibodies

This panel of 52 antibodies represented a rather formidable challenge and we chose to focus our efforts on those antibodies that bound to the Raji cell line. In preliminary screening by indirect immunofluorescence we found that 33/52 B cell antibodies showed demonstrable binding to Raji cells. Two of these antibodies (B47 and B48) were shown to recognize the gp45/55/65 complex defined by monoclonal antibody BA-1, and are discussed in more detail elsewhere (this volume, Chapter 18). The other 31 antibodies were tested by RIP and SDS–PAGE using ^{125}I-labeled Raji cell NP-40 lysates. The results in Fig. 13.3 show the identification of five distinct B cell surface molecules identified by groups of monoclonal antibodies. The antibodies shown in this figure each precipitated their respective antigens in at least two independent experiments. The 220-Kd molecule recognized by antibodies B10, B50, and B51, and designated B220, is probably identical to the high-molecular-mass B cell form of the common leukocyte antigen previously defined with monoclonal antibody F8-11-13 (10). The 140-Kd molecule recognized by antibodies B9, B33, B35, and B41 is the B2 molecule originally defined with anti-B2 (11,12). Workshop

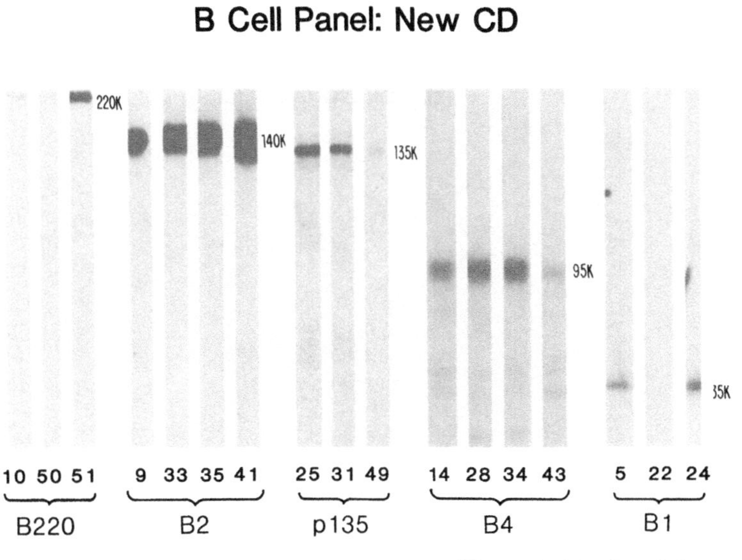

Fig. 13.3. SDS–PAGE and autoradiograph of ^{125}I-labeled Raji cell lysates immunoprecipitated with B cell panel antibodies.

antibodies B9 and B33 were duplicates of anti-B2. A panel of three antibodies (B25, B31, B49) precipitated a molecule, designated p135, that runs slightly faster than B2. This molecule has, to our knowledge, not been previously described. Antibodies B14, B28, B34, and B43 precipitated a 95-Kd molecule. Workshop antibodies B14 and B34 were duplicates of anti-B4 (14). Finally, a 35-Kd molecule was precipitated by antibodies B5, B22, and B24. In this experiment the B22 antibody gives a poorly resolved precipitate. Workshop antibody B5 was anti-B1.

In addition to the potentially new clusters defined in Fig. 13.3, several other antibodies which did not fall into convenient clusters are shown in Fig. 13.4. Workshop antibodies B1 and B23 both precipitated a 29, 34-Kd complex, reminiscent of Ia molecules. Other antibodies and molecules recognized were B8, 80 Kd; B17, 45/140 Kd; B29, 90 Kd; B42, 70 Kd; and B52, 45 Kd doublet. Workshop antibody B52 was 41.H16, and the molecular mass shown in Fig. 13.4 is similar to that reported by Zipf *et al.* (14).

Antibodies tested that did not yield resolvable immunoprecipitates using ^{125}I-labeled Raji cells included B4, B12, B27, B36, B44, B45, and B46. These antibodies were subsequently tested using [^{35}S]methionine-labeled Raji cells and no specific immunoprecipitates were seen. Explanations for our inability to precipitate cell surface molecules recognized by these monoclonal antibodies include: 1) low-titer/low-affinity antibody, 2) a cell surface protein with inaccessible tyrosine or low methionine content, or 3) a glycolipid antigen.

We (3) and others (15) have previously demonstrated that a number of monoclonal antibodies recognizing antigens on human B lineage cells also bind to renal parenchyma. We therefore tested each of the antibodies

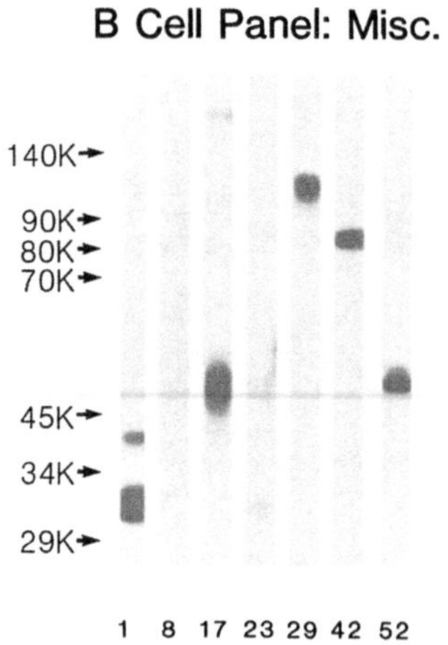

Fig. 13.4. SDS–PAGE and autoradiograph of ^{125}I-labeled Raji cell lysates immunoprecipitated with B cell panel antibodies.

Table 13.1. Reactivity of mature and fetal kidney with B cell workshop antibodies.

	Mature		Fetal	
Cluster	Glomerulus	Tubules	Glomerulus	Tubules
B220		+[a]	+	+
B2	±[b]	+		+
p135		+/−[c]		+/−[c]
B4		+	+	+
B1	±	±		+

[a] Distinct pattern of reactivity observed.
[b] Some antibodies react, but reactivity is ill-defined.
[c] Distinct pattern of reactivity observed with 1 of 3 antibodies.

shown in Fig. 13.3 for reactivity with frozen sections of normal mature and fetal kidney using indirect immunofluorescence. The results of these studies are summarized in Table 13.1. Antibodies from all five clusters showed some degree of reactivity with mature or fetal (21 week) kidney. Most notably, fetal tubular epithelial cells were positive with all five groups. We should emphasize that data in Table 13.1 represent conservative interpretations, and, since the antibodies were not rigorously titered, we may have missed some reactivity patterns (no symbol).

Discussion

As evidenced by the size of this volume, there continues to be tremendous interest in studying the surface of human lympho-hematopoietic cells with monoclonal antibodies. It seems appropriate to briefly discuss our data from the standpoint of the two antibody panels under study.

Examination of the antibodies in the leukemia cell panel which recognize cell surface molecules expressed on NALM-6 cells indicates we have, in great part, recapitulated the past (1). Fifteen of the 21 antibodies recognize p24 (CD9), CALLA (CD10), or the transferrin receptor. Obviously, there are other cell surface molecules expressed on NALM-6 (and other pre-B ALL) such as Class I, Class II, B4 (13), and the gp45/55/65 molecular complex recognized by BA-1 (16). It is apparent, however, that conventional mouse immunizations with intact B cell precursor ALL cells (i.e., those with rearranged heavy- and light-chain genes) are not yielding antibodies recognizing "new" specificities. The immunodominance of some of the aforementioned molecules, particularly Class II, undoubtedly represents a major obstacle.

In contrast to the leukemia cell panel, the B cell panel has revealed the presence of several published and unpublished specificities, at least six of which will form the basis for new clusters of differentiation. Our own data (Fig. 13.3 and this volume, Chapter 18) identify B1, B2, B4, p135, B220, and gp45/55/65, and we only studied about 60% of the total B cell panel. Furthermore, as shown in Fig. 13.4, there are at least five other B cell-associated antigens that we identified with a single antibody. Taken together, it appears that the genome of a human B lymphocyte encodes a *minimum* of 11 cell surface proteins. To no surprise, the major challenge before us is to unravel the function these cell surface molecules subserve to B lymphocytes. As discussed elsewhere in this volume, this area is under active investigation.

As previously reported with monoclonal antibodies BA-1, BA-2, and BA-3 (3), the epitopes (? molecules) recognized by B cell Workshop antibodies are highly conserved on human renal parenchyma (Table 13.1). This relationship continues to defy explanation, but may hold clues for unraveling the function of these molecules.

Summary

We have characterized cell surface molecules recognized by the leukemic cell/B cell panel of monoclonal antibodies. We found that the leukemic cell panel recapitulates, in great part, the results of the First Workshop (predominant specificities = CD9, CD10, transferrin receptor). A *minimum* of six new CD groups will emerge from the B cell panel, including: B220, B2, p135, B4, B1, and gp45/55/65.

Acknowledgments. This work was supported by grants CA-31685 and RR-05385 from the National Institutes of Health. T.W. LeBien is a Scholar of the Leukemia Society of America. J.L. Platt is supported by a Clinician Scientist Award from the American Heart Association.

References

1. LeBien, T.W., J.G. Bradley, D.R. Boué, J.L. Platt, A.F. Michael, and J.H. Kersey. 1984. B cells and kidneys: A "B + CALLA" workshop analysis. In: *Leucocyte typing,* A. Bernard, L. Boumsell, J. Dausset, C. Milstein, and S.F. Schlossman, eds. Springer-Verlag, Berlin, Heidelberg, pp. 346–353.
2. Abramson, C., J.H. Kersey, and T.W. LeBien. 1981. A monoclonal antibody (BA-1) primarily reactive with cells of human B lymphocyte lineage. *J. Immunol.* **126:**83.
3. Platt, J.L., T.W. LeBien, and A.F. Michael. 1983. Stages of renal ontogenesis identified by monoclonal antibodies reactive with lymphohematopoietic differentiation antigens. *J. Exp. Med.* **157:**155.
4. LeBien, T.W., D.R. Boué, J.G. Bradley, and J.H. Kersey. 1982. Antibody affinity may influence antigenic modulation of the common acute lymphoblastic leukemia antigen *in vitro. J. Immunol.* **129:**2287.
5. LeBien, T.W., S.J. Pirruccello, R.T. McCormack, and J.G. Bradley. 1985. p24 and p26, structurally related cell surface molecule identified by monoclonal antibody BA-2. *Molecular Immunol.,* in press.

6. Sutherland, R., D. Delia, C. Schneider, R. Newman, J. Kemshead, and M. Greaves. 1981. Ubiquitous cell-surface glycoprotein on tumor cells is proliferation-associated receptor for transferrin. *Proc. Natl. Acad. Sci. U.S.A.* **78:**4515.
7. Trowbridge, I.S., and M.B. Omary. 1981. Human cell surface glycoprotein related to cell proliferation is the receptor for transferrin. *Proc. Natl. Acad. Sci. U.S.A.* **78:**3039.
8. Pesando, J.M., K.J. Tomaselli, H. Lazarus, and S.F. Schlossman. 1983. Distribution and modulation of a human leukemia-associated antigen (CALLA). *J. Immunol.* **131:**2038.
9. Lebacq-Verheyden, A.-M., A.-M. Ravoet, H. Bozin, D.R. Sutherland, N. Tidman, and M.F. Greaves. 1983. Rat AL2, AL3, AL4, and AL5 monoclonal antibodies bind to the common acute lymphoblastic leukemia antigen (CALLA gp100). *Int. J. Cancer* **32:**273.
10. Dalchau, R., and J.W. Fabre. 1981. Identification with a monoclonal antibody of a predominantly B lymphocyte-specific determinant of the human leukocyte common antigen. *J. Exp. Med.* **153:**753.
11. Nadler, L.M., P. Stashenko, R. Hardy, A. van Agthoven, C. Terhorst, and S.F. Schlossman. 1981. Characterization of a human B cell-specific antigen (B2) distinct from B1. *J. Immunol.* **126:**1941.
12. Oettgen, H.C., P.J. Bayard, W.V. Ewijk, L.M. Nadler, and C.P. Terhorst. 1983. Further biochemical studies of the human B-cell differentiation antigens B1 and B2. *Hybridoma* **2:**17.
13. Nadler, L.M., K.C. Anderson, G. Marti, M. Bates, E. Park, J.F. Daley, and S.F. Schlossman. 1983. B4, a human B lymphocyte-associated antigen expressed on normal mitogen-activated, and malignant B lymphocytes. *J. Immunol.* **131:**244.
14. Zipf, T.F., G.J. Lauzon, and B.M. Longenecker. 1983. A monoclonal antibody detecting a 39,000 M.W. molecule that is present on B lymphocytes and chronic lymphocytic leukemia cells but is rare on acute lymphocytic leukemia blasts. *J. Immunol.* **131:**3064.
15. Metzgar, R.S., M.J. Borowitz, N.H. Jones, and B.L. Dowell. 1981. Distribution of common acute lymphoblastic leukemia antigen in nonhematopoietic tissues. *J. Exp. Med.* **154:**1249.
16. Pirruccello, S.J., and T.W. LeBien. 1985. Monoclonal antibody BA-1 recognizes a novel human leukocyte cell surface sialoglycoprotein complex. *J. Immunol.* **134:**3962.

CHAPTER 14

Biochemical Analysis of Antigens Recognized by Workshop B Series Antibodies, Using "Western Blotting"

B.B. Cohen, Marion Moxley, Patricia Elder, K. Guy, and C. Michael Steel

Introduction

"Western blotting" is the name given to a technique whereby antigenic proteins are separated by sodium dodecyl sulfate–polyacrylamide gel electrophoresis (SDS–PAGE) (1), then transferred electrophoretically from the gel to a sheet of cellulose nitrate. The positions of the antigens, and hence their approximate molecular weights, can be determined by "probing" the sheet with the corresponding antibody which is either labeled itself (with a radioisotope, for example, or an enzyme) or which can be detected indirectly by a second (labeled) probe (2).

Not all determinants remain antigenically intact after solubilization and electrophoresis so that the distribution of epitopes detected by "Western blotting" may differ from that in the source material in its normal biological state. Nevertheless, the method is useful for the investigation of cell surface antigens by means of monoclonal antibodies and, by the same token, can contribute to the analysis of the precise specificities of the antibodies themselves (2–5).

Materials and Methods

Monoclonal Antibodies

All 52 antibodies of the Workshop B series were screened, as described below, and 16 were then selected for more extensive investigation. The samples supplied were diluted 1 : 20 or, in some later tests, 1 : 40 (to give final dilutions of 1 : 200 and 1 : 400) with RPMI 1640 culture medium containing 5% fetal calf serum (FCS) + 5% horse serum (HS). Positive control monoclonal antibodies were CR3.43 (anti-HLA-class II β chain), DA6.147 (anti-HLA-class II α chain), and PD7.26 (anti–leukocyte common antigen). CR3.43 and PD7.26 were generously supplied by Dr D.Y. Mason, Oxford. DA6.147 and the negative control antibody (mouse monoclonal anti-cortisol) were produced in the authors' laboratory. All control reagents were used as ascites fluids in the same dilutions as the Workshop antibodies.

Cell Preparations

The cultured lines and leukemic cells listed in Table 14.1 were used as sources of antigen.

Cells were lysed in NP-40 buffer as described elsewhere (3, 16) using 10^7 cells/ml. The lysates were clarified by centrifugation and adjusted for electrophoresis by adding 4× strength sample buffer (1). Some preparations were loaded directly onto polyacrylamide gels in 0.1% SDS, others received the conventional treatment of boiling in 2% SDS plus 5% mercaptoethanol to separate them into their constituent polypeptide chains. Further reference will be made to these alternative treatments in the Results and Discussion section.

Detection of Epitopes

Cellulose nitrate transfers of solubilized proteins separated by SDS–PAGE were prepared for probing with the Workshop antibodies as described previously (1–5). After soaking in 5% bovine serum albumin (BSA) to saturate nonspecific protein binding sites, the sheets were cut into strips, each of which was incubated with one of the Workshop or

Table 14.1. Cells from which lysates were prepared for "western blotting."[a]

B cells

1. B lymphoblastoid cell lines (EB virus-transformed). Pool of 15 lines from authors' laboratory.
2. Burkitt's lymphoma lines. Pool of 12, including EB_4 and Ramos which are EB virus-negative.
3. Myeloma lines. Pool of RPMI 8226 and U266 B1 (6,7).
4. B chronic lymphocytic leukemia. Blood lymphocytes from one patient (<5% E-rosettes) used both before and after 90-hr culture in 100 ng/ml TPA to induce differentiation and enhance antigen (8–10).

T cells

5. T lymphoid lines MOLT-4 and CCRF.CEM pooled (11,12).
6. T cell chronic lymphocytic leukemia. Blood lymphocytes from one patient. >95% E-rosettes, OKT4 + sero-negative for human T cell leukemia virus.

Null cells

7. K562 cell line from chronic myeloid leukemia (13).
8. Reh cell line from acute lymphoblastic leukemia (14).

[a] Phenotypic characteristics of the cell lines have been published in Ref. 15 and binding of Workshop B series antibodies to intact cells from all the above categories is reported in Chapter 4 of this volume.

control antibodies for a minimum of three hours, with constant agitation. The strips were washed and any bound monoclonal Ig was detected either directly by incubating in ^{125}I-labeled sheep anti–mouse Ig (NEN., used at 0.2 μCi/ml in 5% BSA) or indirectly with rabbit anti–mouse Ig (Miles Yeda) followed by [^{125}I]staphylococcal protein A (Amersham International, used at 0.1 μCi/ml in PBS).

The strips were reassembled and autoradiographed for 1–2 days. Appropriate radiolabeled molecular weight markers were also run in a small well alongside the cell lysates in each gel, transferred simultaneously to the cellulose nitrate sheets, and visualized by autoradiography.

Results and Discussion

Screening of the complete set of Workshop antibodies was carried out in two stages. In the first, the antigen preparations were "boiled" and "unboiled" lysates of pooled B lymphoblastoid lines. Boiling in SDS plus mercaptoethanol splits proteins into their individual polypeptide chains and also ensures that they migrate exactly according to their molecular weights. This procedure should therefore give precise size data on antigenic polypeptides. However the expression of some epitopes may depend upon the interaction of two or more peptide chains and would be lost when these are separated, while the antigenicity of other determinants may be destroyed by heat. Blotting onto unboiled lysates may thus reveal some antigenic bands not detected with conventionally treated preparations though their molecular weights cannot be determined so accurately since, in general, the mobility of proteins in the gel is affected to some degree by omission of the SDS/mercaptoethanol/boiling step.

At this stage, a band or bands could be detected on one or both autoradiographed blots from 19 of the antibodies. There was some correlation with the ability of the antibodies to bind to intact cells (15) in that 13 of the 25 which gave moderate to strong binding were positive on "Western blotting," compared with only six out of 27 "weak" or "negative" binders. However, many of the autoradiographed bands were very faint. For the second stage of the screen, therefore, boiled lysates from a range of cell types were used and individual monoclonals were incubated with transfers from one or more of the cell preparations, selection being guided by the results of the intact cell binding assay (15) as well as by the outcome of the first stage of the blotting screen. In both stages, bound monoclonal Ig was detected with ^{125}I-labeled sheep anti–mouse Ig. Combining the two stages now yielded positive results from 26 of the Workshop antibodies.

As expected, different antibodies showed selectivity for antigens in lysates from different cell types. Some, such as B15, B20, and B50 appeared to react with a large number of polypeptides in a single lysate while B10 and possibly B45 and B20 seemed to recognize antigens of different molecular weights in B and T cells.

In order to investigate these phenomena further, preparations from four sources (pooled B cell lines, B-CLL induced with TPA, Reh cells, and T-CLL) were run in adjacent parallel tracks in a series of SDS gels. After transfer, the cellulose nitrate sheets were cut into broad strips, each spanning the four gel tracks plus markers. These were incubated with one of six selected Workshop monoclonals, followed by rabbit anti–mouse Ig and finally [^{125}I]protein A. Because the original gel pattern did not have to be reconstructed from narrow strips, these blots allowed very accurate comparisons to be made between antigens detected in different cells (Fig. 14.1). This series of experiments was carried out using both boiled and unboiled lysates.

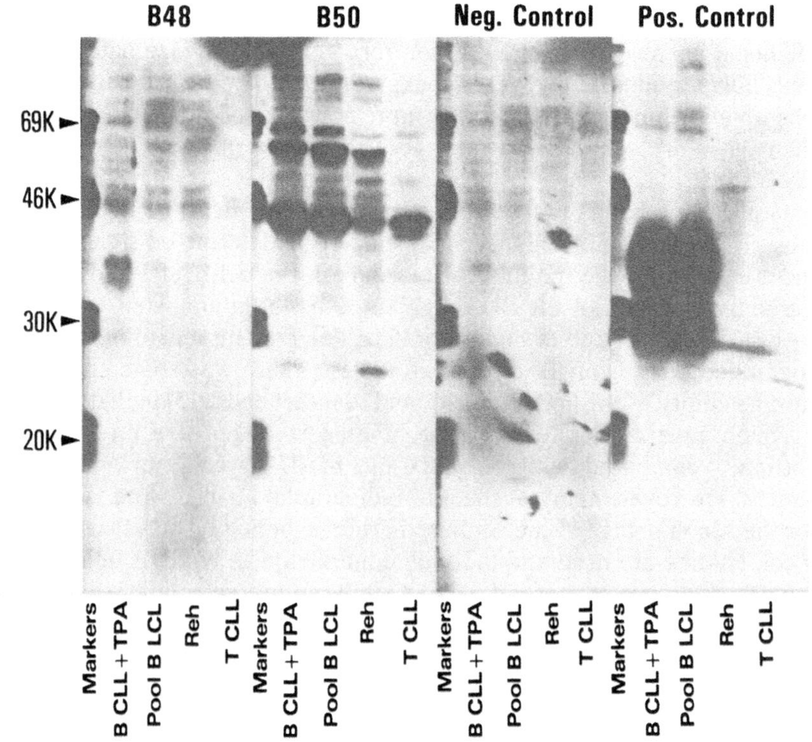

Fig. 14.1. "Western blotting"; boiled lysates from four different cell types in parallel tracks, incubated with Workshop antibodies B48 and B50 plus positive and negative control antibodies.

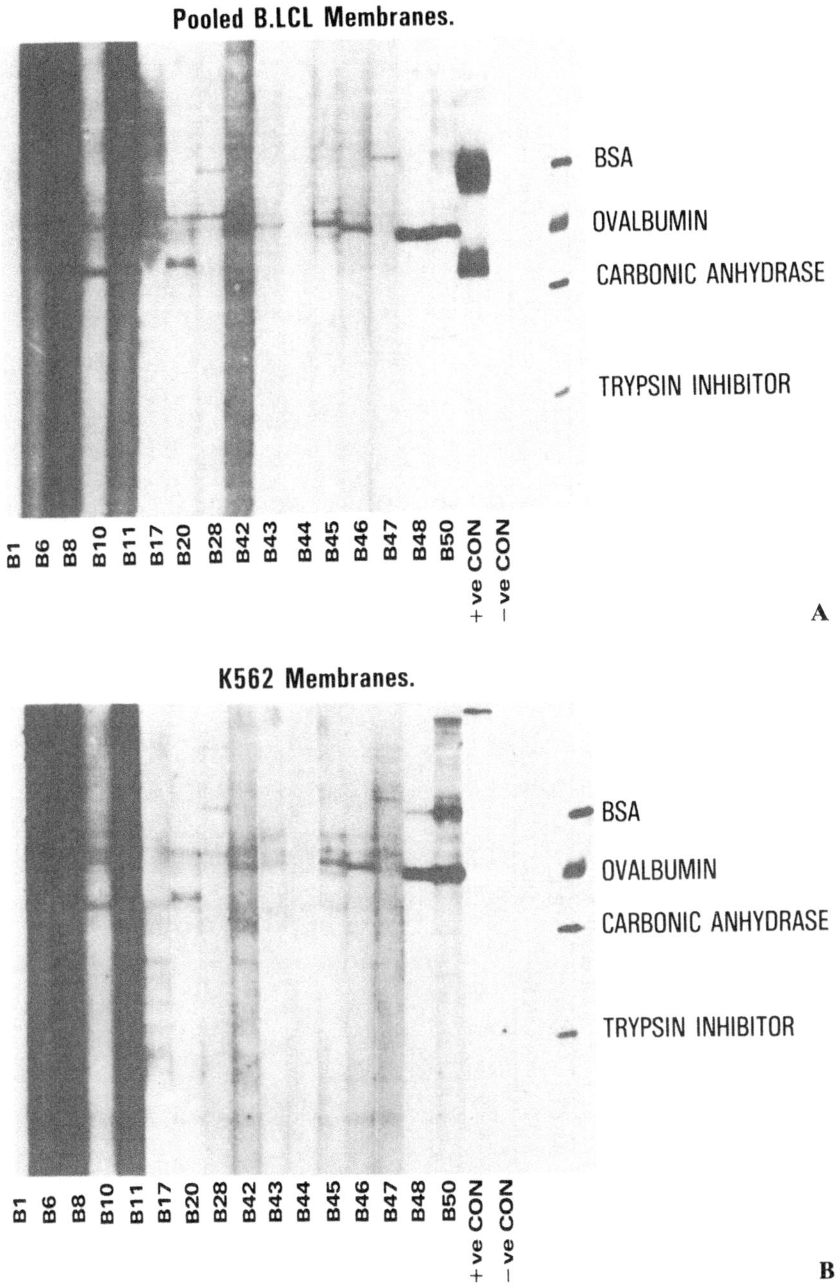

Fig. 14.2. "Western blotting"; lysates from pooled B lymphoblastoid cells (A) or null cell lines K562 (B), transferred after SDS–PAGE to cellulose nitrate. Strips incubated with Workshop or control Antibodies as shown and binding visualized with ^{125}I-labeled sheep anti–mouse Ig.

Table 14.2. Positive western blotting results from workshop B series antibodies.[a]

	Target cell lysate[b]								
	"Boiled" lysates						"Unboiled" lysates		
Antibody	B-LCL pool	B-CLL + TPA	B-CLL, no TPA	B.L. pool	T-CLL	Null cell lines	B-LCL pool	B-CLL + TPA	Null cell lines
B1	70						50		
B4	55								
B5							68		
B6	70								
B8					12, 46		12		
B9					11				
B10	30, 40, 70				68		33, 46	31	33, 46, 48
B11	8								
B15		28, 46, (55)	28, 46, (55)	28, 46, 55	28, 46		60		
B17	12, 30, 68, 70					60 (myeloma)	46, 48, 70, 75, 80		Multiple <20, 27, 32
B18		(28) 46	28, 46						
B20	30, 33, 35, 37, 39, 45	(30) 33, 35, 39, 45			30, 33, 35, 42, 67	30, 33, 35, 37, 39, 45	34 (doublet), 46, 48,	34, 40, 48	34 (doublet), 48,
B28							48, 70		48, 70
B31					(200)				

B32	68								
B37	(200)								
B40	(7)								
B42	(200)						46, 48		26, 46, 48
B43							46, 48		46, 48, 55
B44				(150)			68		
B45	43, (200)	43		43, 55	(12), 43, 150	43	40, 46, 47, 48, 55		35, 46, 47, 48, 55
B46				18, 45			46, 48		46, 48, 55
B47							<u>70</u>		<u>70</u>
B48	<u>43</u>	28, <u>35</u>	(35)				44	44, 69	44, 69
B50	25, 28, 30, 38, 43, 55, 60, 68, 100, 150, 200	25, 43, 50, 58, 68, several bands >70		25, 43, 48, 50, 58, 68, >70, (200)	42, 43, 68, (46), (50)	25, 43, 46, 48, 50, 58, (68), two bands >70	26, 45, 50, 69, 78, (<200)	26, 48, 69, 75	48, 69, <200
B51				(200)					

[a] Values are approx. molecular weights (Kd) of bands visualized. Parentheses enclose weak bands. Very strong bands are underlined.
[b] B-LCL, Pool of B lymphoblastoid cell lines; B-CLL, T-CLL: B and T chronic lymphocytic leukemia, respectively; BL, pool of Burkitt's lymphoma cell lines. Details of target cells are given in Table 14.1.

Finally, 14 of the Workshop antibodies were incubated with transfers from unboiled lysates from a similar range of cell types substituting K562 for Reh as the source of Null cells. On this occasion the second antibody was directly labeled as in the screening phase. The ability of several of the monoclonals to react with two or more polypeptides in a single lysate was confirmed (Fig. 14.2). The results from the entire series of experiments are presented together in Table 14.2.

The results of "Western blotting" analysis are clearly not to be interpreted in isolation from other data which should become available on these reagents. There are technical problems with antigens of low abundance and with antibodies of low affinity. Obviously, the cleaner the background in control blots the more confident can be the identification of weak bands. However, in this respect we found little if any advantage in the use of indirect [^{125}I]protein A labeling as compared to directly labeled second antibody.

The correlation between strength of binding, to intact cells and positive results on "Western blotting" is far from exact. However this is to be expected since the former method detects only those epitopes which are exposed on the cell surface while a lysate can contain many antigens inaccessible on the intact cell. Even when intact cells have been used in the immunization protocol, monoclonal antibodies may be produced which "see" internal targets as, for example, in the case of the anti-DR α-chain reagent DA6.147 (3–5).

The present analysis is not exhaustive. Repeated assays on an even wider range of cell lysates would be required to identify consistent epitopes and to estimate with more precision the molecular weights of the polypeptides which bear them. Nevertheless, the data on several of the Workshop antibodies appear to be reproducible and give a biochemical dimension to antigenic differences between B, T, and Null cells.

References

1. Laemmli, U.K. 1970. Cleavage of structural proteins during the assembly of the head of bacteriophage T4. *Nature* **277**:80.
2. Towbin, H., T. Staehelin, and J. Gordon. 1979. Electrophoretic transfer of proteins from polyacrylamide gels to nitrocellulose sheets: procedure and some applications. *Proc. Natl. Acad. Sci. U.S.A.* **76**:4350.
3. Cohen, B.B., D.L. Deane, V. van Heyningen, K. Guy, D. Hutchins, M. Moxley, and C.M. Steel. 1983. Biochemical variation of human Ia-like antigens detected with monoclonal antibodies. *Clin. Exp. Immunol.* **53**:41.
4. Deane, D.L., B.B. Cohen, M. Moxley, and C.M. Steel. 1983. Detection of Ia epitopes by monoclonal antibody blotting. *Immunol. Lett.* **6**:93.
5. Cohen, B.B., D.L. Deane, and M. Moxley. 1984. Analysis of antibody specificity by Western blotting. *Disease Markers* **2**:135.
6. Matsuoka, Y., G.E. Moore, Y. Yagi, and D. Pressman. 1967. Production of

free light chains by a haemopoietic cell line derived from a patient with multiple myeloma. *Proc. Soc. Exp. Biol. Med.* **125:**1246.
7. Nilsson, K. 1971. Characteristics of established myeloma and lymphoblastoid cell lines derived from an E myeloma patient. A comparative study. *Int. J. Cancer* **7:**380.
8. Totterman, T.H., K. Nillson, and C. Sundstrom. 1980. Phorbol ester-induced differentiation of chronic lymphocytic leukemia cells. *Nature* **288:**176.
9. Guy, K., V. van Heyningen, E. Dewar, and C.M. Steel. 1983. Enhanced expression of human 1a antigens by chronic lymphocytic leukemia cells following treatment with 12-*O*-tetradecanoylphobol-13-acetate. *Eur. J. Immunol.* **13:**156.
10. Caligaris-Cappio, F., G. Janossy, D. Campana, M. Chilosi, L. Bergin, R. Foa, D. Delia, M.C. Guibellino, P. Preda, and M. Gobbi. 1984. Lineage relationship of chronic lymphocytic leukemia and hairy cell leukemia: Studies with TPA. *Leuk. Res.* **8:**567.
11. Minowada, J., T. Ohunuma, and G.E. Moore. 1973. Rosette-forming human lymphoid cell lines. 1. Establishment and evidence of origin from thymus-derived lymphocytes. *J. Natl. Cancer Inst.* **49:**891.
12. Foley, G.E., H. Lazarus, S. Farber, B.G. Uzman, B.A. Boone, and R.E. McCarthy. 1965. Continuous culture of human lymphocytes from peripheral blood of a child with acute leukemia. *Cancer* **18:**522.
13. Lozzio, C.B. and B.B. Lozzio. 1975. Human chronic myelogenous leukemia cell line with positive Philadelphia chromosome. *Cancer* **45:**321.
14. Rosenfeld, C., A. Goutner, C. Choquet, A.M. Venuat, B. Kayibanda, J.L. Pico, and M.F. Greaves. 1977. Phenotypic characteristics of a unique Non-T, non-B acute lymphoblastic leukemia cell line. *Nature* **267:**841.
15. Minowada, J. 1978. Markers of human leukemia–lymphoma cell lines reflect haematopoietic cell differentiation. In: Human lymphocyte differentiation: Its application to cancer. INSERM symposium No. 8, B. Serrou and C. Rosenfeld, eds. Elsevier/North Holland, Amsterdam, pp. 337–344.
16. Jones, P.J. 1980. Selected methods in cellular immunology. B.B. Mishell and S.M. Shiigi, eds. Freeman, San Francisco, p. 398.

CHAPTER 15

Human B Cell Antigens Detected by the Workshop Antibodies: A Comparison of Serological and Immunochemical Patterns

Keizo Horibe and Robert W. Knowles

A number of cell surface molecules expressed on human B cells have been identified by murine monoclonal antibodies (mAbs) in addition to the conventional markers (1–4). These new reagents have contributed to the characterization of B cell differentiation antigens and their functional properties. In the First Workshop, two B-related antigens, CD9 and CD10, were well characterized and defined as clusters of differentiation (5). In the present study, the 72 mAbs included in the B cell/leukemia protocol of this workshop have been analyzed serologically and immunochemically to characterize the B cell antigens which they recognize and to define new clusters of differentiation.

Materials and Methods

Target Cell Panel

The cell panel used in the serological analysis consisted of 20 cells: fresh peripheral blood mononuclear leukocytes; non-T cell fraction separated by nylon-wool column method (PBL non-T), $n = 3$; T cell fraction (PBL-T), $n = 1$; frozen B-chronic lymphocytic leukemia (B-CLL) cells, $n = 2$; cultured hematopoietic cell lines, $n = 14$. All cell lines were obtained from Dr. Bo Dupont and were maintained as described previously (6,7).

Monoclonal Antibodies

W6/32 (anti-HLA-A,B,C framework), obtained from Dr. W.F. Bodmer, was used in addition to the 72 Workshop antibodies. A murine hybridoma clone producing PL13 (B19) was established from the same fusion described previously (7).

Serological Assays

A mixed hemadsorption (MHA) assay and a complement-dependent microcytotoxicity (c-cytotoxicity) assay were performed as described previously (7). The Workshop antibodies were tested at two dilutions, 1 : 10 and 1 : 500, of the provided antibody solution (already diluted at 1 : 10). The results obtained by the MHA assay combined with those by the c-cytotoxicity assay are presented.

Immunoprecipitation and Electrophoresis on SDS Gels

Two B cell lines, PLH (an EBV-transformed B lymphoblastoid cell line) and Raji (a Burkitt's lymphoma-derived cell line), were used for the immunochemical analysis of the 72 Workshop antibodies. Four other B cell lines, Daudi (a Burkitt's lymphoma-derived cell line), SK-LY-18 (a B-lymphoma-derived cell line), NALL-1 (a null-ALL-derived cell line), and BALL-1 (a B-ALL-derived cell line) were also used for some antibodies. Cell surface proteins were labeled with ^{125}I using lactoperoxidase and glucose oxidase, and lysed with Nonidet P-40 as described previously (8). Immunoprecipitation was performed using 10 μl of the antibody solutions provided for the Workshop. *S. aureus* (IgG SORB, enzyme center, Medford, MA) precoated with rabbit anti–mouse IgG antibodies was used for all mAbs tested. SDS–polyacrylamide gel electrophoresis was performed under reducing conditions as described previously (8).

Results

The immunochemical analysis of the 72 Workshop antibodies was performed using ^{125}I-surface-labeled lysates from several B cell lines. The molecular weights (M.W.) of cell surface molecules recognized by 49 out of 72 Workshop antibodies were obtained using at least one cell lysate as shown in Table 15.1. The serological patterns of the 72 Workshop antibodies on the target cell panel are shown in Tables 15.2–15.5. Ten distinct clusters were found by combining the immunochemical and serological results.

Table 15.2 demonstrates the patterns of reactions obtained by the mAbs detecting the four clustered pan-B antigens (125 Kd, 87 Kd, 35 Kd, and 35 Kd/32 Kd). Five antibodies which failed to immunoprecipitate detectable amounts of antigen were included in the 35-Kd cluster because of their similar serological patterns. The term "pan-B" is used when the antibody reacts with more than 80% of PBL non-T cells. Table 15.3 demonstrates the serological patterns of other clustered antigens (140 Kd, 45 Kd, 80 Kd, 24 Kd, 100 Kd, 90 Kd/70 Kd). Two antibodies which failed to immunoprecipitate detectable antigens were included in the 100-Kd cluster because

Table 15.1. Molecular weights (Kd) of the antigens detected by the workshop antibodies using immunoprecipitation on SDS gels.

Workshop mAb	PLH	Raji	Others[a]	Workshop mAb	PLH	Raji	Others
B1	—	35/32		B38	—	—	24(S)
B2	—	75	80/30(D)	B39	60–80/45/35	45	
			80(N, B)	B40	125	125	
B3	—	75	80/30(D)	B41	140	140	
			80(N, B)	B42	—	75	80/30(D)
B4	—	—					80(N, B)
B5	35	35	35(D)	B43	87	87	
B6	—	—		B44	—	—	
B7	125	125		B45	ND	28	28(S)
B8	—	—		B46	—	68	
B9	ND[b]	140		B47	—	—	
B10	—	—		B48	—	—	
B11	60–80/45/35	45		B49	125	125	
B13	—	—		B50	45/32	70/45/32	
B14	—	87		B51	—	—	
B15	—	—		B52	40	40	
B16	—	—		L1	—	—	
B17	160/43	160/43		L2	—	100	
B18	—	—	—(S)	L3	90/70	90/70	
B19	60–80/45/35	—		L4	—	—	
B20	—	—		L6	220	220	
B21	—	—	40(S)	L7	—	—	
B22	—	—		L8	35/32	35/32	
B23	45/32	45/32		L9	160	160	
B24	—	35		L10	—	100	
B25	125	125		L11	—	100	
B26	—	26		L12	220	220	
B27	—	—		L13	90/70	90/70	
B28	—	87		L14	—	100	
B29	85	85		L15	—	—	
B30	—	—		L16	24	24	70/24(S)
B31	125	125		L17	87	87	
B32	—	—		L18	24	24	70/24(S)
B33	ND	140		L19	180/90/70	180/90/70	
B34	—	87		L20	90/70	90/70	
B35	140	140		L21	—	—	
B36	—	—		L22	—	—	24(S)
B37	45	—					

[a] D, Daudi; N, NALL-1; B, BALL-1; S, SK-LY-18.
[b] ND, not determined.

of their similar serological patterns. Table 15.4 shows the serological patterns of the unclustered structurally defined antigens. Table 15.5 describes the serological data obtained from a series of mAbs which could not be serologically classified with any cluster and failed to immunoprecipitate detectable antigens.

Table 15.2. Serological patterns of clustered pan-B antigens.

M.W.	125 Kd	87 Kd	35 Kd	35 Kd/32 Kd
Antigen/prototype		B4	B1	HLA-DR
PBL non-T	++[a]	++	++	++
PBL-T	–	–	–	–
B-CLL	–	++	++	++
PLH(EBV-transformed)	++	++	++	++
SWEIG(EBV-transformed)	++	++	++	++
Raji(Burkitt's)	++	++	++	++
Daudi(Burkitt's)	++	++	++	++
SK-LY-18(B-lymphoma)	++	++	++	–
SK-LY-16(B-lymphoma)	++	++	++	++
BALL-1(B-ALL)	++	++	++	++
NALL-1(Null-ALL)	++	++	++	++
NKL-2(Null-ALL)	+	++	–	++
HUT78(cutaneous-T)	–	–	++[c]	++
HSB-2(T-ALL)	–	–	–[d]	–
HPB-ALL(T-ALL)	–[b]	–	–	–
SKO-007(myeloma)	–	–	–	–
U 937(monocytoid)	–	–	–	–/+
Workshop mAbs	B7 B25 B31 B40 B49	B14 B28 B34 B43 L17	B5 B24 (B4)[e] (B6) (B22) (B30) (B36)	B1 L8

[a] ++, 80%–100%; +, 40%–80%; w, 10%–40%; –, 0%–10%.
[b] B49, weakly positive.
[c] B4, B30, and B36, weakly positive.
[d] B5, weakly positive.
[e] Antibodies shown in parentheses failed to immunoprecipitate detectable amounts of antigen, but reacted with a serological pattern similar to that of the assigned cluster.

Pan-B Antigens

Four pan-B antigens including the HLA-DR antigen were clustered using the Workshop antibodies (Table 15.2). The immunoprecipitation patterns of the 125-Kd and 87-Kd clusters on SDS gels are shown in Fig. 15.1(A). The 125-Kd molecule was usually accompanied by slightly smaller molecules, possibly due to differences in glycosylation. The 87-Kd molecule was clearly visible as a single band and was easily distinguished on 7.5% SDS gels from the 85-Kd molecule detected by B29. These two new clusters, 125 Kd and 87 Kd, were found to have similar serological patterns except for their expression on B-CLL cells. The 87-Kd molecule appears to be the same molecule detected by B4 based on its similar serological and immunochemical patterns (3), while the 125-Kd molecule appears to be newly identified in the present study.

Serological patterns of the other two pan-B antigens, 35-Kd and 35-Kd/32-Kd molecules, were distinct from the 125-Kd and 87-Kd clusters with

Table 15.3. Serological patterns of other clustered B cell-associated antigens.

M.W.	140 Kd	45 Kd	80 Kd	24 Kd	100 Kd	90 Kd/70 Kd
Antigen/prototype	B2			CD9	CD10	
PBL non-T	–/w[a]	–/w	w/+	–/w	–	–
PBL-T	–	–	–	–	–	–
B-CLL	–	++	–	–	–	–
PLH	–	++	w	–	–	w/+
SWEIG	+	++	–	+[d]	–	+/++
Raji	++	++	–	++[d]	++	+/++
Daudi	–	–	++	–	++	+/++
SK-LY-18	–	+	–	++	++	–
SK-LY-16	–	w	++	–	++	++
BALL-1	–	–	++	–	–	+/++
NALL-1	–/w	w	++	w/+	w	+/++
NKL-2	–	w	–[c]	++[d,e]	–	w/+
HUT78	–	–[b]	–[c]	–	–	++
HSB-2	–	–[b]	–	–	–	++
HPB-ALL	–	–	–[c]	–	w/++	++
SKO-007	–	–	–	–	–	w/+
U 937	–	+	–	–	–	–/w
Workshop mAbs	B9	B11	B2	L16	L2	L3
	B33	B19	B3	L18	L10	L13
	B35	B39	B42	L22[e]	L11	L19
	B41			B38[d]	L14	L20
					(L15)[f]	
					(L21)	

[a] ++, 80%–100%; +, 40%–80%; w, 10%–40%; –, 0%–10%.
[b] B11, moderately positive.
[c] B42, weakly positive.
[d,e] B38 and L22, negative.
[f] Antibodies shown in parentheses failed to immunoprecipitate detectable amounts of antigen, but reacted with a serological pattern similar to that of the assigned cluster.

regard to the reactivity of HUT78 (a cutaneous T-lymphoma cell line). Moreover, the 35-Kd cluster was discriminated from the 35-Kd/32-Kd cluster by the difference in reactivity to SK-LY-18 and NKL-2 (a null-ALL derived cell line). Immunoprecipitation patterns of B5 and B24 compared to B1 and L8 (anti-HLA-DR) antibodies are shown in Fig. 15.1(B). The 35-Kd molecule appears to be equivalent to the B1 antigen because of its similar serological and immunochemical patterns (2).

Other Clustered B Cell-Associated Antigens

Six clusters other than the pan-B clusters were identified by the immunochemical and serological analysis (Table 15.3). These antigens were weakly or not expressed on PBL non-T cells and each showed a distinct serological pattern on the target cell panel.

Strong reactivity of the anti-140-Kd mAb was only found with Raji cells in the cell panel. Immunoprecipitation patterns of this antigen are shown

Table 15.4. Serological patterns of unclustered immunochemically identified antigens.

Workshop mAb	Pan-B B52	B-associated B45	B-associated B37	B-associated B23	B-associated B50	B-associated B26	B-associated B21	B-associated B46	B-associated L6	Pan-leukocyte B29	Pan-leukocyte L12	Pan-leukocyte B17	Pan-leukocyte L9
M.W.	40 Kd	28 Kd	45 Kd	45 Kd/ 32 Kd	45 Kd/ 32 Kd	26 Kd	40 Kd	68 Kd	220 Kd	85 Kd	220 Kd	160 Kd/ 43 Kd	160 Kd
PBL non-T	++[a]	+	w	−/w	−/w	−	−	−/w	−	−/w	++	++	++
PBL-T	−	−	−	−	−	−	−	−	−	−	++	++	++
B-CLL	++	++/−[b]	−	−	−	−	−	−	−	++	++	++	−
PLH	++	+	+	++	−	+	−	−	−	++	++	++	++
SWEIG	++	+	+	+	−	+	−	w	−	++	++	++	++
Raji	++	+	−	++	++	++	w	+	++	++	++	++	++
Daudi	++	+	+	+	−	−	−	−	++	++	++	+	++
SK-LY-18	w	−	w	−	−	−	+	−	++	−	++	++	++
SK-LY-16	+	−	+	+	−	−	+	−	++	++	++	+	++
BALL-1	++	++	w	+	+	++	++	−	−	++	++	++	++
NALL-1	++	++	+	w	++	++	+	−	+	++	−	++	++
NKL-2	−	++	−	+	−	−	+	−	−	−	w	w	−
HUT78	−	−	+	+	−	+	−	−	−	++	++	+	++
HSB-2	−	−	−	−	−	−	−	−	−	+	++	++	++
HPB-ALL	−	−	+	−	−	−	−	−	w	−	−	+	++
SKO-007	−	−	w	−	−	−	−	−	−	++	++	−	−
U 937	++	−	−	−	−	+	−	−	−	++	−	+	++

[a] ++, 80%–100%; +, 40%–80%; w, 10%–40%; −, 0%–10%.
[b] One B-CLL was positive, but another was negative to B45.

Table 15.5. Serological patterns for other workshop antibodies.[a]

Workshop mAb	B-associated												T-associated
	B44	B47	B48	B51	L4	B8	B10	B13	B15	B16	B18	B27	L7
PBL non-T	+/++[b]	++	+/++	++	++	w/+	–/w	–	w	–/w	w	–/w	–
PBL-T	–	–	–	–/+	–	–	–	–	–	–	–	–	–
B-CLL	++	++	++	–	–	++	–	–	–	+	–	–/++[c]	–
PLH	++	–	–	+	w	w	w	w	–	–	–	–	–
SWEIG	++	–	–	w	+	++	–	w	–	–	–	–	–
Raji	++	+	–	++	++	++	++	–	–	+	–	++	–
Daudi	++	–	–	++	–	–	–	–	–	++	–	+	–
SK-LY-18	+	++	++	+	++	++	+	–	+	+	+	++	–
SK-LY-16	+	–	–	–/w	+	++	–	–	–	++	–	–	–
BALL-1	++	++	++	++	–	+	+	–	–	+	+	+	–
NALL-1	++	++	++	++	–	+	+	–	+	+	+	+	w
NKL-2	++	++	++	+	++	++	–	–	+	–	+	+	–/w
HUT78	–	–	–	–	–	–	–	w	–	–	–	++	w
HSB-2	–	–	–	–	+	–	–	–	–	w	–	–	++
HPB-ALL	–	–	–	–	–	–	–	–	–	–	–	–	+
SKO-007	++	–	–	–	–	–	–	–	–	–	–	–	–
U 937	–	–	–	++	–	–	++	–	–	–	–	–	–

[a] B20, B32, and L1 did not react to any target cells tested.
[b] ++, 80%–100%; +, 40%–80%; w, 10%–40%; –, 0%–10%.
[c] One B-CLL was negative, but another was positive to B27.

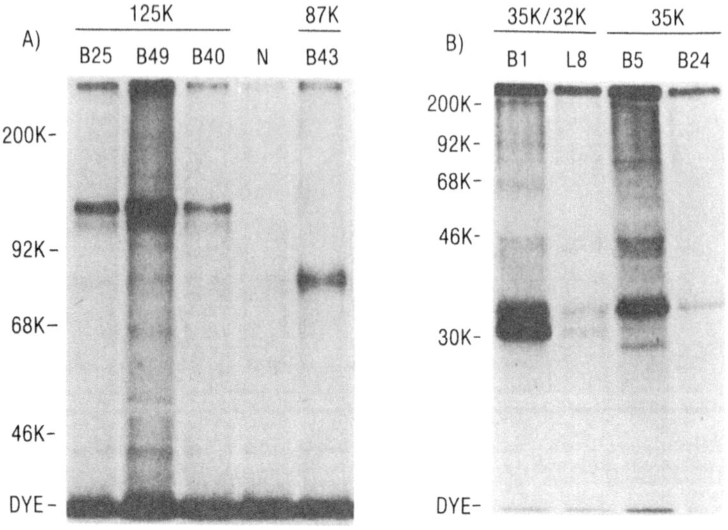

Fig. 15.1. Immunoprecipitation of clustered pan-B antigens. (A) Immunoprecipitation patterns of the 125-Kd and 87-Kd clusters on 7.5% SDS gels were obtained from lysates of ^{125}I-surface-labeled Raji cells. "N" means no mAb control. Molecular weights are indicated based on the mobility of ^{14}C-labeled standard proteins (Amersham) run on each gel. (B) Immunoprecipitation patterns of the 35-Kd/32-Kd and 35-Kd clusters on 12.5% SDS gels were obtained from Daudi lysates.

in Fig. 15.2(A). This cluster appears to be equivalent to the B2 antigen (2).

Three antibodies which detect the same 45-Kd antigen formed another cluster. Their immunoprecipitation patterns using a PLH lysate are shown in Fig. 15.2(B). B11 and B39 showed somewhat stronger bands than B19 (PL13) (B11, data not shown). This heterogeneity may be due to heterogeneity in the titer or affinity of these antibodies. These antibodies commonly immunoprecipitated a broad band ranging in size from 60 Kd to 80 Kd and 35-Kd bands in addition to the 45-Kd major band. When compared with the HLA-class I antigens, detected by W6/32, and the HLA-class II antigens, detected by the Workshop antibody L8, the 45-Kd molecule was found to be distinct from the HLA-A,B,C heavy chains and not to be associated with β_2-microglobulin. The 35-Kd bands may be HLA-class II antigens but are different from the molecules detected by the Workshop antibody L8 [Fig. 15.2(B)]. Furthermore, the 45-Kd molecule detected by PL13 (B19) was found to have a more acidic isoelectric point than the HLA-class I heavy chain detected by W6/32 on two-dimensional isoelectric focusing (IEF)–SDS gels (data not shown). Two-dimensional gel patterns of B11 and B39 were found to be identical to that of PL13 (B19) (data not shown). Serologically this antigen was strongly ex-

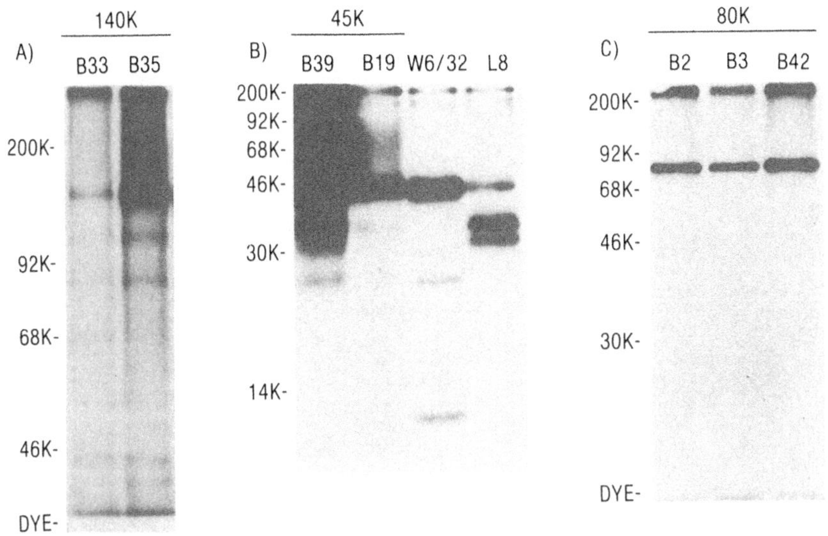

Fig. 15.2. Immunoprecipitation of clustered B cell-associated antigens. (A) Immunoprecipitation patterns of the 140-Kd cluster on 7.5% SDS gels were obtained from lysates of ^{125}I-surface-labeled Raji cells. (B) Immunoprecipitation patterns of the 45-Kd cluster, the HLA-class I antigens detected by W6/32, and the HLA-class II antigens detected by a Workshop antibody, L8, on 15% SDS gels were obtained from PLH lysates. (C) Immunoprecipitation patterns of the 80-Kd cluster on 12.5% SDS gels were obtained from NALL-1 lysates.

pressed on EBV-transformed cell lines and less strongly on some B-lymphoma-derived cell lines.

B2, B3, and B42 were also found to form a cluster serologically and immunochemically. Serological reactivity of this cluster was restricted to some leukemia/lymphoma cell lines derived from the B cell lineage, in addition to moderate reactivity with PBL non-T cells. The immunoprecipitation pattern of this cluster from the NALL-1 lysates is shown in Fig. 15.2(C). There is some heterogeneity in the size of this 80-Kd molecule among the several cell lines tested (data not shown).

CD9 and CD10 antigens, which were defined in the First Workshop (5), were also found to form clusters in this workshop. They showed characteristic serological patterns in the target cell panel. CD9 was expressed on limited leukemia/lymphoma cell lines of the B cell lineage, while CD10 was expressed on one T cell line, HPB-ALL, in addition to leukemia/lymphoma cell lines of the B cell lineage (Table 15.3). Their immunoprecipitation patterns are shown in Fig. 15.3(A) and 15.3(B). The 70-Kd band, in addition to the 24-Kd band, was found on the immunoprecipitation patterns from SK-LY-18 lysates with L16 and L18 Workshop antibodies.

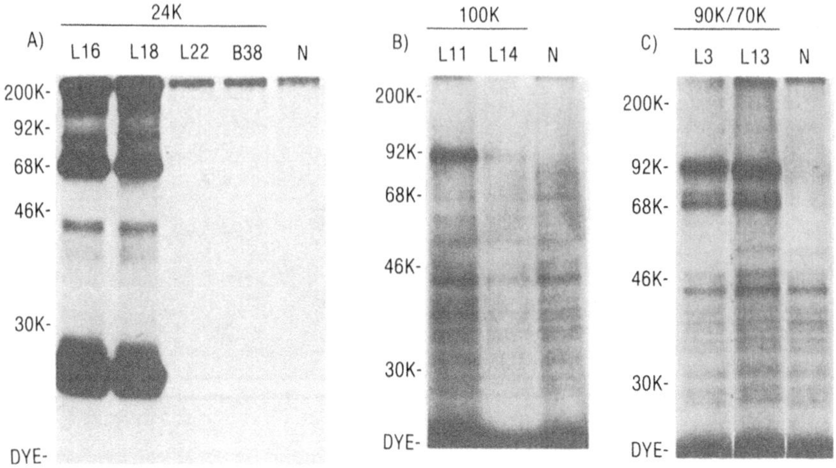

Fig. 15.3. Immunoprecipitation of other clustered B cell-associated antigens. (A) Immunoprecipitation patterns of the 24-Kd cluster on 12.5% SDS gels were obtained from lysates of ^{125}I-surface-labeled SK-LY-18 cells. "N" means no mAb control. (B) Immunoprecipitation patterns of the 100-Kd cluster on 10% SDS gels were obtained from Raji lysates. (C) Immunoprecipitation patterns of the 90-Kd/70-Kd cluster on 10% SDS gels were obtained from Raji lysates.

On the other hand, L22 and B38 showed only a very small amount of the 24-Kd band and serologically showed weaker reaction to positive target cells than L16 and L18. It is possible that the antigens detected by L22 or B38 are different from the antigen detected by L16 or L18, but it appears likely that this is simply due to lower titer or lower affinity of L22 and B38.

One additional antigen with subunits of 90 Kd and 70 Kd also formed a cluster [Fig. 15.3(C)]. This antigen was expressed on both B and T cell lines, but was not expressed on mature cells, such as PBL or B-CLL cells. Some heterogeneity was also found in this cluster. L20 was found to immunoprecipitate only a 70-Kd molecule from Raji lysates. Moreover, L19 was found to immunoprecipitate an additional band, 180 Kd. These findings may be simply due to the difference in titer or affinity among the antibodies. This antigen appears to be expressed predominantly on activated cells.

Unclustered B Cell-Associated Antigens

Thirteen antigens other than clustered antigens identified by immunochemical analysis were found to have unique serological patterns (Table 15.4). Nine of the thirteen antigens seemed to show serological patterns restricted to the B cell lineage. B23 and B50 immunoprecipitated 45-Kd and 32-Kd molecules, but their serological patterns were distinct from each other. Likewise, L6 and L12 immunoprecipitated 22-Kd molecules, but the serological pattern of L12 was pan-leukocyte reactive and that of L6 was very similar to that of CD10. B52 and B21 immunoprecipitated a similar-sized 40-Kd molecule from different cell lysates, but their serological patterns were distinct from each other. B17 and L9 shared similar serological patterns and both immunoprecipitated a 160-Kd band, but an additional strong band at 43 Kd was seen for B17. B29 immunoprecipitated an 85-Kd band and its serological pattern seemed to be restricted to activated cells. B37 immunoprecipitated a 45-Kd band similar to the 45-Kd cluster, but serologically reacted with a distinct pattern. B26, B45, and B46 also immunoprecipitated single bands, but did not cluster with the other antibodies.

The antibodies listed in Table 15.5 in addition to those listed in parentheses in Tables 15.2 and 15.3 failed to immunoprecipitate any detectable antigen. The serological patterns of B47 and B48 were very similar to each other. B15 and B18 also showed similar serological patterns.

The ten clusters identified in the present study are summarized in Table 15.6. Even when the cell panel was reduced to six cells, a distinct serological pattern for each of the ten clusters could be identified. Two T cell lines were useful in order to serologically distinguish the 35-Kd, 35-Kd/32-Kd, 100-Kd, and 90-Kd/70-Kd clusters from the others.

Discussion

The purpose of the present study was to compare the antigens detected by the Workshop antibodies included in the B cell/leukemia protocol of this workshop and to classify them into clusters of antibodies detecting the same antigens. Two types of analysis, the serological phenotyping and the identification of the M.W. of each antigen were performed in parallel. By using a cell panel selected to represent various stages of hematopoietic differentiation, the antigen expression at each stage was determined. A distinct pattern was found to be characteristic for each antigen forming a cluster of differentiation. A good correlation was found combining the serological phenotyping and the identification of the M.W. of the antigen. As a result, seven clusters in addition to CD9, CD10, and HLA-DR were identified in the present study using the Workshop antibodies included in the B cell/leukemia protocol. Three clusters identify pan-B antigens, three clusters identify other B cell-associated antigens, and one cluster identifies an antigen expressed predominantly on activated cells.

The pan-B 125-Kd antigen was found to be expressed on all cells of the B cell lineage except a myeloma-derived cell line and the two samples of B-CLL cells tested.

The pan-B 87-Kd antigen, which appears to be equivalent to the B4

Table 15.6. Summary of the clustered B cell antigens.

M.W.	125 Kd	87 Kd	35 Kd	35 Kd/32 Kd	140 Kd	45 Kd	80 Kd	24 Kd	100 Kd	90 Kd/70 Kd
Antigen/prototype		B4	B1	HLA-DR	B2			CD9	CD10	
PBL non-T	+	+	+	+	−/w	−/w	w	−/w	−	−
B-CLL	−	+	+	+	−	+	−	−	−	−
Raji	+	+	+	+	+	+	−	+	+	+
Daudi	+	+	+	+	−	−	+	−	+	+
SK-LY-18	+	+	+	−	−	+	−	+	+	−
NKL-2	+	+	−	+	−	w	−	+	−	+
HUT78	−	−	+	+	−	−	−	−	−	+
HPB-ALL	−	−	−	−	−	−	−	−	+	+

antigen (3), is expressed on all cells of the B cell lineage except the myeloma-derived cell line.

SK-LY-18 (a B-lymphoma-derived cell line) was found to lack the expression of any detectable amounts of the "DR" subpopulation of HLA-class II antigens but does express other class II molecules related to the DQ (DC) or DP (SB) antigens, using serological absorption and immunochemical analysis (unpublished observation). This cell line was useful in serological classification of the pan-B antigens including the HLA-DR antigens. The pan-B 35-Kd antigen was serologically distinct from 125-Kd and 87-Kd clusters with regard to its lack of reactivity with NKL-2 (a null-ALL derived cell line). Therefore, this antigen may be absent at the extremely early stage of B cell differentiation. Additionally, since HUT78 (a cutaneous T-lymphoma derived cell line) expresses this antigen in addition to the HLA-DR antigen, it may be possible that the 35-Kd antigen is closely related to HLA-class II antigens.

The 140-Kd antigen, which is equivalent to the B2 prototype antigen, is only expressed at a limited stage of the B cell lineage. Recently this antigen was found to be identical to the C3d receptor (CR2) (9,10) and to be equivalent to the EBV receptor (11). Further investigation into the relationship between the EBV and C3 receptors and B cell activation should prove interesting.

The 45-Kd antigen showed a serological distribution and M.W. similar to EBVCS (12), B532 (13), and B-last-1 (14). They may be closely related to each other, although a direct comparison has not been reported. They appear to be differentiation antigens expressed on transformed B lymphoblasts and strongly expressed on EBV-transformed cells. BB-1 is also an antigen related to activated cells and strongly expressed on EBV-transformed cells (15), but its M.W. might be smaller than 45 Kd, and FITC-conjugated BB-1 was not blocked by either EBVCS or B532 (16). A Workshop antibody, B37, also immunoprecipitated a 45-Kd molecule, but its serological pattern could be distinguished from that of this cluster. PL13 (B19) was assigned to the 45-Kd cluster. This antigen was found to be distinct from HLA-class I antigens by comparison of their two-dimensional patterns of IEF–SDS gels.

Another B cell-associated cluster, the 80-Kd antigen, was found in the present study. The molecular size and serological pattern suggest that it may be equivalent to μ-chain of IgM. It is interesting that the serological pattern of this cluster was complementary to that of the 45-Kd cluster in the B cell lineage.

Although the serological distribution of the 90-Kd/70-Kd cluster included cells of the T cell lineage, its specificity seems to be restricted to activated cells. There was heterogeneity in this cluster for both serological and immunochemical patterns.

The combination of a serological analysis and an immunochemical analysis proved extremely powerful in defining the T cell clusters of differentiation in the First Workshop (17), and the present study is another example of the advantage of combining these techniques. Seven new clusters of differentiation in addition to the previously defined CD9, CD10, and HLA-DR antigens have been identified using the 72 B cell Workshop antibodies. This should provide a basis for further studies which can be performed with this extensive panel of monoclonal antibodies to B cell-associated antigens.

Summary

Seven new clusters in addition to CD9, CD10, and HLA-DR were identified from the 72 mAbs in the B cell/leukemia protocol of this workshop. They consisted of three pan-B antigens (125 Kd, 87 Kd, 35 Kd), three other B cell-associated antigens (140 Kd, 45 Kd, 80 Kd), and one cluster (90 Kd/70 Kd) for activated cells. Each cluster showed a distinct serological pattern and a selected cell panel with only six target cells could be used to distinguish each cluster.

Acknowledgments. We wish to thank Dr. Bo Dupont for his generous support and encouragement throughout this study, and Ms. Louise Rozos for preparation of the manuscript. This work was supported by grants CA-22507, CA-23766, CA-33050, and CA-08748 from the U.S. Public Health Service and a grant from the Xoma Corporation, San Francisco, CA, U.S.A.

References

1. McKenzie, I.F.C., and H. Zola. 1983. Monoclonal antibodies to B cells. *Immunol. Today* **4**:10.
2. Nadler, L.M., K.C. Anderson, M. Bates, E. Park, B. Slaughenhoupt, and S.F. Schlossman. 1984. Human B cell-associated antigens: expression on normal and malignant B lymphocytes. In: *Leucocyte typing,* A. Bernard, L. Boumsell, J. Dausset,C. Milstein, and S.F. Schlossman, eds. Springer-Verlag, Berlin, Heidelberg, p. 354.
3. Nadler, L.M., K.C. Anderson, G. Marti, M. Bates, E. Park, J.F. Daley, and S.F. Schlossman. 1983. B4, a human B lymphocyte-associated antigen expressed on normal, mitogen-activated, and malignant B lymphocytes. *J. Immunol.* **131**:244.
4. Wang, C.Y., W. Azzo, Al-Katib, N. Chiorazzi, and D.M. Knowles II. 1984. Preparation and characterization of monoclonal antibodies recognizing three distinct differentiation antigens (BL1, BL2, BL3) of human B lymphocytes. *J. Immunol.* **133**:684.
5. Bernard, A., L. Boumsell, and C. Hill. 1984. Joint report of the First International Workshop on Human Leucocyte Differentiation Antigens: By the investigators of the participating laboratories: B2 protocol. In: *Leucocyte typ-*

ing, A. Bernard, L. Boumsell, J. Dausset, C. Milstein, and S.F. Schlossman, eds. Springer-Verlag, Berlin, Heidelberg, p. 61–81.

6. Naito, K., R.W. Knowles, F.X. Real, Y. Morishima, K. Kawashima, and B. Dupont. 1983. Analysis of two new leukemia-associated antigens detected on human T-cell acute lymphoblastic leukemia using monoclonal antibodies. *Blood* **62:**852.
7. Horibe, K., N. Flomenberg, M.S. Pollack, T.E. Adams, B. Dupont, and R.W. Knowles. 1984. Biochemical and functional evidence that an MT3 supertypic determinant defined by a monoclonal antibody is carried on the DR molecule on HLA-DR7 cell lines. *J. Immunol.* **133:**3195.
8. Knowles, R.W., and W.F. Bodmer. 1982. A monoclonal antibody recognizing a human thymus leukemia-like antigen associated with β_2 microglobulin. *Eur. J. Immunol.* **12:**676.
9. Iida, K., L.M. Nadler, and V. Nussenzweig. 1983. Identification of the membrane receptor for the complement fragment C3d by means of monoclonal antibody. *J. Exp. Med.* **158:**1021.
10. Weis, J.J., T.F. Tedder, and D.T. Fearon. 1984. Identification of a 145,000 Mr membrane protein as the C3d receptor (CR2) of human B lymphocytes. *Proc. Natl. Acad. Sci. U.S.A.* **81:**881.
11. Fingeroth, J.D., J.J. Weis, T.F. Tedder, J.L. Strominger, P.A. Biro, and D.T. Fearon. 1984. Epstein–Barr virus receptor of human B lymphocytes is the C3d receptor CR2. *Proc. Natl. Acad. Sci. U.S.A.* **81:**4510.
12. Kintner, C., and B. Sugden. 1981. Identification of antigenic determinants unique to the surfaces of cells transformed by Epstein–Barr virus. *Nature* **294:**458.
13. Slovin, S.F., D.M. Frisman, C.D. Tsoukas, I. Royston, S.M. Baird, S.B. Wormsley, D.A. Carson, and J.H. Vaughan. 1982. Membrane antigen on Epstein–Barr virus-infected human B cells recognized by a monoclonal antibody. *Proc. Natl. Acad. Sci. U.S.A.* **79:**2649.
14. Thorley-Lawson, D.A., R.T. Schooley, A.K. Bhan, and L.M. Nadler. 1982. Epstein–Barr virus superinduces a new human B cell differentiation antigen (B-LAST 1) expressed on transformed lymphoblasts. *Cell* **30:**415.
15. Yokochi, T., R.D. Holly, and E.A. Clark. 1982. B lymphoblast antigen (BB-1) expressed on Epstein–Barr virus-activated B cell blasts, B lymphoblastoid cell lines, and Burkitt's lymphomas. *J. Immunol.* **128:**823.
16. Clark, E.A., and T. Yokochi. 1984. Human B cell and B cell blast-associated surface molecules defined with monoclonal antibodies. In: *Leucocyte typing,* A. Bernard, L. Boumsell, J. Dausset, C. Milstein, and S.F. Schlossman, eds. Springer-Verlag, Berlin, Heidelberg, p. 339.
17. Horibe, K., R.W. Knowles, K. Naito, Y. Morishima, and B. Dupont. 1984. Analysis of T lymphocyte antibody specificities: comparison of serology with immunoprecipitation patterns. In: *Leucocyte typing,* A. Bernard, L. Boumsell, J. Dausset, C. Milstein, and S.F. Schlossman, eds. Springer-Verlag, Berlin, Heidelberg, p. 212–223.

CHAPTER 16

Further Evidence that the Human Differentiation Antigen p24 Possesses Activity Associated with Protein Kinase

Theodore F. Zipf, Gamil R. Antoun, Gilles J. Lauzon, and B. Michael Longenecker

A fundamental premise underlying the study of differentiation is that the individual molecules on the surface of a cell, at a particular stage of maturation, have a functional role. This role may involve either the maintenance of the individual cell or the regulation of its interaction with other cells and possibly the microenvironment. Significant clarification of the nature of the molecular composition of the cell surface during commitment to lineage and differentiation has been provided by investigations utilizing murine monoclonal antibodies (mAbs). The determination of the functional role played by these molecules has recently been addressed.

The nature of protein phosphorylation, catalyzed by protein kinases (PKs), has been studied extensively. There are suggestions that this process may play a major role in both inter- and intracellular regulation (1,2). Protein kinases have been shown to be on the surface of a variety of mammalian cells but, in spite of their rather ubiquitous distribution, their exact function remains unknown (3–7). Recently, several oncogene products have also been shown to have PK activity (8–12).

The p24 molecule (13,14) first described by Kersey *et al.* has now been shown to be on a wide variety of different tissues. Recently, we have demonstrated that immune complexes containing p24 from Nalm-6 cells have an associated PK activity. This work utilized mAb 50H.19 (15) which has been shown to completely inhibit the binding of antibodies BA-2 (13,14) and SJ-9A4 (16,17) on p24^{+} cell lines.

In this report we present data which extend our previous findings on the protein kinase activity associated with p24.

Materials and Methods

Somatic Cell Hybridization and Human Cell Lines

mAb 50H.19 was generated following immunization of a BALB/c mouse with the human melanoma cell line, MEL-T (15). Selected hybridomas were subsequently maintained by i.p. injection of approximately 1×10^6 cells into BALB/c mice which had been primed with Pristane (Aldrich Chemical Co., Milwaukee, WI). The origins of the cell lines and the culture techniques utilized in this work have been described previously (15,18).

Assay for PK Activity

The assay for PK activity (8,19) was performed in the presence of casein (Sigma Chemicals, St. Louis, MO) and consisted of a 15-min incubation at 37°C in a 100-μl reaction mixture containing 20 m*M* Tris–HCl, pH 7.6, 5 m*M* $MgCl_2$, 5 μg casein, 0.1% NP-40, and 10 μCi [γ^{32}-P]ATP (New England Nuclear, Boston, MA). Other reactions were performed in which 10-μg phosvitin (Sigma) or calf thymus histone (Sigma) replaced casein in the reaction mixture. The reaction was terminated by adding 0.5 ml of 10% trichloroacetic acid (TCA) at 4°C and the mixture was allowed to stand for 1 hr at 4°C. The precipitated proteins were then pelleted by centrifugation at 12,000 × *g* for 10 min. The pellets were washed three more times at 4°C with 10% TCA and twice with acetone. They were then dissolved by a 3-min incubation at 95°C in 100 μl of sample buffer (20) and polyacrylamide gel electrophoresis (PAGE) was performed.

Radiolabeling, Affinity Chromatography, Immunoprecipitation, and Gel Electrophoresis

Prior to immunoprecipitation, intact cells radioiodinated with ^{125}I (Amersham Ltd., Oakville, Ontario) were solubilized, precleared, and incubated with immunoadsorbent Sepharose C1-4B gel (Pharmacia Fine Chemicals, Dorval, Quebec) which had mAb covalently attached. These procedures have been described in detail previously (18). Sodium dodecyl sulfate–polyacrylamide gel electrophoresis (SDS–PAGE) was carried out with a 4% stacking gel and a 7 to 17% acrylamide separating gel (140 × 120 × 1.5 mm). Antibody BA-2 was donated by Dr. Tucker LeBien (University of Minnesota, Minneapolis, MN), SJ9A-4 by Dr. Joseph Mirro (St. Jude Children's Research Hospital, Memphis, TN), and L-243 (21) by Dr. Ronald Levy (Stanford University, Stanford, CA).

For the demonstration of the co-migration of the PK activity and the 50H.19-associated antigen, one set of experiments utilized isoelectric focusing (IEF) as described by O'Farrell (22). Radioiodinated and unlabeled cells were used in these experiments. Gels 10 cm in length were prepared in 0.5-cm-ID glass tubing and Pharmalyte pH 3–10 (Pharmacia) was used at a final concentration of 5%. Voltage was applied to the gels at 400 V for 6 hr followed by 500 V for 1 hr. The gels were then removed from the tubes, cut in 0.5 cm-long sections, and ^{125}I radioactivity was measured (LKB Rack Gamma II Counter, LKB, Bromma, Sweden). For the assay of PK activity, gel sections from the unlabeled cells were equilibrated with 20 m*M* Tris–HCl, pH 7.6, for 2 hr at 22°C and incubated in PK reaction mixture for 2 hr at 37°C; the reaction was terminated by the addition of 100 μl of SDS buffer. Aliquots of 10 μl from each incubation were applied to 0.45-μm pore size filters (Millipore, type HA, Millipore Corp. Bedford, MA) and subsequently washed with 50 ml of 10% TCA. These were placed in vials containing 10 ml of toluene-PRO and ^{32}P radioactivity was measured (LKB Rack Beta Liquid Scintillation Counter).

Lectin Affinity Chromatography

Detergent-solubilized extracts from 2×10^7 radioiodinated cells were washed and suspended in buffer consisting of 5 m*M* Tris–HCl, pH 7.5, 200 m*M* NaCl, and 1 m*M* each of $MgCl_2$, $CaCl_2$, and $MnCl_2$. These were then applied at 20°C to 1-ml columns of Sepharose C1-4B covalently linked to Con A (Pharmacia) and the unbound fraction obtained. After a further wash with 100 ml of buffer, the bound protein was eluted with buffer containing 0.2 *M* mannose.

Results

We have previously shown that PK activity co-migrates with radioiodinated p24 molecule in both PAGE and IEF experiments. This result has been confirmed several times and the data of a further set of IEF experiments are shown in Fig. 16.1.

Under the conditions described above, the p24 molecule from lactoperoxidase (LPO)-labeled Nalm-6 cells was bound to the Con A–Sepharose columns and specifically eluted with mannose. Figure 16.2 shows the behavior with PAGE of the radioiodinated 50H.19-associated antigen which was immunoprecipitated from whole cell lysates and from the bound and unbound fractions of the Con A chromatography. The results of PK assays performed using antigen bound to Sepharose gel with covalently attached 50H.19 or L-243 are shown in Fig. 16.3. PK activity in phosphorylating exogenous casein was associated with antigen immunoprecipitated using 50H.19 from the Nalm-6 cell lines, but was absent in 50H.19 immunoprecipitates of the RPMI 8402 line which demonstrates very weak expression of p24 (16). Further immunoprecipitates prepared from the Nalm-6 line using the anti-HLA-DR antibody L-243 were devoid

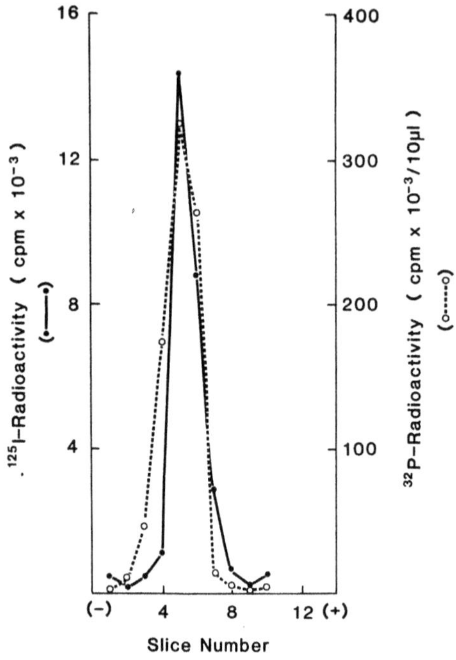

Fig. 16.1. Co-migration of radioiodinated 50H.19-associated antigen and PK activity. 50H.19 immunoprecipitates prepared from detergent-solubilized lysates corresponding to 2.5×10^6 radioiodinated Nalm-6 cells or to 2×10^7 unlabeled cells were loaded on separate gel rods and resolved by IEF. ^{125}I radioactivity and ^{32}P transfer to exogenous casein was determined on 0.5-cm sections of the corresponding gels.

of PK activity using casein as substrate. Figure 16.3 also demonstrates that the PK activity which was bound to 50H.19 was absent in the fraction from the Nalm-6 lysates which did not bind to Con A but was clearly present in the fraction eluted with mannose. Collectively, these results provide further evidence that the PK activity is associated with the p24 molecule.

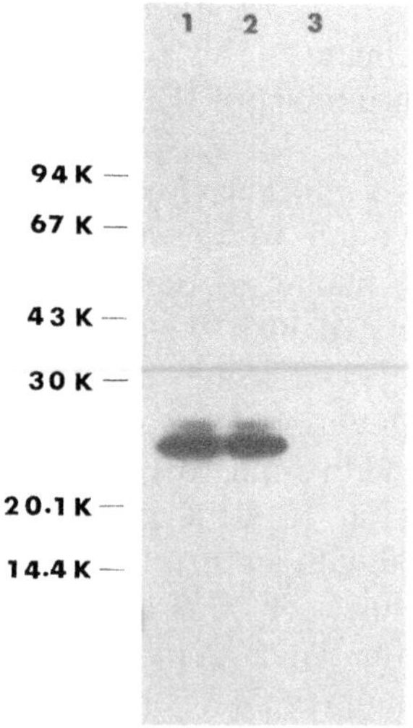

Fig. 16.2. Binding of 50H.19-associated antigen to Con A lectin. Detergent-solubilized lysates equivalent to 5×10^6 radioiodinated Nalm-6 cells were applied to a Con A affinity gel. 50H.19 immunoprecipitates were prepared from the wash-through fractions (lane 3) or the specific mannose-eluted fractions (lane 2) and analyzed by SDS–PAGE. 50H.19 immunoprecipitates prepared from unfractionated lysates corresponding to 2.5×10^6 radioiodinated cells were also subjected to SDS–PAGE (lane 1).

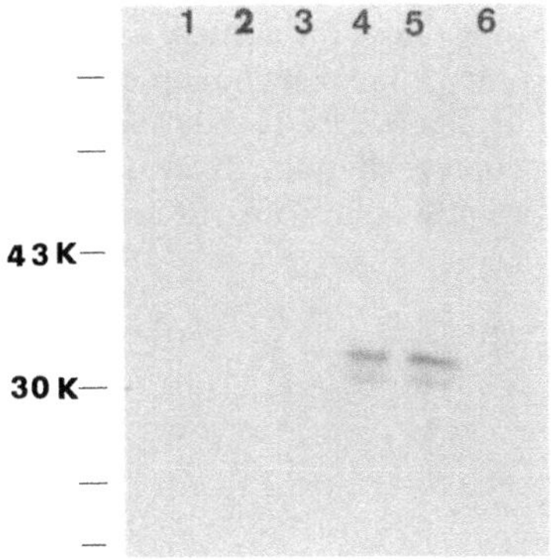

Fig. 16.3. Association of PK activity with the 50H.19-associated antigen. Immunoprecipitates utilizing antibodies 50H.19 and L-243 were prepared from unlabeled, detergent-solubilized extracts of 2×10^7 Nalm-6 or RPMI 8402 cells, either fractionated or unfractionated on Con A affinity gels. PK activity was assessed using antibody affinity gel immunoprecipitates and the results shown are for the PK activity in the following incubations: lane 1, Sepharose gel with covalently bound 50H.19 in the absence of cell extract; lane 2, in L-243 immunoprecipitates prepared from unfractionated RPMI 8402 cell extracts; lane 4, in 50H.19 immunoprecipitates prepared from unfractionated Nalm-6 cell extracts; lane 5, in 50H.19 immunoprecipitates of the portion of the Nalm-6 cell extracts which bound to the Con A affinity gel and was specifically eluted with mannose; lane 6, in 50H.19 immunoprecipitates of the portion of the Nalm-6 extracts which did not bind to the Con A affinity gel. PK activity is represented by radiolabeled casein with bands at 30 Kd and 33.6 Kd.

Association of PK activity with the p24 molecule from other tissues and cell lines has been investigated. Among the mature elements of peripheral blood, only platelets have been shown to bear significant amounts of p24 on their surface. This molecule was precipitated from enriched preparations of platelets and the PK activity assessed; the results, shown in Fig. 16.4, clearly demonstrate that casein phosphorylating activity is associated with the 50H.19 immunoprecipitates. Likewise, PK activity was associated with immunoprecipitates obtained using lysates of the 50H.19$^+$ colorectal adenocarcinoma cell line, LoVo. The CALLA$^-$, HLA-DR$^+$ phenotype of acute lymphoblastic leukemia usually bears the P24 molecule on its surface (16). We have demonstrated the presence of this molecule on the blasts from a child with this immunophenotype and have shown that PK activity in phosphorylating casein is associated with immunoprecipitates of the p24 molecule.

Further studies of malignant tissues were carried out using ascites from adult patients. In one patient with metastatic breast carcinoma, the p24 molecule was demonstrated on LPO radioiodinated cells obtained from the ascitic fluid; correspondingly, activity phosphorylating exogenous ca-

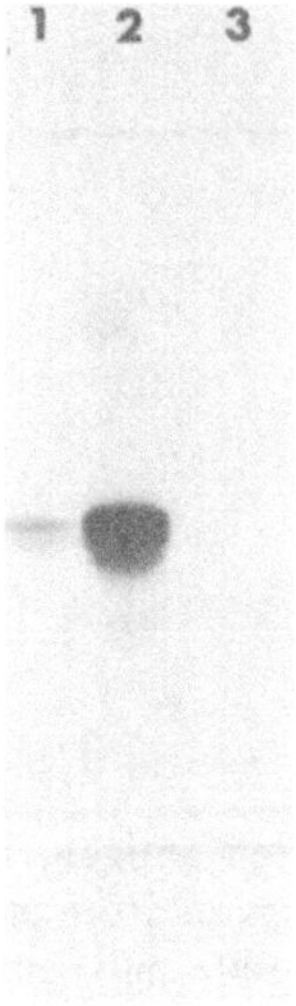

Fig. 16.4. PK activity associated with the 50H.19-associated antigen immunoprecipitated from extracts of 1×10^7 Nalm-6 cells (lane 1) and platelets (lane 2). Lane 3 shows the result of the assay using the Sepharose gel with covalently bound 50H.19 in the absence of cell extract.

sein was associated with the 50H.19 immunoprecipitates prepared from the malignant cells. In a second patient with disseminated ovarian carcinoma, the p24 molecule was not detectable in immunoprecipitates of LPO radioiodinated malignant cells; assays performed on 50H.19 immunoprecipitates from this specimen failed to show any phosphorylation of the exogenous casein.

We have studied the selectivity of the PK activity associated with 50H.19 immunoprecipitates with respect to exogenous protein substrate. Casein was clearly the preferred substrate; phosvitin and calf thymus histone phosphorylation were not detected in our assay (data not shown).

The 41H.16-associated antigen is a 39-Kd glycoprotein which appears on the cell surface near the time of presentation of surface immunoglobulin (18). Immunoprecipitates from lysates of the $41H.16^+$ B-lymphoblastoid cell line RPMI 8392 were devoid of PK activity using casein as substrate. This result suggests that this molecule does not have associated protein kinase activity.

Discussion

The finding that oncogene products have PK activity (8,12) has stimulated interest in protein phosphorylation. The role of this process, especially that which is specific for tyrosine residues (9,11), in malignant transformations may be significant (10). However, it appears unlikely that p24 is an oncogene product. The *ras* p21 oncogene product, if cross-reactive with the 50H.19 antigen, would migrate to roughly the same region of the gels in our experiments. However, p21 is unable to phosphorylate exogenous protein acceptors (23) such as casein. In addition, it has been shown that p24 phosphorylates serine and threonine but not tyrosine (24). There are other known oncogene products which are associated with tyrosine-specific PK activity (11,25,26) but have molecular weights which would not be confused with the antigen detected by mAb 50H.19.

Membrane proteins with associated PK activity have also been shown to act as receptors for epidermal growth factor (27), platelet-derived growth factor, (28) and insulin (29). Ectoprotein kinases have been described on a wide variety of cells including rabbit polymorphonuclear leukocytes (30) and the murine macrophage-like cell line J774 (3). In many of these reports the enzyme phosphorylates serine and threonine but not tyrosine.

The p24 cell surface molecule has a tissue distribution which extends from immature, dividing cells to those at the end-stage of differentiation (16,17,31,32). Previously, we demonstrated that the antigen precipitated by 50H.19 had associated PK activity when the Nalm-6 line was used as a source of p24. In this report we confirm and extend these findings. Although the evidence is strongly suggestive, it does not prove unequivocally that the PK activity is intrinsic to the p24 molecule. Efforts are underway to further purify this molecule to facilitate more detailed studies.

Summary

The molecule referred to as p24 is detected by several monoclonal antibodies which have been described previously. It was originally described as a B cell differentiation antigen but has subsequently been shown to have a much broader tissue distribution. We have previously reported the demonstration of protein kinase activity associated with the p24 molecule. In this report we present data to further confirm these findings and extend the tissue distribution of the protein kinase activity associated with p24.

Acknowledgments. This work was supported by grants from the Alberta Heritage Savings Trust Fund—Applied Research—Cancer, the National Cancer Institute of Canada, and the Medical Research Council of Canada. Dr. Longenecker is a research associate of the National Cancer Institute of Canada.

References

1. Rubin, C.S., and O.M. Rosen. 1975. Protein phosphorylation. *Annu. Rev. Biochem.* **44:**831.
2. Greengard, P. 1978. Phosphorylated proteins as physiologic effectors. *Science* **199:**146.

3. Amano, F., T. Kitagawa, and Y. Akamatsu. 1984. Protein kinases activity on the cell surface of a macrophage-like cell line, J774.1 cells. *Biochim. Biophys. Acta* **803:**163.
4. Kang, E.E., R.E. Gates, T.M. Chiang, and A.H. Kang. 1979. Ectoprotein kinase activity of the isolated rat adipocyte. *Biochem. Biophys. Res. Commun.* **86:**769.
5. Kubler, D., W. Pyerin, and V. Kinzel. 1982. Protein kinase activity and substrates at the surface of intact HeLa cells. *J. Biol. Chem.* **257:**322.
6. Maken, N.R. 1979. Phosphoprotein phosphatase activity at the outer surface of intact and transformed 3T3 fibroblasts. *Biochim. Biophys. Acta* **585:**360.
7. Mastro, A.M., and E. Rozengurt. 1976. Endogenous protein kinase in outer plasma membrane of cultured 3T3 cells. *J. Biol. Chem.* **251:**7899.
8. Collett, M.S., and R.L. Erikson. 1978. Protein kinase activity associated with the avian sarcoma virus src gene product. *Proc. Natl. Acad. Sci. U.S.A.* **75:**2021.
9. Collett, M.S., A.F. Purchio, and R.L. Erikson. 1980. Avian sarcoma virus-transforming protein $pp60^{src}$ shows protein kinase activity for tyrosine. *Nature* **285:**167.
10. Courtneidge, S.A., A.D. Levinson, and J.M. Bishop. 1980. The protein encoded by the transforming gene of avian sarcoma virus ($pp60^{src}$) and a homologous protein in normal cells ($pp^{proto\text{-}src}$) are associated with the plasma membrane. Proc. Natl. Acad. Sci. U.S.A. **77:**3783.
11. Hunter, T., and B.M. Sefton. 1980. Transforming gene product of Rous sarcoma virus phosphorylates tyrosine. *Proc. Natl. Acad. Sci. U.S.A.* **77:**1311.
12. Shih, T.Y., A.G. Papageorge, P.E. Stokes, M.O. Weeks, and E.M. Skolnick. 1980. Guanine nucleotide-binding and autophosphorylating activities associated with the $p21^{src}$ protein of Harvey murine sarcoma virus. *Nature* **287:**686.
13. Kersey, J.H., T.W. LeBien, C.S. Abramson, R. Newman, R. Sutherland, and M. Greaves. 1981. A human leukemia-associated and lymphopoietic progenitor cell surface structure identified with monoclonal antibody. *J. Exp. Med.* **153:**726.
14. Newman, R.A., D.R. Sutherland, T.W. LeBien, J.H. Kersey, and M.F. Greaves. 1982. Biochemical characterisation of leukemia-associated antigen p24 defined by the monoclonal antibody BA-2. *Biochem. Biophys. Acta* **701:**318.
15. MacLean, G.D., J. Seehafer, A.R.E. Shaw, M.W. Kieran, and B.M. Longenecker. 1982. Antigenic heterogeneity of human colorectal cancer cell lines analyzed by a panel of monoclonal antibodies: I. Heterogenous expression of Ia and HLA-like antigenic determinants, *J. Natl. Cancer Inst.* **69:**357.
16. Komada, Y., S.C. Peiper, S.L. Melvin, D.W. Metzgar, G.H. Tarnowski, and A.A. Green. 1983. A monoclonal antibody (SJ-9A4) to p24 present on common ALLs, neuroblastomas and platelets-I. Characterization and development of a unique radioimmunometric assay. *Leuk. Res.* **7:**487.
17. Komada, Y., S.C. Peiper, S.L. Melvin, B. Tarnowski, and A.A. Green. 1983. A monoclonal antibody (SJ-9A4) to p24 present on common ALLs, neuroblastomas and platelets-II. Characterization of p24 and shedding *in vitro* and *in vivo*. *Leuk. Res.* **7:**499.
18. Zipf, T.F., G.J. Lauzon, and B.M. Longenecker. 1983. A monoclonal antibody detecting a 39,000 m.w. molecule that is present on B lymphocytes and

chronic lymphocytic leukemia cells but is rare on acute lymphocytic leukemia blasts. *J. Immunol.* **131**:3064.
19. Griffin, J.D., D. Spangler, and D.M. Livingston. 1979. Protein kinase activity associated with simian virus 40 T antigen. *Proc. Natl. Acad. Sci. U.S.A.* **76**:2610.
20. Shackelford, D.A., L.A. Lampson, and J.L. Strominger. 1981. Analysis of HLA-DR antigens by using monoclonal antibodies: recognition of conformational differences in biosynthetic intermediates. *J. Immunol.* **127**:1403.
21. Lampson, L.A., and R. Levy. 1980. Two populations of Ia-like molecules on a human B cell line. *J. Immunol.* **125**:293.
22. O'Farrell, P.H. 1975. High resolution two dimensional electrophoresis of proteins. *J. Biol. Chem.* **250**:4007.
23. Papageorge, A., D. Lowy, and E.M. Skolnick. 1982. Comparative biochemical properties of p21 *ras* molecules coded for by viral and cellular *ras* genes. *J. Virol.* **44**:509.
24. Seehafer, J., B.M. Longenecker, and A.R.E. Shaw. 1984. Human tumor cell membrane glycoprotein associated with protein kinase activity. *Int. J. Cancer* **34**:815.
25. Eckhart, W., M.A. Hutchinson, and T. Hunter. 1979. An activity phosphorylating tyrosine in polyoma T antigen immunoprecipitates. *Cell* **18**:925.
26. Witte, O.N., A. Dasgupta, and D. Baltimore. 1980. Abelson murine leukaemia virus protein is phosphorylated *in vitro* to form phosphotyrosine. *Nature* **283**:826.
27. Zenisek, S.C., and J.A. Fernandez-Pol.1982. Modulation of protein phosphorylation in human colon adenocarcinoma cell membrane preparations by epidermal growth factor *in vitro*. *Int. J. Cancer* **29**:277.
28. Ek, B., B. Westermark, A. Wasteson, and C. Heldon. 1982. Stimulation of tyrosine-specific phosphorylation by platelet derived growth factor. *Nature* **295**:419.
29. Kasuga, M., Y. Fujita-Yamaguchi, D.L. Blither, and C.R. Kahn. 1983. Tyrosine-specific protein kinase activity is associated with the purified insulin receptor. *Proc. Natl. Acad. Sci. U.S.A.* **80**:2137.
30. Emes, C.H., and N. Crawford. 1982. Ectoprotein kinase activity in rabbit peritoneal polymorphonuclear leukocytes. *Biochim. Biophys. Acta* **717**:98.
31. Ash, R.C., J. Jansen, J.H. Kersey, T.W. LeBien, and E.D. Zanjoni. 1982. Normal human pluripotential and committed hematopoietic progenitors do not express the p24 antigen detected by monoclonal antibody BA-2: Implications for immunotherapy of lymphocytic leukemia. *Blood* **60**:1310.
32. Jones, N.H., M.J. Borowitz, and R.S. Metzgar. 1982. Characterization and distribution of a 24,000 molecular weight antigen defined by a monoclonal antibody (DU-ALL-1) elicited to common acute lymphoblastic leukemia (cALL) cells. *Leuk. Res.* **6**:449.

CHAPTER 17

Clustering of Anti-Leukemia and Anti–B Cell Monoclonal Antibodies

Anne-Marie Ravoet and Anne-Marie Lebacq-Verheyden

Introduction

Determination of molecular weight (M.W.) of antigens has become an essential step in the analysis of the specificity of monoclonal antibodies (mAbs).

We used the following approach:

1. An appropriate target cell was found by studying the reactivity of the mAbs with normal and malignant cells. At the same time, these studies allowed clustering of mAbs according to their specificity.
2. Binding of mAbs to *Staphylococcus aureus* (S.A.) was checked.
3. Target cells bearing high amounts of the antigen were radiolabeled and membrane proteins were extracted and used for immunoprecipitation.
4. M.W. of the antigen was determined after electrophoresis of the immunoprecipitate and detection of radiolabeled polypeptides by autoradiography.
5. When possible, epitopes defined by mAbs of the same cluster were compared by competition studies.

Materials and Methods

Antibodies

Upon arrival, the 1-ml samples of mAbs were supplemented with 2.5 m*M* phenylmethyl sulfonyl fluoride (PMSF). All were mouse mAbs, except L14, L15, L16, B15, and B18 which were rat mAbs. Second-layer antibodies were affinity-purified rabbit anti–mouse Ig (anti–rat Ig) antibodies (1).

Cells

Mononuclear cells (PBL) and buffy coat cells were purified from peripheral blood of healthy donors on Ficoll–Paque or Dextran 150, respectively. Tonsil cell samples were enriched in B lymphocytes by rosetting with AET-treated sheep erythrocytes. Malignant cells were obtained from patients with cALL (Tdt^+, DR^+, $SmIg^-$, E^-, $CALLA^+$), lymphoblastic lymphoma (DR^+, $SmIg^+$, K^+, E^-, $CALLA^-$), lymphocytic lymphoma (DR^+, $SmIg^+$, λ^+, E^-), or immunoblastic lymphoma (DR^+, $SmIg^+$, K^+, E^-). Cell lines were derived from non-T, non-B ALL (Nalm-6, KM3), CML-LyBC (Nalm-1), Burkitt's lymphoma (Daudi), EBV-transformed B lymphocytes (IM9), myeloma (ARH-77), T-ALL (Jurkat, MOLT-4), CML-BC (K562), APML (HL-60), neuroblastoma (CHP100/212), small-cell lung carcinoma (LICR-BRU-OC1/OC2, NCI-H69), and melanoma (DES). These cell lines were graciously provided by M. Greaves, G. Janossy, C. Rosenfeld, B. Van Camp, J. Kemshead, and M. Symann.

Radiobinding Assays

5×10^5 cells were incubated for 30 min at 4°C in polyvinyl microtiter plates in the presence of mAb (1/200) in 50-μl Eagles medium. After 3 washes, saturating amounts of tritium-labeled second-layer antibodies (4 μg/well, 600,000 cpm/μg protein) were added. After a 30-min incubation at 4°C, cells were washed 3 times and pellet-associated radioactivity was measured.

Surface Labeling of Cells

Iodination of cells was performed using 5-U lactoperoxidase and 1-mCi [^{125}I]iodine/10^7 cells and *in situ*-generated H_2O_2 (5 U glucose oxidase and 0.5 mg glucose/10^7 cells).

Extraction of Proteins and Fractionation on lentil Lectin

As described in Ref. 2.

Immunoprecipitation

In order to clear cell extracts of protein A-reacting molecules, 200 μl of a 10% suspension of S.A. were added per ml extract. After a 30-min incubation, the bacteria were removed by centrifugation. For mAbs which bound to S.A., 0.25 ml of cleared cell extract were supplemented with 3-μl ascites. After 30 min at 4°C, 50 μl of a 10% suspension of S.A. were added and incubation was continued for 30 min. For non-S.A. binders, 1-μl ascites and 100-μl S.A., coated with rabbit anti–mouse Ig (anti–rat Ig) antibodies, were used. Washing of immunoprecipitate, elution of antigens, and SDS–PAGE were performed as described previously (1).

Purification and Labeling of mAbs

AL2 was purified and tritium-labeled as described (1). AL6-containing ascites were pooled. Proteins precipitated at 45% ammonium sulfate were further separated by a 24-hr electrophoresis at 4 V/cm on Pevicon in barbital buffer, pH 8.6. AL6 (50 μg) was labeled with 0.5-mCi Bolton-Hunter reagent (430,000 cpm/μg protein) or using 1 mCi [^{125}I]iodine and 0.2 μg chloramine T (35,700 cpm/μg).

Results

Clustering of mAbs According to Reactivity Pattern

Binding of the 21 anti-leukemia and the 52 anti–B cell mAbs to different normal and malignant cells was measured by radiobinding assay. Reactivity patterns are illustrated in Figs. 17.1 and 17.2. It should be emphasized 1) that intensity of labeling depends on total amount of antigen in the sample (at least if mAb is present in saturating amounts), and hence on density of antigen, on surface area of the cells, and on proportion of cells labeled, and 2) that a result < 3000 cpm can indicate a negative reaction as well as a weak reaction or a reaction with a small subset of cells. Three mAbs of the L protocol and 24 mAbs of the B protocol gave no intense reaction with any of the cell samples tested (not shown).

Anti-leukemia antibodies are grouped into six clusters. L2, L10, L11, L14, L15, and L21 have an anti-CD10-like reactivity pattern. L4, L16, L18, and L22 have an anti-CD9-like reactivity pattern. The labeling in PBL and buffy coat samples is probably on platelets. In contrast to L16, L22 labels neuroblastoma cell lines better than hematologic cells. This could point to a slight difference in epitope specificity. L8 has a typical anti-HLA-DR reactivity pattern. L3, L13, L19, and L20 react with cell lines of different lineages. L9 shows preferential reactivity with myeloid cell lines. L6 reacts with all cell samples tested, except for the melanoma line. L12, although labeling brightly most of the hematologic cells, does not label the solid tumor cell lines.

Among the anti–B cell mAbs, B1 and B12 have an anti-HLA-DR reactivity pattern similar to the one obtained for L8. Four mAbs (B15, B18, B47, B48) exhibit labeling of non-T, non-B lymphoblasts, most B cells tested, K562, neuroblastomas, and small-cell lung carcinoma lines. The reactivity pattern of B47 is similar to that reported for BA1 (3). The reactivity of B15 and B18, evaluated by immunofluorescence, always correlates with the reactivity of BA1, but B15 and B18 consistently show

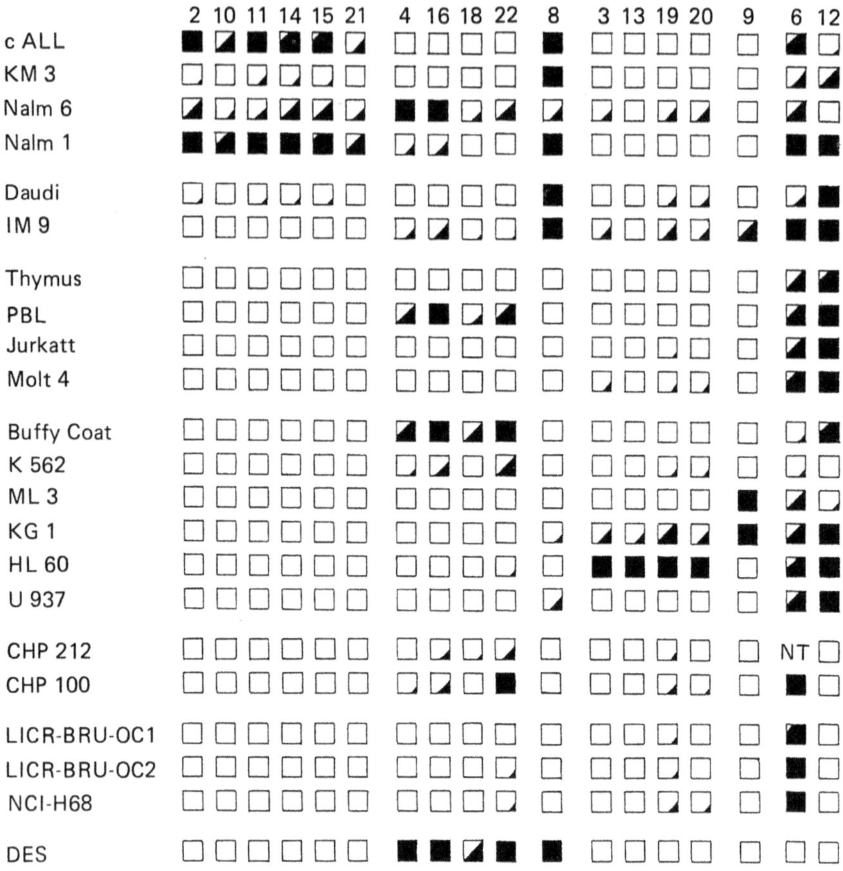

Fig. 17.1. Binding of anti-leukemia mAbs to some hematologic and solid tumor cell samples. Radioactivity measured was □ < 3000 cpm; ◪ 3000 cpm; ◪ 8000 cpm; ■ > 15000 cpm.

weaker labeling of immature and mature B cells than of non-T, non-B lymphoblasts and K562. Hence the epitopes recognized are probably different, although related. B38 has an anti-CD9 reactivity pattern.

The bulk of mAbs with reactivity restricted to immature and mature B cells can be split up into two major groups. mAbs exhibiting positive reaction with Daudi and no reaction with IM9 mostly also react with B-lymphomas, while the $IM9^+$, $Daudi^-$ group exhibits no significant reaction with B-lymphomas. Only B27 reacts with Daudi and IM9 cells. Tight clusters are formed by B5, B22, and B24, by B11, B19, and B39 and by B23 and B29. Reactivity of B4, B10, B44, and B50 is not restricted to B cells. B3, B7, B8, and B45 recognize widely distributed antigens.

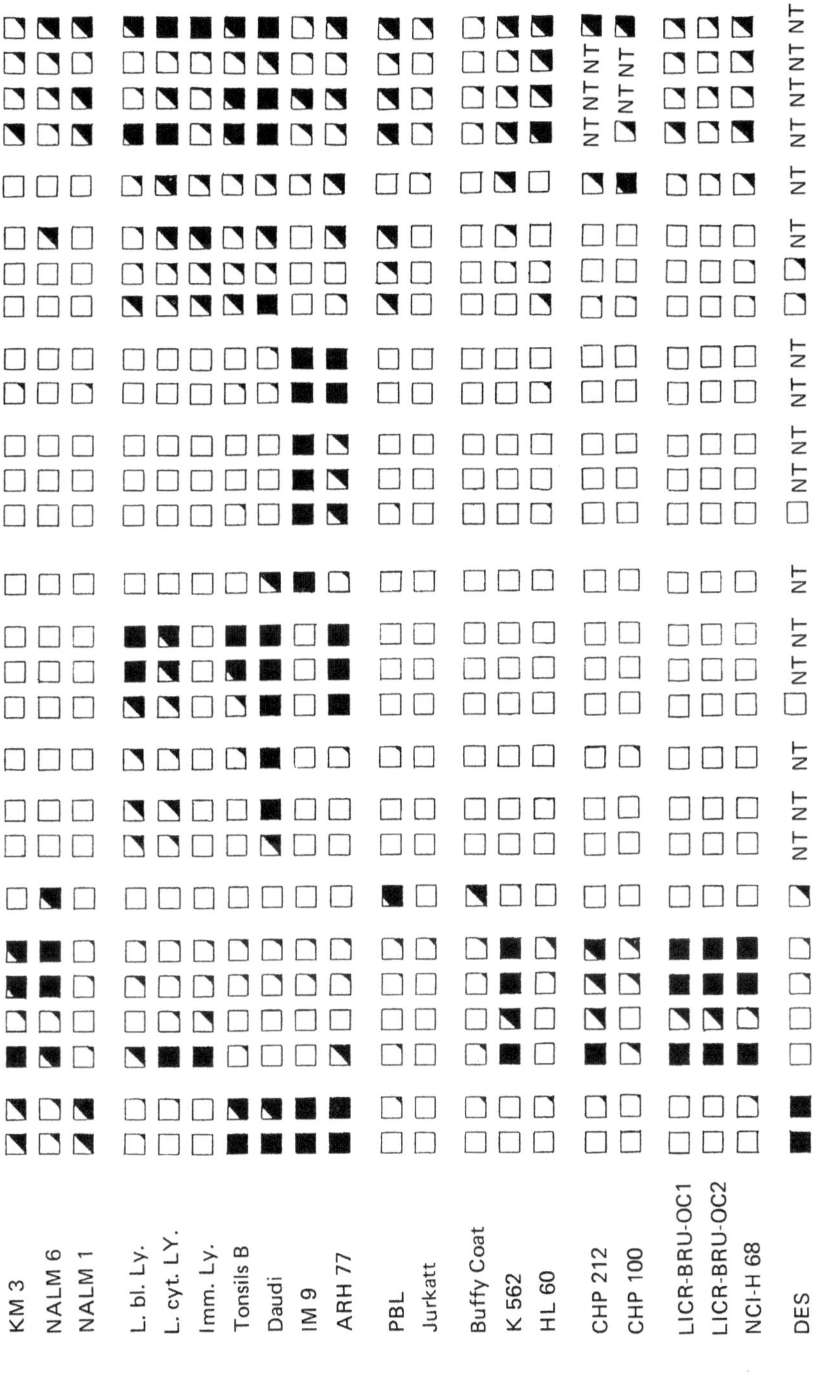

Fig. 17.2. Binding of anti–B cell mAbs to some hematologic and solid tumor cell samples. See Fig. 17.1 for definition of symbols.

Clustering of mAbs According to M.W. of the Precipitated Antigens

All mAbs of the CD10 cluster, except L15, precipitate the expected gp100 CALLA (Fig. 17.3). Among the anti-CD9 mAbs, precipitation of p24 could be evidenced for L4, L16, and L18 (Fig. 17.3). Epitopes recognized were compared by competition studies, using ^{3}H-AL2 (=L14) or ^{125}I-AL6 (=L16) as radioactive probes for the CD10 and the CD9 clusters, respectively (Figs. 17.4 and 17.5). All six anti-CD10 mAbs displace ^{3}H-AL2, indicating that they all recognize the same or closely related epitopes. The negative control, L1, and the anti-HLA-DR mAb, L8, do not interfere with binding of AL2. ^{125}I-AL6 is more easily displaced by our reference AL6 than by the Workshop AL6 (L16) indicating partial inactivation of the latter. L4, L18, and B38, as well as BA2 (4) compete for binding with AL6, while L22 does not. The anti-CD10 mAb L14 and normal mouse serum have no effect. Hence, at least two different epitopes are recognized by the anti-CD9 mAbs.

L8 precipitates the expected gp31/37 HLA-DR dimer (Fig. 17.6). L9 precipitates polypeptides with M.W. of 105 and 160 Kd when KG1 is the target cell, or a single 160-Kd polypeptide if the ML3 cell line is used. This protein is retained on lentil lectin, indicating the presence of mannose or glucose residues (not shown). Antigen recognized by L13 and L19 mi-

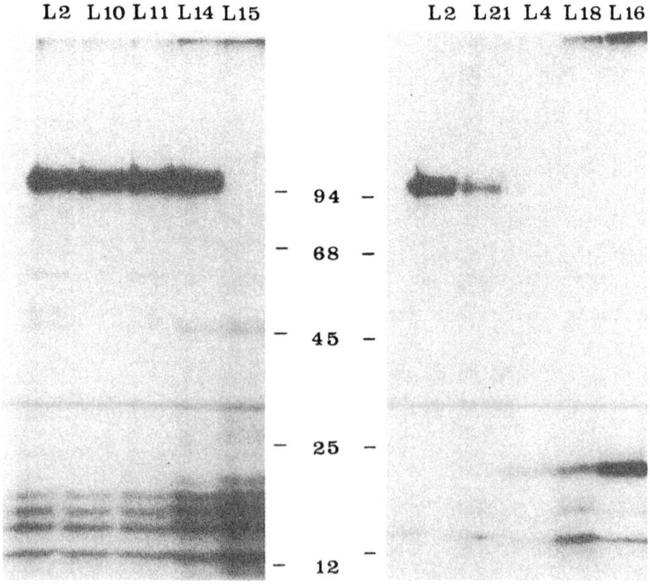

Fig. 17.3. SDS–PAGE and autoradiography of antigens immunoprecipitated from ^{125}I-Nalm-6 cell extracts with mAbs of the CD10 and CD9 clusters.

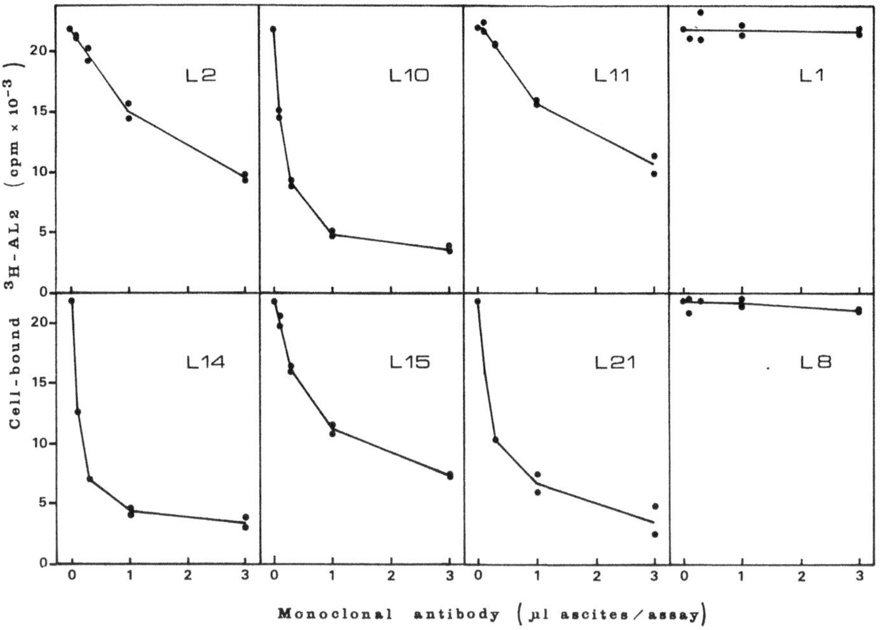

Fig. 17.4. Competitive binding of ^{3}H-AL2 and anti-CD10 mAbs. 2.5×10^6 cells of a cALL patient were incubated for 2 hr at 0°C with 80,000 cpm ^{3}H-AL2 and the indicated amounts of ascites. After 3 washes, pellet was solubilized in 10% Triton X-100 and radioactivity was measured.

grates as a 97-Kd molecule under reducing conditions. For L3 and L20 no specifically precipitated antigen was detected.

Antigen of B15 (=AL1a) and B18 can be visualized after ^{35}S[methionine] incorporation. The polymeric target molecule is built up of 220-Kd, 180-Kd, and, in the case of B15, 110-Kd subunits (Fig. 17.7). In the same experiment, no antigen was detected using B47 or B48. In its nonreduced form, the AL1 antigen does not enter a 7% polyacrylamide lattice. The antigen is not *N*-glycosylated. Indeed, when methionine labeling occurs in the presence of tunicamycin, no modification in M.W. is noted, while for CALLA, M.W. shifts from 100 Kd in the absence of tunicamycin to about 80 Kd in the presence of the inhibitor (Fig. 17.8). In addition, the AL1 antigen is not retained on lentil lectin. In competition studies, B15 and B18 completely displace ^{125}I-AL1a, while addition of B47 or of BA1 results in partial displacement of ^{125}I-AL1 (not shown).

Among mAbs brightly labeling Daudi cells, B42 precipitates a polymeric protein with subunits of 29, 34, and 90 Kd (Fig. 17.9). B17 precipitates a 47 Kd structure (reducing and nonreducing conditions). None of the mAbs of the B5, B24 cluster precipitates any antigen, regardless of the type of labeling used. It was unexpected to find that B3, a mAb with broad

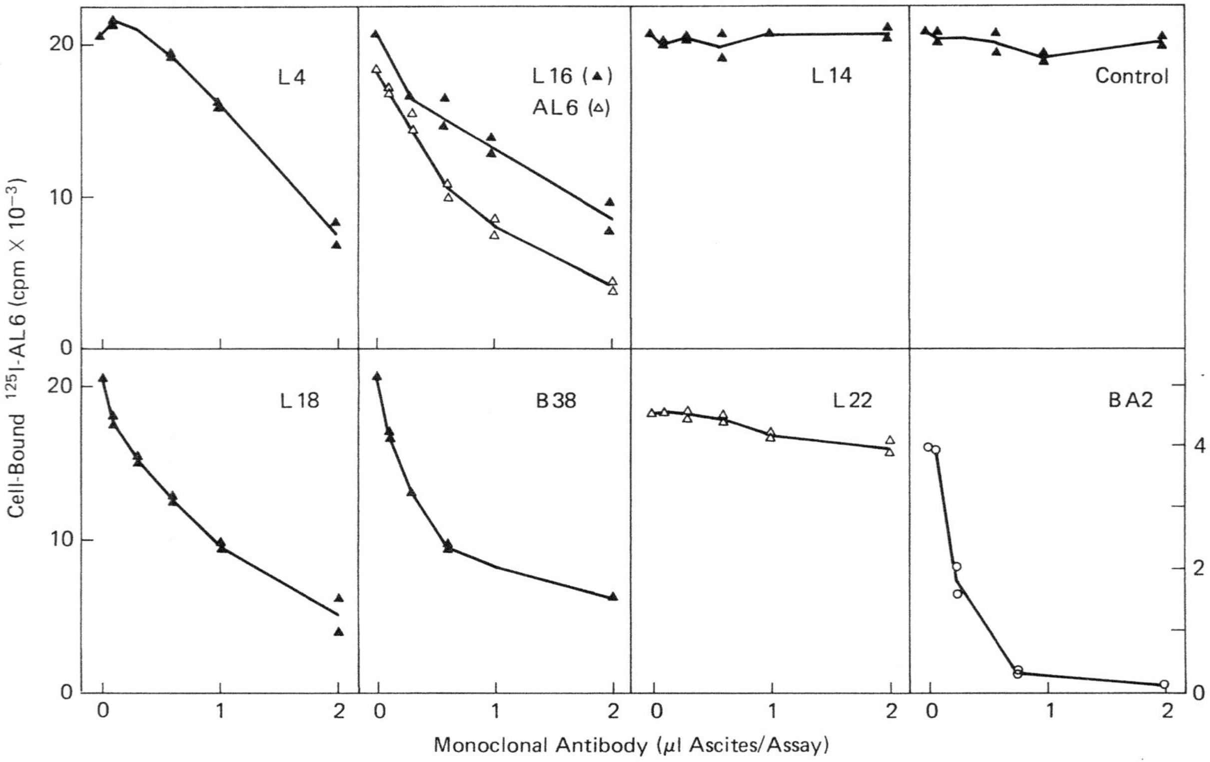

Fig. 17.5. Competitive binding of ^{125}I-AL6 and anti-CD9 mAbs. 1.5×10^6 Nalm-6 cells were incubated for 2 hr at 0°C with 430,000 cpm (▲,△) or 18,000 cpm (○) ^{125}I-AL6 and the indicated amounts of ascites. After 3 washes, pellet-associated radioactivity was measured.

Fig. 17.6. SDS–PAGE and autoradiography of antigens immunoprecipitated with anti-leukemia mAbs from ^{125}I-IM9 (L8), ^{125}I-KG1 (L9,c), and ^{125}I-HL-60 (L20, L19, L13, L3, c) cell extracts.

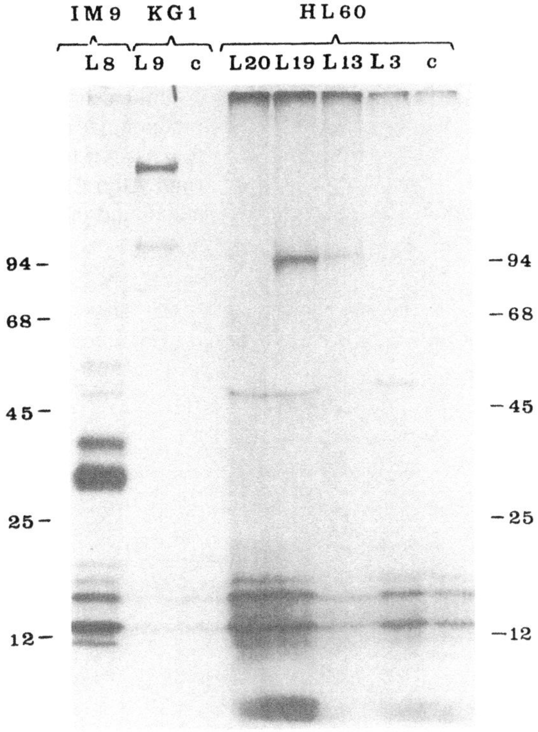

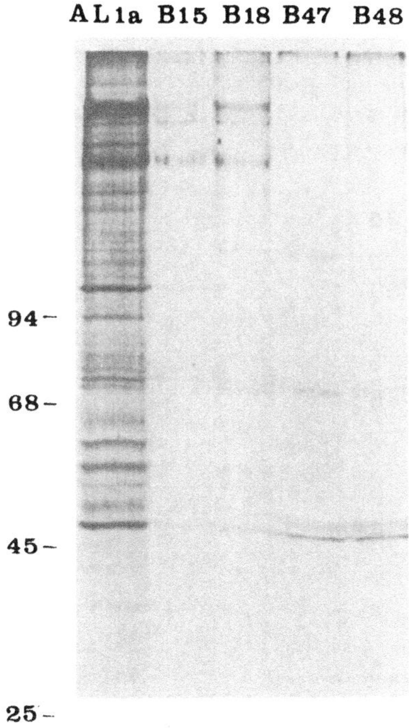

Fig. 17.7. SDS–PAGE and autoradiography of antigens immunoprecipitated with BA1-like mAbs. Cryopreserved cells from a cALL patient were thawed and incubated overnight at 37°C in methionine-free Eagles medium (5 × 10^6 cells/ml) supplemented with [^{35}S]methionine (60 μCi/ml) and 10% fetal calf serum.

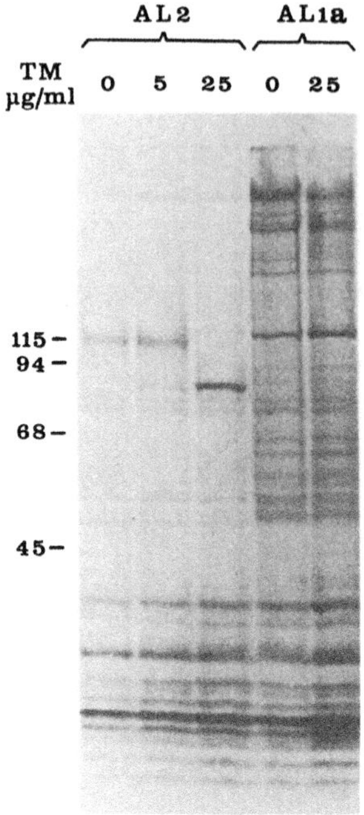

Fig. 17.8. Effect of tunicamycin on M.W. of antigens precipitated with AL1a (B15) and AL2 (L14). Nalm-6 cells (10^7/ml) were incubated at 37°C in methionine-free Eagles medium supplemented with 0, 5, 25 μg tunicamycin/ml and 10% fetal calf serum. After 1 hr, [^{35}S]methionine (0.1 mCi/ml) was added and incubation was continued for 4 hr.

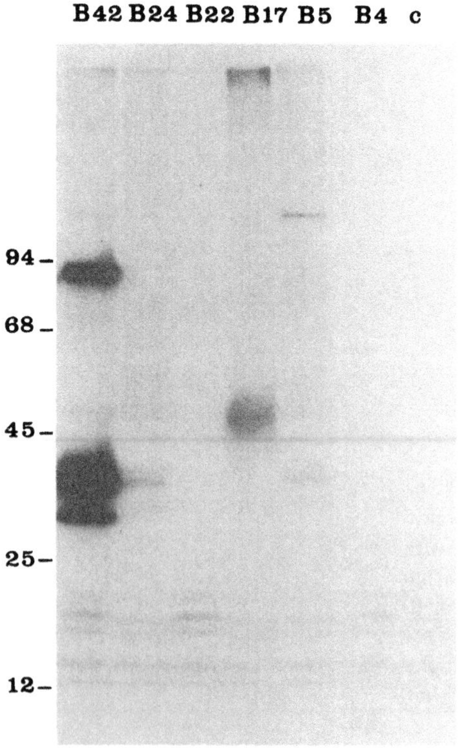

Fig. 17.9. SDS–PAGE and autoradiography of antigens immunoprecipitated from ^{125}I-Daudi cells with B cell-specific mAbs.

specificity, precipitates a molecule similar to the B-restricted p29/34/90 molecule, defined by B42 (Fig. 17.10). When these p29/34/90 antigens are compared to HLA-DR precipitated by B1 and L8, clearly distinct mobilities are observed. Moreover, after three immunoprecipitations with B3, cell extract is depleted of p29/34/90, but still contains HLA-DR (not shown). The additional band obtained at 72 Kd possibly arises from proteolysis (Fig. 17.10).

Among mAbs reacting brightly with IM9 cells, all three mAbs of the B11, B19, B39 cluster precipitate a molecule built up of a subunit of 48 Kd and a subunit with heterogenous M.W. extending from 72 to 100 Kd (Fig. 17.11). Neither B27 nor B23 or B29 precipitated any molecule.

Discussion

Clustering of mAbs according to their reactivity patterns was corroborated by biochemical studies. One exception was noted with B3 and B42: although precipitating the same molecule, B3 has a broader specificity than B42. Whether this discordance is due to nonspecific adsorption of B3 to the cells or the wells in radiobinding assay, or to the presence of two mAbs in the B3 ascites is not known. Probably due to inactivation, almost 50% of the mAbs of the B protocol did not react with any of the 15 cell samples tested.

Most of the anti-leukemia mAbs are against well-known differentiation antigens. L4, L16, L18, and B38 define one epitope on the CD9 (nTnB,

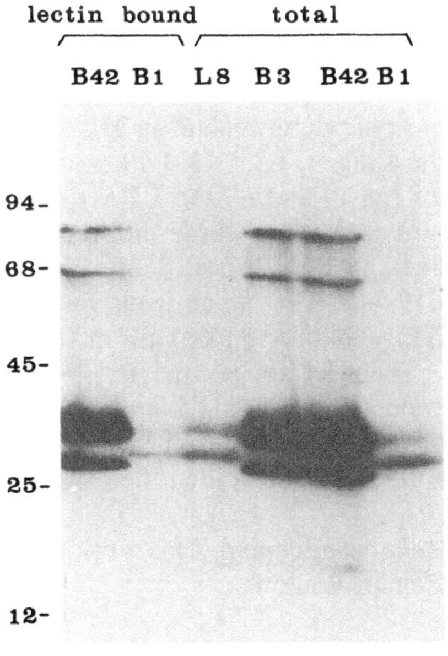

Fig. 17.10. Comparison of p29,34,90 and HLA-DR. Total extract or the lentil lectin-bound fraction of ^{125}I-Daudi cells were immunoprecipitated with the stated mAb.

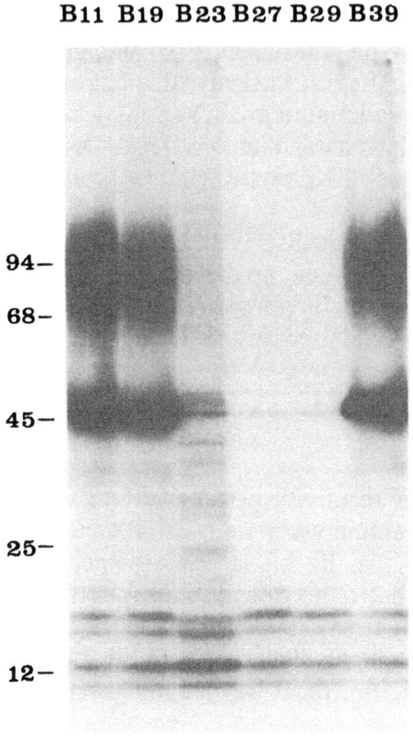

Fig. 17.11. SDS–PAGE and autoradiography of antigens immunoprecipitated from 125I-IM9 cell extracts with B cell-specific mAbs.

p24) antigen. L22, although closely related to this cluster, recognizes a different epitope. L2, L10, L11, L14, L15, and L21 all compete for the same epitope on CD10 (nTnB, gp100). L8, B1, and B12 recognize a monomorphic determinant on HLA-DR (gp31,37). According to the M.W. of the antigen, L3, L13, L19, and L20 could recognize the transferrin receptor (gp90) and L9 the LFA-1 molecule (gp105,160).

We identified three different B-restricted antigens: 1) gp29,34,90, defined by B42; 2) p47, defined by B17; 3) p48, 72–100, recognized by B11, B19, and B39. No antigens were detected with B27, with mAbs of the B5, B22, B24 cluster, or with mAbs of the B23, B29 cluster.

The AL1 antigen (p110 ?,180,220) defined by B15 and B18 is not only expressed on non-T lymphoblasts, but also on solid tumor cells. Related to these are the BA1-like mAbs, B47 and B48, for which a different antigen has been identified (see this volume, Chapter 18).

Acknowledgment. This work was supported by grant no. 818543 from the Region Wallone.

References

1. Lebacq-Verheyden, A.M., A.M. Ravoet, H. Bazin, D.R. Sutherland, N. Tidman, and M.F. Greaves. 1983. Rat AL2, AL3, AL4 and AL5 monoclonal antibodies bind to the common acute lymphoblastic leukemia antigen (CALLA gp 100). *Int. J. Cancer* **32:**273.
2. Sutherland, D.R., J. Smart, P. Niaudet, and M.F. Greaves. 1978. Acute lymphoblastic leukaemia associated antigen. II. Isolation and partial characterization. *Leuk. Res.* **2:**115.
3. Abramson, C., J. Kersey, and T. LeBien. 1981. A monoclonal antibody (BA1) reactive with cells of human B lymphocyte lineage. *J. Immunol.* **126:**83.
4. Kersey, J.H., T.W. LeBien, C.S. Abramson, R. Newman, R. Sutherland, and M. Greaves. 1981. p24: A human leukemia-associated and lymphohemopoietic progenitor cell surface structure identified with monoclonal antibody. *J. Exp. Med.* **153:**726.

CHAPTER 18

A Structurally Novel Human B Cell Surface Molecule

Samuel J. Pirruccello and Tucker W. LeBien

Introduction

Monoclonal antibody BA-1 was produced in this laboratory by immunization of mice with the pre-B human leukemic cell line NALM-6 (1). BA-1 binds to cells at multiple stages of B cell development, including B cell precursors, but does not bind to plasma cells or normal and malignant T cells. BA-1 is also reactive with granulocytes and several non-hematopoietic tissues, most notably adult and fetal kidney (2). Biochemical characterization of the BA-1 antigen has been hampered by our inability to reproducibly radiolabel the molecule using conventional protein labeling techniques. We have, however, recently defined the BA-1 antigen as a three-chain, nondisulfide-linked glycoprotein complex of 45, 55, and 65 kilodaltons (gp45/55/65) (3) by cell surface radiolabeling of galactosyl or sialosyl residues with $^{3}H\text{-}NaBH_4$ (4,5). We therefore undertook an investigation of the B cell Workshop antibodies to identify other monoclonal antibodies which recognize gp45/55/65. This was accomplished by immunofluorescent screening of the B cell Workshop panel followed by radioimmunoprecipitation and SDS–polyacrylamide gel electrophoresis of those antibodies exhibiting a serologic profile similar to BA-1. We report here the finding of two Workshop antibodies which recognize a molecular complex that appears similar to gp45/55/65 by SDS–PAGE. We also include a brief summary of the known biochemical characteristics of gp45/55/65.

Materials and Methods

Immunofluorescence

Indirect immunofluorescent staining of human leukemic cell lines was performed as previously described (1), utilizing Workshop antibodies at the requested final dilution of 1:250. Examination of stained cells was performed on a Zeiss fluorescent microscope equipped with Ploem epiillumination. Fluorescence intensity was rated 0 to 3+, with BA-1 staining of HPB-NULL cells rated 3+, and control ascites rated 0. The intensity rating reflected the brightness of individual cells when 50% or more of the cell population bound a given antibody. Human leukemic cell lines (6) used included the Burkitt's lymphoma cell line Raji, the erythroleukemia cell line K562, and the pre-B acute lymphoblastic leukemia (ALL) cell line HPB-NULL. Chronic lymphocytic leukemia (CLL) cells cryopreserved in 10% DMSO were obtained through the Cell Marker Laboratory, Department of Laboratory Medicine and Pathology. A single peripheral blood specimen obtained from a CLL patient at the time of diagnosis was thawed in RPMI-1640 containing 5% FBS for use in the screening panel.

Radiolabeling

Cell surface labeling of galactosyl residues with neuraminidase/galactose oxidase/^{3}H-$NaBH_4$ was performed essentially according to Gahmberg (4) as previously described (7). HPB-NULL and NALM-6 cells were used for surface radiolabeling of galactosyl residues. Labeled cells were lysed in 0.5% NP-40 at 1×10^8 cells/ml as previously described (7).

Radioimmunoprecipitation (RIP) and Sodium Dodecyl Sulfate–Polyacrylamide Gel Electrophoresis (SDS–PAGE)

These experiments were conducted according to methods we have previously described in detail (8). Briefly, 100 μl of radiolabeled lysate were incubated with 50 μl of BA-1 ascites, W6/32 (American Type Culture Collection, Rockville, MD), MOPC IgM (Bionetics, Charleston, SC), or the appropriate Workshop antibody, overnight at 4°C. Antigen–antibody complexes were precipitated by incubation with 50 μl of 10% protein A–Sepharose CL-4B (Pharmacia, Uppsala, Sweden), precoated with 25 μl of rabbit anti–mouse IgM (Miles Scientific, Naperville, IL), at 4°C for one hr. The complexes were eluted by boiling in SDS sample buffer and electrophoresed under reducing conditions on a 12.5% polyacrylamide gel. The rabbit anti–mouse IgM cross-reacted with mouse IgG, and precipitated all eight Workshop antibodies used, as evidenced by monoclonal banding of immunoglobulin light chains on the coomassie blue-stained gels.

Results and Discussion

Immunofluorescent Screening

Immunofluorescent screening of the B cell Workshop antibodies utilized a panel of four different cell targets; HPB-NULL, K562, Raji, and CLL cells. These cells were chosen in order to identify the Workshop antibodies showing serologic identity to BA-1 prior to RIP and SDS–PAGE. Only those Workshop antibodies reactive with all four cell targets were used for immunoprecipitation experiments. As shown in Table 18.1, eight B cell Workshop antibodies fulfilled the serologic criteria. These included antibodies B1, B14, B31, B34, B44, B47, B48, and B49. Most notable were antibodies B14, B47, B48, and B49 which reacted with 50% of K562 cells. This staining pattern was similar to BA-1, and suggested antigen homology among these four Workshop antibodies.

RIP and SDS–PAGE

The eight Workshop antibodies identified by immunofluorescent screening were utilized for RIP and SDS–PAGE of galactose oxidase/^{3}H-$NaBH_4$-radiolabeled HPB-NULL and NALM-6 lysates. W6/32, which recognizes a monomorphic determinant on all HLA-A,B,C molecules, was used as a positive control. A myeloma MOPC IgM was used as a negative control.

Initial RIP and SDS–PAGE of the eight workshop antibodies using radiolabeled HPB-NULL lysates revealed that B47 and B48 precipitated a two-chain complex of 45 and 55 Kd which co-migrated with the 45- and 55-Kd components of gp45/55/65 recognized by BA-1. These bands were present after four weeks of fluorographic exposure, but were insufficient for photographic demonstration. Of the other six antibodies, only two gave discernible precipitates. Workshop antibody B1 precipitated a two-chain glycoprotein of 28 and 33 Kd, consistent with HLA-DR, and Workshop antibody B14 precipitated a 90-Kd glycoprotein. Three of the remaining four antibodies were subsequently identified as recognizing antigens distinct from gp45/55/65 by surface labeling with ^{125}I-lactoperoxidase (see this volume, Chapter 13). These included B31 and B49 (M_r 130 Kd), and B34 (M_r 90 Kd). B44 gave no precipitates with either labeling technique. RIP and SDS–PAGE with B47 and B48 was subsequently repeated using ^{3}H-$NaBH_4$-labeled NALM-6 lysate (Fig. 18.1). Both antibodies again precipitated a 55-Kd glycoprotein that co-migrates with the 55-Kd component of gp45/55/65. In addition, a 65-Kd component could be seen visually in the B47 lane but did not photograph. Neither antibody precipitated a clearly discernible 45-Kd chain from NALM-6. This may reflect lower affinity, lower antibody titers, or both. It is also possible that B47 and B48 are selectively precipitating only one or two chains of the three-chain complex. All three chains are accounted for by combining the data from the HPB-NULL and NALM-6 precipitations with B47. Obviously, more rigorous, sequential precipitations would be required to definitively prove that B47 and B48 recognize the identical gp45/55/65 recognized by BA-1.

As summarized in Table 18.2, and reported in detail elsewhere (3), gp45/55/65 is quite novel in both its structure and labeling properties. None of the components can be surface-labeled with [^{125}I]lactoperoxidase

Table 18.1. Serologic/structural comparison of B cell workshop antibodies with BA-1.

	Cells tested				RIP & SDS–PAGE
Ab	Raji	CLL	K562	HPB-NULL	gp45/55/65
BA-1	++[a]	+++	++[b]	+++	+
B1	+++	+++	+	++	–
B14	+++	+	++[b]	++	–
B31	++	+	+	+	–
B34	++	+	+	+	–
B44	++	+++	+	+++	–
B47	++	+++	+[b]	++	+[c]
B48	++	+++	+[b]	+++	+[c]
B49	++	+	+[b]	++	–

[a] + Bright, ++ brighter, +++ brightest by indirect immunofluorescent staining.
[b] 50% of cell population showed positive staining.
[c] All three chains were not consistently precipitated.

or biosynthetically with [^{35}S]methionine, or a cocktail of several ^{3}H- or ^{14}C-amino acids. Surprisingly, however, the molecule (or possibly the epitope) is quite pronase sensitive on the cell surface. SDS–PAGE under reducing and nonreducing conditions shows that none of the three chains in the complex are disulfide-linked. There are 5-kilodalton (Kd) shifts in

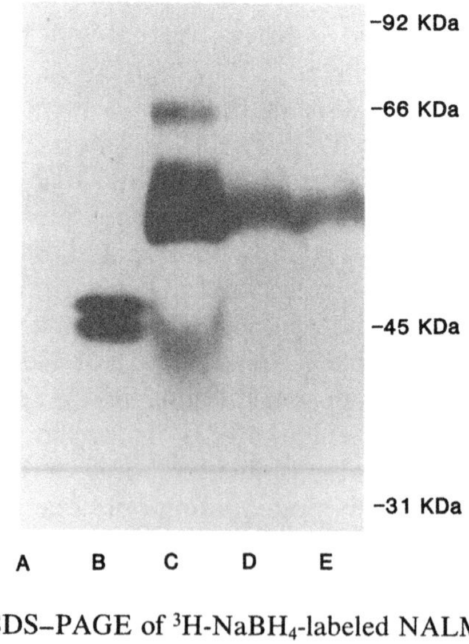

Fig. 18.1. RIP and SDS–PAGE of ^{3}H-NaBH$_4$-labeled NALM-6 lysates. Lane A, MOPC IgM (negative control); lane B, W6/32 (positive control); lane C, BA-1; lane D, Workshop antibody B47; lane E, Workshop antibody B48. Radiolabeled precipitates were run on a 12.5% SDS polyacrylamide gel under reducing conditions. M_r markers are indicated at the side of the gel. Exposure time = 4 weeks.

Table 18.2. Biochemical properties of gp45/55/65.

SDS–PAGE		Radiolabeling properties				IEF
M_r reduced (Kd)	M_r nonreduced (Kd)	Galactose	Sialic acid	[^{35}S]Met	[^{125}I]lactoperoxidase	pI
45	45	+	+	−	−	neutral
55	50	+	+	−	−	acidic
65	60	+	−	−	−	?

the molecular mass of the 55- and 65-Kd components on going from reducing to nonreducing conditions, suggesting the presence of intrachain disulfide linkages. The molecular mass shifts, together with the pronase sensitivity, are consistent with primary and secondary protein structure. Our inability to biosynthetically or surface radiolabel amino acid residues in the molecular complex suggests that the protein cores of all three components may be quite small. Alternatively, the molecules may have an unusual amino acid content. By surface labeling with periodate/^{3}H-$NaBH_4$ (5), the 55-Kd component is seen to contain sialic acid residues and has an acidic pI on isoelectric focusing. The 45-Kd component also contains sialic acid residues but focuses near neutrality, suggesting the presence of basic amino acids in the protein backbone. The 65-Kd component does not appear to label with periodate/^{3}H-$NaBH_4$ and the isoelectric point after labeling of galactosyl residues has not yet been identified.

In sequential immunoprecipitation experiments (3) we have shown that BA-1 and monoclonal antibody OKB2 (9) both recognize gp45/55/65. In this study we have identified two additional Workshop antibodies, B47 and B48, which also recognize gp45/55/65. This is consistent with a preliminary report which suggested that B47 and B48, designated HB-8 and HB-9 by Tedder *et al.* (10), are serologically identical to BA-1. It is probable that the monoclonal antibody αBL1, recently described by Wang *et al.* (11), recognizes gp45/55/65. This is predicted based on the remarkable similarities in serologic distribution between αBL1 and BA-1, and the inability to precipitate an antigen with αBL1 after surface labeling with [^{125}I]lactoperoxidase.

We have been able to precipitate gp45/55/65 from several hematopoietic cell sources, including two pre-B human leukemic cell lines (HPB-NULL and NALM-6), a newly diagnosed ALL, tonsillar lymphocytes, and peripheral blood granulocytes (3). By immunofluorescence the BA-1 *epitope* is also detected on other tissues including adult and fetal kidney (2), neuroblastoma cells (12), and a human gastrointestinal adenocarcinoma cell line SW480 (unpublished observations). It is currently unclear whether BA-1 is recognizing a common epitope on disparate molecules or identical antigens in these disparate tissues. This question can be addressed by surface labeling with galactose oxidase/^{3}H-$NaBH_4$ and SDS–

PAGE. Information about the actual tissue distribution of gp45/55/65 may yield some clues to the biologic function of this novel cell surface molecule.

The gp45/55/65 molecular complex recognized by BA-1 is structurally distinct from previously described human B cell-associated antigens, including those recognized by monoclonal antibodies: anti-B1 (30 Kd) (13,14), anti-B2 and HB-5 (140 Kd) (10,14), anti-B4 (40, 80 Kd) (15), anti-BL2 (68 Kd) (11), anti-BL3 (105 Kd) (11), OKB1 (168 Kd) (9), OKB4 (87 Kd) (9), OKB7 (175 Kd) (9), BB-1 (37 Kd) (16), B-LAST-1 (45 Kd) (17), F8-11-13 (220 Kd) (18), 41H.16 (39 Kd) (19) as well as newly defined antigens that were part of this workshop (see this volume, Chapters 12 and 13). Many laboratories are actively pursuing the function these molecules subserve to human B cells. Although some progress is being made (20,21) most of the aforementioned molecules have no assignable function as of this writing.

Summary

Monoclonal antibody BA-1 was produced by immunization of mice with the pre-B human leukemic cell line NALM-6. After cell surface radiolabeling of galactosyl residues with ^3H-$NaBH_4$, BA-1 consistently precipitates a three-chain, nondisulfide-linked, glycoprotein complex of 45, 55, and 65 kilodaltons (gp45/55/65). This antigen is distinct from previously described B cell antigens and is quite novel in its structure and radiolabeling properties. Other antibodies which recognize gp45/55/65 include OKB2 and Workshop antibodies B47 and B48. The function of gp45/55/65 is currently unknown.

Acknowledgments. This work was supported by grants CA-31685 and RR-05385 from the National Institutes of Health. T.W. LeBien is a Scholar of the Leukemia Society of America.

References

1. Abramson, C.S., J.H. Kersey, and T.W. LeBien. 1981. A monoclonal antibody (BA-1) reactive with cells of human B lymphocyte lineage. *J. Immunol.* **126:**83.
2. Platt, J.L., T.W. LeBien, and A.F. Michael. 1983. Stages of renal ontogenesis identified by monoclonal antibodies reactive with lymphohematopoietic differentiation antigens. *J. Exp. Med.* **157:**155.
3. Pirruccello, S.J., and T.W. LeBien. 1985. Monoclonal antibody BA-1 recognizes a novel human leukocyte cell surface sialoglycoprotein complex. *J. Immunol.* **134:**3962.
4. Gahmberg, C.G. 1976. External labeling of human erythrocyte glycoproteins. Studies with galactose oxidase and fluorography. *J. Biol. Chem.* **251:**510.

5. Gahmberg, C.G., and L.C. Anderson. Selective radioactive labeling of cell surface sialoglycoproteins by periodate-tritiated borohydride. *J. Biol. Chem.* **252:**5888.
6. Minowada, J., G. Janossy, M.F. Greaves, T. Tsubota, B.I. Sahai Srivastava, S. Morikawa, and E. Tatsumi. 1978. Expression of an antigen associated with acute lymphoblastic leukemia in human leukemia–lymphoma cell lines. *J. Natl. Cancer Inst.* **60:**1269.
7. LeBien, T.W., J.G. Bradley, and B. Koller. 1981. Preliminary structural characterization of the leukocyte cell surface molecule recognized by monoclonal antibody TA-1. *J. Immunol.* **130:**1833.
8. LeBien, T.W., D.R. Boué, J.G. Bradley, and J.H. Kersey. 1982. Antibody affinity may influence modulation of the common acute lymphoblastic leukemia antigen *in vitro*. *J. Immunol.* **129:**2287.
9. Mittler, R.S., M.A. Talle, K. Carpenter, P.A. Rao, and G. Goldstein. 1983. Generation and characterization of monoclonal antibodies reactive with human B lymphocytes. *J. Immunol.* **131:**1754.
10. Tedder, T.F., L.T. Clement, and M.D. Cooper. 1983. Use of monoclonal antibodies to examine differentiation antigens on human B cells. *Fed. Proc.* **42:**415A.
11. Wang, C.Y., W. Azzo, A. Al-Katib, N. Chiorazzi, and D.M. Knowles II. 1984. Preparation and characterization of monoclonal antibodies recognizing three distinct differentiation antigens (BL1, BL2, BL3) on human B lymphocytes. *J. Immunol.* **133:**684.
12. Kemshead J.T., J. Fritschy, U. Asser, R. Sutherland, and M.F. Greaves. 1982. Monoclonal antibodies defining markers with apparent selectivity for particular hematopoietic cell types may also detect antigens on cells of neural crest origin. *Hybridoma* **1:**109.
13. Stashenko, P., L.M. Nadler, R. Hardy, and S.F. Schlossman. 1980. Characterization of a human B lymphocyte-specific antigen. *J. Immunol.* **125:**1678.
14. Nadler, L.M., P. Stashenko, R. Hardy, A. van Agthoven, C. Terhorst, and S.F. Schlossman. 1981. Characterization of a human B cell specific antigen (B2) distinct from B1. *J. Immunol.* **126:**1941.
15. Nadler, L.M., K.C. Anderson, G. Marti, M. Bates, E. Park, J.F. Daley, and S.F. Schlossman. 1983. B4, a human B lymphocyte-associated antigen expressed on normal, mitogen-activated, and malignant B lymphocytes. *J. Immunol.* **131:**244.
16. Yokochi, T., R.D. Holly, and E.A. Clark. 1982. B lymphoblast antigen (BB-1) expressed on Epstein Barr-activated B cell blasts, B lymphoblastoid cell lines, and Burkitt's lymphoma. *J. Immunol.* **128:**823.
17. Thorley-Lawson, D.A., R.T. Schooley, A.K. Bhan, and L.M. Nadler. 1982. Epstein Barr virus superinduces a new human B cell differentiation antigen (B-LAST-1) expressed on transformed lymphoblasts. *Cell* **30:**415.
18. Dalchau, R., and J.W. Fabre. 1981. Identification with a monoclonal antibody of a predominantly B lymphocyte-specific determinant of the human leukocyte common antigen. *J. Exp. Med.* **153:**753.
19. Zipf, T.F., G.J. Lauzon, and B.M. Longenecker. 1983. A monoclonal antibody detecting a 39,000 M.W. molecule that is present on B lymphocytes and chronic lymphocytic leukemia cells but is are on acute lymphocytic leukemia blasts. *J. Immunol.* **131:**3064.

20. Iido, K., L. Nadler, and V. Nussenzweig. 1983. Identification of the membrane receptor for the complement fragment C3d by means of a monoclonal antibody. *J. Exp. Med.* **158:**1021.
21. Wiess, J.J., T.F. Tedder, and D.T. Fearon. 1984. Identification of a 145,000 Mr membrane protein as the C3d receptor (CR2) of human B lymphocytes. *Proc. Natl. Acad. Sci. U.S.A.* **81:**881.

CHAPTER 19

Human Neutrophils Synthesize Different Forms of the Common Acute Lymphoblastic Leukemia Antigen

Robert T. McCormack, J. Garrett Bradley, and Tucker W. LeBien

Introduction

The common acute lymphoblastic leukemia antigen (CALLA) is a nonintegral, 100-kilodalton (Kd) membrane glycoprotein (1), initially determined to be expressed predominantly on acute lymphoblastic leukemia (ALL) cells (2). Recently, the expression of CALLA has been extended to include several other cell types, including neutrophils (3,4). Anti-CALLA monoclonal antibodies precipitated a molecule from ^{125}I-labeled neutrophils that had a M_r slightly larger than that of the form precipitated from ALL cells (3,4). Furthermore, the reactivity was mostly limited to the mature neutrophil, with only slight reactivity being detected with the less mature stages of neutrophil development (4). To date, however, it has not been determined if neutrophils actually synthesize CALLA, or, for example, passively adsorb it from plasma.

Using FACS analysis, we identified six leukemic cell Workshop monoclonal antibodies that bound to neutrophils and also precipitated a CALLA-like molecule. Furthermore, we have used the anti-CALLA monoclonal antibody BA-3 (5) to study the synthesis and expression of CALLA by normal human neutrophils.

Materials and Methods

Cells

Granulocytes from normal donors were harvested from the pellets of Ficoll–Hypaque gradients (6) and freed of contaminating erythrocytes by hypotonic lysis. Final preparations contained >95% neutrophils. The established human B cell precursor leukemia cell line, Nalm-6, was also used.

Antibodies

Workshop antibodies L2, L6, L10, L11, L14, L15, and L21 were selected based on their specificity for leukemic cell CALLA (this volume, Chapter 13). BA-2 and BA-3 were produced in this laboratory as previously described (5,7). W6/32, a monoclonal antibody that recognizes an epitope on HLA-A,B,C molecules (8), was produced from hybridoma cells obtained from the American Type Culture Collection. Control ascitic fluid was obtained by injecting BALB/c mice with non-antibody-secreting hybridomas or the parent (NS-1) myeloma cell line.

Immunofluorescence assays

Antibody binding was analyzed by indirect immunofluorescence using the fluorescence-activated cell sorter (FACS-IV, Becton Dickinson, Mountainview, CA) as previously described (5).

Radiolabeling of CALLA, Radioimmunoprecipitation, and Sodium Dodecyl Sulfate–Polyacrylamide Gel Electrophoresis (SDS–PAGE)

Neutrophils were biosynthetically labeled using [^{35}S]methionine. Cells were suspended to 2–4 × 10^6/ml in RPMI 1640 formula 78-0404 (Gibco, Grand Island, NY) methionine-free media containing 10% dialyzed, heat-inactivated human serum, 1% Penn-Strep (Gibco), and 1% glutamine (Gibco). 100 μCi of [^{35}S]methionine (1200 Ci/mmole, New England Nuclear, Boston, MA) per 20 × 10^6 cells were added and cultures were incubated for 8 hr at 37°C in 5% CO_2/95% air, with gentle vortexing every hour.

Neutrophils at 5 × 10^7/ml in phosphate-buffered saline (PBS) were surface radiolabeled with 2.5 mCi of ^{125}I (New England Nuclear, Boston, MA) using the lactoperoxidase method as previously described (5).

Following radiolabeling, cells were washed in PBS and lysed in 0.1 *M* Tris buffer, pH 8.1, containing 0.9% NaCl, 0.5% NP–40 (Nonidet P-40, Particle Data Laboratories, Elmhurst, IL), 2 m*M* phenylmethyl sulfonyl fluoride (PMSF, Sigma) and 1% aprotinin (Sigma). The lysates were centrifuged at 20,000 × *g* and absorbed with protein A–Sepharose (Pharmacia Fine Chemicals, Piscataway, NJ). Precleared lysates were either used immediately or frozen at −70°C.

50–100 μl of radiolabeled lysates were mixed with 20 μg of BA-3 or CAF and incubated overnight at 4°C. Antigen–antibody complexes were precipitated with a 10% protein A–Sepharose solution in lysis buffer and processed as previously described (5). Radiolabeled antigens were resolved by SDS–PAGE on 12.5% gels using the Tris glycine buffer system described by Laemmli (9). ^{125}I-labeled proteins were visualized by autoradiography of dried gels using Kodak X-Omat XAR-5 film and intensifier screens, and were stored at −70°C for exposure. ^{35}S-labeled proteins were visualized by treating the gels with 1 *M* sodium salicylate (Aldrich, Milwaukee, WI) according to the method of Chamberlain (10). Dried gels were placed on Kodak X-Omat XAR-5 film and stored at −70°C for exposure.

Results and Discussion

Workshop antibodies that precipitated a 100-Kd molecule from the surface of Nalm-6 cells (see this volume, Chapter 13) were tested for binding activity against normal human neutrophils. As shown in Fig. 19.1, Work-

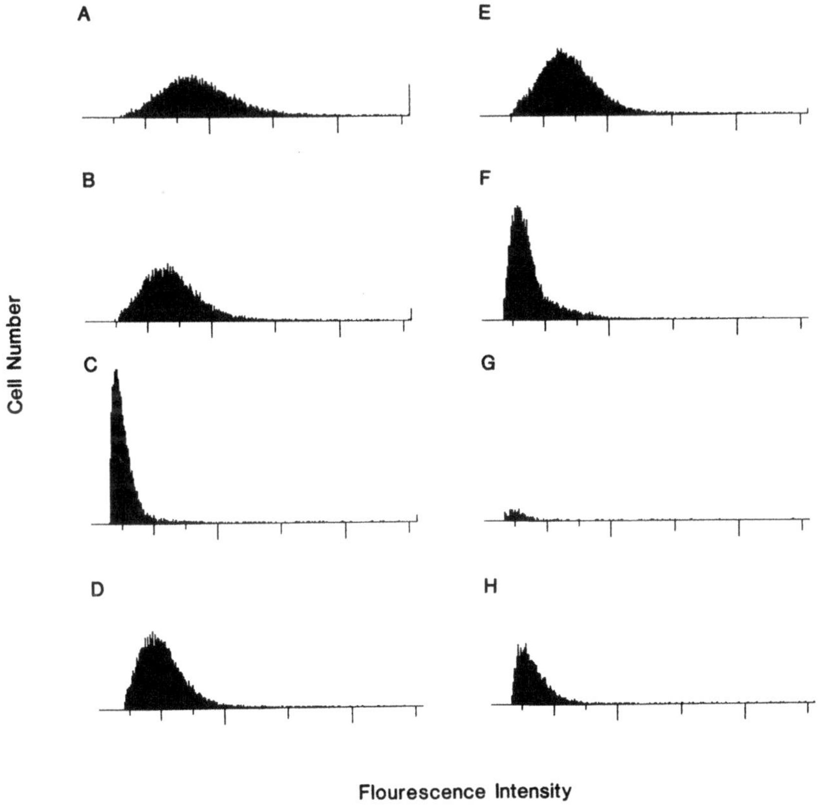

Fig. 19.1. FACS analysis of anti-CALLA Workshop antibodies tested against normal neutrophils. The *x*-axis indicates relative fluorescence intensity and the *y*-axis indicates relative cell number. All figures are corrected for background using control ascitic fluid. (A) BA-3 ascites, positive control; (B) L2; (C) L6; (D) L10; (E) L11; (F) L14; (G) L15; (H) L21.

shop antibodies L2, L6, L10, L11, L14, and L21 showed appreciable binding to neutrophils. Percent positives ranged from 50% (L21) to 86% (L11), with a mean percent positive of 72 ± 13. Antibody L15 is a rat IgM anti-CALLA reagent (AL3) (11) which showed weak binding to neutrophils when tested by FACS analysis [Fig. 19.1(G)]. In our hands L15 also reacted poorly with CALLA-positive leukemic cells, and it is possible that our FITC goat anti–mouse Ig reagent did not recognize rat IgM antibodies.

Figure 19.2 shows the results of an experiment with human neutrophils biosynthetically labeled with [^{35}S]methionine. Using monoclonal antibody BA-3, we were able to precipitate a CALLA-like molecule with an apparent molecular mass between 95 Kd and 110 Kd (Fig. 19.2, lane 4). As expected, neutrophils synthesized Class I MHC molecules as evidenced by immunoprecipitation with antibody W6/32 (Fig. 19.2, lane 2). We were unable to precipitate p24 from neutrophils with monoclonal antibody BA-2 (Fig. 19.2, lane 3). Neutrophils usually show weak reactivity with BA-2. Our failure to detect synthesis could indicate an extremely low level of synthesis, or the weak reactivity detected by FACS analysis could result from contamination with platelets, which are p24 positive and known to adhere to neutrophils.

When Nalm-6 cells and human neutrophils were surface labeled with ^{125}I, the neutrophil form of CALLA was seen to have a M_r greater than the Nalm-6 form (data not shown). However, our initial experiments led us to believe that this difference in M_r between Nalm-6 cells and neutrophils was not constant, but appeared to vary depending on the source of donor neutrophils. We therefore isolated neutrophils from six normal donors,

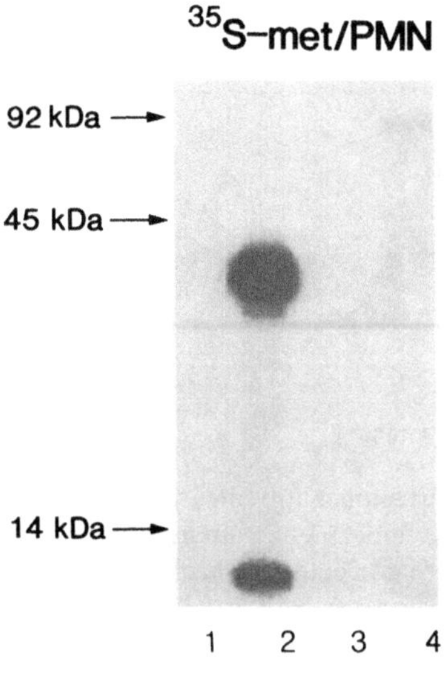

Fig. 19.2. SDS–PAGE and fluorograph of normal neutrophils biosynthetically labeled with [^{35}S]methionine. Lane 1, CAF; lane 2, W6/32; lane 3, BA-2; lane 4, BA-3.

radiolabeled them with ^{125}I, immunoprecipitated the NP-40 lysates with BA-3, and analyzed the immunoprecipiated antigens on a 12.5% polyacrylamide gel. Figure 19.3 shows the results of one such experiment. At least two forms of neutrophil CALLA can be seen. Both forms have a M_r greater than CALLA isolated from Nalm-6 cells, and vary from each other by about 6 Kd, depending on the gel conditions.

Although CALLA was initially believed to be expressed only on non-B, non-T ALL cells and their nonmalignant counterparts in normal bone marrow (12), reports have recently appeared describing the occurrence of CALLA on several other diverse cell types (13–16). The reports by Cossman *et al.* (3) and Braun *et al.* (4) showed that normal mature human neutrophils also express CALLA. However, it had not been determined whether normal neutrophils actually synthesize CALLA or, for example, passively adsorb it from the plasma. The latter phenomenon is not a trivial possibility and has recently been reported for human platelets, which were shown to adsorb Class I molecules from the plasma (17).

We have shown that Workshop antibodies that precipitate a 100-Kd molecule from Nalm-6 cells also bind to human neutrophils when tested by FACS analysis, confirming earlier reports (3,4). Furthermore, using BA-3, we were able to immunoprecipitate a molecule from neutrophils biosynthetically labeled with [^{35}S]methionine—providing unequivocal evidence that neutrophils synthesize CALLA. Normal mature neutrophils undergo little protein synthesis (18), consistent with the relative paucity of both ribosomes and endoplasmic reticulum. However, we found that 8-hr incubation at 37°C was sufficient to detect not only CALLA synthesis, but Class I heavy- and light-chain synthesis as well.

Our initial investigations led us to believe that the differences in M_r

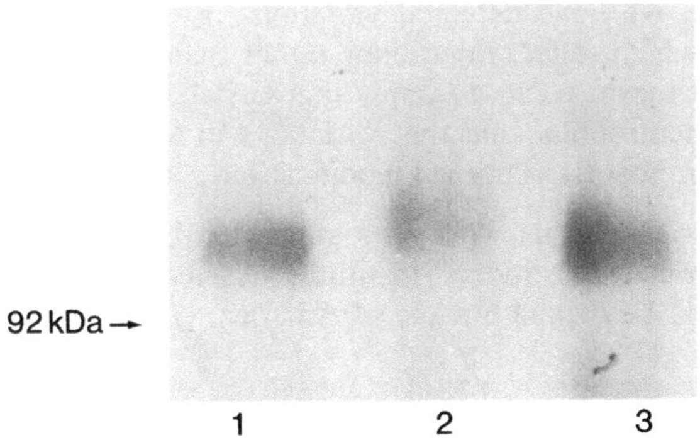

Fig. 19.3. SDS–PAGE and autoradiograph of ^{125}I-labeled normal neutrophil lysates immunoprecipitated with BA-3. Lanes 1 and 3, two different normal donors with faster-migrating form of CALLA. Lane 2, third donor with slower-migrating form of CALLA.

between neutrophil and leukemic cell CALLA were not invariable. Using neutrophils isolated from several normal donors, we were able to identify at least two forms of neutrophil CALLA, both with a greater M_r than the leukemic cell form. Using various concentrations of polyacrylamide gels, we were able to discern a difference of about 6 Kd between the two forms. To date, we have been able to detect the faster-moving form on neutrophils from 75% of normal donors.

We have also been investigating whether the variations in M_r of neutrophil CALLA are unique to this cell type, or whether these variations exist with other CALLA-positive cells. Evidence from our laboratory shows the latter to be true; in data to be reported elsewhere we have detected at least three forms of CALLA on leukemic cells isolated from patients with ALL (19).

Newman *et al.* reported that CALLA contains about 25% carbohydrate (1). It is likely, therefore, that the M_r differences seen in the various forms of neutrophil and leukemic cell CALLA can best be explained by post-translational modifications (i.e., glycosylation). We are currently exploring this possibility with the aid of various glycosidic enzymes.

Finally, since we have shown that neutrophils synthesize CALLA, we reason that CALLA subserves some functional role to neutrophils. Neutrophils are specialized cells whose triad of functions—motility, phagocytosis, and microbial killing—lend themselves readily to *in vitro* testing. Experiments are currently in progress to address these possibilities.

Summary

Human neurophils have recently been shown to express CALLA, a 100-Kd glycoprotein. We have extended this observation by providing unequivocal evidence that neutrophils synthesize the CALLA molecule. Furthermore, we have detected variations in the M_r of CALLA from neutrophils which reflect interdonor rather than intradonor differences. We interpret these results to imply that CALLA subserves a particular function to neutrophils, and that variations in M_r are best explained by differences in post-translational modification.

Acknowledgments. This work was supported by grants CA-31685 and RR-05385 from the National Institutes of Health. T.W. LeBien is a Scholar of the Leukemia Society of America.

References

1. Newman, R., R. Sutherland, and M. Greaves. 1981. A biochemical characterization of a cell surface antigen associated with acute lymphoblastic leukemia and lymphocyte receptors. *J. Immunol.* **126:**2024.
2. Greaves, M.F., G. Brown, N. Rapson, and T. Lister. 1975. Antisera to acute lymphoblastic leukemia cells. *Clin. Immunol. Immunopathol.* **4:**67.
3. Cossman, J., L. Neckers, W. Leonard, and W. Greene. 1983. Polymorphonu-

clear neutrophils express the common acute lymphoblastic leukemia antigen. *J. Exp. Med.* **157:**1064.
4. Braun, M., P. Martin, J. Ledbetter, and J. Hansen. 1983. Granulocytes and cultured human fibroblasts express common acute lymphoblastic leukemia-associated antigens. *Blood* **61:**718.
5. LeBien, T., D. Boue, J. Bradley, and J. Kersey. 1982. Antibody affinity may influence antigenic modulation of the common acute lymphoblastic leukemia antigen *in vitro*. *J. Immunol.* **129:**2287.
6. Boyum, A. 1968. Isolation of mononuclear cells and granulocytes from human blood. Isolation of mononuclear cells, and of granulocytes by combining centrifugation and sedimentation at 1g. *Scand. J. Clin. Lab. Invest.* **21**(Suppl. 97)**:**77.
7. Kersey, J., T. LeBien, C. Abramson, R. Newman, R. Sutherland, and M. Greaves. 1981. p24: A human hematopoietic progenitor and acute lymphoblastic leukemia-associated cell surface structure identified with monoclonal antibody. *J. Exp. Med.* **153:**726.
8. Barnstable, C., W. Bodner, G. Brown, G. Galfre, C. Milstein, A. Williams, and A. Ziegler. 1978. Production of monoclonal antibodies to Group A erythrocytes, HLA and other human cell surface antigens—New tools for genetic analysis. *Cell* **14:**9.
9. Laemmli, U. 1970. Cleavage of structural proteins during the assembly of the head of bacteriophage T4. *Nature* **227:**680.
10. Chamberlain, J. 1979. Fluorographic detection of radioactivity in polyacrylamide gels with water-soluble fluor, sodium salicylate. *Anal. Biochem.* **9:**132.
11. Lebacq-Verheyden, A.-M., A.-M., Ravoet, H. Bozin, D.R. Sutherland, N. Tidman, and M.F. Greaves. 1983. Rat AL2, AL3, AL4, and AL5 monoclonal antibodies bind to the common acute lymphoblastic leukaemia antigen (CALLA gp100). *Int. J. Cancer* **32:**273.
12. Greaves, M., D. Delia, G. Janossy, N. Rapson, J. Chessells, M. Woods, and G. Prentice. 1979. Acute lymphoblastic leukaemia associated antigen. IV. Expression on non-leukaemic 'lymphoid' cells. *Leuk. Res.* **4:**15.
13. Metzgar, R., M. Borowitz, N. Jones, and B. Dowell. 1981. Distribution of common acute lymphoblastic leukemia antigen in nonhematopoietic tissues. *J. Exp. Med.* **154:**1249.
14. Platt, J., T. LeBien, and A. Michael. 1983. Stages of renal ontogenesis identified by monoclonal antibodies reactive with lymphohemopoietic differentiation antigens. *J. Exp. Med.* **157:**155.
15. Pesando, J., K. Tomaselli, H. Lazarus, and S. Schlossman. 1983. Distribution and modulation of human leukemia-associated antigen (CALLA). *J. Immunol.* **131:**2038.
16. Keating A., C.K. Whalen, and J.W. Singer. 1983. Cultured marrow stromal cells express common acute lymphoblastic leukaemia antigen (CALLA): implications for marrow transplantation. *Brit. J. Haematol.* **55:**623.
17. Blumberg, M., D. Masel, T. Mayer, P. Horam, and J. Heal. 1984. Removal of HLA-A,B antigens from platelets. *Blood* **63:**448.
18. Wade, B., and G. Mandell. 1983. Polymorphonuclear leukocytes: Dedicated professional phagocytes. *Am. J. Med.* **74:**686.
19. McCormack, R.T., and T.W. LeBien. 1985. Structure/function studies of the common acute lymphoblastic leukemia antigen (CALLA/CD10) on human neutrophils. Submitted for publication.

IV. Immunohistochemical Analysis of B Cell/Leukemia Panel Monoclonal Antibodies

CHAPTER 20

Immunohistochemical Analysis of Monoclonal Anti-B Cell Antibodies

D.Y. Mason, H. Ladyman, and K.C. Gatter

Introduction

The development of immunohistochemical methods for labeling tissue sections with monoclonal antibodies has provided a powerful means of analyzing these reagents, both during the initial screening and cloning steps necessary for the production of monoclonal antibodies (1), and also when investigating the specificity of established antibodies. The present report describes the immunocytochemical reaction patterns (on normal and neoplastic lymphoid tissue and on non-lymphoid tissues) of the monoclonal antibodies included in the B cell panel of the Workshop. The data presented in this report were compiled before the Workshop, but the results have been subsequently retabulated so as to correlate the immunohistological patterns obtained with the molecular weight (where known) of the antigens recognized.

Materials and Methods

Immunohistochemical staining of the majority of tissues was performed as described previously, using immunoperoxidase techniques applied to acetone-fixed frozen sections of human tissue (2). All antibodies were tested initially at two dilutions (1/250 and 1/1000) and then used subsequently at whichever of these dilutions gave the strongest reaction. For a few samples the APAAP immuno-alkaline phosphatase technique (3) was used in place of the immunoperoxidase method. Samples were obtained via the Surgical Histology Department of the John Radcliffe Hospital, Oxford and comprised specimens of normal tissue (e.g., skin, kidney, liver) and of biopsy material from cases of lymphoproliferative disorders. These latter cases had been previously classified by conventional histological criteria and by immunophenotyping on cryostat sections.

Staining was also performed on blood smears from a case of common acute lymphoblastic leukemia. Staining in this instance was performed by the APAAP immuno-alkaline phosphatase technique (4,5).

Results

Initial Screening of Antibodies

Of the 52 antibodies studied, 8 gave no reaction on sections of tonsil tissue (see Table 20.1). A further 18 either reacted very weakly with B cell areas or else gave reactions which indicated that they were not selectively reactive with B cells (see Table 20.1). None of the antibodies in any of these groups was investigated further.

Antibodies Selectively Reacting with B Cells

The remaining 26 antibodies all reacted with lymphoid follicles in tonsil sections, and were analyzed further by staining normal tissues and biopsies from cases of lymphoproliferative disorders. The patterns obtained are summarized in Table 20.2.

Anti-p35 (CD20)

Two of the three monoclonal antibodies in this category (B22 and B24) gave only weak labeling of B cell follicles when initially screened against tonsil sections (see Table 20.1) and were therefore not studied further. The third reagent, B5 [antibody B1 from Nadler *et al.* (6)] stained both

Table 20.1. Antibodies in the B cell panel which were not investigated further: Summary of their immunohistological reactions on tonsil sections.

Negative reactions
B1, B16, B27, B30, B32, B37, B46, B48
Weak staining of B cell areas
B8, B10, B22, B24
Staining of squamous epithelium alone
B13, B38
Staining of squamous epithelium plus other elements
B15, B18, B26, B29, B50
Miscellaneous staining reactions (vessels, fibers, etc.)
B2, B12, B20, B23, B44, B45, B47

mantle zone lymphocytes in B cell areas of lymphoid tissue and also showed a meshwork pattern of reactivity in germinal centers (Fig. 20.1, page 253). This latter pattern was closely similar to that obtained previously with antibodies recognizing dendritic reticulum cells (7,8).

No non-lymphoid tissue was stained by reagent B5, and the majority of B cell neoplasms (with the exception of the single case of common ALL tested) were labeled (see Table 20.2). However, the staining of B cells in normal lymphoid tissue was relatively weak (when compared with the results obtained using other pan-B antibodies, notably in categories CD19 and CD22—see below), in keeping with the authors' previous experience (D.Y. Mason and H. Stein, unpublished).

The two anti-p35 antibodies which gave weak reactions on initial screening (B22 and B24) were subsequently tested by the more sensitive APAAP technique on tonsil sections (3) and gave essentially identical reactions to reagent B5.

Anti-p45 (CD23)

The three reagents in this category (B11, B19, and B39) gave identical reactions. The majority of mantle zone lymphocytes were clearly labeled (Fig. 20.1). However, there was variation in the intensity of labeling between different cells, with the result that, at higher dilutions of these antibodies, a mosaic pattern of mixed positive and negative (or weak) cells (reminiscent of the labeling pattern of anti-Ig light-chain antibodies on these cells) was seen.

Within the germinal centers of lymphoid follicles these three antibodies also showed a selective pattern of cell staining, in that a meshwork was seen, restricted to certain areas (usually the upper "light" zones). This appearance suggested that these reagents recognize a subpopulation of dendritic reticulum cells. Germinal center lymphoid cells (e.g., centroblasts and centrocytes) were unlabeled.

When tested against neoplastic B cells the three reagents in this group labeled B cell CLL but gave weak or negative reactions with the majority of other B cell lymphoproliferative disorders. Labeling of probable normal dendritic reticulum cells was seen in many sections, although it was evident in this tissue (as in normal germinal centers) that only a subpopulation of these cells was detected.

Anti-p95 (CD19)

The three antibodies in this category gave essentially identical reactions; in reactive lymphoid tissue both mantle zone and germinal center lymphoid cells were stained (Fig. 20.1), and all B cell lymphoproliferative disorders gave positive reactions.

In germinal centers coarse labeling in a meshwork pattern was visible (Fig. 20.1), although this was not as intense as the staining with antibodies

Table 20.2. Immunocytochemical labeling by antibodies showing selective reactivity with B cells.[a,b]

	Normal tissues							B cell lympholiferative disorders			
	Ton.	LNM	Spl.	Kid.	Liv.	Br.	Lu.	CLL	ALL	HCL	LBL
Antibody to p35 (CD20)											
5 (B1)	+/D	–	–	–	–	–	–	+	–	+	+
Antibodies to p45 (CD23)											
11 (MHM6)	(+/D)	–	+	–	–	–	–	+	–	–/D	+
19 (PL-13)	(+/D)	–	+	–	–	–	–	+	–	–	–
39 (Blast-2)	(+/D)	–	+	–	–	–	–	+	–	–/(D)	–
Antibodies to p95 (CD19)											
14, 34 (B4)	+/D	?wk	+	–	–	–	–	+	+	+	+
28 (HD37)	+/D	?wk	+	–	–	–	–	+	+	+	+
43 (AG7)	+	–	+	–	–	–	–	+	+	+	+
Antibodies to p135 (CD22)											
7 (29-110)	+	–	+	–	–		–	wk	+	+	?wk
25 (HD6)	+	±	+	–	–	–	–	wk	+	+	+
31 (HD39)	+	–	+	–	–	–	–	±	+	+	+
40 (SJ10)	+	–	+	–	–	–	–	–	?+	+	+
49 (SHCL-1)	+	–	+	–	–	–	–	wk	–	+	wk
Antibodies to p140 (CD21)											
9, 33 (B2)	+/D	+	+	(+)	–	–	–	+/D	–	–/D	±
35 (BL-13)	+/D	+	+	–	(+)	–	–	+/D	–	–/D	±
41 (HB5)	+/D/E	+	+	–	+	–	?wk	+/D	–	–/D	wk
Antibodies to μ chain											
3 (29–132)	+	–	+	BG	BG	–	BG	+	?+	wk	+
42 (B-7)	+/BG	BG	+/BG	BG	BG	BG	BG	+	BG	+	+/BG
Anti-p220 (anti-leukocyte common)											
51 (HB11)	(+)/E	–	+	(+)	+	+	+	±	–	+	–
Miscellaneous or unknown specificities											
4 (HH1)	+/?D	?+	+	–	–	–	–	+	–	+	+
6 (NUB1)	(+)	–	(+)	–	–	–	–	(+)	–	+	–
17 (HD28)	+	–	+	–	–	–	–	+	–	+	+
21 (SHCL-2)	+	+	+/RP	–	+	+	(wk)	wk/TZ	+	+	+
36 (BL-14)	+	?wk	+	–	–	–	–	+	–	+	wk
52 (41H16)	(+)/M	+	(+)/RP	+	+	+	(?M)	+	–	+	–/M

[a] Results in parentheses indicate that only a subpopulation of cells were labeled. Positive reactions on *tonsil* and *spleen* tissue refer to the reactivity of B cell areas. Positive reactions on *kidney, breast,* and *lung* refer to the labeling of epithelium, with the exception of B21 and B52 on these tissues (which label presumed macrophages, including Kupffer cells in the liver). Results on *lymphoma* tissue show the neoplastic cell labeling before the stroke, additional reactions after it (e.g., –/D indicates that neoplastic cells were unreactive but that normal dendritic reticulum cells were labeled).

[b] *Abbreviations:* ALL, Acute lymphoblastic leukemia; BG, background or interstitial staining (which often prevented assessment of specific cellular labeling); Br, breast; CB/CC, centroblastic/centrocytic lymphoma (equivalent to follicular lymphoma); CC, centrocytic lymphoma (equivalent to diffuse follicle center cell lymphoma); CLL, chronic lymphocytic leukemia; D, dendritic reticulum cells; Diff NHL, diffuse non-Hodgkin's lymphoma; E, epithelium (in tonsil); HCL, hairy cell leukemia; Kid, kidney; LBL, lymphoblastic lymphoma (Burkitt type); Liv, liver; LNM, lymph node medulla; Lu, lung; M, probable macrophages; ND, no data; RP, red pulp (in spleen); Spl, spleen; Ton, tonsil TZ, T cell zone (in tonsil); wk, weak reactivity.

to p140 (see below). Because of the strong staining of adjacent germinal center cells it was difficult to assess whether or not this reactivity represented staining of dendritic reticulum cells. However, coarse staining of intercellular material was seen with these antibodies in a number of diffuse B cell lymphomas (in areas where dendritic reticulum cells were absent), raising the possibility that antigen shed from the surface of B cells was detected.

B cell lympholiferative disorders									
CC	CB/ CC1	CB/ CC2	CB/ CC3	Diff NHL1	Diff NHL2	Diff NHL3	Diff NHL4	Diff NHL5	Summary of reactions
+	+	+	+	+	+	+	+	+	B cells & DRC
+	(D)	wk/(D)	–	–	wk	–	–	(+)	(DRC & B cells)
ND	(D)	(D)	–	±/(D)	wk	–	–	(+)	(DRC & B cells)
+	(D)	–/(D)	–	–	–	–	–	(+)	(DRC & B cells)
+	+	+	ND	+	+	+	+	+	B cells & ?DRC
+	+	+	+	+	+	+	+	+	B cells
+	+	+	+	+	+	+	+	+	B cells
?wk	+	+	+	+	+	wk	+	+	B cells
+	+	+	+	+	+	+	+	+	B cells
–	+	+	wk	+	+	+	+	+	B cells
+	+/wk	+	wk	+	–	+	+	ND	B cells
+	+/wk	+	wk	+	+	+	?	+	B cells
+/D	–/D	–/D	–/D	–/D	+	–	–	–	DRC & (B cells)
+/D	wk/D	–/D	–/D	wk/D	+	–	?	–	DRC & (B cells)
+/D	(wk)/D	–/D	–/D	wk/D	+	–	wk	–	DRC, (B cells), epithelium
ND	(+)	–	+	+	–	–	–	–/BG	IgM pattern
+	+	–	+	+	–	–	?	–/BG	IgM pattern
±	+/wk	–	ND	+	±	–	–	+	(B cells,) epithelium
+	+	+	+	+	?+	–	+	+	B cells & ?DRC
+	wk	+	–	wk	–	–	–	?wk	(B cells)
+	+	+	+	+	wk	+	+	+	B cells
+	+/TZ	+	–	+	+/M	+	+	+	B cells & macrophages
+	+	+	+	+	±	?±	+	wk	B cells
+	+	+	+	+	–/M	M	M	–/M	(B cells) & macrophages

Anti-p135 (CD22)

The five reagents in this category reacted with B cells in both the mantle zones and germinal centers of secondary lymphoid follicles (Fig. 20.1). They also reacted with many different categories of B cell lymphoproliferative disorders (including the single case of ALL recorded).

Anti-p140 (CD21)

The three antibodies in this category gave similar reactions (Fig. 20.1), the most striking of which was an intense staining in lymphoid follicles of presumptive dendritic reticulum cells (this interpretation being based on the coarse meshwork-like pattern of reactivity). B cells in the mantle zones of lymphoid follicles were less strongly stained.

The majority of B cell neoplasms gave negative or weak reactions, although the one case of CLL and a diffuse B cell lymphoma gave stronger reactions. In many sections of B cell lymphomas normal dendritic reticulum cells were strongly stained.

Reagents B9 and B33 [both antibody B2 (9)] stained renal tubular cells in a coarse granular pattern, localized towards the luminal aspect of these cells. Reagents B35 and B41 stained biliary epithelial cells in the liver.

Other Antibodies

Six other antibodies, which did not recognize B cell antigen clusters, or which did not have clearly defined molecular targets, also reacted selectively with B cells in tissue sections. They gave a variety of different patterns on normal and neoplastic samples (see Table 20.2), and no two antibodies appeared to be identical. It was of interest that reagent B6 (antibody NUB1) showed selectivity within the B cell lineage in that mantle zone cells were labeled, but not germinal centers (although the antigen appeared on some neoplasms of germinal center origin) (Table 20.2).

Reagent B21 (antibody SHCL-2) was also of interest in that it reacted with probable macrophages in many tissues, including splenic red pulp, T zones of lymph nodes, and liver (Kupffer cells). Reagent B52 (antibody 41H16) also appeared to react with macrophages in many tissues although its pattern of labeling was different from that of B21 (e.g., sinusoidal lining cells rather than Kupffer cells were labeled in the liver). Furthermore, B52 tended to label selectively mantle zone cells but not germinal centers.

Anti–μ Chain and Anti-p220

The probable specificities of two antibodies which were subsequently shown (by immunoprecipitation) to detect μ heavy chain were predicted on the basis of their characteristic immunohistological labeling reactions (i.e., staining of mantle zone lymphocytes, meshwork staining of germinal center immune complexes, extensive staining of collagen and connective tissue, etc.).

One antibody against a high molecular weight antigen (p220), presumed to belong to the leukocyte common group of glycoproteins (reagent B51), reacted with B cell areas in lymphoid tissue, but also stained epithelial cells in many organs.

Discussion

p95 and p135

Two groups of reagents (anti-p95 and anti-p135) emerged from this study as being good pan B cell antibodies suitable for detecting both normal and neoplastic B cells by immunohistochemical procedures. It is of interest that other studies in this Workshop (summarized in Chapter 1 of this volume) indicate that p135 is a restricted B cell marker, and that activation of B cells leads to its disappearance from the cell surface. The tissue section staining results reported in this paper suggest, in contrast, a relatively broad pattern of distribution among B cells and also that germinal center cells (presumed to be the physiological equivalent of B cells activated *in vitro*) express the antigen. Furthermore immunoenzymatic studies of a large series of B cell neoplasms (to be reported elsewhere) have confirmed the findings from the present study that the antigen is expressed by many B cell lymphomas and leukemias.

One possible explanation of this discrepancy is that immunocytochemical staining of tissue sections or cell smears reveals intracytoplasmic antigen which is not expressed on the cell surface membrane. There is recent evidence from Campana *et al.* (10), and also data presented at the Workshop by Dorken *et al.*, indicating that the p135 molecule is initially expressed by B cells as an intracytoplasmic molecule and then emerges on the cell surface during B cell maturation.

p35

The p35 molecule was reported in this Workshop as being a broadly expressed B cell marker. Whilst the present immunocytochemical studies confirm earlier reports by Nadler *et al.* (6) that the p35 molecule is present on many different B cell types, we have found in the present and in previous studies that the molecule is a relatively poor marker of B cell neoplasms, principally because of weak immunoenzymatic reactivity. It also appeared in the present study to label dendritic reticulum cells in germinal centers. In previous studies of antibody B1 (11,12) similar labeling reactions have been reported.

p45

The antibodies against the p45 molecule appeared, on the basis of their immunocytochemical reactivity pattern, to be identical to antibody Tül (13) and probably also to antibody B-532 (12). The immunocytochemical labeling reactions obtained in this present investigation, and previously by Stein *et al.* (14) using antibody Tül, agree with the results summarized at this Workshop, although the frequency of cases of chronic lymphocytic leukemia stained for this antigen is higher by the immunocytochemical technique.

It is of interest that studies summarized in the present Workshop indicate that the p45 molecule is absent from normal circulating B cells, since it is expressed [as also noted by Hofman *et al.* (12)] on at least a major population of mantle zone cells (the tissue equivalent of these cells) (see Results). It is not at present clear whether this indicates loss of antigen when these cells enter the circulation or possibly intracytoplasmic expression. It may also be noted that the antigen was found in the present study to be absent from germinal center lymphoid cells (although present on a subpopulation of dendritic reticulum cells—see Results) since *in vitro* activation experiments reported at the Workshop indicate that the antigen can be induced on peripheral blood cells by stimulation with anti-Ig or with Epstein–Barr virus.

p140

The strong expression of p140 [C3d receptor (15)] by probable dendritic reticulum cells is in keeping with the report by Bhan *et al.* (11) of "intracellular" labeling observed in germinal centers using antibody B2. However, it may be noted that it remains to be formally proved (e.g., by studying isolated dendritic reticulum cells) that this meshwork pattern of staining does indeed represent dendritic reticulum cell labeling. More importantly there is a possibility that the p140 antigen may be acquired by dendritic reticulum cells from other cell types (e.g., B cells). There is a precedent for this in that immune complexes in germinal centers show a labeling pattern very similar to that of integral dendritic reticulum cell antigens, despite the fact that the complexes are of exogenous origin. Future studies, including the investigation of "burnt out" B cell follicles which are depleted of B cells (such as are seen in cases of angioim-

---

Fig. 20.1. Immuno-alkaline phosphatase labeling of tonsil tissue with monoclonal anti–B cell antibodies (acetone-fixed cryostat sections).

CD19 (*p95*)*:* The reactions of Workshop antibody B28 are shown at low and high power. Note strong staining of B cell follicles and also of scattered cells outside these areas. T cell areas (T) are unstained. In the high-power view the edge of a follicle abutting against the squamous epithelium (E) of the tonsil is seen. Note strong labeling of mantle zone (MZ) lymphocytes and also extensive labeling within the germinal center (GC), some of which may represent dendritic reticulum cell or extracellular labeling.

CD20 (*p35*)*:* Labeling of a B cell follicle with Workshop antibody B5 shows relatively weak labeling of both the germinal center (GC) and mantle zone (MZ) areas. Note the meshwork pattern of staining within the germinal center.

CD21 (*p140*)*:* A B cell follicle is strongly labeled with Workshop antibody B9. The strongest labeling is seen as a coarse meshwork pattern within the germinal center.

DC22 (*p135*)*:* Staining is seen at low and high power with Workshop antibody B25. B cell follicles are strongly stained. In the high-power view labeling of both germinal center and mantle zone cells is seen, together with scattered extrafollicular B cells within the squamous epithelium (E).

CD23 (*p45*)*:* The characteristic labeling reaction of antibodies within this group is shown by staining with Workshop antibody B11. There is labeling of both mantle zone cells (which show some variability in staining intensity) and also a dense meshwork localized in one region of the germinal center (the so-called "light zone"). In the high-power view the negative reaction of many of the cells in the germinal center is clearly seen.

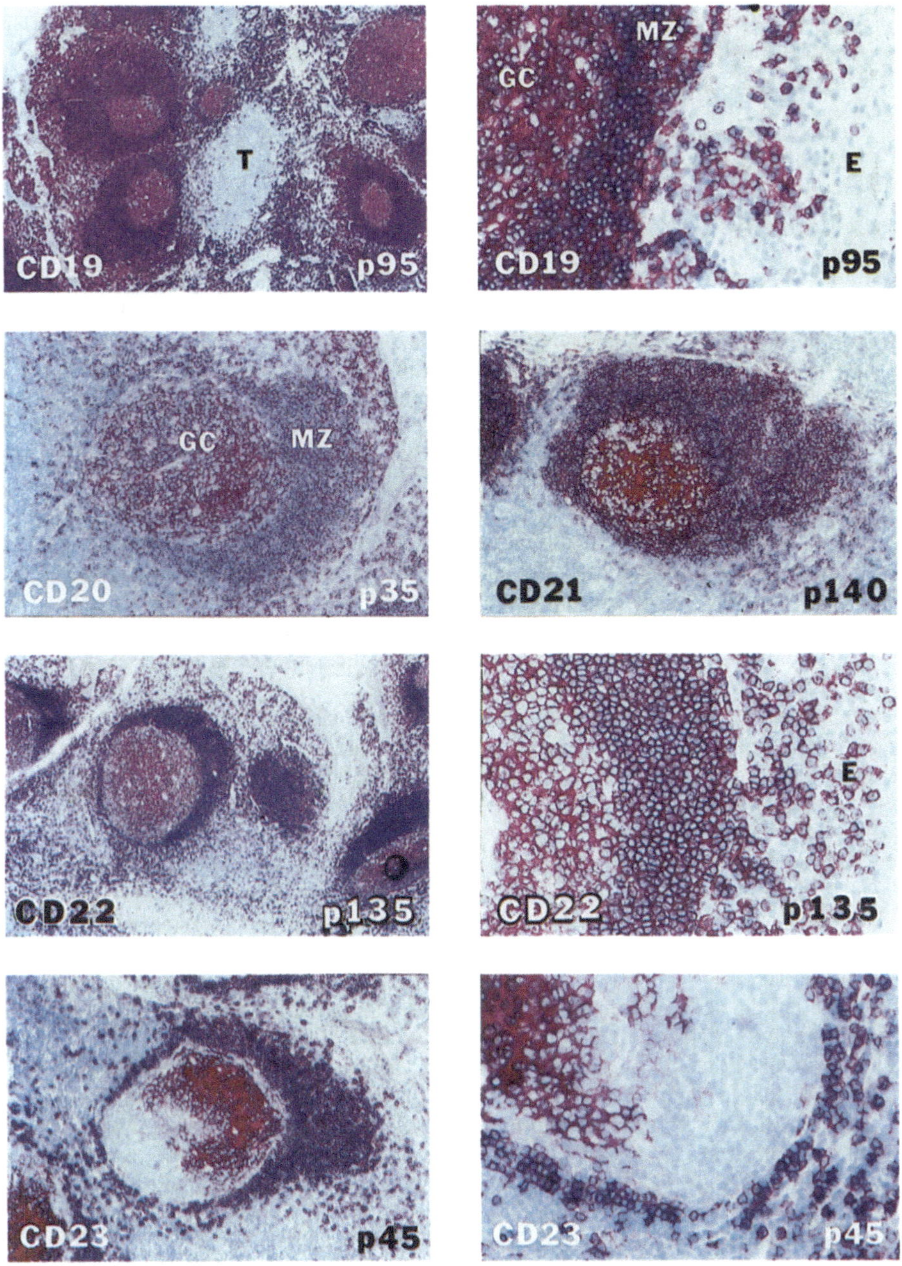
T
CD19
p95
MZ
GC
E
CD19
p95
GC
MZ
CD20
p35
CD21
p140
CD22
p135
E
CD22
p135
CD23
p45
CD23
p45

munoblastic lymphadenopathy), would be informative since if the p140 molecule was then no longer detectable it would provide evidence that the antigen was not the product of the dendritic reticulum cells themselves.

Acknowledgments. This work was supported by grants from the Leukaemia Research Fund and the Wellcome Trust.

References

1. Naiem, M., J. Gerdes, Z. Abdulaziz, C.A. Sunderland, H. Stein, and D.Y. Mason. 1982. The value of immunohistological screening in the production of monoclonal antibodies. *J. Immunol. Methods* **50:**145.
2. Gatter, K.C., B. Falini, and D.Y. Mason. 1984. The use of monoclonal antibodies in histopathological diagnosis. In: *Recent advances in histopathology,* Vol 12, P. Anthony and R. MacSween, eds. pp. 35–67. Churchill Livingstone, Edinburgh London, Melbourne New York.
3. Cordell, J.L., B. Falini, W.N. Erber, A.K. Ghosh, Z. Abdulaziz, S. MacDonald, K.A.F. Pulford, H. Stein, and D.Y. Mason. 1984. Immunoenzymatic labelling of monoclonal antibodies using immune complexes of alkaline phosphatase and monoclonal anti-alkaline phosphatase (APAAP complexes). *J. Histochem. Cytochem.* **32:**219.
4. Moir, D.J., A.K. Ghosh, Z. Abdulaziz, P.M. Knight, and D.Y. Mason. 1983. Immunoenzymatic staining of haematological samples with monoclonal antibodies. *Brit. J. Haemat.* **55:**395.
5. Erber, W.N., A.J. Pinching, and D.Y. Mason. 1984. Immunocytochemical detection of T and B cell populations in routine blood smears. *Lancet* **i:**1042.
6. Nadler, L.M., J. Ritz, R. Hardy, J.M. Pesando, S.F. Schlossman, and P. Stashenko. 1981. A unique cell surface antigen identifying lymphoid malignancies of B cell origin. *J. Clin. Invest.* **67:**134.
7. Naiem, M., J. Gerdes, Z. Abdulaziz, H. Stein, and D.Y. Mason. 1983. Production of a monoclonal antibody reactive with human dendritic reticulum cells and its use in the immunohistological analysis of human lymphoid tissue. *J. Clin. Pathol. 36:*167.
8. Gerdes, J., H. Stein, D.Y. Mason, and A. Ziegler. 1983. Human dendritic reticulum cells of lymphoid follicles: Their antigenic profile and their identification as multinucleated giant cells. *Virch. Arch. B* **42:**161.
9. Nadler, L.M., P. Stashenko, R. Hardy, A. van Agthoven, C. Terhorst, and S.F. Schlossman. 1981. Characterization of a human B cell-specific antigen (B2) distinct from B1. *J. Immunol.* **126:**1941.
10. Campana, D., G. Janossy, M. Bofill, L.K. Trejdosiewicz, D. Ma, A.V. Hoffbrand, D.Y. Mason, A-M. Lebacq, and H. Forster. 1985. Human B cell development 1. Phenotypic differences of B lymphocytes in the bone marrow and peripheral lymphoid tissue. *J. Immunol.* **134:**1524.
11. Bhan, A.K., L.M. Nadler, P. Stashenko, R.T. McCluskey, and S.F. Schlossman. 1981. Stages of B cell differentiation in human lymphoid tissue. *J. Exp. Med.* **154:**737.
12. Hofman, F.M., E. Yanagohara, B. Bryne, R. Billing, S. Baird, D. Frisman, and C.R. Taylor. 1983. Analysis of B-cell antigens in normal reactive lymphoid tissue using four B-cell monoclonal antibodies. *Blood* **62:**775.

13. Ziegler, A., H. Stein, C. Müller, and P. Wernet. 1981. Tül: A monoclonal antibody defining a B cell subpopulation—usefulness for the classification of non-Hodgkin's lymphomas. In: *Leukaemia markers,* W. Knapp, ed. Academic Press, pp. 113–116.
14. Stein, H., J. Gerdes, and D.Y. Mason. 1982. The normal and malignant germinal centre. *Clinics in Haematology* **11:**531.
15. Iida, K., L. Nadler, and V. Nusenzweig. 1983. Identification of the membrane receptor for the complement fragment C3d by means of a monoclonal antibody. *J. Exp. Med.* **158:**1021.

CHAPTER 21

Analysis of B and L Workshop Antibodies on Sections of Normal and Neoplastic Lymphoid Tissue and Cell Lines

Ian C.M. MacLennan, Paul D. Nathan,
Gerald D. Johnson, Mahmood Khan,
Léonie Walker, and Noel R. Ling

Introduction

B lymphocyte subgroups can be identified by phenotypic analysis of isolated lymphocyte preparations; further information about changes in these cells on activation can be gained by culture with antigen or polyclonal B cell mitogens. Most of the markers applied to isolated cells can also be used in tissue sections. Immunohistology allows several different cell types to be screened at the same time. Sections also enable functional subsets of cells to be identified by their location in the tissues. In addition to general T and B cell areas a number of different B cell compartments can be identified. These include:

1. The small lymphocyte mantle of follicles which contain recirculating B cells (Figs. 21.1 and 21.2) (1);
2. Germinal centers with mixtures of centroblasts, centrocytes, follicular dendritic cells, and tingible body macrophages [Figs. 21.1(A) and 21.2(A)] (2);
3. The marginal zone B cells of the spleen which do not recirculate and which, unlike recirculating follicular B cells, do not express surface IgD (Fig. 2.2) (3).

In this report we describe the reactivity of the B and L series of monoclonal antibodies submitted to this workshop. This analysis has used both isolated cells in the form of lymphoblastoid cell lines, together with tissue sections. The range of substrates used has been chosen to cover a broad spectrum of B cell maturation and differentiation. Both normal and neoplastic tissues have been included.

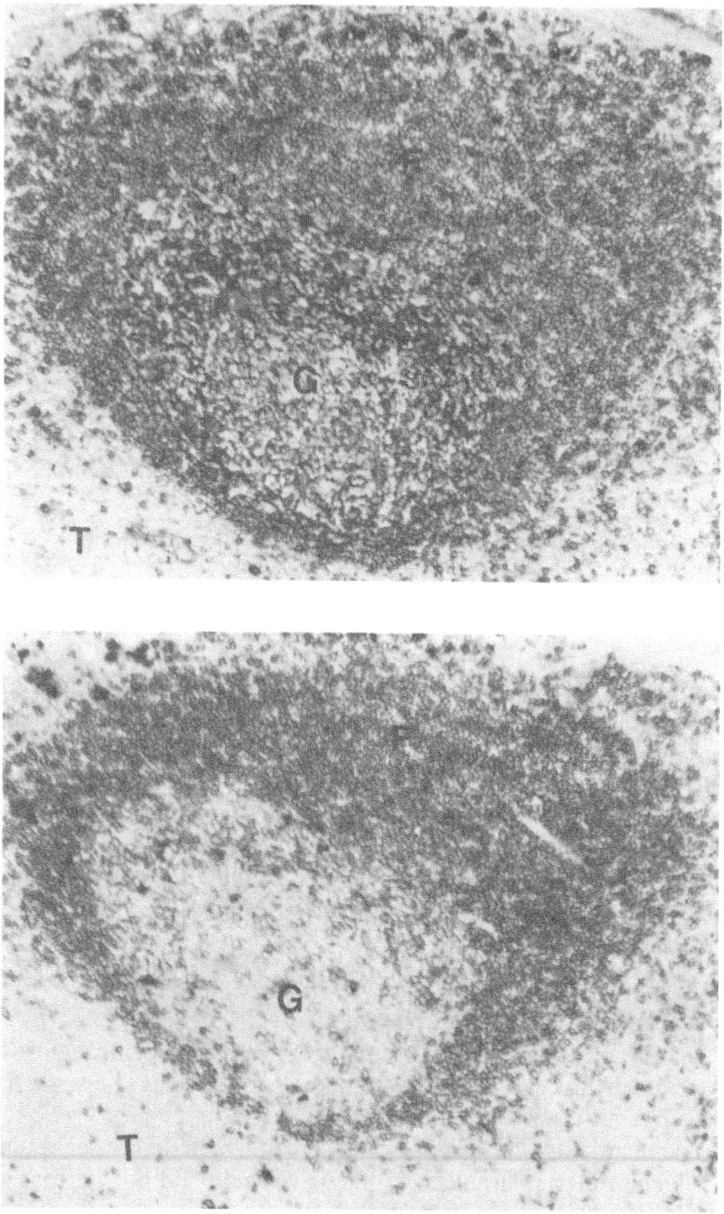

Fig. 21.1. Serial frozen sections of human tonsil. (A) Stained to reveal IgM. F = Follicular mantle; G = germinal center with IgM immune complex shown on follicular dendritic cells; T = T zone. (B) Stained to reveal IgD symbols as in (A).

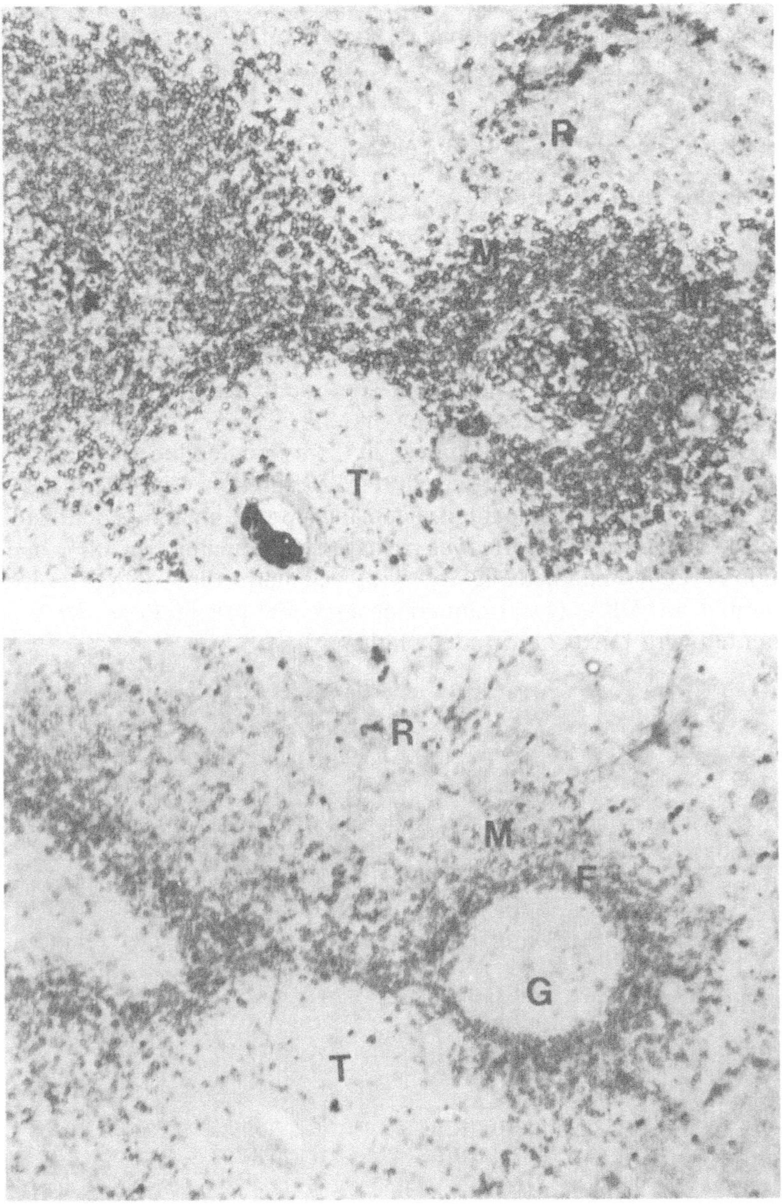

Fig. 21.2. Serial frozen sections of human spleen. (A) Stained to reveal IgM. R = Red pulp; T = T zone of periarteriolar lymphocytic sheath; G = germinal center; M = marginal zone. (B) Stained to reveal IgD. The $SIgM^+$ IgD^+ cells of the follicular mantle, F, are seen. Occasional IgD^+ B cells in the marginal zone are probably in transit.

The results obtained demonstrate the usefulness of immunohistological screening. They also reveal phenotypic features of cultured cells which were not identified in sections of any of the normal or neoplastic tissues examined.

Materials and Methods

Immunohistology

The tissues used for screening are shown in Table 21.1. In all cases they were snap-frozen in liquid nitrogen and cryostat sections made. Sections were mounted on four- or twelve-spot Teflon-coated multispot slides (Hendley and Essex), air-dried, and fixed in acetone at 4°C for 20 min for immunoperoxidase and for 5 min at 20°C for immunofluorescence. Workshop monoclonal antibodies used at 1/250 dilution were identified with FITC or peroxidase-conjugated sheep anti–mouse immunoglobulin (G, A, and M). Peroxidase activity was revealed with diaminobenzidine and the sections counterstained lightly with haematoxalin, dehydrated, and mounted in DPX (4). Immunofluorescence preparations were wet-mounted with DABCO to reduce fading (5).

Table 21.1. Cells identified by immunohistological assessment of acetone-fixed frozen sections.[a]

Substrate	Technique	B Cells identified	T Cells	Antigen-presenting cells	
Normal spleen	IF, IP	RFB, MZB, GCB, PC	TZT	TZIDC	FDC
Normal tonsil	IF, IP	RFB, GCB, PC	TZT	TZIDC	FDC, Langerhan cells[b]
CLL spleen 1	IP	CLL B			
2	IP	CLL B			
3	IP	CLL B			
CB.CC lymph node	IP	CB/CC			
HCL spleen 1	IP	RFB, HCL	TZT		
2	IP	RFB, HCL	TZT		
3	IP	HCL			
Reactive lymph node		RFB, GCB, PC	TZT	TZIDC	FDC

[a] Abbreviations: CLL, Chronic lymphocytic leukemia; CB/CC, centroblastic/centrocytic lymphoma; HCL, hairy cell leukemia (Fig. 21.5); RFB, small recirculating B cells of follicular mantle; MZB, static B cells of splenic marginal zone (Fig. 21.2); GCB, B cells of germinal centers (Figs. 21.1 and 21.2); PC, plasma cells; TZT, T zone T cells (Figs 21.1 and 21.2); TZIDC interdigitating cells of T zones (Fig. 21.3); FDC, Follicular dendritic cells (Figs. 21.1 and 21.2) IP, indirect immunoperoxidase using peroxidase-conjugated sheep anti–mouse immunoglobulin; IF, indirect immunofluorescence using fluorescein-conjugated sheep anti–mouse immunoglobulin.

[b] Langerhans cells = intraepithelial dendritic cells.

Table 21.2. Cell lines used in workshop antibody analysis.

Name	Cell of origin
SMS/SB	Pre-B ALL
EB4	Burkitt
Raji	Burkitt
Daudi	Burkitt
Ed-1	B?
ARH-77	B (Ref. 6)
WM-1	Waldenström's macroglobulinemia
RPMI-8226	Plasmacytoma
HSB-2	T-LCL
MOLT-4	T-LCL
HL-60	Promyelocytic leukemia

Cell Lines

The cell lines used and their characteristics are shown in Table 21.2. Reaction with monoclonal antibodies was revealed by indirect immunofluorescence as above using cells prepared in two ways:

1. Live cells in suspension (surface staining).
2. Cells air-dried and acetone-fixed for 5 min at 20°C on 12-spot multispot slides (surface and cytoplasmic staining). Preparations were mounted in DABCO as above before reading by fluorescence microscopy.

Results

Workshop antibodies are divided into eleven groups based on screening procedure results. These are described below. Tables 21.3–21.9 give details of the reactivity of antibodies by group. Cell line reactivity in these tables is shown without discriminating between the results achieved with live as opposed to acetone-fixed cells. In general, similar results were achieved with both techniques. However, with nine of the antibodies major differences were detected between reactions on cell line cells tested in these two ways. Details of these differences are shown in Table 21.10.

Group 1: Antibodies Reacting with Most B Cells in the Follicular Mantle, Marginal Zone, and Germinal Centers (Table 21.3)

These antibodies also reacted with hairy cells, CLL cells, and CB/CC lymphoma cells. They were free of T cell reactivity but identified all B cell lines except the "pre–B cell" line. The failure to show any reactivity with follicular dendritic cells (FDC) or plasma cells indicates that they are not anti-IgM. The reactivity with germinal centers and marginal zone cells

Table 21.3. Group 1. Antibodies reacting with most B cells in the follicular mantle, marginal zone, and germinal centers.

Workshop no.	Small follicular B	Marginal zone B	Germinal center B	FDC	IDC T zone	Langerhans cells	Non-B lymphocytes	CB/CC B	CB/CC FDC
B4	++	+	+	?	–	+	–	+	–
B17	+	+	±	–	–	–	–	+	–
B22	+	+	+	–?	–	–	–	+	–
B24	+++	+	+	–	–	–	some T?	+	–?
B36	++	+	+	–	–	–	some T?	+	–?

Workshop no.	HCL	CLL B	Pre-B	EB4	Raji	Daudi	Ed-1	ARH	WM-1	8226	HSB-2	MOLT-4	HL-60
B4	+	++	–	+	+	+	+	–	+	–	–	–	–
B17	++	+	–	+	++	++	++	++	+	±	–	–	±
B22	++	+	–	+++	++	+++	+++	+++	+++	++	–	–	–
B24	++	+	–	+	+	+	+	±	±	±	–	–	–
B36	++	++	occ +	++	+	++	++	++	++	++	–	–	–

Table 21.4. Group 2. Antibodies reacting with most B cells in the follicular mantle and germinal center together with interdigitating cells in T zones but showing relatively weak activity with marginal zone B cells.

Workshop no.	Small follicular B	Marginal zone B	Germinal center B	FDC	IDC T zone	Langerhans cells	Non-B lymphocytes	CB/CC B	CB/CC FDC	HCL	CLL
B1	most +	–	++	?	++	+	–	++	?	+	+
B12	++	±	+	–?	++	++	–	±	?	+	±

Workshop no.	Reactive LN germinal center B	Pre-B	EB4	Raji	Daudi	Ed-1	ARH	WM-1	8226	HSB-2	MOLT-4	HL-60
B1	++	+	++	+	++	++	++	++	rare ++	–	++	–
B12		–	++	+	+++	+++	+++	+++	rare ++	–	++	–

Table 21.5. Group 3. Antibodies reacting with most B cells in the follicular mantle and marginal zone but lacking reactivity with germinal center B cells.

Workshop no.	Small follicular B	Marginal zone B	Germinal center B	FDC (tonsil)	IDC T zone	Langerhans cells	Non-B lymphocytes	CB/CC B	CB/CC FDC
B7	++	++	±	–	–	–	–	±	–
B21	+	+	–	–	–	–	–	+	–
B25	+	+	±	–	–	–	–	–	–
B31	+	+	±	–	–	–	?	+	?
B40	+	+	±	–	–	–	–	±	–
B49	++	++	±	–	–	–	–	±	–

Workshop no.	HCL	CLL B	Pre-B	EB4	Raji	Daudi	Ed-1	ARH	WM-1	8226	HSB-1	MOLT-4	HL-60
B7	+++	+	++	+	±	+++	++	++	±	±	–	–	–
B21	++	+	±	++	+++	++	++	++	occ +	–	–	–	–
B25	++	++	+	±	+	+	±	+	–	–	–	–	–
B31	+++	++	++	++	++	++	+	+	++	±	–	–	–
B40	+++	++	–	–	±	±	±	±	–	–	–	–	–
B49	++	++	±	±	±	+	+	+	±	+	–	–	–

Table 21.6. Group 4. Antibodies reacting with B cells in follicular mantle and marginal zones together with FDCs.

Workshop no.	Small follicular B	Marginal zone B	Germinal center B	FDC	IDC T zone	Langerhans cells	Non-B lymphocytes	CB/CC B	CB/CC FDC
B3	++ (most)	++ (some)	?	+	−	−	−	−	−
B9	++	++	?	++	−	−	−	−?	++
B28	+	+	?	+	−	−	−	−	−
B33	++	++	?	++	−	−	±	−?	++
B34	+	+	?	+	−	−	−	±	−?
B35	+	+	−	++	−	−	−	−?	++
B41	±	+	?	++	−	−	−	+	++
B42	+	+ (not all)	?	++	−	+ (some)	−	−	−

Workshop no.	HCL	CLL B	Pre-B	EB4	Raji	Daudi	Ed-1	ARH	WM-1	8226	HSB-2	MOLT-4	HL-60
B3	+	++x2 (±)x1	−	+	+	+	+ (occ)	−	+	−	−	−	−
B9	−	±	−	−	−	±	−	+	−	−	−	−	−
B28	++	++	+	±	++	+	++	+	+	−	−	−	−
B33	±	+	−	±	++	±	++	++	+	±	−	−	−
B34	+	−	++	+	+++	++	+++	++	+++	±	−	−	−
B35	−	+	−	−	++	+	+	+++	++	+	−	−	−
B41	−	−	−	−	−	±	+ (occ)	+	+	−	−	−	−
B42	−	++x2 −x1	±	−	−	+	+ (occ)	−	−	−	−	−	−

Table 21.7. Group 6. Antibodies reacting with B cells in CB/CC lymphoma and FDCs in tonsil.[a]

Workshop no.	Small follicular B	Marginal zone B	Germinal center B	FDC	IDC T zone	Langerhans cells	Non-B lymphocytes	CB/CC B	CB/CC FDC
L4	–	–	?	+	–	?	–	++	?
L18	–	–	?	+	–	?	–	+	–
L22	–	–	–	+	–	–	–	++	?

Workshop no.	HCL	CLL B	Pre-B	EB4	Raji	Daudi	Ed-1	ARH	WM-1	8226	HSB-2	MOLT-4	HL-60
L4	–	–	–	–	–	–	–	–	–	+	–	–	–
L18	–	–	+	+	–	–	+	–	–	++	–	+	–
L22	–	–	+	+	–	±	+	–	–	++	–	+	+
				(occ)[b]			(occ)						

[a] Also: Stratified epithelium ++, arteries ++, splenic sinusoids ++.
[b] occ = Occasional.

Table 21.8. Group 7. Antibody reacting with follicular mantle B cells in tonsil but not spleen, and with T zone IDC and IDC surrounding splenic white pulp.

Workshop no.	Small follicular B	Marginal zone B	Germinal center B	FDC	T zone IDC	IDC red pulp	Non-B lymphocytes	CB/CC B	CB/CC FDC	HCL	CLL B
B27	–	–	–	–	–spleen +tonsil +lymph node	++	–	–	–	++	+x2 –x1

Workshop no.	Reactive LN germinal center	Pre B	EB4	Raji	Daudi	Ed-1	ARH	WM-1	8226	HSB-2	MOLT-4	HL-60
B27	±	–	±	+	++	++	+	++ (rare)	–	–	–	–

Table 21.9. Group 8. Antibodies not reacting with lymphocytes in tissues but having reactivity with some cell line cells.

Workshop no.	Pre-B	EB4	Raji	Daudi	Ed-1	ARH-77	WM-1	8226	HSB-2	MOLT-4	HL-60
B18											
Suspension	++	+ (occ)[a]	–	–	–	–	–	–	–	–	–
Multispot	–	(+)			±	–	–	–	–	–	–
B32											
Suspension	–	–	++	+	++	–	++	±	–	–	–
Multispot	–	±		–	–	occ +	–	+	–	–	–
B38											
Suspension	–	++	–	–	–	occ +	–	occ +	–		–
Multispot	++	+		–	–	–	–	±	–	–	–
L13											
Suspension	–	±	–	–	–	–	–	–	–	–	–
Multispot	–	++		(+)	occ +	±	+	+++	+	+++	+
L19											
Suspension	occ +	++	+++	+++	++	+++	+++	++	+++	++	++
Multispot	–	+++		50% +	+++	++	++	+++	–	30% ++	++
L20											
Suspension	++	++	++	++	++	++	+++	+++	++	+++	20% ++
Multispot	±	+++		++	occ +	++	+++	+++	++	+++	++

[a] occ = Occasional.

Table 21.10. Antibodies showing preferential reactivity with the surface of live cell line cells (membrane fluorescence) or acetone-fixed cell spreads.

	Pre-B ALL	Burkitt lymphoma			B-LCL	B-Lymphoma		P.C.	T-LCL		Myel.
Workshop no.	SMS/SB	EB4	Raji	Daudi	Ed-1	ARH-77	WM-1	8226	HSB-2	MOLT-4	HL-60
B7											
S[a]	–	–	–	–	–	+/–	+/–	–	–	–	–
Cy	++	+		+++	++	++	+/–	(+)	–	–	–
B12											
S	–	–	+	–	–	–	–	–	–	–	–
Cy	–	++		+++	+++	+++	+++	+++	–	++	–
B22											
S	–	+++	++	+++	+++	+++	+++	++	–	–	–
Cy	–	–	–	+	–	+	–	–	–	–	–
B31											
S	++	++	++	++	+	+	++	+/–	–	–	–
Cy	–	–	–	–	–	–	–	–	–	–	–
B34											
S	++	+	+++	++	+++	++	+++	+/–	–	–	–
Cy	–	–	–	–	–	–	–	–	–	–	–
B35											
S	–	–	++	+	+	+++	++	+	–	–	–
Cy	–	–	–	–	–	–	–	–	–	–	–
B36											
S	+	++	+	++	++	++	++	++	–	–	–
Cy	–	–	–	–	–	–	–	–	–	–	–
B45											
S	–	–	–	–	–	–	–	–	–	–	–
Cy	+++	+++		+++	++	+++	+++	+++	++	++	++
B50											
S	–	–	–	–	–	–	–	–	–	–	–
Cy	+++	+++		+++	+++	+++	+++	+++	+++	+++	+++

[a] S = Surface staining; Cy = cytoplasmic staining.

excludes the possibility that they are anti-IgD. In general these antibodies form a homogeneous group and have a "pan-B" reactivity.

Group 2: Antibodies Reacting with Most B Cells in the Follicular Mantle and Germinal Center Together with Interdigitating Cells in T Zones But Showing Relatively Weak Activity with Marginal Zone B Cells (Table 21.4)

The reaction of this group with T zone interdigitating cells (Fig. 21.3) and Langerhans cells is reminiscent of anti-Class II MHC antigens. However, the relative lack of reactivity against CLL and marginal zone B cells makes this unlikely. No reactivity with T cells in tissue sections was apparent although both antibodies reacted with the T-LCL MOLT-4.

Group 3: Antibodies Reacting with Most B Cells in Follicular and Marginal Zones but Lacking Reactivity against Germinal Center B Cells (Table 21.5)

This group differs from group 1 by not showing clear reactivity against normal or neoplastic germinal center B cells (Fig. 21.4). Also most antibodies of this group, unlike group 1 antibodies, react with the pre-B cell line.

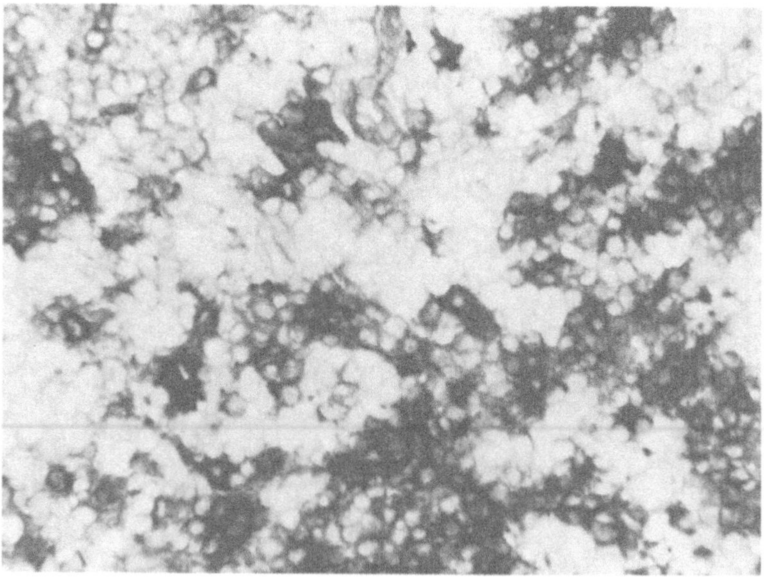

Fig. 21.3. Interdigitating cells, IDC. Stained with monoclonal B1 in the T zone of normal human spleen. The IDC cells are stained black. Some of the lymphocytes associated with the IDC are $B1^+$; most of the surrounding lymphocytes in the T zone are $B1^-$.

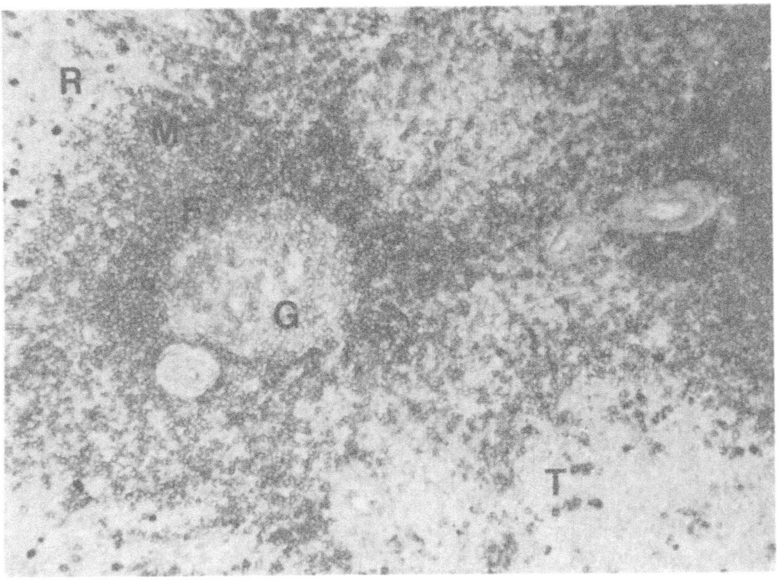

Fig. 21.4. Section of normal human spleen stained with monoclonal B49. Symbols as in Fig. 21.2.

Group 4: Antibodies Reacting with B Cells in Follicular Mantle and Marginal Zone and Follicular Dendritic Cells (Table 21.6)

This very heterogeneous group is discussed in detail in Chapter 23 of this volume. Only half the antibodies in this group reacted with FDC in neoplastic tissues. Many of these antibodies, unlike the antibodies of groups 1 and 3, fail to react with hairy cells (Fig. 21.5).

Group 5: Antibodies Reacting with Follicular Dendritic Cells But Not with Normal or Neoplastic Lymphocytes in Sections

These and FDC antibodies (B11, B19, B29 and B39) and their reaction with cell line cells are discussed in detail in Chapter 23 of this volume.

Group 6: Antibodies Reacting with Follicular Dendritic Cells in Tonsil and with CB/CC Lymphoma B Cells (Table 21.7)

These antibodies appear to form a homogeneous group. They react strongly with the B cells of a patient with centroblastic centrocytic lymphoma and also with germinal center B cells in a patient with reactive lymph node hyperplasia. They all reacted with the cell line RPMI-8226 which has some plasmablast features. However, there was no evidence of reactivity with tissue plasma cells. These antibodies also react with stratified squamous epithelium and a component in the walls of blood vessels.

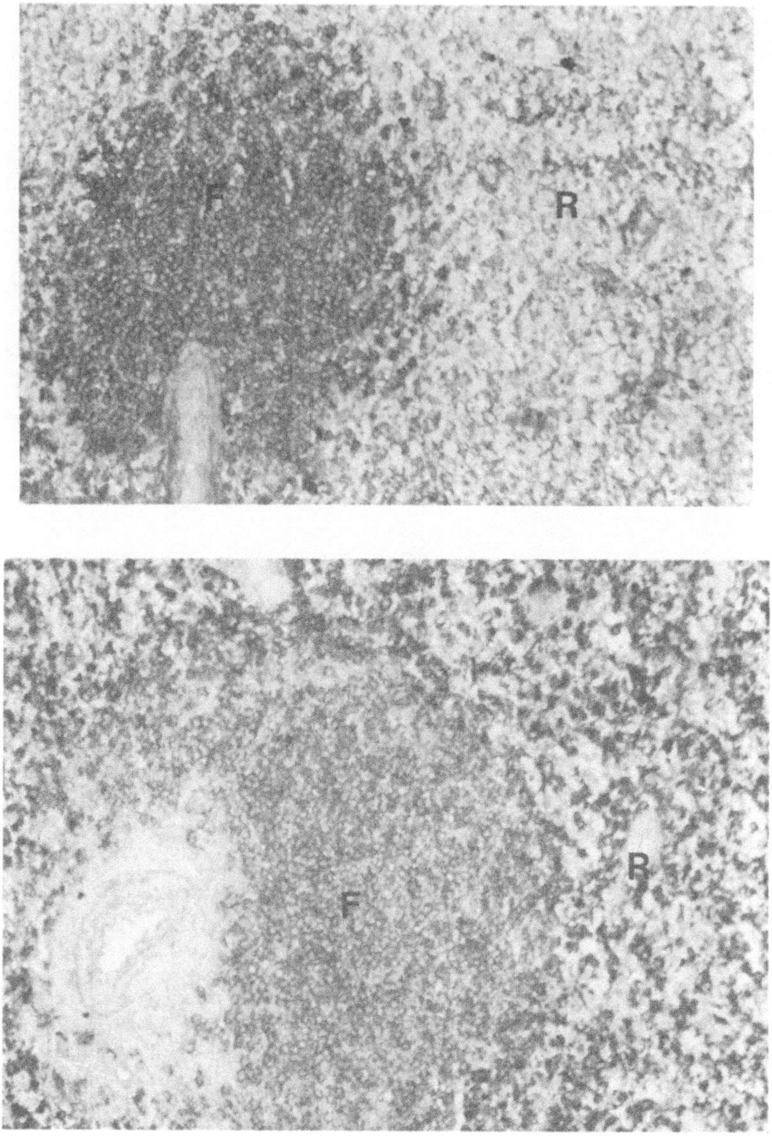

Fig. 21.5. Sections of spleen from a patient with hairy cell leukemia. (A) Stained with monoclonal antibody B42. Normal follicular (F) B cells are stained; hairy cells in the red pulp (R) are not. (B) Stained with monoclonal B49. Both follicular B cells and hairy cells in the red pulp are stained.

Group 7: Antibody Reacting with Follicular Mantle B Cells in Tonsil But Not Spleen and Also Revealing T Zone IDC and IDC Surrounding the White Pulp of Spleen (Table 21.8)

The reaction of this antibody with IDC in that part of the red pulp of the spleen adjacent to the white pulp is most unusual (Fig. 21.6) These IDC appear to be surrounded by lymphocytes. Further study of these cells is clearly required.

Group 8: Antibodies Not Reacting with Lymphocytes in Tissues But Having Reactivity with Some Cell Line Cells (Table 21.9)

The strong reactivity of L19 and L20 with most of the cell line cells (including T-LCLs and HL-60) was striking. Other antibodies were more restricted, reacting principally with the pre-B line (B18) or the plasma cell line (L13). The nature of the antigens strongly expressed on cell line cells but poorly on normal lymphocytes is not known. It does indicate, however, that some specificities are missed if screening is purely histological.

Although not reacting with lymphocytes in sections, antibodies L13, L19, and L20 appeared to show reactivity with tingible body macrophages together with a vascular component. However, this group of antibodies showed strong reactivity with some or all of the LCLs.

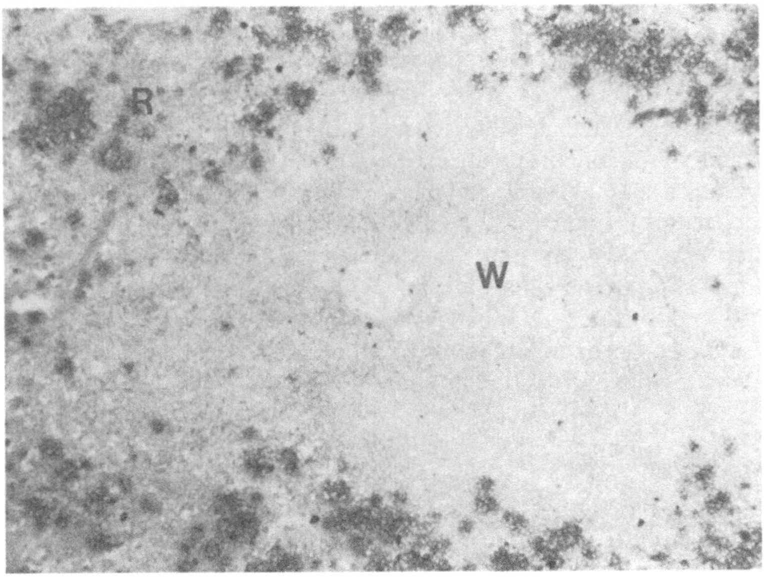

Fig. 21.6. Section of normal human spleen stained with B27. Interdigitating cells in the area of red pulp (R) adjacent to the white pulp (W) are revealed.

Group 9: Antibodies Reacting with Both T and B Lymphocytes

This group consists of antibodies B10, B20, B44, B45, B47, B48, B50, B51, L8, L9, and L12.

Group 10: Samples Containing Antibodies of Two Distinct Isotypes—B2, L6, L17

The isotype of the antibodies was only determined in a proportion of the samples. This finding is discussed in detail in Chapter 23 of this volume.

Group 11: Antibodies Having Relatively Little or No Reaction on Any of the Substrates Tested

This group consists of antibodies B5, B6, B8, B13, B14, B15, B16, B23, B26, B30, B37, B43 and B46, B52, L1, L2, L3, L7, L10, L14, L15, L16, and L21.

Discussion

Although the Workshop B and L series of monoclonal antibodies included specificities which covered the entire range of B cells it is interesting that none of them appeared to identify plasma cells. This emphasizes the dramatic phenotypic change which B cells undergo on terminal differentiation to plasma cells. It is already recognized that surface membrane immunoglobulin and class II MHC antigens are lost from plasma cells. This may reflect a loss of receptors for external regulatory signals in a highly specialized end cell which has a finite life span. It is unwise to draw too many firm conclusions from the present limited study as some of the reactions seen may prove to be nonrepresentative on further experience. However, it is clear that many of the antibodies do identify molecules which are selectively expressed on B cells or subsets of B cells. The identification of molecules which are expressed on putative antigen-presenting cells as well as B cells is intriguing. It will be important to study these molecules in greater detail to determine if they have a role in initiating and regulating B cell maturation induced by antigen.

References

1. Nieuwenhuis, P., and W.L. Ford. 1976. *Cell. Immunol.* **23**:254.
2. Stein, H., J. Gerdes, and D.Y. Mason. 1982. *Clinics in Haematology* **11**:531.
3. MacLennan, I.C.M., D. Gray, D.S. Kumararatne, and H. Bazin. 1982. *Immunology Today* **3**:305.

4. Hoffman-Fezer, G., H. Rodt, M. Eulitz, and S. Thierfelder. 1976. *J. Immunol. Methods* **13**:261.
5. Johnson, G.D., R.S. Davidson, K.C. McNamee, G. Russell, D. Goodwin, and E.J. Holborow. 1982. *J. Immunol. Methods* **55**:231.
6. Burk, K.H., B. Drewinko, J.M. Trujillo, and M.J. Ahearn. 1978. *Cancer Res.* **38**:2508.

CHAPTER 22

Immunohistological Analysis of Tissue Specificity of the Fifty-two Workshop Anti–B Lymphocyte Monoclonal Antibodies

G. Pallesen

Introduction

For the Second International Workshop on Human Leukocyte Differentiation Antigens the 52 monoclonal antibodies in the anti–B cell panel were investigated, attaching particular importance to the demonstration of the tissue localization of antigens recognized by the antibodies, and to the selection of antibodies that might be valuable in immunohistological techniques. Because of limitations in our laboratory resources and in time, only such Workshop antibodies that were selected by a screening procedure were selected for more extensive study in the search for antibodies recognizing B cell differentiation stages, and in the search for clustering of antibody reactivity patterns. For this purpose, frozen biopsies of B cell lymphomas of well-established histological subcategories which had already been extensively phenotyped with other monoclonal antibodies were used. In addition, a few cases of T cell lymphomas, Hodgkin's disease, and polymorphic large-cell Ki-1-positive lymphoma ("malignant histiocytosis") were also included in the extended study. Finally, all the selected anti–B cell antibodies were tested for reactivity in sections of formalin-fixed and paraffin-embedded tissue. The results are presented in the following.

The limitations in resources and time did not, however, permit a systematic search for the occurrence of antigens in various human nonlymphoid tissues, as defined by staining by the individual monoclonal antibodies. In all probability this is a very important phase in the investigation of new antibodies if therapeutic immunotoxic application of such antibodies is considered.

Materials and Methods

The source of tissue biopsies, their preparation and storage, and the immunohistological method employed in this laboratory (using acetone–chloroform fixation of frozen tissue sections and a 3-layer immunoperoxidase method) have been published previously (1,2).

Formalin-fixed, paraffin-embedded tissue sections were dewaxed, treated with methanol and H_2O_2 to block endogenous peroxidase, then stained with a 3-layer immunoperoxidase method following trypsinization.

The present investigation was organized in the following way:

1. Frozen tissue sections of a normal (reactive) lymph node were stained by each of the 52 workshop anti–B cell antibodies diluted 1/500.
2. In cases with weak or no staining of the B-zone lymphocytes, repeat stainings were performed using 1/50 dilution.

If no or only weak staining of the B-zone lymphocytes was obtained following the second procedure, the individual antibodies were excluded from further immunohistological investigation. Also excluded were antibodies which exhibited a very broad staining pattern not confined to B-zone lymphocytes. Finally, antibodies which apparently reacted with HLA-DR (staining germinal centers, lymphocytic mantle zone, interdigitating reticulum cells, and endothelial cells) were excluded.

The extended study of the remaining antibodies comprised staining of cryostat tissue sections from the following tissues:

1. Kidney—to disclose reactivity with the C3b receptor of glomeruli;
2. Thymus—to detect reactivity with thymic cells;
3. Skin—to detect reactivity with Langerhans cells. Epidermis was included in the tissue biopsies from two of the selected lymphomas;
4. Granulomas of epitheloid histiocytes—such were present in one case of hairy cell leukemia in splenic tissue;
5. Twenty cases of B cell lymphomas in which previous immune phenotyping had been performed with a wide panel of monoclonal antibodies. The lymphomas were selected in such a way that lymphomas of both low- and high-grade malignancy, and lymphomas of early, intermediate, and late B lymphocyte differentiation stages were represented. The following histological types were selected for the study: lymphoma corresponding to B-type chronic lymphatic leukemia (B-CLL) [2], hairy cell leukemia [2], centrocytic lymphoma [2], centroblastic/centrocytic lymphoma [1 nodular and 1 diffuse type], immunocytoma [2], plasmacytoma [1], anaplastic centrocytic lymphoma [2], centroblastic lymphoma [1 nodular, 1 diffuse], B-type immunoblastic lymphoma [1 with and 1 without plasmacytic differ-

entiation], lymphoblastic lymphoma, Burkitt's type [1], and lymphoblastic lymphoma, unclassified [2; one was immunophenotyped as pre-pre-B-type, the other as B-type]. The principles of the Kiel classification were employed. For translation to the Working Formulation, see Rilke and Lennert (3);

6. Four cases of T cell lymphomas (lymphoblastic lymphoma of convoluted-cell type, mycosis fungoides, pleomorphic T cell lymphoma, and T-type immunoblastic lymphoma), 1 case of Hodgkin's disease, mixed cellularity, and 1 case of so-called malignant histiocytosis [recently shown in most cases to be neoplasms derived from activated, Ki-1 and Tü69 positive T lymphocytes (4)] were also included in the study to investigate cross-reactivity with non-B cell lymphomas.

For purposes of comparison with Workshop antibodies, the staining results obtained with three commercially available pan-B-lymphocyte antibodies were included in the study of tissues in 5) and 6). They comprised DAKO-pan-B (=To15) from Dakopatts, anti-Leu-12 from Becton-Dickinson, and anti-B1 from Coulter Electronics.

Results and Discussion

Screening of Workshop Anti–B Cell Monoclonal Antibodies

The staining of normal lymph node with the 52 anti–B cell antibodies diluted 1/500 or eventually 1/50 resulted in the exclusion of 27 antibodies from the extended study. The reasons are given in Table 22.1. The weak staining produced by nine of the antibodies was too weak (borderline reactivity) to make these antibodies interesting for immunohistological techniques like the present. Of the seven antibodies showing a broad staining pattern, some apparently stained T lymphocytes or a fraction of

Table 22.1. Reasons for the exclusion of 27 antibodies from further investigation following the immunohistological staining of normal lymph node cryostat tissue sections by all 52 anti–B cell workshop monoclonal antibodies.

No staining of B-zone lymphocytes	B13, B16, B18, B23, B26, B30, B32, B37, B38, B46
Too weak staining of B-zone lymphocytes	B2, B5, B8, B15, B20, B22, B29, B36, B48
Broad staining pattern not confined to B-zone lymphocytes, and/or severe background staining	B10, B12, B27, B42, B47, B50, B51
Other reasons	B1[a]

[a] The staining pattern in lymph nodes was similar to that of anti-HLA-DR antibodies.

them in addition to B lymphocytes whereas others (B42, B47, and B50) produced an unclean background since they stained most structures of the tissue. Finally, B1 was excluded since it apparently recognized HLA-DR antigen. It stained germinal centers, lymphocytes of the mantle zone, interdigitating reticulum cells in the T-zone, and endothelial cells. In another experiment B1 was shown to be reactive also with epidermal Langerhans cells.

Staining of Cells Other Than B-Zone Lymphocytes by Selected Antibodies

The typical staining patterns obtained with the 25 Workshop antibodies selected for the extended part of the study in cells other than B-zone lymphocytes are shown in Table 22.2. Seven antibodies recognized dendritic reticulum cells of the lymphatic follicle [Fig. 22.1 (A) and (B); Fig. 22.3], and four antibodies stained endothelial cells. (Thirteen of the excluded antibodies: B1, B8, B12, B15, B23, B26, B29, B32, B38, B42, B47, B48, and B50 also stained endothelium of the postcapillary venules of lymph nodes.) None of the antibodies stained C3b receptors of glomeruli. Langerhans cells of the skin were stained rather weakly by B52 which was unreactive with endothelium; therefore the antigen recognized by B52 could not be HLA-DR. Seven antibodies stained granulomatous epitheloid histiocytes with varying intensity. Three antibodies, B9, B33, and B52, showed reactivity with thymic structures. B9 and B33 produced a dot-like cytoplasmic and probably nonmenbranous staining in a fraction of thymic medullary cells whereas B52 stained a fraction of thymic epithelial cells, particularly in the cortical region.

Staining of B Cell Lymphomas Using Selected Workshop Antibodies

The results obtained from staining 20 B cell lymphomas with the 25 selected Workshop antibodies and the three pan-B-cell reagents are shown in Table 22.3.

Table 22.2. Structures other than B-zone lymphocytes that were stained by the 25 monoclonal workshop antibodies included in the more extensive study.

Follicular dendritic reticulum cells	B9, B11, B19, B33, B35, B39, B41
Endothelial cells	B3, B4, B44, B45
C3b receptor (glomeruli)	None
Langerhans cells (skin)	B52
Epitheloid histiocytes	B3, B17, B24, B41, B43, B45, B52
Thymic cells	B9[a], B33[a], B52[b]

[a] Dot-like cytoplasmic staining in a fraction of the thymic medullary cells.
[b] Staining of a fraction of thymic epithelial cells, mainly cortical.

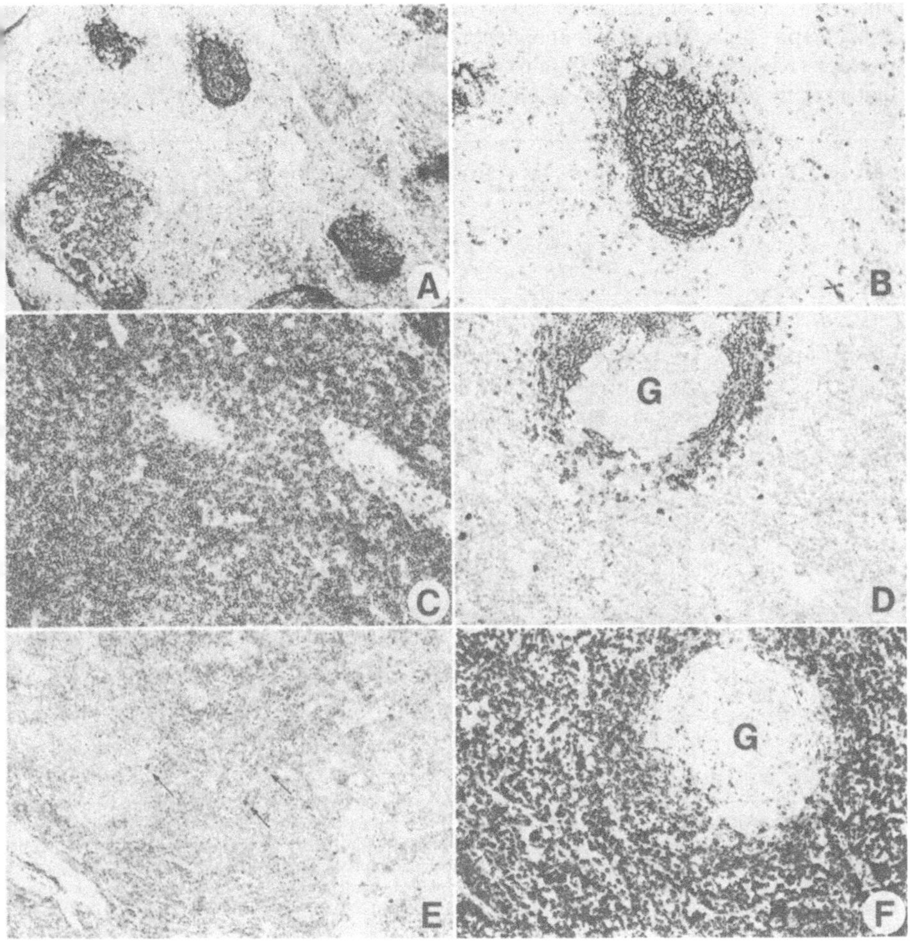

Fig. 22.1. Immunoperoxidase stainings using Workshop anti–B cell monoclonal antibodies that recognize follicular dendritic cells, or may discriminate between B-type chronic lymphocytic leukemia (B-CLL) and hairy cell leukemia: (A) reactive lymph node stained by B11. The dendritic reticulum cells are clearly seen with this antibody; (B) in higher magnification it is seen that B11 stains not only the dendritic meshwork of the follicle but also a large fraction of the mantle zone lymphocytes like monoclonal antibody Tül; (C) like Tül, B11 stains the neoplastic lymphocytes of B-CLL strongly whereas vessels remain unstained; (D) splenic tissue in hairy cell leukemia stained with B11. The hairy cells are unstained as are the histiocytes of the epitheloid granuloma (G) which is surrounded by the strongly stained remnants of a follicle; (E) B-CLL in lymph node stained with B31. Apart from three single and strongly stained lymphocytes (arrows) which are probably residual benign B lymphocytes, the diffuse cellular infiltrate remains unstained; (F) on the other hand, B31 stains hairy cells strongly and does not react with the histiocytes of the epitheloid granuloma (G). (A, ×18; all others, ×45).

Table 22.3. Staining of neoplastic cells by 25 selected workshop anti–B cell monoclon antibodies in 20 previously immunophenotyped B cell lymphomas of well-defined histology (Kiel classification). In addition, the staining results obtained with three commercially available pan B cell antibodies (DAKO-pan-B, anti-Leu-12, anti-B1) are shown.[a,b]

Histology	B3	B4	B6	B7	B9	B11	B14	B17	B19	B21	B24	B25	B28
B-CLL	+/−	2+	−	−	2+	3+	2+	1+	3+	1+	−	(+)	2+
B-CLL	+/−	2+	−	−	1+	3+	2+	1+	3+	1+	−	(+)	2+
HCL	+/−	3+	−	3+	−	−	3+	3+	−	3+	2+	3+	3+
HCL	+/−	2+	−	3+	−	−/+	3+	3+	−/+	3+	1+	3+	3+
CC	3+	1+	1+	1+	(+)	−	2+	1+	−	3+	(+)	2+	2+
CC	2+	2+	−	1+	1+	−	1+	2+	−	3+	(+)	1+	2+
CB/CC, diffuse	2+	1+	1+	3+	3+	−	3+	1+	−	3+	−	3+	3+
CB/CC, nodular	3+	2+	2+	1+	+/−	2+	2+	1+	2+	3+	−	2+	2+
IC	2+	2+	(+)	3+	2+	−	3+	1+	−	3+	−	3+	3+
IC	3+	3+	1+	1+	2+	−	3+	(+)	−	2+	−	2+	3+
Plasmacytoma	−	−	−	−	−	−	−	−	−	−/+	−	−	−
CC, anapl.	−	3+	−	2+	1+	−	1+	1+	−	3+	−	1+	2+
CC, anapl.	3+	−	(+)	2+	2+	−	1+	(+)	−	2+	−	2+	1+
CB, diffuse	3+	−	−	−	−	−	1+	−	−	1+	−	−	2+
CB, nodular	2+	(+)	(+)	1+	1+	−/+	2+	−	−/+	2+	−	2+	3+
B-IB without plasm.	−	(+)	−	3+	−/+	−	2+	1+	−	2+	−	2+	2+
B-IB with plasm.	−	−	−	−	−	−	−	−	−	−	−	−	−
LB, Burkitt type	2+	(+)	−	(+)	−	−	(+)	1+	−	1+	−	(+)	1+
LB, unclassified[c]	−	−	−	(+)	−/+	−	2+	−	−	2+	−/+	(+)	2+
LB, unclassified[d]	3+	2+	−	2+	1+	−	2+	1+	−	(+)	−	2+	2+

[a] Abbreviations: B-CLL = lymphoma corresponding to B-type chronic lymphatic leukemia; B-IB without plasm B-type immunoblastic lymphoma without plasmacytic differentiation; B-IB with plasm. = B-type immunobla lymphoma with plasmacytic differentiation; CB = malignant lymphoma, centroblastic; CB/CC = malignant ly phoma, centroblastic/centrocytic; CC = malignant lymphoma, centrocytic; CC, anapl. = malignant lymphor anaplastic centrocytic; HCL = hairy cell leukemia; IC = malignant lymphoma, immunocytoma; LB = malign lymphoma, lymphoblastic; ND = not determined.

[b] Degree of staining: − = no staining; (+) = very weak (borderline) staining; 1+ = weak staining; 2+ = moder staining; 3+ = strong staining; +/− = partial staining with occurrence of positive and negative tumor cells, howev with predominance of positive tumor cells; −/+ = partial staining with predominance of negative tumor cells.

[c] Immunological phenotype = pre-pre-B-type.

[d] Immunological phenotype = B-type.

Eleven antibodies (B3, B4, B14, B21, B25, B28, B31, B34, B43, B49, and B52) stained the majority of the lymphomas and might thus be denoted pan-B-cell reagents. Except for two antibodies (B21, B52), the pan-B antigens recognized by these antibodies were not expressed in the plasmacytic stage. The most efficient reagents to detect the B-cell nature of these lymphomas were B14, B21, B28, B34, B43, B49, and B52 which were comparable with the three commercial pan-B-cell reagents.

Antigens expressed by late differential B lymphocyte stages (plasmacytoma, immunoblastic lymphoma with plasmacytic differentiation) were recognized by three Workshop antibodies, B21, B45, and B52. The ex-

3	B34	B35	B39	B40	B41	B43	B44	B45	B49	B52	DAKO-pan-B	Anti-Leu-12	Anti-B1 (Coulter)
	1+	2+	2+	–	3+	2+	–	–	(+)	3+	1+	3+	2+
	1+	2+	3+	–	2+	2+	–	–	(+)	2+	1+	2+	2+
	3+	–	–	3+	–	2+	+/–	3+	3+	3+	3+	3+	3+
	3+	–	–/+	3+	–	2+	+/–	3+	3+	3+	3+	3+	3+
	1+	(+)	–	1+	1+	2+	+/–	+/–	2+	2+	3+	2+	3+
+	1+	–	–	(+)	–	1+	1+	2+	1+	2+	3+	3+	1+
	3+	2+	–	1+	3+	2+	–/+	–	2+	3+	3+	3+	1+
	2+	1+	2+	1+	1+	2+	–/+	–	2+	2+	3+	3+	2+
	3+	1+	–	3+	1+	2+	+/–	–/+	3+	3+	3+	3+	3+
	1+	–	–	–/+	+/–	2+	1+	–/+	2+	3+	2+	3+	2+
	–	–	–	–	–	–	–	1+	–	–/+	–	–	–
	2+	2+	–	1+	2+	2+	–	–	2+	–	3+	2+	1+
	1+	2+	–	1+	2+	1+	1+	+/–	3+	2+	3+	2+	3+
	1+	–	–	–	–	1+	1+	2+	ND	2+	1+	3+	2+
–	2+	–/+	–/+	1+	2+	2+	–/+	–	2+	3+	3+	3+	2+
)	2+	2+	–	2+	2+	2+	–	3+	2+	–/+	3+	2+	3+
	–	–	–	–	–	–	+/–	1+	–	–/+	–	–	–
	(+)	–	–	(+)	–	1+	–	(+)	1+	1+	3+	2+	3+
+	2+	–/+	–	–/+	–/+	1+	–	(+)	1+	–/+	1+	2+	1+
+	2+	1+	–	1+	1+	2+	–	–	2+	(+)	2+	2+	2+

pression in these lymphomas was rather weak, or partial. However, the staining pattern obtained with B45 suggests that this antibody reacts with an antigen predominantly expressed by late B lymphocytes since it was not expressed, or only weakly, in early B lymphocyte malignancies (B-lymphoblastic lymphomas, B-CLL).

Antigens that were preferentially expressed by early B lymphocyte malignancies were not detected using this panel of lymphomas.

It is seen (Table 22.3) that the antigen detected by B6 is mainly expressed by neoplastic germinal center cells. In normal lymph node, B6 stained the germinal center cells stronger than the mantle zone lymphocytes. This suggested that an antigen characteristic of lymphocytes of an intermediate differential stage like centroblasts and centrocytes was recognized by B6. In a later study, completed after the submission of results for the Workshop, the expression of this antigen in a larger panel of B cell lymphomas was investigated. These results did not support the original findings since the antigen was often not expressed by lymphomas of the germinal center type. B6 might, however, be an interesting antibody in

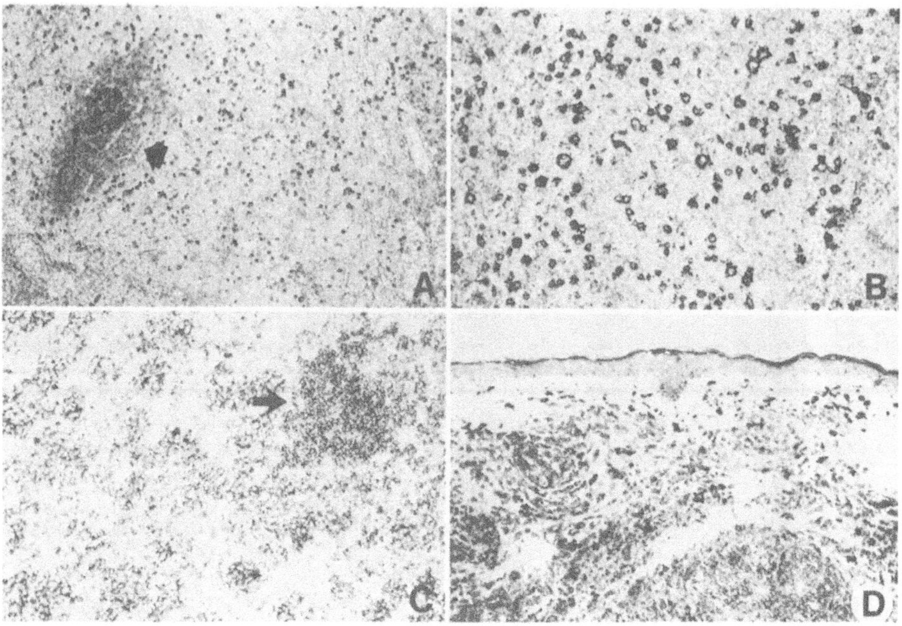

Fig. 22.2. Examples of non-B cell staining obtained with Workshop anti–B cell monoclonal antibodies: (A) lymph node with Hodgkin's disease, mixed cellularity, stained by B11. In low magnification the strong staining of a residual follicle (arrow) is seen as well as numerous large and strongly stained cells in the diffuse infiltrate; (B) same case as (A) in higher magnification. Hodgkin's and Reed–Sternberg cells are stained very clearly with antibodies B11, B19, and B39; (C) same case of Hodgkin's disease, stained with B17. A residual positively stained follicle is seen (arrow). In the diffuse infiltrate many more cells are stained than with B11 but the staining is weaker. Probably both Hodgkin's and Reed–Sternberg cells, as well as macrophages, are stained; (D) skin infiltrated with T-type immunoblastic lymphoma stained with B52. This T cell lymphoma was stained by five of the workshop anti–B cell antibodies: B6, B17, B24, B41, and B52. (A, ×18; others, ×45).

view of the restricted occurrence of the corresponding antigen, and should be studied in more depth. At present there seems to be no monoclonal antibody that selectively reacts with B lymphocytes of the germinal center type.

An immunological discrimination between B-CLL and hairy cell leukemia was possible using several of the workshop monoclonal antibodies [Fig. 22.1(C)–(F)]. Seven antibodies (B9, B11, B19, B33, B35, B39, and B41) stained B-CLL but not hairy cell leukemia whereas the reverse staining pattern was obtained with six antibodies (B7, B24, B31, B40, B44, and B45). The seven cluster-forming antibodies staining B-CLL but

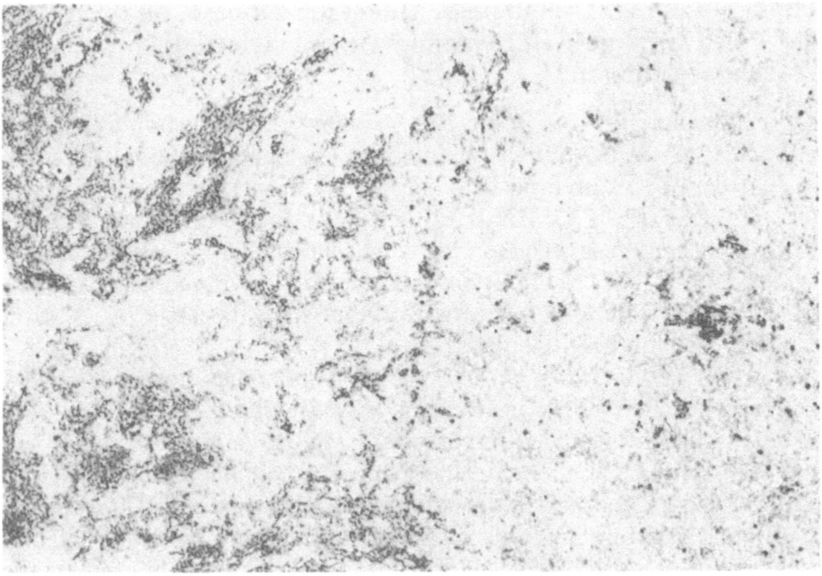

Fig. 22.3. An example of the application of a monoclonal Workshop antibody that recognizes follicular dendritic cells. In this case of a peripheral and pleomorphic T cell lymphoma, the staining by B9 discloses a peculiar proliferation of the dendritic reticulum cells (left part of picture) which make a widespread loose meshwork associated with proliferation of vessels. This phenomenon is not rare in peripheral T cell lymphomas and may suggest a close and probably functional association between neoplastic T lymphocytes and non-neoplastic B cell areas. (×46).

not hairy cell leukemia were the same ones that stained dendritic reticulum cells and mantle zone lymphocytes in normal lymph nodes. This pattern bears a resemblance to that obtained with monoclonal antibody Tül (4). In this cluster it was possible to identify subclusters of antibodies: B9 and B33 were the only antibodies of this group that did also react with thymic cells (Table 22.2). Analyzing the staining results of Table 22.3 it becomes clear that at least two subclusters are formed. Rather similar but not identical stainings were obtained by B9, B33, B35, and B41 which differed slightly from those obtained by B11, B19, and B39. The difference between the two groups became obvious when tissue from Hodgkin's disease was stained: B11, B19, and B39 stained Hodgkin's and Reed–Sternberg cells similarly, but differently from B9, B33, B35, and B41. In all the tissues stained for this study, B11, B19, and B39 have produced identical stainings, and they most probably recognize the same epitope.

Staining of T Cell Lymphomas, Hodgkin's Disease, and Ki-1-Positive Large-Cell Lymphoma Using Selected Workshop Antibodies

The results are shown in Table 22.4. None of the workshop monoclonal antibodies stained the thymic lymphoblastic lymphoma of convoluted-cell type or two of the three post-thymic lymphomas. However, five anti–B cell antibodies (B6, B17, B24, B41, and B52) stained an immunologically well-established case (Lyt3$^+$, T3$^+$, UCHT1$^-$, T6$^-$, Leu3a$^+$, Tü68$^-$, Tü33$^+$, Tü71$^-$, Tü14$^-$) of T-type immunoblastic lymphoma of the skin [Fig. 22.2(D)] which was selected for this study because of its strong reactivity with anti-B1 (Coulter).

A surprisingly high number of Workshop antibodies stained Hodgkin's and Reed–Sternberg cells in Hodgkin's disease tissue (Table 22.4). Some of the antibodies produced rather weak staining confined to tumor cells (and B lymphocytes) whereas others apparently stained macrophages as well [Fig. 22.2(C)]. Some of the antibodies (B11, B19, and B39) produced a particularly strong and clear staining of the tumor cells in Hodgkin's disease [Fig. 22.2(A) and (B)]. As stated above these three antibodies apparently recognize the same epitope and show a Tül-like staining pattern. However, the staining of Hodgkin's disease tissue discloses that

Table 22.4. Staining of neoplastic cells by the 25 selected workshop anti–B cell monoclonal antibodies and three commercially available pan B cell antibodies in T cell lymphomas, Hodgkin's disease, and "Ki-1-positive large-cell lymphoma" (Malignant histiocytosis sensu auctorum).

Histology		Antibodies with positive staining	Antibodies with negative staining
T cell lymphomas[a] (n = 4)	LB-conv.[b]	Anti-B1 (Coulter)	All others, DAKO-pan-B, anti-Leu-12
	Myc. fung.	None	All
	T-pleom.	None	All
	T-IB	B6, B17, B24, B41, B52, anti-B1 (Coulter)	All others, DAKO-pan-B, anti-Leu-12
Hodgkin's disease (mixed cellularity) (n = 1)		B3, B6, B9, B11, B14, B17, B19, B21, B24, B28, B33, B34, B35, B39, B41, B43, B44, B45, B52, anti-B1 (Coulter)	B4, B7, B25, B31, B40, B49, DAKO-pan-B, anti-Leu-12
Ki-1-positive large-cell lymphoma[c] (n = 1)		B3, B6, B11, B17, B19, B21, B24, B39, B41, B52	B4, B7, B9, B14, B25, B28, B31, B33, B34, B35, B40, B43, B44, B45, B49, DAKO-pan-B, anti-Leu-12 anti-B1 (Coulter)

[a] the cases were previously phenotyped using a wide panel of monoclonal antibodies.
[b] Abbreviations: LB-conv. = malignant lymphoma, lymphoblastic, convoluted-cell type; Myc. fung. = mycosi fungoides; T-pleom. = pleomorphic T cell lymphoma; T-IB = T-type immunoblastic lymphoma.
[c] This case was shown to be T lymphocyte-derived by multiple marker analysis.

they nevertheless are distinct from Tül since this antibody does not stain Hodgkin's and Reed–Sternberg cells.

The case of Ki-1-positive large-cell lymphoma was stained by 10 of the Workshop anti–B cell antibodies (Table 22.4). These tumors were usually considered cases of malignant histiocytosis because of their morphology and because they are usually alpha-1-antitrypsin and 1-antichymotrypsin positive and also show strong diffuse cytoplasmic staining for acid phosphatase and nonspecific esterase, like macrophages. Recent studies have brought evidence, however, that they represent neoplasias of activated lymphocytes, usually T lymphocytes, since they express activation antigens (Ki-1^+, Tü69^+, Tac^+, Tü35^+) (5). The present case was phenotypically T-derived ($Lyt3^+$, $T3^-$, $UCHT1^+T6^-$, $Leu3a^+$, Tü68^-, Tü33^+, Tü71^+, Tü14^-).

Staining of Formalin-Fixed and Paraffin-Embedded Tissue

To screen reactivity with antigens in routinely processed histological sections, a biopsy with Hodgkin's disease was stained since it contained residual benign B and T cell areas as well as Hodgkin's and Reed–Sternberg cells. Only two of the selected Workshop antibodies produced staining at all. B44 stained the cytoplasm of endothelial cells and scattered large cells, probably a fraction of macrophages, but did not stain B lymphocytes and Reed–Sternberg cells. B45 stained almost all cells including lymphocytes and Reed–Sternberg cells.

Resource Consumption

For this study about 900 immunohistological sections were produced representing the work of one laboratory technician in 50 working days, or of two technicians in five weeks as was the case in our preparations for the Workshop. To allow more in-depth study it is recommended that more time be provided between the receipt of antibodies and the submission of results for the Third International Workshop.

Acknowledgments. This work was supported by the Danish Cancer Society (grant No. 20/80) and the Danish Medical Research Council (grant No. 12-4528). I thank Ms. B. Hermansen and Ms. A. Fuglsang for their technical contribution, Ms. A. Holmehave for typing the manuscript, and Mr. S. Shapiro for the photographic work.

References

1. Pallesen, G., M. Madsen, and S. Schifter. 1983. Immune marker expression in 53 lymphomas of high-grade malignancy. *Histopathol.* **7**:841.
2. Pallesen, G., P.C.L. Beverley, E.B. Lane, M. Madsen, D.Y. Mason, and H. Stein. 1984. The nature of non-B/non-T lymphomas. An immunohistological

study on frozen tissues from 24 cases using monoclonal antibodies. *J. Clin. Pathol.* **37**:911.

3. Rilke, F., and K. Lennert. 1981. Perspective of the Kiel Classification in relation to other recent classifications of non-Hodgkin's lymphomas, with special reference to the Working Formulation. In: *Histopathology of non-Hodgkin's lymphomas (based on the Kiel Classification).* K. Lennert, ed. In collaboration with H. Stein. Springer-Verlag, Berlin, Heidelberg, New York, p. 112–118.
4. Ziegler, A., H. Stein, C. Müller, and P. Wernet, 1981. Tü1: A monoclonal antibody defining a B cell subpopulation—usefulness for the classification of non-Hodgkin's lymphomas. In: *Leukaemia markers,* W. Knapp, ed. Academic Press, London, pp. 113–116.
5. Stein, H., D.Y. Mason, J. Gerdes, N. O'Connor, J. Wainscoat, G. Pallesen, B. Falini, G. Delsol, H. Lemke, and K. Lennert. The expression of the Hodgkin's disease associated antigen Ki-1 in reactive and neoplastic lymphoid tissue. Evidence for a relationship between Reed–Sternberg cells and polymorphic large cell lymphomas. *Blood* (in press).

CHAPTER 23

Reactivity of Monoclonal Antibodies of the B and L Series with Follicular Dendritic Cells in Tissue Sections and with Lymphoblastoid Cell Lines

Gerald D. Johnson, Ian C.M. MacLennan,
Noel R. Ling, Debbie L. Hardie,
Paul D. Nathan, and Léonie Walker

Introduction

The B cell and leukemia Workshop antibodies were tested on tissue sections by immunofluorescence and immunoenzyme staining. This identified 16 which showed unequivocal reactivity with follicular dendritic cells (FDC) in germinal centers (GC). It was observed that patterns of FDC staining obtained with different antibodies were not identical—some gave extensive staining of the GC and others a restricted pattern involving fewer FDC or possibly a smaller part of each FDC. The antibodies were therefore compared by double immunofluorescence (using conjugates specific for mouse immunoglobulin isotypes) with two established anti-FDC monoclonal antibodies (mAbs), and a third anti-FDC antibody produced in this Department, all of which give completely coincident staining of FDC when paired with each other.

Additional differences between Workshop FDC-reactive antibodies were revealed by studying their reactivity with normal and neoplastic lymphocytes in sections, with FDC from patients with malignant lymphoma, and with various cell lines. It is concluded that multiple antigens are involved in the reactions of different Workshop antibody preparations with FDC and comprise at least four subsets on the basis of the findings on tissue sections.

In the course of this study the immunoglobulin isotypes of 22 of the Workshop preparations were determined by immunofluorescent staining with isotype-specific conjugates. In four of the 22, this revealed the presence of antibodies associated with two different isotypes. In three of these, the two isotypes gave different patterns of cell reactivity.

Materials and Methods

Tissues

Surgical specimens of tonsil, normal spleen, and a lymph node and three spleens from patients with low-grade non-Hodgkin's lymphoma which contained abundant FDC as identified by the mAb R4/23 (*vide infra*) were snap-frozen and stored at −70°C. Cryostat sections cut at 6 μm were air-dried and fixed in acetone for 5 min at room temperature.

Cell Lines

The lymphoblastoid and other cell lines used were stained in suspension and as air-dried acetone-fixed cell spreads.

Immunofluorescence Procedure

Standard techniques were employed as previously described (1). The final dilution of the Workshop antibodies was 1 : 250 in the screening tests and 1 : 100 in the double-staining procedure. Mouse immunoglobulin isotypes of Workshop antibodies were identified by staining with monospecific conjugates (Seward, U.K.). Comparison of FDC staining with that obtained with R4/23 (2) (kindly supplied by Professor H. Stein) and BL13 (3) (kindly supplied by Dr. J. Brochier) was made by double fluorescence—sections were first treated with a mixture of Workshop antibody and defining antibody of different isotype and then with a mixture of fluorescein and rhodamine conjugates of appropriate isotype specificities. A third anti-FDC mAb (2BF11) having the same isotype as R4/23 gave completely coincident staining when paired with BL13.

Immunoperoxidase Technique

Sections were treated with the mAbs followed by sheep anti–mouse Ig peroxidase conjugate and developed with diaminobenzidine.

Results

The screening tests of the mAbs comprising the B and L Workshop series performed on sections of tonsil and normal spleen by immunofluorescence and immunoperoxidase staining identified 33 preparations that showed some reactivity with FDC in germinal centers. Of these 16 giving unequivocal staining were selected for further study. Most of these gave extensive staining of the germinal centers but three appeared to react with only a few cells (Fig. 23.1). When compared with defining antibodies of appropriate isotype specificity it was found that the staining pattern obtained with antibodies giving extensive GC staining coincided completely with the pattern obtained with the defining antibodies; the restricted pattern involved a subset of the cells reactive with defining antibody. The three subset-reactive antibodies gave no staining of about one third of the germinal centers in normal tissue that were stained by defining antibody, suggesting a *germinal center subset* specificity. However, this may have been a section-sampling artifact since the unstained GC tended to be smaller than those containing positive cells—i.e., these antibodies might be reacting with the more central parts of each FDC rather than staining all of some FDC but no part of others. The mouse Ig isotypes and results of the paired staining experiments are shown in Table 23.1.

The anti-FDC antibodies were further tested on sections of a lymph node from a patient with centroblastic/centrocytic lymphoma and on cells from cultured cell lines. Results were analyzed in relation to the findings on tissue sections (Table 23.2). It will be seen that all the antibodies which gave the *restricted* FDC-staining pattern also stained FDC in the malignant lymphoma, failed to stain lymphocytes in sections of normal tissue,

Table 23.1. Isotype specificity and FDC reactivity on tissue sections.

mAb	Mouse Ig isotype	Comparison with defining antibodies
B series		
2	G1[a]	Identical
3	G1	Identical
9	M	Identical
11	G1	Restricted
19	G1	Restricted
28	G1	Identical
29	G2b	Identical
33	M	Identical
34	G1	Identical
35	G1	Identical
39	G1	Restricted
41	G2a	Identical
42	G1	Identical
L series		
4	G3	Identical
17	G1[b]	Identical
18	M	Identical
Defining antibodies		
R4/23[c]	M	Identical to BL13
BL13[d]	G1	Identical to R4/23 & 2BF11
2BF11	M	Identical to BL13

[a] FDC-nonreactive G2b anti-lymphocyte antibody also present.
[b] FDC-nonreactive G3 anti-lymphocyte antibody also present.
[c] Ref. 2.
[d] Ref. 3.

A

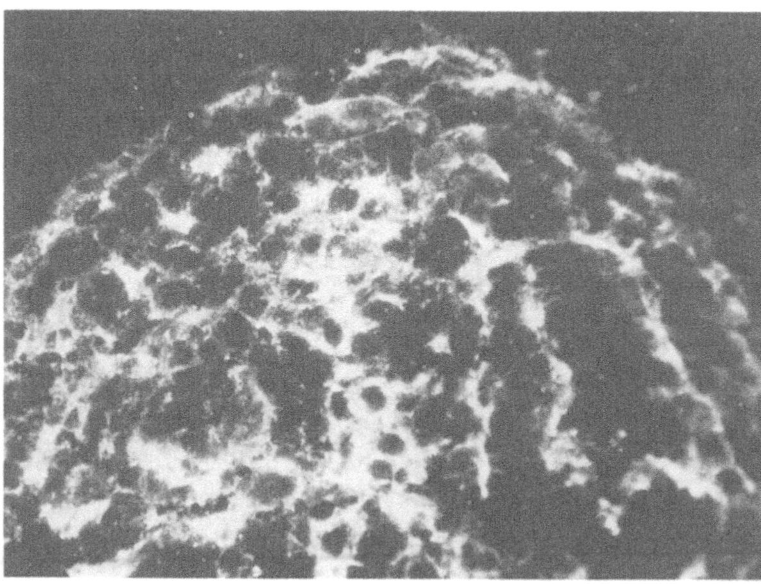

B

Fig. 23.1. Double immunofluorescence staining of tonsil sections. (A) Defining mAb BL13 demonstrated with rhodamine-labeled anti–mouse IgG1. (B) Workshop mAb B9 demonstrated with fluorescein-labeled anti–mouse IgM, same field as (A), showing completely coincident FDC staining.

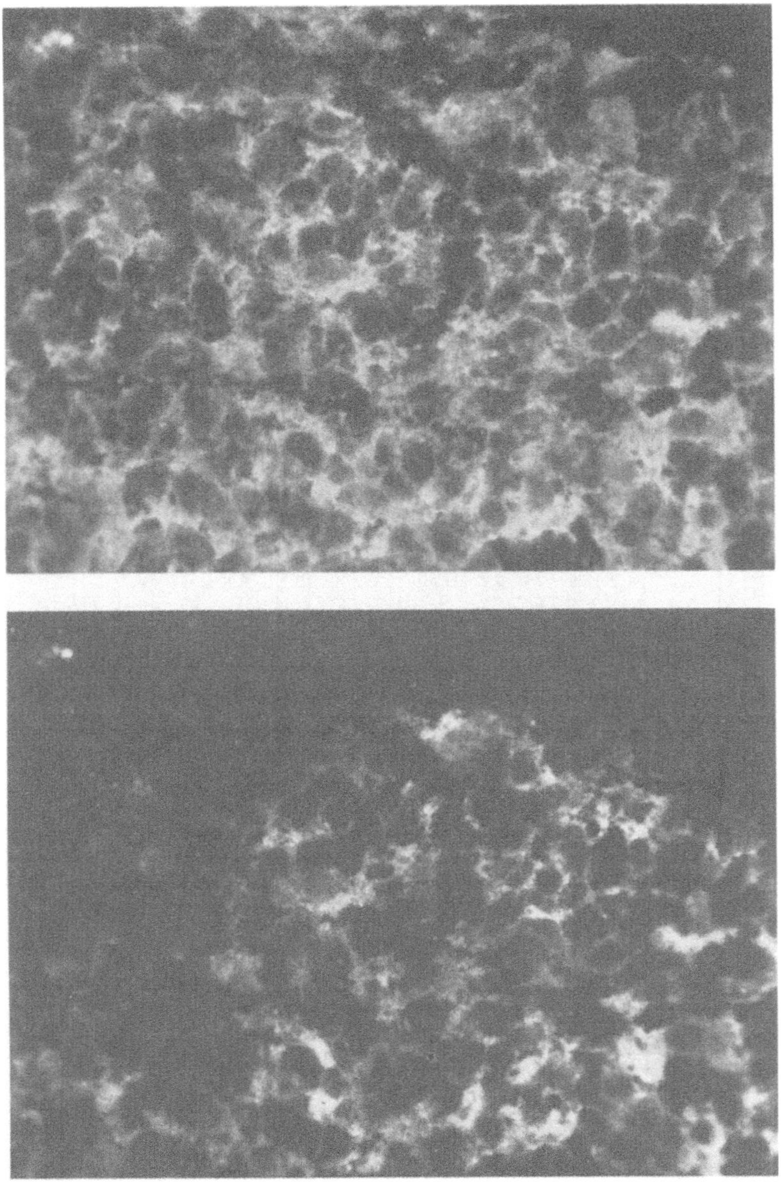

C

D

Fig. 23.1. (*Continued*) (C) Defining mAb R4/23 demonstrated with rhodamine-labeled anti–mouse IgM. (D) Workshop mAb B11 demonstrated with fluorescein-labeled anti–mouse IgG1, same field as (C), showing restricted pattern.

Table 23.2. Reactivity of anti-FDC workshop mAbs with tissue sections and cell lines.

Workshop reactivity with cells in tissue sections					Reactions with cells of B and other cell lines[e]										
					Pre-B ALL	Burkitt			B-LCL	Late B		Plasma cell	T-LCL		Myeloid
Workshop no.	Tonsil FDC[a]	Lymphocytes[b]	Lymphoma FDC[c]	Assay[d]	SMS/SB	EB4	Raji	Daudi	Ed-1	ARH-77	WM-1	RPMI-8226	HSB-2	MOLT-4	HL-60
B3	Total	+	–	S	–	rare±	–	–	–	–	±	–	–	–	–
				M	–	–	ND[f]	+++	++	–	+++	–	–	–	–
B28	Total	+	–	S	+	±	++	+	++	+	+	–	–	ND	–
				M	±	–	ND	±	–	±	–	–	–	–	–
B34	Total	+	–	S	++	+	+++	++	+++	++	+++	±	–	ND	–
				M	–	–	ND	–	–	–	–	–	–	–	–
B42	Total	+	–	S	–	–	–	–	–	–	–	–	–	–	–
				M	±	–	ND	+	occ+[g]	–	–	–	–	–	–
B41	Total	+	+	S	–	–	–	–	occ+	+	+	–	–	ND	–
				M	–	–	ND	±	–	–	occ+	–	–	–	–
B9	Total	+	+	S	–	–	–	±	–	+	–	–	–	–	–
				M	–	–	ND	±	occ+	+	–	–	–	–	–
B33	Total	+	+	S	–	–	++	–	++	++	+	±	–	ND	–
				M	–	±	ND	±	–	±	–	–	–	–	–
B35	Total	+	+	S	–	–	++	+	+	+++	++	+	–	ND	–
				M	–	–	ND	–	–	–	–	–	–	–	–
L4	Total	–	Doubtful	S	–	–	–	–	–	–	–	–	–	–	–
				M	–	–	ND	–	–	–	–	+	–	–	–
L18	Total	–	Doubtful	S	50%+	–	–	–	rare+	–	–	50%++	–	20%+	–
				M	+	+	ND	–	occ+	–	–	+++	–	–	–
B29	Total	–	Doubtful	S	–	+	++	+	++	+	–	++	–	ND	–
				M	–	–	ND	–	occ±	+	–	+	–	–	–
B11	Subset	–	+	S	–	–	+	–	+	++	+	±	–	–	–
				M	–	–	ND	–	+++	+++	++	rare+	–	–	–
B19	Subset	–	+	S	–	+	+	(+)	++	+++	+++	++	–	ND	–
				M	–	–	ND	–	+++	+++	+++	+	–	–	–
B39	Subset	–	+	S	–	–	–	–	++	+	+	±	–	–	–
				M	–	–	ND	–	+	+	±	–	–	–	–

[a] Compared with reactivity of defining antibodies.
[b] B cells in tonsil and spleen.
[c] Lymph node (centroblastic/centrocytic lymphoma).
[d] S = Surface staining; M = multispot technique—acetone-fixed cell-spreads.
[e] Intensity of immunofluorescence shown by the majority of cells is given.
[f] ND: Not done.
[g] occ: Occasional.

but gave positive reactions with the B-lymphoblastoid line Ed-1 and the two late B lines ARH-77 and WM-1. The results obtained with Workshop mAbs giving *completely coincident* staining with that obtained with defining antibodies showed marked heterogeneity and included three other antibodies giving unequivocal staining of B cell lines which did not stain lymphocytes in normal tissue (B29, L4, L18). All seven antibodies which did not stain FDC in the malignant lymphoma lymph node also showed no staining of FDC in the neoplastic GC in sections of spleen from a patient with low-grade non-Hodgkin's lymphoma; B3, B42, and L4 were also tested on spleens from two further cases of low-grade non-Hodgkin's lymphoma which showed abundant R4/23-positive areas; FDC were again not revealed by these antibodies. Four patterns of reactivity could be identified by analysis of the results obtained on tissue sections, as Table 23.2 demonstrates, but they showed no correlation with the staining of the cell lines—apart from the pattern associated with the FDC-subset-reactive antibodies noted above. A fifth pattern of reactivity on sections, viz., pan-FDC-positive on normal and malignant lymphoma tissues and negative on normal B lymphocytes, was seen only with the defining antibody R4/23 (Table 23.3).

Four mAbs of the B series (11, 19, 29, and 39) and two of the L series (4 and 18) which bound to FDC showed little or no binding to B cells (follicular or marginal zones) in sections of lymphoid tissue. Only one (L4) failed to react significantly with any of the lymphoblastoid cell panel—apart from a positive reaction with fixed RPMI-8226 cells—thus showing a high degree of specificity for FDC. L18 was unusual in binding to cells of the pre-B line and strongly to cells of the plasma cell line RPMI-8226. It also reacted with MOLT-4—the only mAb of the group which reacted with any of the non-B cell lines.

An incidental observation arising from the isotype-specific staining experiments was the demonstration in individual Workshop preparations of antibodies giving different staining patterns associated with two isotype specificities. Two of the preparations—B2 and L17—contained anti-lymphocyte, FDC-nonreactive antibodies of different isotype from that of the FDC-reactive antibody they contained. B12 was found to have FDC reactivity associated with both IgG3 and IgM fractions. A fourth preparation

Table 23.3. Five distinct epitopes on FDC.

Antibodies	B cells	FDC tonsil Total	FDC tonsil Part	FDC lymphoma Total	FDC lymphoma Part
R4/23	−	+	−	+	−
B3, 28, 34, 42	+	+	−	−	−
B9, 33, 35, 41	+	+	−	+	−
L4, 18, 29	−	+	−	?	−
B11, 19, 39	−	−	+	−	+

included in the 22 selected for determination of isotype specificity showed broad lymphocyte reactivity in both IgG2b and IgM fractions (L6); distinctive isotype-specific features were, however, also apparent in the tissue sections, including staining of large vessel endothelium and basal epithelial cells.

Discussion

Thirteen of the 51 B series preparations tested and three of the 22 L series gave unequivocal staining of FDC in germinal centers in sections of tonsil and normal spleen. When these were compared with established anti-FDC mAbs 13 gave an identical pattern but three showed less extensive reactivity. Although these antibodies are designated here as reacting with follicular dendritic cells we have not formally shown binding to cells. Rather the binding is follicular and shows a pattern which resembles that of immune complex localized on FDC. The "pan-FDC" monoclonal antibodies show a more extensive staining pattern than that normally associated with immune complex on FDC. These pan-FDC show staining between lymphocytes in the follicular mantle as well as in germinal centers. FDC in centroblastic/centrocytic lymphoma rarely have associated immune complex. This may be because the active transport mechanism involved in localizing immune complex on FDC is unable to operate in neoplastic tissue (4). It is possible that those antibodies which show FDC in normal but not neoplastic tissue are revealing antigens evoked by the presence of immune complex.

The method of analysis we employed—isotype-specific double immunofluorescence—is a powerful procedure for studying sites of *coincident* reactivity since the contribution of the individual reactants can be separately determined by fluorescence microscopy with the specific filter combinations. The availability of two reference anti-FDC mAbs of different isotypes enabled all the Workshop FDC-reactive preparations to be paired with a defining antibody. The paired antibodies were applied simultaneously as mixtures. Coincident staining of antigen in this system may be detecting different isotypes having a similar distribution. However, two antibodies of comparable affinity detecting the same epitope will also give coincident staining by this technique. This contrasts with the effect of *sequential* application of antibodies when the antibody applied first may specifically block or sterically hinder the attachment of the second.

The analysis of the other findings reported here suggests that multiple epitopes are involved in the reactions showing complete coincidence with the established anti-FDC antibodies. However the three antibodies giving the restricted FDC-staining pattern all showed the same profile of reactivity when tested on other cells and tissues. Unfortunately, paired isotype staining of these three antibodies with each other was not possible because they all have the same isotype—IgG1. The results obtained on the various tissue sections indicate that none of the pan-FDC-reactive Workshop antibodies has identical specificity to that shown by the defining antibody R4/23.

Our results confirm the value of screening monoclonal antibodies on sections of lymphoid tissues. This allows an extensive variety of cell types in characteristic sites to be tested at one time. The incidental finding that a significant proportion of the antibody preparations (three of 22 tested) showed different patterns of reactivity associated with two immunoglobulin isotypes underlines the need for caution in interpreting results on samples whose monoclonality has not been formally confirmed.

References

1. Johnson, G.D., and Holborow, E.J. Preparation and use of fluorochrome conjugates. In: *Handbook of experimental immunology,* 4th Edition, D.M. Weir and L.A. Herzenberg, eds. Blackwells, Edinburgh and Oxford, Chap. 31. (in press).
2. Naiem, M., J. Gerdes, Z. Abdulaziz, H. Stein, and D.Y. Mason. 1983. Production of a monoclonal antibody reactive with human dendritic reticulum cells and its use in the immunohistological analysis of lymphoid tissue. *J. Clin. Pathol.* **36:**167.
3. Brochier, J., D. Schmitt, E. Yonish-Rouach, G. Codier, and C. Viac. 1984. Use of tissue distribution studies to determine the specificity of monoclonal anti-lymphocyte antibodies. In: *Leucocyte typing,* A. Bernard, L. Boumsell, J. Dausset, C. Milstein, and S.F. Schlossman, eds. Springer-Verlag, Berlin, Heidelberg, pp. 465–469.
4. Gray, D., D.S. Kumararatne, J. Lortan, M. Khan, and I.C.M. MacLennan. 1984. Relation of intra-splenic migration of marginal zone B cells to antigen localisation on follicular dendritic cells. *Immunology* **52:**659.

CHAPTER 24

The Staining of a Panel of Routine Diagnostic Tissue Biopsies with the Workshop "L" Series Antibodies

David B. Jones, Karen M. Britten, and Dennis H. Wright

Introduction

Clinical schemes for the classification of lymphoma assume a relationship between the normal development of lymphoreticular cells *in vivo* and their neoplastic counterparts (1). The staining patterns of lymphomas are therefore representative of normal tissue from which they are derived (2). In this study we have prepared a panel of normal and neoplastic lymphoid tissue and have attempted to establish predominant staining patterns for the Workshop antibodies in frozen section.

Materials and Methods

Workshop antibodies were received as a 1/25 dilution of ascitic fluid. The antibodies were diluted in Tris-buffered saline (TBS), pH 7.6, to the recommended dilution of 1/250 at the time of screening. Individual antibodies are identified by their Workshop number. Table 24.1 presents the tissue panel employed in this investigation. Tissue was selected from routine diagnostic biopsy material on the basis of previous histological diagnosis. This was confirmed using a monoclonal antibody panel on tissue sections and also by phenotypic analysis of cells isolated from fresh tissue biopsies.

Immunohistochemical staining was performed by a modification of the method of Stein *et al.* (2). 6-μm cryostat sections were dried at room temperature and stored at $-20°C$ over dessicant until required for staining. On the day of staining, sections were fixed in dry acetone for 15 min at room temperature and washed in TBS. Antibodies were then applied to the sections which were incubated for 30 min and then washed twice in TBS. A second layer of peroxidase-conjugated rabbit anti–mouse Ig (Dako; diluted 1/80 in TBS) was then applied for a further 30 min with two more washes. The reaction product was then developed by the application of diaminobenzidine (3). The sections were counterstained with hematoxylin and mounted for viewing.

Table 24.1. Tissue panel employed in this study.

Patient	Diagnosis	Abbreviation
M.B.	T cell lymphoma	T
J.G.	T cell lymphoma	T
R.K.	Lymphoblastic lymphoma, T	T-LBL
E.L.	Reactive	R
D.J.	Sarcoid	SARC
T.	Reactive	R
H.H.	Follicular center cell lymphoma follicular	FCC CB/CC F
S.L.	Follicular center cell lymphoma follicular	FCC CB/CC F
B.L.	Follicular center cell lymphoma follicular	FCC CB/CC F
M.H.	B cell lymphoma, centrocytic diffuse	FCC CC DIFF
M.E.	B cell lymphoma, centrocytic diffuse	FCC CC DIFF
F.E.	B cell lymphoma, centroblastic diffuse	FCC CB DIFF
W.T.	B cell lymphoma, centroblastic diffuse	FCC CB DIFF
F.R.	Diffuse large cell lymphoma (histiocytic?)	LLDIF
E.B.	Lymphocytic lymphoma	LC
D.W.	Reactive	R
J.T.	Reactive	R
I.C.	Reactive	R

Results

The staining patterns of the various L series antibodies on the frozen tissue panel are presented in Table 24.2. The predominant staining patterns of the antibodies provided may be summarized as follows.

Antibodies L1 and L15 were not reactive with either neoplastic or reactive lymphoid tissue. No particular staining pattern could be identified for antibodies L10 and L11. L12 stained all lymphoreticular cells present in all sections; capillary endothelium and connective tissue cells were negative (Fig. 24.1, Fig. 24.2). The staining observed with L6 also suggested a leukocyte common specificity but the staining observed was weaker than that seen with L12. After staining with L6, lymphoblastic lymphoma cells were negative though residual lymphocytes were positive in the same section (Fig. 24.3).

Six antibodies (L4, L7, L16, L18, L21, L22) showed a similar reactivity. The predominant staining pattern seen was of vessel walls and connective tissue in lymph nodes (Fig. 24.4) with occasional positivity of macrophages and a few lymphoid cells. Follicle centers showed weak staining with a reticular pattern in some sections and in the case of L22 the squamous epithelium of the tonsil was heavily stained.

Antibody L9 stained the T cell areas of reactive lymph nodes, T cell lymphoma cells (Fig. 24.5), and residual lymphocytes (Fig. 24.6) in B cell lymphoma. T-LBL cells were positive (Fig. 24.7). Macrophages were also positive with this antibody. Staining of B cell tumors was seen with L8 and L17, though the intensity varied; residual cells in T cell lymphomas were also positive with these antibodies. Residual macrophages and occasional endothelial staining was also seen with L8 (Fig. 24.8).

Table 24.2. Predominant reactivity of L series antibodies on human lymph node biopsy material.

	L1	L2	L3	L4	L6	L7	L8
T	Negative	Macrophages	Macrophages	Vessel walls	Tumor cells	Negative	Residual cells
T		Macrophages	Macrophages	Vessel walls	Tumor cells	Negative	Residual cells
LBL		Tumor cells	Macrophages	Vessel walls	Residual lymphocytes	Negative	Macrophages
R		Macrophages	Macrophages	Vessel walls	Lymphocytes	Negative	Some lymphocytes
SARC		Macrophages, granulomas	Macrophages, granulomas	Vessel walls	Lymphocytes, macrophages	Negative	Some lymphocytes
R		Cells in blood vessels	Follicular and interfollicular macrophages	Weak vessel walls and endothelium	Lymphocytes, macrophages	Weak vessel walls and endothelium	Some lymphocytes
CB/CC Fol	Negative	Weak positivity in follicles	Macrophages	Vessel walls, connective tissue	Positive tumor cells, interfollicular cells	Negative	Neoplastic cells, some cells in blood vessel walls
FCC-CC DIF X2	Negative	—	—	Vessel walls, connective tissue	Positive tumor cells	Negative	Positive tumor cells Positive residual cells
FCC-CB DIF	Negative	—	Residual macrophages, ?DRC+	Vessel walls, connective tissue	Positive tumor cells	Negative	Weak positive tumor cells
LL-DIF	Negative	Tumor cells	Tumor cells	Vessel walls, connective tissue	Tumor cells	Weak vessel walls and endothelium	Tumor cells and some residual cells
LC	Negative	Occasional macrophages	Occasional macrophages	Vessels	Positive	Weak vessels	Lymphocytes
R x3	Negative	Interfollicular macrophages weak positive	As L2	Vessels, weak staining in follicles	Leukocytes	Weak vessels	Heavy reticular staining of follicles, interfollicular cells, endothelium

Table 24.2. (*Continued*)

	L9	L10	L11	L12	L13	L14	L15
T T	Tumor cells positive	Epithelioid cells	Weak epithelioid cells	Leukocytes	Macrophages+	—	Negative
LBL	Weak+	Tumor cells	Tumor cells	Leukocytes	Residual macrophages	Positive tumor	Negative
R	T cell areas positive	—	—	Leukocytes	Macrophages	Weak staining of epithe-lioid cells	Negative
SARC	Lymphs, granulomas	—	—	Leukocytes	Macrophages & granulomas		Negative
R	T cell pattern	Weak in granuloma	Weak in granuloma	Leukocytes	Macrophages		Negative
CB/CC Fol x3	Residual T cells	Weak positive in neoplastic follicles		Leukocytes	Residual cells outside tumor areas		Negative
CC DIF x2	Residual T cells	—	—	Leukocytes	—	Weak macrophages	Negative
CB DIF x2	Residual T cells	—	—	Leukocytes	Some weak staining of tumor cells	Weak macrophages	Negative
LL DIF	Residual cells	Positive	Positive	Leukocytes	Macrophages		Negative
LC	Residual cells			Leukocytes	Occasional macrophages	—	Negative
R x3		Weak positive in follicles	As L10 but weaker	Leukocytes	Strong macrophage staining	Weak macrophages	Negative

	L16	L17	L18	L19	L20	L21	L22
T	Endothelium	—	Connective tissue and vessels	Macrophages	Macrophages	Connective tissue and vessels	Vessels
T		Residual cells					
LBL	Endothelium	Very weak tumor cell	Connective tissue and vessels	Macrophages	Macrophages	Negative	Vessels
R	Endothelium	Some lymphs	Connective tissue and vessels	Negative	Macrophages	Negative	Vessels
SARC	Endothelium	Some lymphs	Connective tissue and vessels	Macrophages++	As L19 but weaker	Negative	Vessels
R	Tonsil epithelium blood vessels	Follicles Stain	Connective tissue and vessels	Endothelium−	As L19 but weaker	Negative	Tonsil epithelium
CB/CC Fol x3	Vessels connective tissue	Neoplastic follicles	Connective tissue and vessels	Strong macrophages	As L19 but weaker	Negative	—
CC DIF x2	Some cells+	— —	Connective tissue and vessels	Strong macrophages	As L19 but weaker	Negative	—
CB DIF x2	—	Weak positive tumor cells	Connective tissue and vessels	Some tumor cells	As L19 but weaker	Negative	—
LL DIF	—	Negative	Connective tissue and vessels		As L19 but weaker	Negative	—
LC	—	Tumor cells positive	Connective tissue and vessels		As L19 but weaker	Negative	—
R x3	—	Follicles and mantle cells	Connective tissue and vessels	Extensive macrophage staining ?DRC ?IDRC some vessels	As L19 but weaker	Negative	—

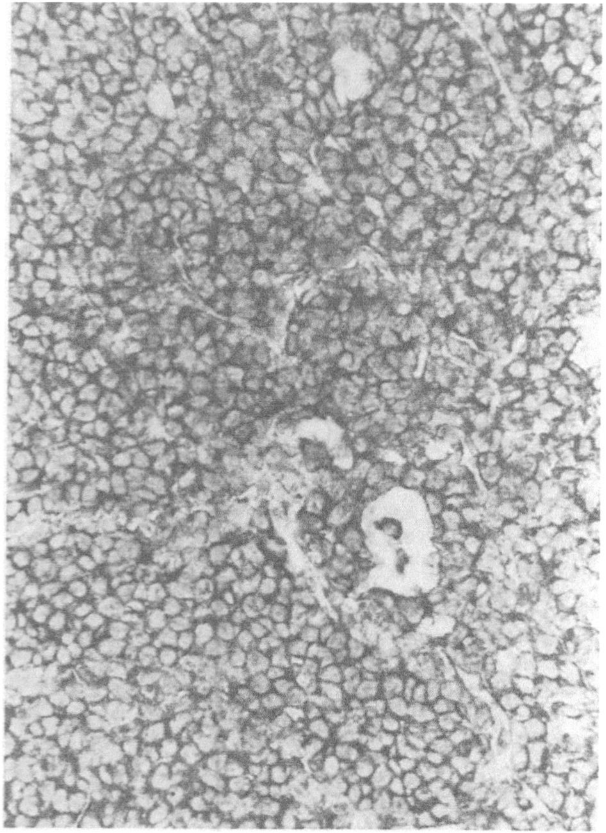

Fig. 24.1. B cell lymphoma (FCC centroblastic diffuse) stained with antibody L12. All lymphoid cells are stained. (×350)

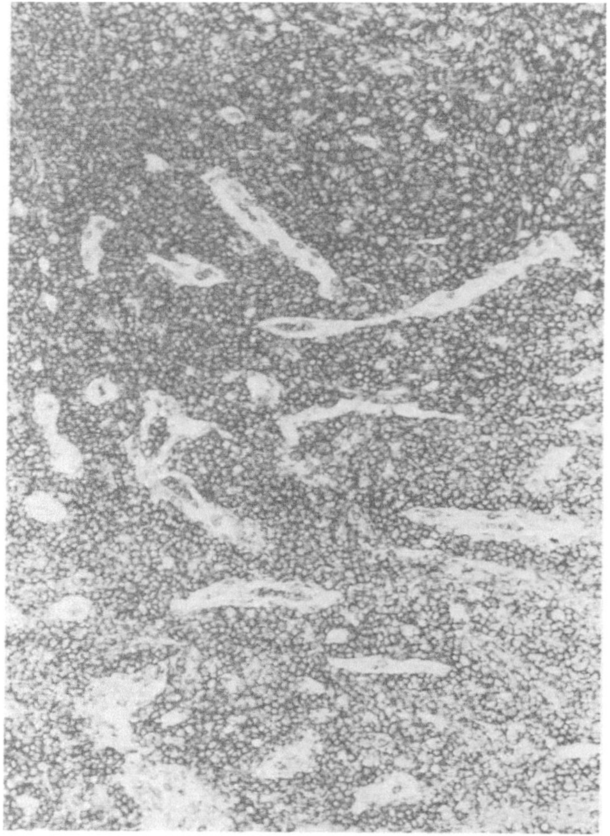

Fig. 24.2. T cell lymphoma stained with L12. Again all lymphoid cells show positive staining. Capillaries are negative. (×140)

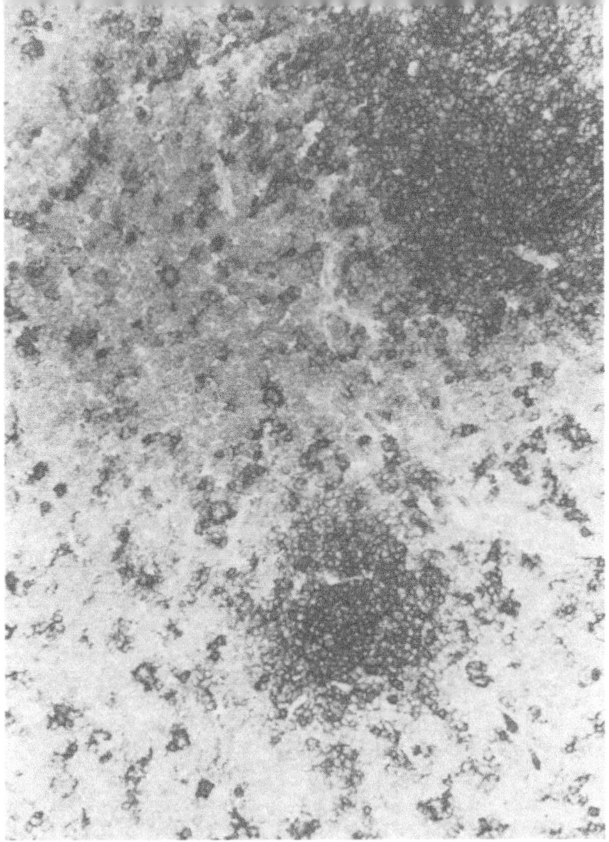

Fig. 24.3. Lymphoblastic lymphoma stained with L6. Residual lymphoid cells stain strongly. (×140)

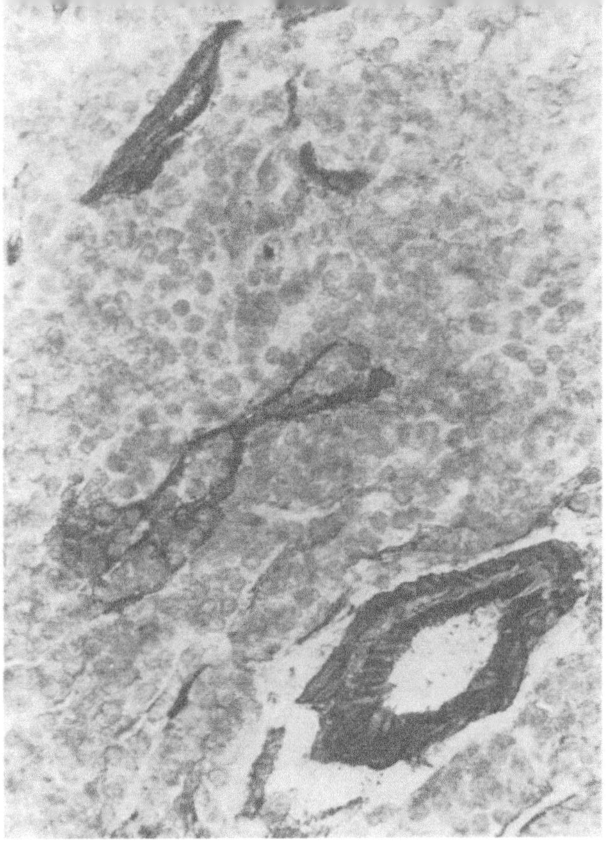

Fig. 24.4. Follicle center cell lymphoma. CB : CC follicular. Extensive staining of vessel walls in a pattern which resembles intermediate filament. (L22; ×350)

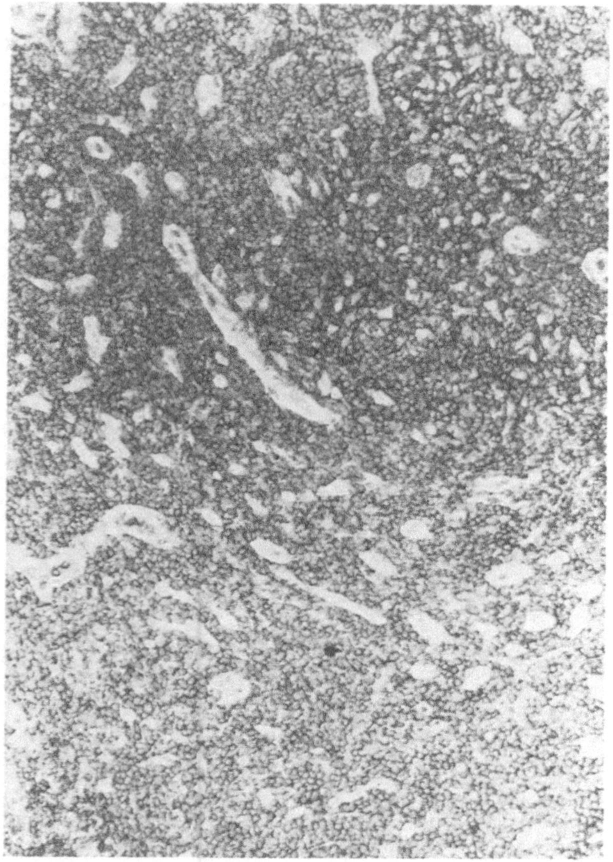

Fig. 24.5. Positive sheet of tumor cells with antibody L9 in a node-

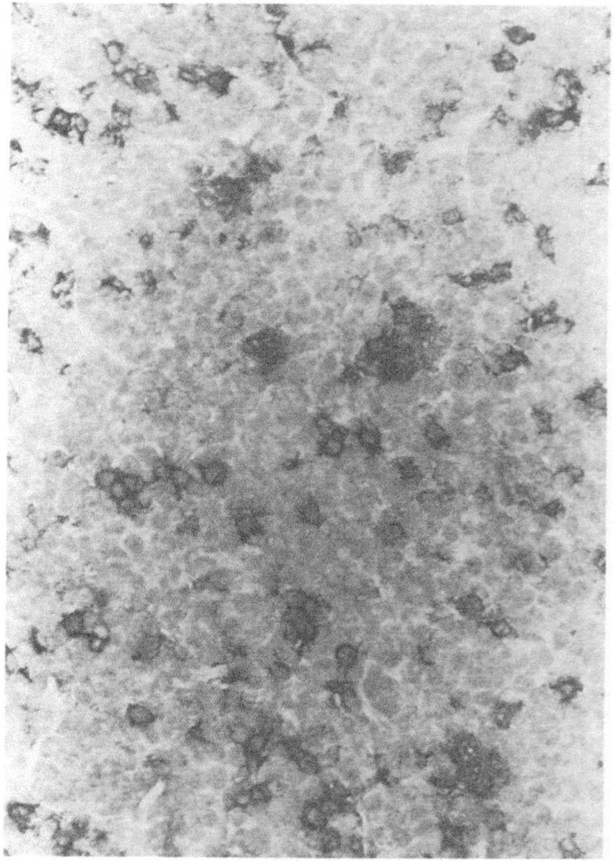

Fig. 24.6. Residual cells in B cell lymphoma stained with L9.

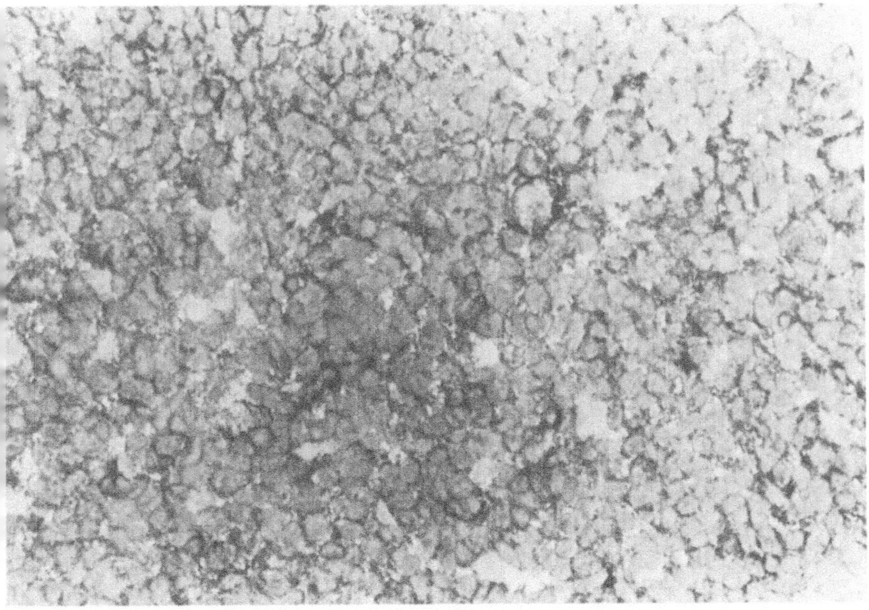

Fig. 24.7. Positive lymphoblastic lymphoma cells with antibody L9. (×140)

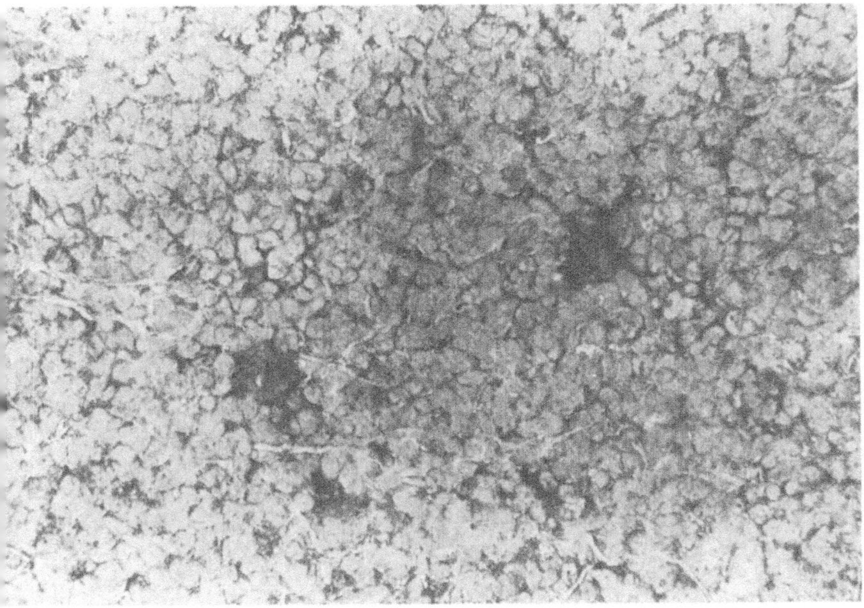

Fig. 24.8. HLA-like pattern of staining. Positive macrophages and B lymphoma cells. (W.T. ×350)

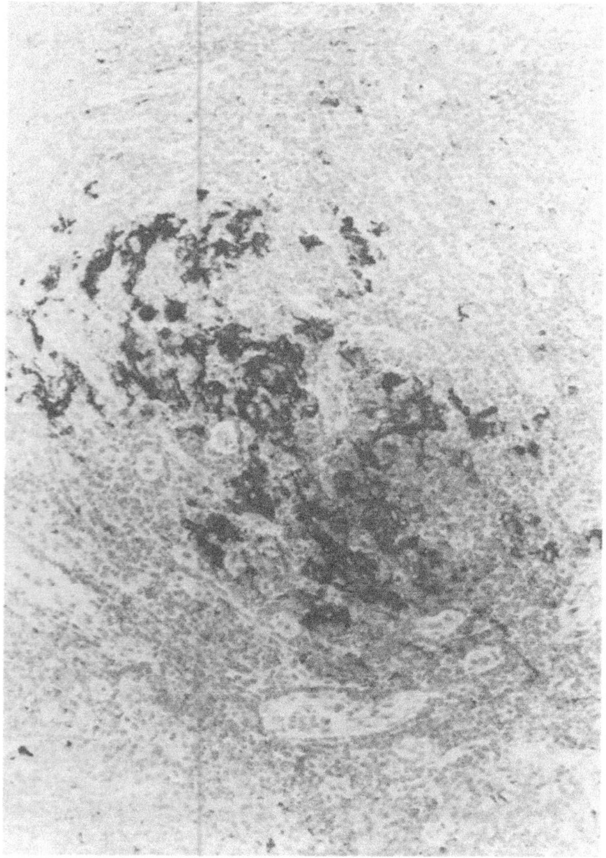

Fig. 24.9. Positive staining of epithelioid macrophages in T cell lymphoma. (L3; ×140)

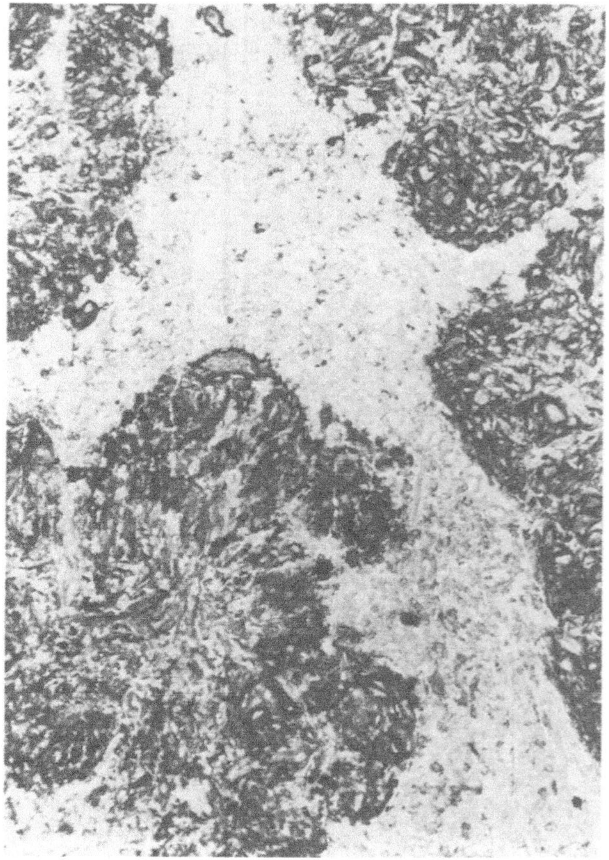

Fig. 24.10. Sarcoid granulomas; strong positive staining. (L19; ×140)

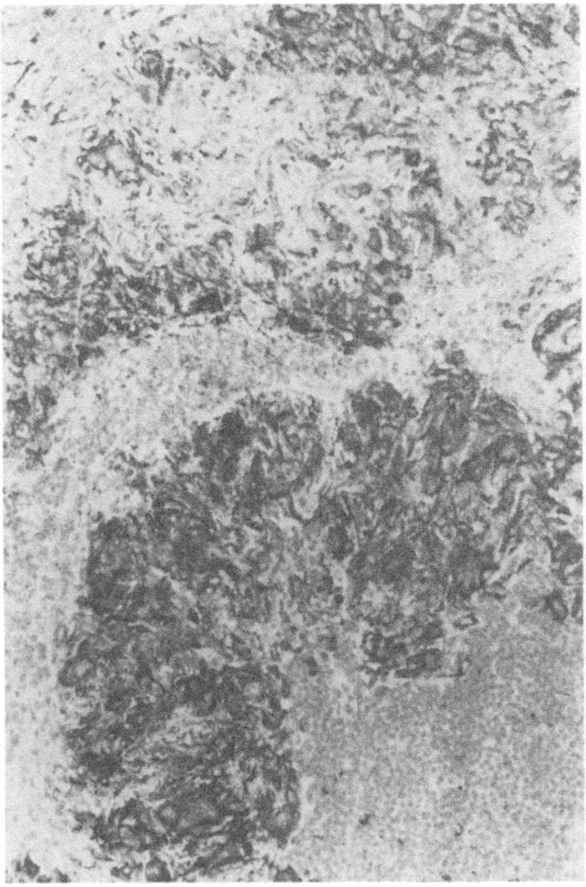

Fig. 24.11. Sarcoid granulomas; strong positive staining. (L3; ×140)

The staining of macrophage populations with L2, L3, L19, and L20 is of particular interest (Figs. 24.9–24.11). The mature cells of sarcoid granulomas stained strongly with L3 and L19. L14 was generally negative though a CALLA-positive T lymphoblastic lymphoma was reactive with this antibody.

Discussion and Conclusion

The antibodies presented in this section of the workshop gave a range of reactivities. Two (L1, L15) were negative, and two (L6, L12) showed a leukocyte common specificity, with L12 giving stronger staining at a titer of 1:250. Capillary endothelium was negative with these antibodies as with other leukocyte common reagents (4). The pattern of staining shown with six antibodies (L4, L7, L16, L18, L21, and L22) was consistent with activity directed towards intermediate filaments (IF) and parallels the pattern seen in this laboratory with monoclonal reagents directed towards cytokeratins (5) or other IF (6).

The intensity of staining varied greatly. One reagent, L9, stained phenotypically confirmed T cell lymphoma and cells in reactive lymph nodes and B cell lymphoma which also stained with our current T cell reagent (7). It would therefore be reasonable to allocate a T cell specificity to this antibody. However, macrophages also appeared to stain with this antibody.

A relative B cell specificity was apparent with L17, which predominantly identified proliferating follicle centers, mantle cells, and follicle center cell lymphomas. It was not possible to subdivide the lymphoma types stained with this reagent but as with other antibodies showing pan-B activity (8) the level of staining of tumor cells was less than that observed on reactive B cells.

The heavy staining of macrophages, particularly the mature cells in granulomas, with L2, L3, L13, and L19 is of particular interest. Many monoclonal antibodies identify monocytic cells but rarely stain histiocytes present in tissue. The level of staining showed by L3 and L19 is therefore of considerable interest.

In other cases the biopsy tissue included in the panel did not allow a predominant staining pattern to be identified. Two other antibodies are worthy of comment however. The reagent L10 identified the tumor cells in lymphoblastic lymphoma though no other particular pattern could be identified. Antibody L8 stained normal and neoplastic B cells, endothelium, residual macrophages, and in some blocks strongly stained follicle centers. This pattern resembles that seen with anti-class II reagents on biopsy tissue (9).

Summary

The Workshop antibodies L1 to L22 were employed at a dilution of 1/250 to stain a panel of frozen tissue biopsies. The panel comprised three T cell neoplasms, six reactive biopsies, and eight biopsies from B cell lymphomas of differing types. Peroxidase-conjugated rabbit anti-mouse antibody was used as the second layer. Two ascitic fluids, L1 and L15, did not stain and two, L6 and L12, stained as leukocyte common antibodies. The staining of all cells with L6 was not of equal intensity to that with L12, with tumor cells and lymphoblastic leukemia showing a weak or negative reaction. The antibody L9 stained all tissues with a T cell pattern, including residual lymphocytes in B cell lymphoma. L17 stained normal follicles, follicle center cell tumors, and lymphocytic lymphoma. Seven of the antibodies provided gave a staining pattern which resembled that obtained in lymphoid tissue with anti–intermediate filament antibodies. The pattern obtained with L8 resembled HLA-class II. L2, L3, and L13 stained macrophages predominantly in all tissues. The sarcoid granulomas were strongly stained with L3 and L13.

Acknowledgments. We acknowledge Miss Joanne Nesbitt for preparation of the manuscript.

References

1. Stein, H. The immunologic and immunochemical basis for the Kiel classification. In: *Malignant lymphomas other than Hodgkins disease,* K. Lennert, ed. S. Karger, p529 Berlin, 1978.
2. Stein, H., A. Bonk, G. Tolksdorf, K. Lennert, H. Rodt, and J. Geides. 1980. Immunological analysis of the organisation of normal lymphoid tissue and non-Hodgkins lymphomas. *J. Histochem. Cytochem.* **78:**746.
3. Graham, R.C., and M.J. Karnovsky. 1966. The early stages of absorption of injected horseradish peroxidase in the proximal tubules of mouse kidney: Ultrastructural cytochemistry by a new technique. *J. Histochem. Cytochem.* **14:**291.
4. Beverly, P. 1982. Application of monoclonal antibodies to typing and isolation of lymphoreticular cells. *Proc. Roy Soc. Edinburgh* **81B:**221.
5. Makin, C.A., I.G. Bobvrow, and W.F. Bodmer. 1984. Monoclonal antibody to cytokeratin for use in routine histopathology. *J. Clin. Pathol.* **37:**984.
6. Lane, E.B., and B. Anderton. 1982. Focus on filaments: embryology to pathology. *Nature* **298:**706.
7. Beverly, P.C.L., and R.E. Callard. 1981. Distructive functional characteristics of human T-lymphocytes defined by E-rosetting and a monoclonal anti T-cell antibody. *Eur. J. Immunol.* **11:**329.
8. Gobbi, M., F. Caligoris Capio, and G. Janossy. 1983. Normal equivalent cells of B-cell malignancies. Analysis with monoclonal antibodies. *Brit. J. Haematol.* **54:**393.
9. Krajewski, A.S., K. Guy, A.I. Dewar, and D. Cossar. 1984. Immunohistochemical analysis of human MHC class II antigens in B-cell non-Hodkins lymphoma. *J Pathol.* **145:**185.

CHAPTER 25

Immunohistochemical Reactivity of Anti–B Cell Monoclonal Antibodies in Thymus, Lymph Node, and Normal Skin

Emilio Berti, Carlo Parravicini, Giorgio Cattoretti, Domenico Delia, Filippo de Braud, and Marco Cusini

Post-medullary differentiation of B cell lineage largely depends on activation of secondary follicles in different lymphoid organs like spleen, tonsils, and lymph nodes. In all these organs the lymphatic follicles show the same basic structure and cytology. The phenotype of follicular lymphocytes is mainly related to different stages of B cell activation, and at present no organ-specific patterns of immunoreactivity have been reported (1). In the present study the immunohistochemical analysis of follicular reactivity of Workshop mAbs was restricted to the lymph node, whereas thymus and normal skin have been included to screen cross-reactivities against T lymphocytes and extra-medullary structures.

Material and Methods

Multiple blocks of normal skin, lymph nodes, and thymus removed from different patients have been employed in this study. Specimens were snap-frozen immediately after surgical removal. 4-μ thick cryostat sections from each specimen were mounted on the same slide and simultaneously processed. After a brief fixation with 3% paraformaldehyde in 0.1 *M* cacodylate buffer, pH 7.2, and washing in phosphate-buffered saline (PBS), pH 7.2, the sections were sequentially incubated for 30 min with 20% normal horse serum in PBS, primary mAbs, biotinylated horse anti–mouse H+L immunoglobulins (Vector Lab), and preformed avidin–biotin–peroxidase complex (ABCPx) (Vector Lab). Reaction was revealed in 3-amino-9-ethylcarbazole. Sections were counterstained with hematoxylin.

Table 25.1. Pan follicular mAbs.

mAbs	Reactivity[a]
B4, 5, 7, 17, 22, 23, 24, 25, 27[b], 30, 31, 47, 48, 36, 40, 49	MZ^+, GC^+: strong cytoplasmic pos.
B14, 21, 28, 34, 43, L17	MZ^+, GC^+: weak, extracellular branching pos.

[a] MZ: Mantle zone; GC: germinal center.
[b] mAbs 27 also stains granulocytes.

Results

Pan Follicular Reactivity (Table 25.1)

Anti–B cell Workshop mAbs of this group display two different types of follicular reactivity. The first type gives an intense cytoplasmic staining of the germinal center (Fig. 25.1), while the second one shows a weak labeling of the germinal center associated with an extracellular branching posi-

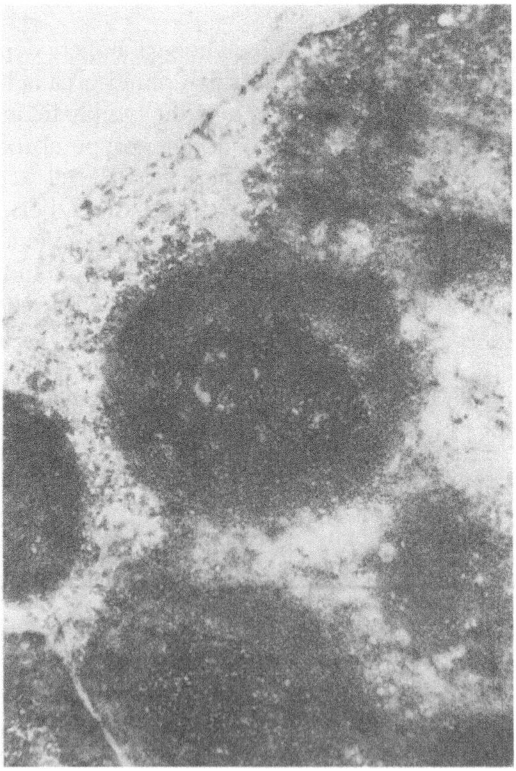

Fig. 25.1. Human lymph node. Pan follicular reactivity displayed by mAb B17. Note the intense staining of germinal center. Hematoxylin counterstain (orig. magn. 100×).

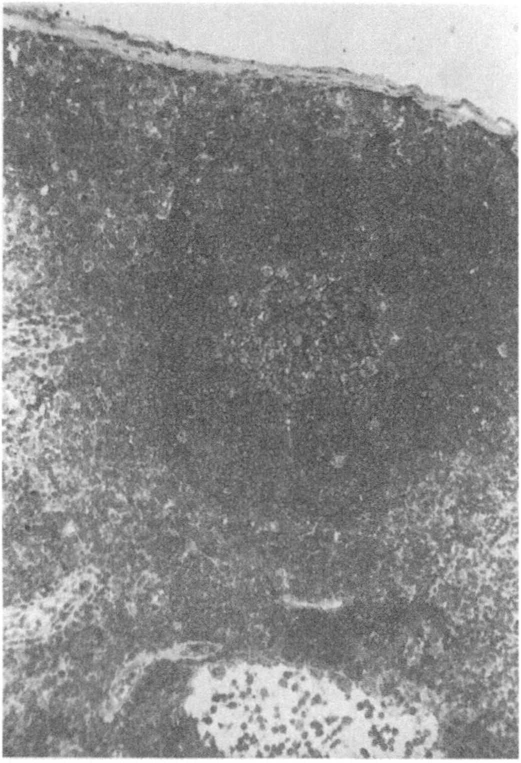

Fig. 25.2. Human lymph node. Pan follicular reactivity displayed by mAb B14. Note extracellular branching positivity into the germinal center. Hematoxylin counterstain (orig. magn. 160×).

tivity (Fig. 25.2). None of the mAbs in this group stain appreciably thymic cells or skin.

GC Reactivity (Table 25.2)

mAbs B11, B19, and B39 give an intense staining of both the germinal center and dendritic reticulum cells of follicles (DRC) (Fig. 25.3) mAbs

Table 25.2. mAbs staining the germinal center only.

mAbs	Reactivity[a]
B11, 19, 39	GC$^+$ strong, cytoplasmic
L4, L16, L18, L22	GC$^+$ weak, extracellular branching pos.[b]
L10, L11, L14, L15, L21	GC$^+$, weak
L2, L3, L13, L19, L20	GC$^+$, weak[c]

[a] GC: Germinal center; Nv: nerves; EG: eccrine glands; MC: monocytoid cells; Ep: epidermis; Mu: muscles; St: stroma.
[b] Also react with Ep, Nv, Mu, EG, St.
[c] Also react with MC.

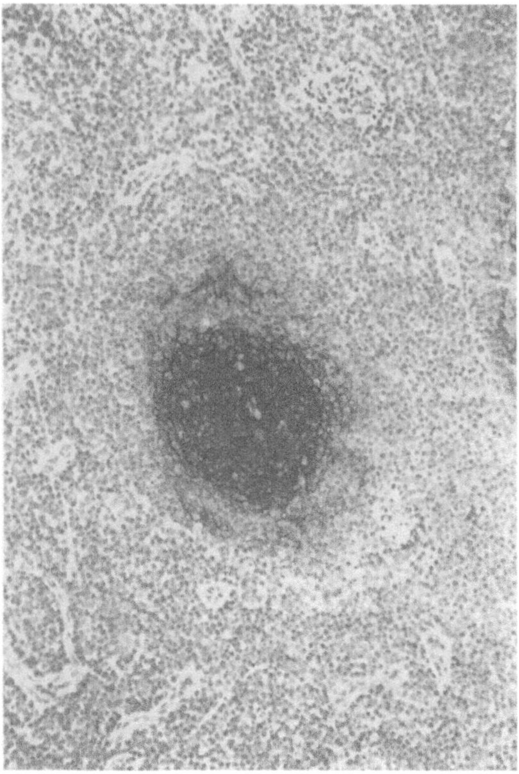

Fig. 25.3. Human lymph node. Intense reactivity of germinal center with mAb B11. Hematoxylin counterstain (orig. magn. 160×).

L2, L3, L13, L19, and L20 react also with monocytoid cells in all the tissues tested. mAbs L4, L16, L18, and L22 stain the germinal center and dendritic reticulum cells of follicles but display also reactivity with reticulum and collagen fibers, vessels, epidermis, etc.

DRC and Sinus Histiocyte Reactivity

mAbs B9, B33, B35, and B41 show the same pattern of follicular immunoreactivity but some of these mAbs also stain vessels, eccrine sweat glands, and elastic fibers (Table 25.3).

Heterogeneous Reactivity

We have found several mAbs with different and miscellaneous reactivities. Some mAbs of this group show an "HLA-DR"-like pattern (B1 and L8); some a "pan-leukocyte" reactivity (L6, L9, L12); some give DRC

Table 25.3. mAbs staining DRC, SH.[a]

	mAbs			
Reactivity	B9	B33	B35	B41
Vs⁺	+	–	–	–
EG⁺	+	+	–	–
EF⁺	+	–	–	–

[a] DRC: Dendritic reticulum cells of follicles; SH: sinus histiocytes; Vs: vessels; EG: eccrine glands; EF: elastic fibers.

and nerve staining (B2, B3, B42); and finally some display not groupable or weak positivity (Table 25.4). The data for mAbs with extrafollicular reactivity only are reported in Table 25.5. Some of the workshops mAbs-18, 26, 45, 46, L1, L7 were consistently negative in the tissue tested.

Discussion and Concluding Remarks

Many of the Workshop mAbs display a selective staining pattern restricted to the B cell lineage. Our results show that at least two classes of antigens—respectively cytoplasmic and membranous in type—can be expressed by B cells in the follicle. Antigens of each type can be shared by the resting lymphocytes of the mantle zone and by the proliferating cells of the germinal center, or can be expressed in the germinal center only.

Table 25.4. Heterogeneous reactivity.

	mAbs									
Reactivity[a]	B6	B10	B12	B15	B20	B44	B48	B50	B51	B52
GC⁺	+	+	+	–	+	+	–	+	+	–
MZ⁺	+	+	+	–	–	–	–	+	+	+
IDL⁺	–	–	+	–	–	–	–	–	–	–
SH⁺	–	–	+	+	+	+	+	–	–	+
DDC⁺	–	–	–	–	–	–	–	+	+	+
HC⁺	–	–	–	–	–	–	–	+	–	+
Ep⁺	–	–	–	–	–	–	–	+	–	–
EG⁺	–	–	–	+	+	+	–	+	–	–
Vs⁺	–	–	–	+	+	+	+	+	+	–
Nv⁺	–	–	–	–	–	–	–	+	–	–
Mu⁺	–	–	–	+	+	–	–	+	–	–
St⁺	–	–	–	–	–	–	–	+	–	+
EF⁺	–	–	–	–	–	–	–	–	–	+

[a] GC: Germinal center; MZ: mantle zone; IDL: interdigitating cell of lymph node; SH: sinus histiocyte; DDC: dendritic dermic cell; HC: Hassal corpuscle; Ep: epidermis; EG: eccrine glands; Vs: vessels; Nv: nerves; Mu: muscle; St. stroma (reactivity with both collagen and reticulum fibers); EF: elastic fibers.

Table 25.5. mAbs without follicular reactivity.

mAbs	Extrafollicular reactivities[a]
B46, L1, L7	—
B29	SH^+, Vs^+
B32, B37	EG^+
B38	Ep^+, Nv^+, EG^+
B13	Vs^+, EG^+, Mu^+, Ep^+
B16	Vs^+

[a] SH: Sinus histiocytes; Vs: vessels; EG: eccrine glands; Nv: nerves; Mu: muscle; Ep: epidermis.

From the group of mAbs specific for the germinal center only, those characterized by a cytoplasmic staining show an irregular pattern of positivity, with a broad morphological spectrum observed among the follicles in the same lymph node. This finding, in our opinion, may be related to the different stages recently described in the development of lymphatic follicles (2).

mAbs with membranous reactivity do not display a morphological pattern significantly related to the functional partition of the germinal center (dark and light zone). Some of these mAbs, however, show an extracellular branching staining possibly due to antigens shared by B cells and DRC.

References

1. Hsu, S.M., and E.S. Jaffe. 1984. Phenotypic expression of B-lymphocytes. 1 Identification with monoclonal antibodies in normal lymphoid tissue. *Am. J. Pathol.* **114**:387.
2. Hsu, S.M., and E.S. Jaffe. 1984. Phenotypic expression of B-lymphocytes. 2 Immunoglobulin expression of germinal center cells. *Am. J. Pathol.* **114**:396.

CHAPTER 26

Report to Second International Workshop on Human Leukocyte Differentiation Markers: Boston, 1984 Tissue Localization of B Cell and Leukemic Reagents (Special Studies)

Norbert Kraft, Peter S. Giddy, Wayne W. Hancock, and Robert C. Atkins

Analysis of B Cell Protocol Reagents; Use of Tissue Section Staining to Subclassify Antibody Groups

Introduction

B cells are most commonly typed with monoclonal antibodies with the use of fluorescent second antibodies applied to cell suspensions. Until recently the most useful reagents for this purpose have been those which identified Dr and CALLA antigens; other antibodies such as B1 and B2 (1) have become prominent since the last workshop on B cell markers but their absolute specificity for B cells in the context of tissue section analysis remains doubtful (2). For the purpose of tissue studies and to investigate the biology of the B cell it is of interest to identify B cell-specific antibodies by the criterion of non-cross-reactivity with tissue elements. The analogy with the well-known T cell markers and the elucidation, in consequence of their discovery, of the T cell antigen receptor should serve as a powerful stimulus in this endeavor.

Methods

Tissues for examination were obtained from surgery or autopsy rooms in fresh condition in saline. They were diced into 2-mm blocks and fixed in paraformaldehyede, lysine, periodate solution as described previously (3). Cell smears were made from suspensions in 50% fetal calf serum, they were air-dried overnight prior to use or stored in dessicant at −80°C.

Smears were fixed in the above-mentioned preservative on the day of use. Fixed sections or smears were incubated with diluted (1/250) antibody for 30 min, washed in phosphate-buffered saline (0.2% gelatin) × 3 and then incubated with peroxidase-conjugated rabbit anti–mouse Ig (Dako) for 30 min. Following 3 washes in the same buffer, H_2O_2 and DAB were added to complete the reaction. The sections were lightly counterstained with hematoxylin to identify nuclei but not cytoplasmic elements in tissues.

Results: Normal Spleen

The initial screening involved testing all reagents on two samples of normal spleen. In addition standard anti-leukocyte antibodies were used to verify the staining reaction (they comprised common leukocyte, Dr common, HLA common, pan T, helper T). In this tissue 32/52 of the B series gave strong reactions with "B" cell areas; five did not stain; three labeled B cells, macrophages, and vascular endothelium but not T cells; three reacted with all tissue elements; eight reacted with macrophages and/or smooth muscle and two labeled interfollicular cells, comprising polymorphs and unidentified cells. See Table 26.1 for a summary and Table 26.3 for a detailed listing.

Leukemic Reagent Tissue Staining

The 22 antibodies in this series were used on the same panel of normal kidney, liver, and lymphoid tissues as were studied for the B cell series. The staining patterns observed are listed in Table 26.2. In summary, every antibody in the series reacted with at least one normal tissue component. 19/22 reacted with kidney, 8/22 reacted with liver, and 12/22 reacted with macrophages in lymphoid tissues.

Table 26.1. Patterns of label on normal spleen with B series reagents.

Pattern	Antibody
No stain	8, 10, 32, 37, 46
All of section	1, 20, 50
Macrophages and/or SM	15, 16, 23, 26, 29, 38, 46, 48
Interfollicular cells	27, 47
B cells & macrophages & endothelium	3, 7, 42
B cell areas	2, 4, 5, 6, 9, 11, 12, 13, 14, 17, 18, 19, 21, 22, 24, 25, 28, 30, 31, 33, 34, 35, 36, 39, 40, 41, 43, 44, 45, 49, 51, 52.

Table 26.2. Normal tissue distribution patterns of antigens defined by leukemic marker antibodies.[a]

	Spleen					Tonsil & RLN				Kidney			Liver	
Antibody	B	MAC	T	P	OTH	GC	M	INT	OTH	GL	TUB	OTH	K	P
1	–	–	–	–	+	–	–	–	+	+	–	+	–	–
2	–	–	–	–	–	–	–	–	–	+	+	–	+	–
3	–	–	–	–	+	+	–	–	–	–	+	–	–	–
4	–	+	–	–	+	+	–	–	+	+	+	+	–	–
6	+	+	+	+	+	+	+	+	+	+	+	–	–	–
7	–	+	–	–	–	–	–	–	–	–	–	–	–	–
8	–	+	–	–	+	+	+	+	+	+	+	+	+	–
9	–	+	–	–	+	+	+	+	+	+	+	+	+	–
10	–	–	–	–	–	–	–	–	–	+	+	+	+	–
11	–	–	–	–	–	–	–	–	–	+	+	+	+	–
12	+	+	+	+	+	+	+	+	–	–	–	+	+	–
13	–	+	–	–	–	+	–	–	–	–	–	+	–	–
14	–	–	–	–	–	–	–	–	–	–	–	+	–	+
15	–	–	–	–	–	–	–	–	–	+	+	+	–	–
16	–	+	–	–	+	+	–	–	+	–	–	–	–	–
17	+	–	–	–	–	+	–	–	–	–	–	–	–	–
18	–	+	–	–	+	+	–	–	+	+	+	–	–	–
19	–	+	–	–	–	+	–	–	+	–	+	–	–	–
20	–	+	–	–	+	+	–	–	+	–	+	–	–	–
21	–	–	–	–	–	–	–	–	–	–	+	+	+	–
22	–	+	–	–	+	+	–	–	+	+	+	–	–	–

[a] Abbreviations: B = B cell area, MAC = macrophage, T = T cell area, P = polymorph., OTH = other tissue elements, GC = germinal center, M = mantle, INT = interfollicular, TUB = tubular, GL = glomerular, K = Kupffer, P = parenchyma, RLN = reactive lymph node.

Labeling of Normal Kidney, Liver, and Lymphoid Tissues with B Cell Series Reagents

All the antibodies defined as B cell reactive from the spleen sections were found on follicles in normal tonsil and reactive lymph nodes. Germinal center and mantle zone label was observed for all these reagents except 5 (no mantle label), 42 (no GC label), and 51 (no follicular label). Mantle zone label was intermittent however and no attempt was made to incorporate it into a subclassification scheme. This group of antibodies could be subdivided on the basis of patterns of labeling observed in these tissues:

1. Antibodies which had a high level of "background reactivity": 3, 7, 42
2. Antibodies which label Kupffer cells: 2, 12, 52.
3. Antibodies which label components of the kidney: 13, 18, 44, 51.
4. Antibodies which label B cell areas and which may label macrophages in tissues (i.e., the remaining 25 reagents of the B cell-reactive group). The degree of macrophage labeling varied from +/– to + and it was not practical to make subdivisions on this basis. Several

Table 26.3. Normal tissue distribution patterns of B cell reagents.[a]

	Spleen					Tonsil & RLN				Kidney			Liver	
Antibody	B	MAC	T	P	OTH	GC	M	INT	OTH	GL	TUB	OTH	K	PA
1	+	+	−	−	+	+	+	+	+	+	+	+	+	−
2	+	−	−	−	+	+	+	+	+	−	−	−	+	−
3	+	+	−	+	+	+	+	−	+	+	+	+	+	+
4	+	+	−	+	−	+	+	+	−	+	−	−	−	−
5	+	−	−	−	+	+	−	+	−	−	−	−	−	−
6	+	+	−	−	−	+	+	+	+	−	−	−	−	−
7	+	+	−	−	+	+	+	+	+	+	+	+	+	+
8	−	−	−	−	−	−	−	−	−	−	−	−	−	−
9	+	+	−	−	−	+	+	+	+	−	−	−	−	−
10	−	−	−	−	−	−	−	−	−	−	−	−	−	−
11	+	+	−	−	−	+	+	+	+	−	−	−	−	−
12	+	+	−	−	−	+	+	−	−	−	−	+	+	−
13	+	+	−	−	−	+	+	+	+	+	+	−	−	−
14	+	−	−	−	−	+	+	+	+	−	−	−	−	−
15	−	+	−	−	−	−	−	−	−	−	+	−	−	−
16	−	+	−	−	+	−	−	−	+	−	−	−	−	−
17	+	+	−	−	+	+	+	+	+	−	−	−	−	−
18	+	+	−	−	−	+	+	−	+	−	+	+	−	−
19	+	−	−	−	−	+	−	−	−	−	−	−	−	−
20	+	+	+	+	+	+	+	+	+	+	+	+	+	+
21	+	+	−	−	−	+	+	+	+	−	−	−	−	−
22	+	−	−	−	+	+	+	+	+	−	−	−	−	−
23	−	+	−	−	+	+	+	+	+	−	+	+	−	+
24	+	+	−	−	−	+	+	+	+	−	−	−	−	−
25	+	+	−	−	−	+	+	+	+	−	−	−	−	−
26	−	−	−	−	+	+	+	+	+	−	+	−	−	−
27	+	+	−	+	+	+	+	+	+	−	+	−	−	−
28	+	−	−	−	−	+	+	+	+	−	−	−	−	−
29	+	+	−	−	−	+	+	−	+	+	+	+	+	−
30	+	−	−	−	+	+	+	−	+	−	−	−	−	−
31	+	−	−	−	+	+	+	−	−	−	−	−	−	−
32	−	−	−	−	−	−	−	−	−	−	+	−	−	+
33	+	−	−	−	+	+	+	−	+	−	−	−	−	−
34	+	−	−	−	+	+	+	+	+	−	−	−	−	−
35	+	−	−	−	+	+	+	−	+	−	−	−	−	−
36	+	−	−	−	+	+	+	+	+	−	−	−	−	−
37	−	−	−	−	−	−	−	−	−	−	−	−	−	−
38	−	−	−	−	+	−	−	−	−	−	+	−	−	−
39	+	−	−	−	+	+	+	−	+	−	−	−	−	−
40	+	−	−	−	+	+	+	+	−	−	−	−	−	−
41	+	+	−	−	−	+	+	+	+	−	−	−	−	−
42	+	+	−	−	+	−	+	+	−	+	+	+	+	+
43	+	+	−	+	+	+	+	−	+	−	−	−	−	−
44	+	+	−	−	−	+	+	+	+	−	+	−	−	−
45	+	+	−	+	+	+	+	+	+	+	+	+	−	+
46	−	−	−	−	−	−	−	−	−	−	−	−	−	−
47	−	+	−	−	−	−	−	−	+	+	+	+	+	−
48	−	+	−	−	−	+	+	+	−	+	+	+	+	−
49	+	−	−	−	−	+	+	+	+	−	−	−	−	−
50	+	+	−	+	+	+	+	+	+	+	+	+	−	+
51	+	−	−	−	−	−	−	+	−	−	+	+	−	−
52	+	+	−	−	+	−	+	−	−	−	−	+	+	−

[a] Abbreviations: B = B cells, MAC = macrophages, T = T cells, P = polymorphs, OTH = other tissue components GC = germinal center, M = mantle, INT = interfolilcular leukocyte, GL = glomerular, TUB = tubules, K = Kupffe cells, PA = parenchyma.

antibodies in this group (9, 33, 35, 41) labeled follicular reticulum cells and frequently displayed a characteristic "lacework" pattern of follicle label in normal and lymphoma tissues.

Labeling of B Cell Lines

The examination of normal tissues had thus led to a subdivision of B cell-reactive antibodies with a majority group of 25. Although several of these had a distinctive pattern of reaction with follicles, this was considered too subjective for regular use and an attempt was made to discriminate between the antibodies on the basis of B cell line staining. The two lines used were Raji (Ig^-), a well-known Burkitt's lymphoma-derived cell and Cess-B (Ig^+), a spontaneous B cell line derived from a patient with myelomonocytic leukemia (4). These staining results are given in Table 26.4 and indicate that this may indeed be a practical way to determine subgroups of B cell reagents. Two other features are incorporated in this table: 1) the antibodies are listed which labeled a large-cell lymphoma which was considered likely to be a pre-B cell because of its lack of reaction with PHM-1 and its positive reaction with a CALLA antibody (PHM-6); 2) the reactivity of the antibodies which labeled this lymphoma with Nalm-6 cells which is a pre-B cell line. All but one (22) of the B cell-reactive antibodies which labeled the lymphoma also stained the Nalm-6 cells. It is of interest that the group of reagents which gave a characteristic pattern of follicle

Table 26.4. Reaction of B workshop reagents with three B cell lines and a pre-B lymphoma lymph node.

Antibody	Nalm-6	Large-cell lymphoma	Raji	Cess-B
B cell reagents				
2, 14, 21, 42	+	+	−	−
3	+	+	+	−
7, 12, 28, 43	+	+	+	+
22	−	+	+	+
4, 5, 6, 11, 17, 19, 24, 25, 28, 30, 31, 34, 36, 39, 43, 45, 49		−	+	+
9, 33, 35, 40, 41		−	+	−
52 (Kupffer +), 44 (tubule +)		−	+	+
13, 18 (tubule +)		−	−	−
51 (tubule +)		−	+	−
Non-B reagents				
1, 20, 23, 50	+	+	+	+
48	+	+	−	−
47	−	+	−	−

labeling (indicating a possible follicular reticulum series of markers) also emerged as a subgroup in this table (Raji$^+$, Cess-B$^-$, lymphoma$^-$, kidney$^-$).

Labeling Patterns with Lymphoma Samples

Ten lymph nodes were screened with the B series reagents. Only the B cell data for four of these are indicated in Table 26.5. The most interesting features of this series have been indicated in Table 26.4, but several other

Table 26.5. Reaction of B cell reagents with four types of lymphoma lymph nodes.[a]

	Nodular poorly differentiated					Large cell diffuse					Small cell nodular					Mixed cell nodular				
Antibody	A	F	M	MA	OTH	A	F	M	MA	OTH	A	F	M	MA	OTH	A	F	M	MA	OTH
1	+					+					+					+				
2		+		+		+					+					+				
3	+					+						+				+				
4	+					+						+					+			
5	+										+					+				
6		+		+		+					+					+				
7	+					+					+					+				
8					+															
9		+					+					+					+			
10	+						+								+					+
11		+					+					+								+
12				+		+						+					+			
13				+																
14				+		+						+					+			
15															+					
16																				
17	+						+					+				+				
18																				
19		+					+						+							+
20					+	+					+					+				
21		+				+									+		+			
22		+		+		+						+				+				
23				+		+								+	+	+				
24	+						+				+					+				
25		+					+					+				+				
26																				
27												+								
28		+				+											+			
29	+								+					+	+	+				
30		+		+			+				+						+			
31		+								+	+					+				
32																				
33		+					+					+					+			
34		+				+					+						+			
35		+					+					+					+			
36	+						+				+					+				
37																				
38																				
39		+					+						+							+
40							+					+				+				
41		+					+					+					+			
42	+					+									+	+				
43		+		+		+						+					+			
44					+					+					+				+	+
45					+	+									+				+	+
46																				
47	+					+			+	+	+									+
48						+					+									+
49		+								+	+					+				
50					+	+									+	+				
51		+								+	+					+				
52		+				+									+					

[a] Abbreviations: A = Most or all cells, F = follicle, M = mantle, MA = macrophage, OTH = other tissue element or scattered or patches of lymphocytes.

points remain. 1) Most of the B-reactive antibodies labeled either the majority of cells in the sections or labeled the follicles, confirming the markers as defining relatively mature stages of B cell differentiation. 2) There were significant discrepancies between samples with the same classification; for example, 18/52 reagents in two cases of small-cell nodular poorly differentiated lymphoma gave different reactions. 3) In the absence of significant morphological features, little information may be gained from many samples other than that all, some, or none of the cells were stained. In the absence of information about the presence of normal cells (preferably obtained by double labeling) definitive answers are often unobtainable. The diversity of staining patterns obtained renders the data impossible to represent in any condensed form except on the sample basis shown in Table 26.5 and the pattern representation in Table 26.6.

Discussion

We have demonstrated site-specific labeling by 25 B cell reagents in lymphoid tissues, thus confirming their potential for studying solid tissue-derived pathological material. The specificity of this series is much

Table 26.6. Pattern comparison of B cell reagents on four different lymphoma lymph nodes (NPD, large-cell diffuse, small-cell nodular, mixed nodular).

Antibody	Pattern[a]	Antibody	Pattern
4	AFFF	9	FFFF
5	A-AA	33	FFFF
6	FAAA	35	FFFF
11	FFMP	40	-FFA
14	-AFF	41	FFFF
17	AFAA	3	FAAA
19	FFMP	7	AAAA
21	FAIF	42	AASS
22	FAFA		
24	AFAA	2	FAIA
25	FFFA	12	OAFA
28	FA-F	52	FAI-
30	FFAF		
31	F-AA		
34	FAAF		
36	AFAA		
39	FFMP		
43	FAFF		
45	PASS		
49	F-FF		

[a] A = All or most leukocytes, F = follicle, – = no label, S = smooth muscle, P = patch of cells, M = mantle, I = extrafollicular, 0 = macrophage.

greater than the one from the previous workshop (2) which contained few reagents suitable for examining solid tissue samples.

There were some differences in the labeling patterns of the B-reactive antibodies on normal tissues which may aid in the subclassification of these reagents. Thus three of the antibodies labeled Kupffer cells and four reacted with kidney tubules. The majority group of B-reactive antibodies however gave similar patterns of labeling on normal tissues and, except for a subgroup which may react with follicular reticulum cells, may be divisible on the basis of reaction with B cell lines. The fortuitous recognition of a probable pre–B cell lymphoma permitted us to define nine antibodies of the B-reactive series as putative pre–B cell markers and eight of these also labeled the pre-B cell line Nalm-6. On the basis of the above criteria we can suggest that the various antibodies recognize determinants on cells over different sections of the B cell differentiation pathway, as indicated in Table 26.4. This approach to the use of cell line and tumor material may eventually indicate a set of highly detailed differentiation steps of the B cell series. It was however obvious from the series of lymphoma samples that the morphological criteria used for their classification did not provide the same degree of detailed differentiation as seems to be available from the use of an extensive panel of specific antibodies.

Two of the antibodies submitted to the Workshop (B3, B7) were irradiated because of customs regulations. They had considerable increases in background labeling reactions compared to the identical samples which had been retained for comparison. It is therefore likely that the category of B-reactive antibodies with high background activity is an artifactual one.

The leukemic markers all labeled one or more normal tissue component; many of them gave strong reactions with kidney tubules. This may not reflect on their usefulness in categorizing cell suspensions and may in fact prove of help in the subclassification of these antibodies and in the interpretation of the significance of the antigenic determinants involved.

References

1. Nadler, L.M., P. Stashenko, R. Hardy, A. van Agthoven, C. Terhorst, and S.F. Schlossman. 1981. Characterization of a human B cell specific antigen (B2) distinct from B1. *J. Immunol.* **126:**1941.
2. Hancock, W.W., N. Kraft, and R.C. Atkins. 1984. Importance of tissue localization in characterization of monoclonal antibodies to human leucocyte antigens. In: *Leucocyte typing,* A. Bernard, L. Boumsell, J. Dausset, C. Milstein, and S.F. Schlossman, eds. Springer-Verlag, Berlin, Heidelberg, p. 475.
3. Hancock, W.W., G.J. Becker, and R.C. Atkins. 1982. A comparison of fixatives and immunohistochemical techniques for use with monoclonal antibodies. Am. J. Clin. Pathol. **78:**825.
4. Bradley, T.R., G.R. Pilkington, M. Garson, G.S. Hodgson, and N. Kraft. 1982. Cell lines derived from a myelomonocytic leukaemia. *Brit. J. Haematol.* **51:**595.

Part V. Expression of B Cell/ Leukemia Panel on Leukemias and Lymphomas

CHAPTER 27

Quantitative Phenotypes of B Chronic Lymphocytic Leukemia B Cells Established with Monoclonal Antibodies from the B Cell Protocol

Philippe Poncelet, Thierry Lavabre-Bertrand, and Pierre Carayon

Introduction

As seen at the First International Workshop on Human Leucocyte Differentiation Antigens (1), comparison of the serological specificities of mAbs is a very cumbersome task which requires testing reagents on a very broad panel of cell types. Flow microfluorimetry (FMF) analysis was used as well as fluorescence microscopy to perform such studies and finally each sample was characterized, for the purpose of statistical comparisons, by a single figure (2). Indeed, FMF-derived histograms are very rich pieces of data, which are only partly summarized by indicating the percentage of reactive cells. In order to get as much significant information as possible from a limited panel of cells, we propose the use of an additional numerical parameter, the mean antigen density of the cell (sub)population(s), obtained by absolute quantitative measurements of the mean amount of mAb molecules bound to the selected cells under saturating conditions.

The method we derived for measuring mAb binding to cell surfaces by standard indirect immunofluorescence (3) offers the invaluable advantage of applicability with mAbs in crude form (ascites, concentrated culture fluids, etc.) such as those provided for this international exchange of reagents. This quantification technique was applied to the study of eight B-CLL cases with the entire B cell and leukemia panel of mAbs. Clusters (or rather tentative groupings) are proposed for these reagents in comparison with known antibodies. We also give a first illustration of the value of antigen density for the study of B cell differentiation as was suggested by Anderson *et al.* (4).

Materials and Methods

Preparation of Cells

Patients who had been previously diagnosed as B-CLL according to classical clinical and hematological features (5) were selected for this study. Surface immunoglobulin typing was made according to standard techniques (6). Blood samples were recovered in heparinized tubes (or normal ones for serum recovery) and the mononuclear cell fraction isolated by centrifugation through Ficoll–Hypaque. Before staining, cells were adjusted to 10^7 cells/ml in PBS buffer containing 0.1% NaN_3 plus 20% autologous serum as requested by the Workshop organizers. The derivation and characterization of CCRF-CEM subclones will be described elsewhere (Poncelet *et al.*, manuscript in preparation). These cell lines were maintained using standard culture procedures and taken in growing log phase for quantification purposes.

Handling of Reagents and Cells

Aliquots of mAbs provided in the B cell and leukemia kits were diluted in PBS containing 0.1% BSA and 0.1% NaN_3 and stored at 4°C throughout the study. The microtitration Micronic® system (Flow Labs., Paris) was used for mAb dilutions storage as well as cell staining. Thus, 96 1-ml tubes could be processed like a microtiter plate. Cells (50 μl, 10^7 cells/ml) and mAb dilutions (50 μl, 2xC) were distributed in 96 Titertek® tubes. Standard immunofluorescence protocol included a first-step incubation for 2 hr, 3 washings (dilution factor, 10^4), a second-step incubation for 45 min, one final wash in PBS without protein, and cell fixation with 1% paraformaldehyde according to Lanier and Warner (7) (0.5 ml on cell pellet). Stored in the cold, cells had been shown to maintain the same fluorescence intensity for more than 2 weeks.

Quantitative Immunofluorescence by Flow Microfluorometry

This quantification method is based on the construction of an internal standard-curve using cell lines which fix a well-known number of T101 mAb molecules under saturating conditions. All the details of its design will be described elsewhere (3). The essential features are the following:

1. Indirect fluorescence intensity is linearly related to the number of mAb molecules bound to the cells, from background level (~1000 molecules) to 500,000 molecules.
2. Cells with various well-known T1 antigen densities can be used as internal standards to build a calibration curve relating the fluorescence intensity to the number of T101 molecules bound at saturation.

3. Using an appropriate anti-mouse Ig reagent, the linear relationship is the same for any other IgG subtype. Thus such a standard-curve can be used with any IgG mAb to determine the number of cell-bound antibody molecules on the basis of the mean fluorescence intensity.

Thus, the only modifications in the indirect immunofluorescence protocol described above required for quantification were the following:

1. For each sample, the last 12 tubes were reserved for biological standards: 5×10^5 cells of each CCRF-CEM subclone described in Fig. 27.1 received 50 μl of either T101 (20 μg/ml) or a negative control (anti-DNP mAb, 20 μg/ml). Two additional places were left for doublets.
2. The same lot of fluorescein isothiocyanate-conjugated $F(ab')_2$ fragments of affinity-purified sheep anti–mouse Ig antibodies (NEI 504, lot no. FPE 215, New England Nuclear, Boston) was used throughout this study. It has been tested for equivalent quantitative staining of cells coated with mAbs of the four different IgG subtypes directed against the same antigen, and also for saturating concentration (dilution = 1/40).
3. During FMF analysis, operated on a FACS IV cell sorter (Becton-Dickinson, Sunnyvale, CA), fluorescent plastic beads of constant fluorescence (ref.: 7766, Polysciences, Warrington, PA) were analyzed at regular intervals to take into account potential modifications in apparatus settings. Results were always expressed in arbitrary units (a.u.) related to their mean fluorescence intensity (peak of microspheres = 1000 a.u.). The biological standards were analyzed once for each new series of analyses, each tube providing enough cells for multiple analysis. Generally, a preset amount of 10,000 cells, gated on the basis of both forward and perpendicular light scattering, was entered in each histogram. Data were stored on floppy disks for subsequent manipulations.

Statistical Analysis of Data

Statistical analysis of data was done as described by Bernard *et al.* (2). In the present case, the antigen density data entered in a DEC-VAX11/750 computer were log-transformed in order to take into account geometric distances (i.e., ratios) rather than arithmetic distances. Negative results were equated to 1000 bound molecules which is currently equivalent to background. Comparison of antibodies was made with the BMDP2M program ("Cluster analysis of cases," BMDP Statistical Software, Inc., Los Angeles, CA, converted for use on DEC-VAX11 computers by Management Science Associates, Inc., Pittsburgh, PA). BMDP1M ("Cluster analysis of variables") was used for comparison of patients. Procedure amalgamation rule was minimum distance.

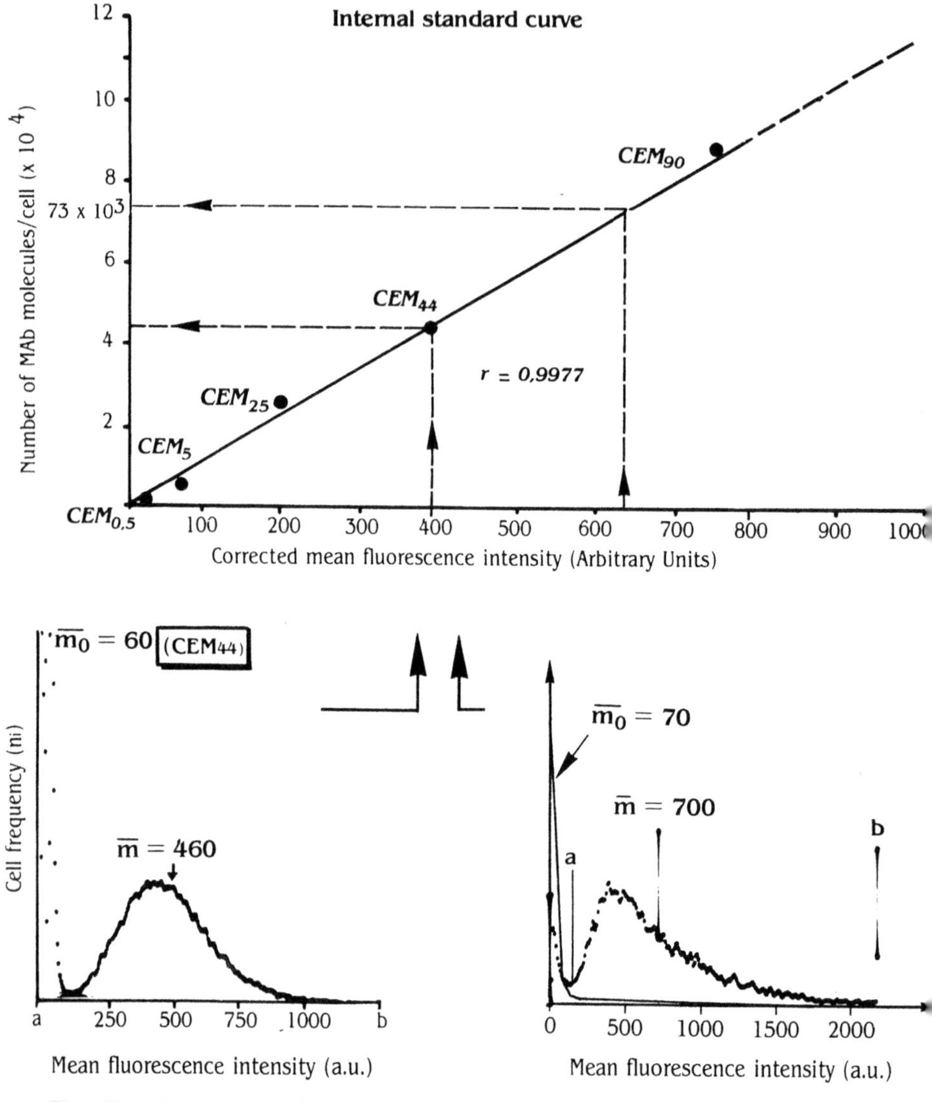

Fig. 27.1. Data processing for quantification by indirect immunofluorescence. Five stable CCRF-CEM subclones were used to construct the standard-curve. Incubated with saturating concentrations of T101 mAb (1 μg/10^6 cells), they repeatedly bound the following mean amounts of T101 molecules/cell: CEM_{90} (90,000), CEM_{44} (44,000), CEM_{25} (25,000), CEM_5 (5000), $CEM_{0.5}$ (500). Corrected mean fluorescence intensities were calculated from FMF histograms according to the following formulas:

$$\overline{m} = \sum_{i=a}^{i=b} i \times ni : \sum_{i=a}^{i=b} ni \quad \text{and} \quad \overline{m}_c = \overline{m} - \overline{m}_0$$

where i = channel number and ni its cell content; $\overline{m}_0$ is the mean fluorescence of the negative control.

Results

Quantitative Immunofluorescence Can Be Applied on a Routine Basis

The 96 tubes processed for each cell sample included 51 anti–B cell mAbs (B12 lacking), 21 mAbs against leukemia-associated antigens (L5 lacking), 12 control mAbs (listed in Table 27.1), and 12 tubes for establishment of the standard-curve. Thus, two cell samples could be processed the same day by one person, further FMF analysis and calculations requiring about two days.

As an example of data processing, the histogram generated on one case (DUCR) with one of the tested mAbs (B17) is presented in Fig. 27.1. The mean fluorescence is calculated on positive cells, i.e., from *a* to *b* limits ($a = 1$ if no clear distinction appears between subpopulations), corrected for background mean fluorescence, and compared to the standard-curve. The latter has been constructed using reference cells (CCRF-CEM subclones selected according to T1 antigen density and established as permanent cell lines) which followed the same treatments (staining, fixation, storage, analysis) at the same time. The equation of the linear regression curve is used to calculate, from the mean fluorescence intensity value, the mean number of mAb molecules bound per cell. Extrapolation can be used for high densities (more than 100,000 molecules/cell) since linearity has previously been checked up to 500,000 bound molecules. When the

Table 27.1. Control mAbs of known specificity.

mAbs	Specificity	References	Source
T101	T1 = CD5 = p67 antigen also present on majority of B-CLL	Royston *et al.* (8)	Hybritech (U.S.A.)
Ba2	Hematopoietic precursor antigen (p24), CD9	Kersey *et al.* (9)	Hybritech (U.S.A.)
BL14 (IOB1)	Normal human B lymphocytes, 100% of B-CLL	Brochier *et al.* (10)	Immunotech (France)
ALB6	p24, equivalent to Ba2, CD9	Boucheix *et al.* (11)	Generous gift, C. Boucheix (Paris)
ALB7, ALB9	Ba1 equivalents	Boucheix *et al.* (11)	Generous gift, C. Boucheix (Paris)
A1b1 (IOT5a)	Anti-CALLA, CD10	Boucheix *et al.* (11)	Immunotech (France)
D1B6 (or D1-12)	Monomorphic HLA-DR	Carrel *et al.* (12)	Generous gift, S. Carrel (Lausanne)
3A1 l.d.	CD7	Haynes *et al.* (13)	A.T.C.C. HB2
Negatives	B cell unrelated	Local control ascites of various subtypes	

origin (mean of the negative control) and the mean density have been positioned, the fluorescence intensity expressed in arbitrary units can be replaced by the antigen density expressed as number of mAb molecules bound per cell, as shown in Fig. 27.2(b).

Correlation of antigen density to the number of mAb molecules bound assumes monovalent binding and thus requires that cells be incubated with saturating doses of mAbs. This was tested on the whole B cell panel

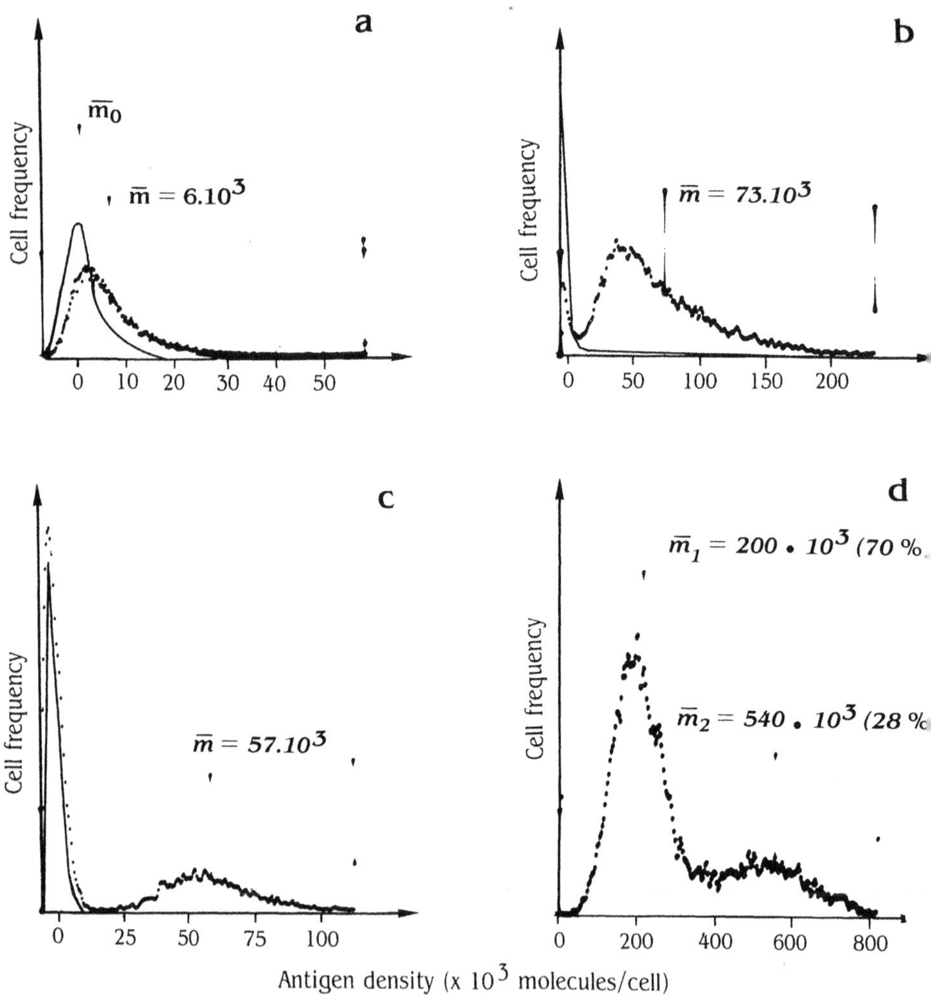

Fig. 27.2. FMF analysis of antigen density on various types of B-CLL-derived histograms. Indirect immunofluorescent stainings of: (a) FARG cells with mAb B39, (b) DUCR cells with mAb B17, (c) DELP cells with mAb B28, (d) BLAI cells with mAb L12. The light lines show the background fluorescence of negative samples.

by comparing fluorescence of cells coated either with the suggested dilution (1/250 final) or a lower one (1/50 final). In such experiments, carried out with two different cases, the following mAbs showed more than 10% decrease in fluorescence intensity from 1/50 to 1/250 dilutions: B1, B5, B15, B30, B47 and L4, L10, L19. This suggests caution in analysis of the related data. In addition, IgM do not fit with the present quantification method. Thus, data obtained with IgM should not be considered as reliable, in terms of antigen density.

Antigen Density Provides Additional Information over the Percentage of Positive Cells

Whenever the distribution of an antigen is heterogeneous and/or its density low, the overlapping of experimental and control histograms gives low real significance to the percentage of positive cells. In the example of Fig. 27.2(a), 33% of the cells could be considered as positive whereas analysis with other mAbs of the same subgroup as B39 (B28, B43; see subgroup 3, Table 27.4) and also other subgroups indicated that this sample was a homogeneous B-CLL cell population, which expressed a moderate density of this particular antigen.

In contrast with Fig. 27.2(a) and (b) which are examples of homogeneous populations, Fig. 27.2(c) and (d) correspond to more complex samples showing two different subpopulations, each one characterized by the indicated value of antigen density and its respective cell content. Figure 27.2(d) is the histogram raised on BLAI cells (28,000 WBC/mm^3, ~ 60% sIg$^+$) with L12. This mAb brightly stained all peripheral blood mononuclear cells in all tested cases. Comparison of the respective percentages of both L12 subpopulations seems interesting. Since in the present case HLA-DR cells were 70%, BL14 cells were 62%, E-rosettes and T11 cells were between 23% and 28%, and 3A1 cells (representing the major part of T lymphocytes) were 20%, it might well be that B leukemia cells are included in the population expressing a mean of 200,000 L12-related molecules whereas contaminating T lymphocytes would then be more brightly stained (~540,000 molecules/cell).

Antigen Density is a Powerful Parameter for Clustering Analysis of Numerous Unknown mAbs

The whole panel of mAbs directed against the B cell and leukemia-associated antigens was studied on blood samples from eight different patients suffering from B chronic lymphocytic leukemia (B-CLL). This panel was completed with the mAbs of known specificity listed in Table 27.1. Since some mAbs gave histograms containing two positive cell populations on certain samples, as shown above for L12, they were computed as two different mAbs. Thus, about 120 × 8 = 960 antigen density values were

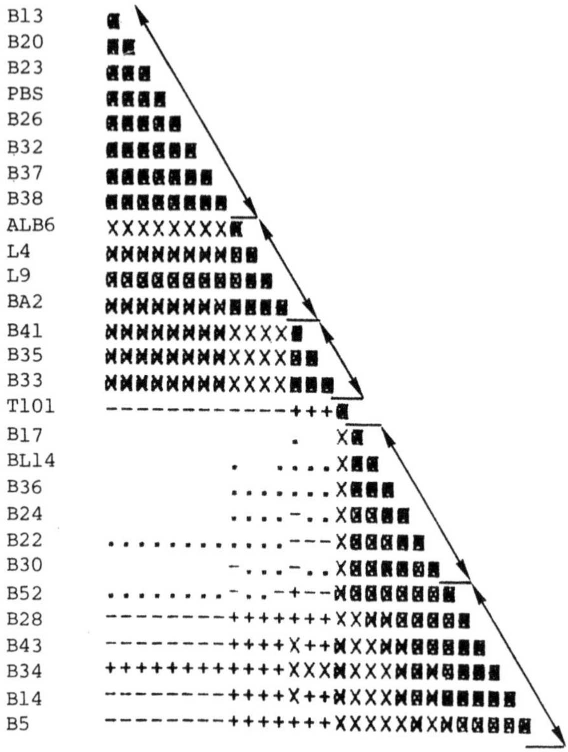

Fig. 27.3. Diagram of distances between antibodies. This correlation (or distance) matrix is constructed with one line and one column for each antibody. The distance between mAb_i and mAb_j is in row i and column j. The distances are represented according to the following scheme: ⊞, less than 0.794; ⊠, from 0.794 to 1.589; ⋈, from 1.589 to 2.009; ×, from 2.009 to 3.761; +, from 3.761 to 4.579; −, from 4.579 to 5.117; ·, from 5.117 to 5.958; blank, greater than 5.958.

processed by the computer. Figure 27.3, provided as an illustration of the presentation of results, is the diagram resulting from the clustering analysis of only 28 selected mAbs. Distances between cases (mAbs) are represented in shaded form: heavy shading indicates small distances (i.e., high correlation). In this example, six different groups can be individualized, comprising mAbs which were consistently negative on B-CLL and also T101 alone. In fact, the 84 mAbs clustered in about 22 subgroups of various sizes, from 1 to 9 mAbs, in addition to the following 100% negative mAbs: B3, 13, 20, 23, 26, 32, 37, 38 and A1b1, L1, 2, 3, 6, 7, 11, 13, 14, 20, 21, 22. Raw data are organized in Tables 27.2 to 27.7. The five largest subgroups are separated in Tables 27.2 to 27.6. Statistical calculations (mean, S.E.M.*) were made on the most tightly related mAbs which

* S.E.M. = Standard error of the mean.

Table 27.2. Subgroup 1 antigen densities.[a]

mAb	Subgroup	CARR	ESTE	FARG	VITI	BLAI	PELE	DELP	DUCR
B4 *	1	91	17	29	114	84	67	110	58
BL14 *	1	83	18	25	93	169	68	115	57
B17 *	1	79	23	31	110	168	70	160	73
B36 *	1	37	15	25	87	135	64	114	64
B22	1	18	16	15	120	60	29	85	40
B24	1	23	11	19	140	90	49	98	64
B30	1	55	9	17	55	107	47	72	52
B5	1	28	8	10	76	3	16	40	15
Mean[b]	1	72.5	18.3	27.5	101	139	67.5	125	63
S.E.M.[b]	1	12	1.7	1.5	6.5	20	1.25	12	6.4

[a] Antigen densities expressed in mean number of antibody molecules bound/cell ($\times 10^3$).
[b] Mean and S.E.M. calculated on marked (*) lines.

Table 27.3. Subgroup 2 antigen densities.

mAb	Subgroup	CARR	ESTE	FARG	VITI	BLAI	PELE	DELP	DUCR
B8M	2	1	1	3	24	8	3	13	4
B9M	2	1	1	2	13	8	1	12	1
B33 *	2	1	1	2	1	16	1	12	1
B35 *	2	1	1	2	1	13	1	18	1
B41 *	2	1	1	3	3	12	2	8	1
Mean[a]	2	1	1	2.3	1.7	13.7	1.3	12.7	1
S.E.M.[a]	2	5.2	0	0	1.2	2.1	0.6	5	0

[a] Mean and S.E.M. calculated on marked (*) lines.

Table 27.4. Subgroup 3 antigen densities.

mAb	Subgroup	CARR	ESTE	FARG	VITI	BLAI	PELE	DELP	DUCR
B14 *	3	16	9	12	16	22	11	40	18
B34 *	3	14	8	11	16	22	12	38	17
B43 *	3	16	12	12	18	20	16	48	19
B28 *	3	24	11	12	17	24	18	57	23
B52	3	27	19	23	38	70	28	71	20
B19	3	12	6	9	1	27	30	42	11
B39	3	9	19	6	1	26	27	45	9
L15M	3	16	3	11	24	27	18	80	1
L17M	3	10	1	10	10	15	9	11	18
Mean[a]	3	17.5	10	11.8	16.8	22	14.3	46	19.3
S.E.M.[a]	3	2.2	0.9	0.25	0.5	0.8	1.7	4.3	1.3

[a] Mean and S.E.M. calculated on marked (*) lines.

Table 27.5. Subgroup 4 antigen densities.

mAb	Subgroup	CARR	ESTE	FARG	VITI	BLAI	PELE	DELP	DUCR
B25	4	1	1	1	1	1	1	35	1
B31	4	1	1	1	1	1	1	23	1
B40	4	1	1	1	1	1	1	25	1
Mean	4	1	1	1	1	1	1	28	1
S.E.M.	4	0	0	0	0	0	0	3.7	0

Table 27.6. Subgroup 6 antigen densities.

mAb	Subgroup	CARR	ESTE	FARG	VITI	BLAI	PELE	DELP	DUC
BA2 *	6	12	1	1	1	1	1	1	9
ALB6 *	6	24	1	1	1	1	1	1	18
L4M *	6	4	1	1	1	1	1	1	10
L16M	6	5	25	1	1	1	1	1	6
L18M *	6	5	1	1	1	1	1	1	6
Mean[a]	6	11.3	1	1	1	1	1	1	10.8
S.E.M.[a]	6	4.6	0	0	0	0	0	0	5

[a] Mean and S.E.M. calculated on marked (*) lines.

we considered as the most representative of each subgroup. This information is provided for easier comparison between patients. Data concerning pairs of mAbs and individual mAbs are assembled in Table 27.7, with indication of subgroup given in the second column. The clustering diagram was only a guide and we always came back to the density and fraction of positive cells data to check similarity of results. However, in the case of several mAbs it was not clear to which subgroup they belonged. MAbs B5 and B52 could be tentatively assigned to either the first

Table 27.7. Antigen density data on paired and individual mAbs.

mAb	Subgroup	CARR	ESTE	FARG	VITI	BLAI	PELE	DELP	DUC
B7	5	1	1	1	7	1	1	30	1
B51M	5	1	1	1	24	1	1	65	1
B46	7	1	1	1	1	1	4	1	1
L10M	7	1	1	1	1	1	2	1	1
B21	8	1	1	6	1	1	1	39	1
B49	8	1	1	3	1	1	1	29	1
B10M	9	34	15	4	164	40	18	100	64
B50	9	9	6	3	100	65	2	105	7
ALB9	12	83	136	60	13	1	49	520	118
ALB7	12	81	130	57	11	1	43	460	126
B47M	13	52	39	30	61	1	62	1	118
B48M	13	37	43	40	58	1	69	1	87
D1B6	15	60	90	46	136	313	210	130	81
L8M	15	58	90	42	100	366	235	94	69
B6	19	9	5	10	12	8	2	20	8
B11	19	9	1	3	2	4	23	37	8
B16	22	5	4	4	2	13	1	14	4
B29	22	3	4	4	1	8	1	4	4
B45M	10	1	4	1	1	1	4	59	1
B44M	11	8	11	5	73	113	60	7	14
L9	14	11	1	1	11	110	1	140	1
L12	16	91	78	70	200	200	170	350	150
T101	17	35	18	29	25	26	20	67	1
B18	20	6	39	4	1	1	3	130	27
B27M	21	1	1	4	1	88	20	1	1

or third group. Since B5 working dilution was found to be subsaturating (cf. this volume, Chapter 1), antigen densities are likely to be underestimated and thus it was finally clustered with the mAbs showing the highest densities, i.e., B14, B17, etc. (subgroup 1). B52 is placed in Fig. 27.3 just between B28 (subgroup 3) and B30 (subgroup 1). By comparison of histograms, it was decided that it could belong to subgroup 3 but it might also correspond to an individual specificity. Associated with subgroup 1, B22 and B24 have very similar reactivities. They might also be directed against another antigen whose expression very often resembles the expression of B114-related antigen. The same remark is applicable to mAbs B19 and B39 which initially clustered with subgroup 3. Also, clustering of L15 and L17 with subgroup 3 is only indicative, the density data being doubtful due to the IgM nature of these antibodies. One case of positivity makes the assignment of L16 to subgroup 6 questionable.

Since such groupings were defined using only one type of target cells (B-CLL), antigen density is shown to be a powerful parameter for serological specificity studies, even more powerful than the percentage of positive cells when working with homogeneous cell populations. This is illustrated in Table 27.8. When comparing mAb pairs B17, B114 (subgroup 1) and B28, B43 (subgroup 3), significant differences can be observed on 8/8 cell samples on the basis of antigen density. On the contrary, significant differences are only apparent for 4/8 cell samples when using the percentage of positive cells. In fact, the fraction of positive cells was specially useful in the cases where B-CLL cells were mixed with normal leukocytes, due to relatively low white blood cell counts (BLAI, DELP). In the majority of reports, such cases would not have been selected for typing studies.

Table 27.8. Compared potencies of antigen density and fraction of positive cells for clustering analysis.

mAb	CARR	FARG	VITI	PELE	ESTE	DUCR	BLAI	DELP
				Antigen density[a]				
B17	79	31	110	70	23	73	168; 29[b]	160; 31[b]
BL14	83	25	93	68	18	57	169; 26[b]	115; 8[b]
B28	24	12	17	18	11	23	24	57
B43	16	12	18	16	12	19	20	48
				Positive cell fraction (%)				
B17	92	95	79	96	98	94	62; 34[b]	24; 62[b]
BL14	94	93	81	93	98	94	62; 32[b]	29; 57[b]
B28	93	96	84	95	42	75	56	34
B43	90	95	86	94	36	84	58	37

[a] Expressed as mean number of antibody molecules bound/cell ($\times 10^3$).
[b] Presence of two positive cell populations.

Potential Interest of Antigen Density for Leukemia Typing

The same type of clustering program which had been used for comparison of reagents was run for comparison of patients. The highest correlation on the basis of the whole matrix of density data was clearly obtained between both CARR and DUCR cases (correlation = 0.94). These patients are the only two out of eight expressing the p24, CD9 antigen (Table 27.6). In addition, densities of p24 appear very similar for these two related cases.

The second finding is that, while not expressing p24 and showing a more different phenotype (correlation CARR/PELE = 0.84, correlation CARR/DUCR = 0.82), PELE shares with CARR and DUCR very similar densities of subgroup 1 and subgroup 3 antigens (Tables 27.2 and 27.4). These are the three out of eight patients whose leukemic cells did not bear detectable amounts of surface immunoglobulins. No other particular clinical or hematological similarity has been noted to date. On the contrary, the lack of T1 (CD5) antigen expression (Table 27.7, subgroup 17) on DUCR cells looks surprising, but it has not yet been possible to test again this reactivity.

Phenotype Modifications are Easily Detected Due to Absolute Measurements of Antigen Density

CARR quantitative phenotype was reestablished 3 months after the first analysis, using mAbs representative of the major subgroups (B114, "1"; B35, "2"; B28, "3"; B25, "4"; A1b6, "6"; B7, "5"; B46, "7"; B21, "8"; B50, "9"; A1b9, "12"; T101, "17"; D1B6, "15"). Among all these reagents, only one showed a variation of antigen density higher than 10%: the number of Ia-like molecules detected by mAb D1B6 rose from 60,000 molecules/cell to 200,000 molecules/cell. This observation illustrates the stability of the B-CLL phenotype, even on a quantitative basis.

Discussion

In looking for serological specificities of mAbs, FMF is a very informative technique. Computer-assisted statistical analysis requires a numerical description of each histogram as already clearly demonstrated (2). Using antigen density as an indicator of mAb reactivity, we have proposed a tentative grouping of the mAbs of the B cell and leukemia kits among seven main subgroups (including negatives which could not be further resolved with B-CLL cells) and nine pairs. Consideration of the organized data suggests that subgroups could be further resolved but this would really require more data on a different type of cells.

The potential value of quantitative phenotypes has been illustrated by the comparison of our B-CLL patients. This revealed that B-CLL could be associated with various differentiation levels, since the two cases with very similar phenotypes were both expressing the CD9 antigen as well as lacking surface immunoglobulins. Since p24 (CD9) has been associated with hematopoietic precursor differentiation stages (9), these two features are suggestive of an earlier differentiation stage for these cases. A third case was similar to these on the basis of subgroup 1 and 3 antigen densities as well as absence of surface immunoglobulins. Since p24 was not expressed, the leukemic cells of the latter case might be arrested a step forward in the B cell differentiation pathway. We also got an indication that the phenotype of a B-CLL could appear very stable on a quantitative basis, as it was reported to be on a qualitative basis (14). Such monitoring studies will be favored by measurements made on an absolute basis, avoiding the preservation of cells (15).

Numerous other applications of antigen density analysis by FMF are envisaged, such as comparison of antigen densities between pathological cells and their supposed normal counterparts, the interest of which was suggested by Anderson *et al.* (4). It might even provide invaluable information for the developing therapeutic approaches based on mAbs or their derivatives (16,17).

Summary

Using a new method of standardization, mAb binding could be quantitated from flow cytometric analysis of samples stained by conventional indirect immunofluorescence. This technique was applied to the study of the whole panel of B cell reagents on B chronic lymphocytic leukemia (B-CLL). Eight B-CLL cases were tested against 84 different mAbs, including an anti-CD5 known to react with such cells, providing for each case an extended quantitative phenotype.

The quantitative information obtained provided:

1. A useful numerical parameter for computer-assisted clustering analysis of tested mAbs.
2. A first insight into the fine heterogeneity of B-CLL differentiation stages.
3. The compared densities of all antigens recognized on B-CLL by the mAbs of the B panel.

Acknowledgments. The authors wish to express their appreciation to G. Derzko for helpful advice in statistical analysis and to G. Lefort (Blood Transfusion Center, Montpellier, France) for initial typing of immunoglobulins. Acknowledgments also go to G. Laurent for stimulating discussions concerning this work, to A. Garcia and M. Planchon for manuscript typing, and C. Pages for illustrations.

References

1. Bernard, A., L. Boumsell, J. Dausset, C. Milstein, and S.F. Schlossman, eds. 1984. *Leucocyte typing*. Springer-Verlag, Berlin, Heidelberg.
2. Bernard, A., L. Boumsell, and C. Hill. 1984. Joint report of the First International Workshop on Human Leucocyte Differentiation Antigens by the investigators of the participating laboratories (data management and statistical analysis). In: *Leucocyte typing*. A. Bernard, L. Boumsell, J. Dausset, C. Milstein, and S.F. Schlossman, eds. Springer-Verlag, Berlin, Heidelberg, p. 122.
3. Poncelet, P., and P. Carayon. Cytofluorometric quantification of cell-surface antigens by indirect immunofluorescence using monoclonal antibodies. *J. Immunol. Methods* in press
4. Anderson, K.C., M.P. Bates, B.L. Slaughenhoupt, G.S. Pinkus, S.F. Schlossman, and L.M. Nadler. 1984. Expression of human B-cell associated antigens on leukemias and lymphomas: a model of human B-cell differentiation. *Blood* **63:**1424.
5. Rundles, R.W. 1983. Chronic lymphocytic leukemia. In: *Hematology,* 3rd Ed., W.J. Williams, E. Beutler, A.J. Erslev, and M.A. Lichtman, eds. Mac Graw Hill Book Company, New York, p. 981–998.
6. Pernis, B., L. Forni, and L. Amante. 1970. Immunoglobulins spots on the surface of rabbit lymphocytes. *J. Exp. Med.* **132:**1001.
7. Lanier, L.L., and L.N. Warner. 1981. Paraformaldehyde fixation of hematopoietic cells for quantitative flow cytometry analysis. *J. Immunol. Methods* **47:**25.
8. Royston, I., J.A. Majda, S.M. Baird, B.L. Meserve, and J.C. Griffiths. 1980. Human T cell antigens defined by monoclonal antibodies. The 65000 dalton antigen is also found on chronic lymphocytic leukemia cells bearing surface immunoglobulins. *J. Immunol.* **125:**725.
9. Kersey, J.H., T.W. LeBien, C.S. Abramson, R. Newman, R. Sutherland, and M. Greaves. 1981. p24. A human leukemia-associated and lymphohemopoietic progenitor cell surface structure identified with monoclonal antibody. *J. Exp. Med.* **153:**726.
10. Brochier, J., D. Schmitt, E. Yonish-Rouach, G. Cordier, and J. Viac. 1984. Use of tissue distribution studies to determine the specificity of monoclonal anti-lymphocyte antibodies. In: *Leucocyte typing,* A. Bernard, L. Boumsell, J. Dausset, C. Milstein, and S.F. Schlossman, eds. Berlin, Heidelberg, p. 465.
11. Boucheix, C., J-Y. Perrot, M. Mirshahi, N. Fournier, M. Billard, F. Giannoni, A. Bernadou, and C. Rosenfeld. 1984. Monoclonal antibodies against acute lymphoblastic leukemia differentiation antigens. In: *Leucocyte typing,* A. Bernard, L. Boumsell, J. Dausset, C. Milstein, and S.F. Schlossman, Berlin, Heidelberg, p. 671.
12. Carrel, S., R. Tosi, M. Grass, M. Tanigaki, A.L. Carmagnona, and R.S. Accolla. 1981. Subsets of human Ia-like molecules identified by monoclonal antibodies. *Mol. Immunol.* **18:**403.
13. Haynes, B.F., G.S. Eisenbarth, and A.S. Fauci. 1980. Human lymphocyte antigens: production of a monoclonal antibody that defines functional thymus-derived lymphocyte subsets. *Proc. Natl. Acad. Sci. U.S.A.* **76:**5829.

14. Enno, A., D. Catovsky, M. O'Brien, M. Cherchi, T.O. Kumaran, and D.A.G. Galton. 1979. "Prolymphocytoid" transformation of chronic lymphocytic leukaemia. *Brit. J. Haematol.* **41:**9.
15. Bernard, A., B. Raynal, J. Lemerle, and L. Boumsell. 1982. Changes in surface antigens on malignant T-cells from lymphoblastic lymphomas at relapse: an appraisal with monoclonal antibodies and microfluorometry. *Blood* **59:**809.
16. Levy, R., and R.A. Miller. 1983. Biological and clinical implications of lymphocyte hybridomas: Tumor therapy with monoclonal antibodies. *Annu. Rev. Med.* **34:**107.
17. Möller, G. ed. 1982. Antibodies as carriers of drugs and toxins in tumor therapy. *Immunol. Rev.* **62:**1.

CHAPTER 28

Analysis of Workshop "L" Series Antibodies: Radioimmunobinding and Biochemical Studies

C.M. Steel, B.B. Cohen, Patricia Elder, Marion Moxley, and Keith Guy

Introduction

Twenty-one of the Workshop "L" series antibodies (L5 was missing) were received in April, 1984. They were thawed and stored at 8°C throughout the study (April–August, 1984). Aliquots were diluted 1:20 for use in both serological and biochemical assays, the diluent being RPMI 1640 medium containing 5% fetal calf serum (FCS) and 5% horse serum (HS) plus 0.02% sodium azide. All the antibodies were tested for the ability to bind to a selected panel of cells comprising established human leukocyte lines and fresh tissue from cases of leukemia or lymphoma. Those which gave positive results were then examined by "Western Blotting" on solubilized membrane glycoproteins, derived from human leukemic cells, which had been run in polyacrylamide gel electrophoresis (PAGE) and electrotransferred to cellulose nitrate sheets.

Materials and Methods

Test Cells

Fresh leukemic bloods and lymphoma biopsy material were provided by colleagues in several clinical departments in Edinburgh whose cooperation is gratefully acknowledged. Leukocytes were recovered by flotation over Ficoll–Hypaque (after teasing out a single cell suspension from the solid tissue in the case of lymphomas). Viability was checked by dye exclusion and only samples with fewer than 10% dead cells were included in the panel.

Long-term cultured lines were maintained in the authors' laboratory either in Ham's F10 medium with 5% FCS or in RPM1 1640 with 5% FCS and 5% HS.

Details of the cell panel are set out in Table 28.1.

Cell suspensions were fixed in 0.125% glutaraldehyde for 5 min at 4°C. An equal volume of 0.15 *M* glycine buffer (pH 7.2) was added for a further 15 min, followed by three washes in phosphate-buffered saline (pH 7.2) containing 1% bovine serum albumin (PBS.BSA). They were stored at 4°C in the same solution, with the addition of 0.02% sodium azide, for up to 2 months. Immediately before use they were washed twice more.

Radioimmunobinding Assay

The diluted Workshop antibodies were distributed in 50-μl aliquots in the wells of round-bottomed Microtest II plates (Cooke microtiter). For each test, three control wells were included, containing, respectively, 50 μl of J5 (anti-CALLA, Coulter, 1:50 of stock Ig solution), H92.1 (monomorphic anti-HLA-class I; undiluted hybridoma supernatant), and MS1 20A ("irrelevant" monoclonal antibody to mouse sperm; undiluted hybridoma supernatant).

Cell suspensions were adjusted to 4 × 10^6/ml in PBS.BSA and 50-μl aliquots (2 × 10^5 cells) added to each well. In a preliminary screen, the incubation period in first antibody was 40 min at room temperature but when eight cell preparations had been tested with all 21 antibodies, only one instance of positive binding was detected (L4 reacted with one sample of acute lymphoblastic leukemia) while the control antibodies J5 and H92.1 gave the expected positive reactions. In all subsequent assays, the first incubation period was increased to 2 hr at 8°C. The cells were washed three times in PBS.BSA by centrifuging the plate, decanting the fluid, and resuspending the pellets on a mechanical shaker. Second antibody was sheep anti–mouse Ig, ^{125}I-conjugated (New England Nuclear) diluted to 1 μCi/ml in culture medium. 50 μl were added to each well and incubation continued for a further 40 min at room temperature. The cells were washed three times, then taken up in 100 μl of PBS.BSA, with 1% sodium dodecyl sulfate, for gamma-counting.

Western Blotting

The principles of Western blotting and their application to the analysis of leukocyte surface antigens with monoclonal antibodies have been described elsewhere (14,15). Briefly, cells were lysed in NP-40 buffer and the lysate run in SDS–PAGE (16). The gel was then overlaid with a sheet of cellulose nitrate and the proteins transferred by horizontal electrophoresis in a commercial Electroblot apparatus (E-C Corporation). Following transfer, the cellulose nitrate sheet was soaked in BSA to saturate protein binding capacity. It was then dried and cut into strips, each of which was incubated with one of the test antibodies. After washing, the strips were exposed first to polyclonal rabbit anti–mouse Ig (Miles), followed by ra-

Table 28.1. Cell panel for testing L series monoclonal antibodies in radioimmunobinding assay.

Long-term cell lines

Non-T, non-B cells

1. Reh. From acute lymphoblastic leukemia (1)[a]
2. K562. From chronic myeloid leukemia (2)[a,b]
3. HL-60. From acute promyelocytic leukemia (3)[b]
4. U 937. From histiocytic lymphoma (4)[b]

B cells

5. EB virus-transformed lines. Pool of 15 lines from authors' laboratory[a,c]
6. Burkitt's lymphoma (BL). Pool of 12 lines, including EB_4 and Ramos which are EB virus negative[a]
7. Myeloma. Pool of RPMI-8226 (λ-secretor) and U266Bl (IgE secretor) (5,6)[a]
8. BLA_1 EB virus-transformed line from acute undifferentiated leukemia (? malignant cells)

Early T cells

9. Pool of MOLT-4 and CCRF-CEM, both from acute lymphoblastic leukemia (7,8)[a]
10. 8402T from acute lymphoblastic leukemia (9)[a]

Fresh leukemia/lymphoma cells

Null cells

1. Acute lymphoblastic lymphoma. Separate samples from 3 adult patients with non-B, non-T phenotype.

B cells

2. B prolymphocytic leukemia. Blood cells from 1 adult patient
3. B follicular lymphoma (lymphoblastic). Groin lymph node from 1 adult patient
4. Hairy cell leukemia. Solid deposit from spleen of 1 adult
5. B chronic lymphocytic leukemia. Blood lymphocytes from two adult patients; samples tested before and after 90 hr of culture in 100-ng/ml TPA to induce HLA-class II and other antigens (11–13). Before induction, DQ antigens were relatively strongly expressed on cells of patient 1 but almost absent from those of patient 2. They increased markedly, in both cases, on TPA exposure.

T cells

6. T cell chronic lymphocytic leukemia. Blood leukocytes from 1 adult patient with lymphocytosis ++ and skin involvement. Cells $OKT3^+$, $OKT4^+$, E-rosette +. No evidence of human T cell leukemia virus.
7. T cell lymphoblastic lymphoma. Samples from two adult patients (1 lymph node, 1 pleural effusion). Both E-rosette, +, $OKT4^+$.

Others

8. Monocytic leukemia. Blood leukocytes from 1 adult
9. Myelo-monocytic leukemia. Blood leukocytes from 1 adult
10. Acute myeloblastic leukemia. Blood leukocytes from 1 adult

[a] Phenotypes of these lines have been reported in detail (Ref. 10). Minor differences are found in the authors' laboratory, e.g., only 8402T now shows high levels of TdT; Reh is now negative for HLA-class II (Ia) antigens.

[b] K562, HL-60, and U 937 can be induced to express differentiated properties, usually of erythroid, myeloid, and monocyte lineages, respectively, but under standard culture conditions may be regarded as poorly differentiated cells.

[c] For one part of the assay, a pool of B cell lines was split into two aliquots, one half "starved" for 9 days so that it was growth-arrested, the other fed every 24 hours and shown to be in log-phase growth at the point of harvest. A comparison of the binding patterns with these two cell preparations should reveal any antigens whose expression is dependent upon cell division.

Table 28.2. Binding of L series antibodies to cell panel.[a]

	Antibody									
	L1	L2	L3	L4	L6	L7	L8	L9	L10	L1
"Null" cells										
ALL 1	−	+	−	+	+	−	+	−	++	+
ALL 2	−	−	−	−	±	−	−	−	±	−
ALL 3	−	−	±	±	±	−	+	±	−	−
Reh	−	−	++	−	++	+	±	±	−	±
K562	−	−	++	−	++	−	−	−	−	−
B cells										
B PLL	−	−	−	−	+	−	±	−	−	−
BLLma[b]	−	−	−	−	+	−	+	+	−	−
HCL	−	−	−	−	±	−	+	±	−	−
B.CLL 1 No TPA	−	−	−	−	+	−	++	−	−	−
B.CLL 1 + TPA	−	−	−	−	+	−	+++	−	−	−
B.CLL 2 No TPA	−	−	−	−	−	−	−	−	−	−
B.CLL 2 + TPA	−	−	−	±	−	−	±	−	−	±
B.L.	−	±	−	−	−	−	+	−	+	+
BLA 1	−	−	++	−	+	±	+++	−	−	−
L.C.L.[c]	−	−	−	−	±	−	++	−	−	−
Myeloma (8226 + U266B1)	−	−	−	−	±	−	++	−	±	−
T cells										
T CLL	−	−	−	−	±	−	−	±	−	−
TLLma 1[d]	−	−	−	−	++	−	±	±	−	−
TLLma 2	−	±	−	−	+++	±	±	++	±	+
8402 T	+	±	+	±	±	−	±	±	++	+
T L.C.L.[e]	−	−	+	−	++	±	±	+	−	−
Others										
Mono L	−	−	−	−	+	−	±	±	−	−
MML	−	−	−	−	+	−	±	+	−	−
U937	−	−	−	−	−	±	±	−	+	±
HL-60	−	−	−	−	±	±	±	++	−	−
AML	−	±	±	±	+	−	±	±	−	−
B-LCL pool										
Log Ph.	−	−	±	−	−	−	+	−	−	−
Static	−	−	−	−	−	−	++	±	−	−

[a] Each binding reaction is scored − to +++ according to the system described in the text.
[b] Lymphoblastic lymphoma.
[c] Lymphoblastoid cell lines.
[d] T lymphoblastic lymphoma.
[e] T lymphoblastoid cell lines (MOLT-4 + CCRF-CEM)

dioiodinated staphylococcal protein A (Amersham), then finally reassembled into a sheet for autoradiography.

Results

The pattern of antibody binding to the cell panel is shown in Table 28.2, using the following scoring system:

Antibody											
L12	L13	L14	L15	L16	L17	L18	L19	L20	L21	L22	J5
++	−	−	−	−	−	+	±	−	+	+	+
+	±	−	−	−	−	−	+	+	−	±	±
++	±	−	+	±	−	±	+	+	+	±	±
+++	++	−	−	−	±	±	+++	+++	±	−	++
+++	+++	−	−	−	−	+	+++	+++	−	±	+
++	−	−	−	−	−	−	−	+	−	−	±
+++	−	−	−	−	−	−	−	±	−	±	±
++	−	−	−	−	±	−	±	±	±	−	±
+++	−	−	−	−	±	−	−	±	−	−	±
++	−	−	−	−	±	−	−	±	−	−	±
+++	−	−	−	−	−	−	−	−	−	−	±
+++	−	−	−	−	−	+	−	+++	±	−	−
++	±	−	−	−	−	−	++	+	±	±	+
−	−	−	−	−	−	−	±	++	−	+	±
+	−	−	−	−	−	−	+	±	−	±	±
±	−	−	−	−	−	−	+	+	−	+	+
+	−	−	−	−	−	−	±	±	−	−	±
+++	+	−	−	−	−	−	+	±	−	+	±
+++	±	−	−	−	−	+	±	±	±	±	+
−	+	+	±	±	−	±	+++	++	+	+	++
+++	+	−	±	−	−	±	+++	+++	−	+	+
+++	±	−	−	−	−	−	++	+	−	−	+
+++	−	−	−	−	−	−	±	+	−	−	+
+++	−	−	−	−	−	±	−	−	±	−	±
++	−	−	−	−	−	−	±	±	−	±	±
+++	+	±	−	−	−	+	++	+	−	±	+
±	±	−	−	−	−	−	++	±	−	−	±
+	−	−	−	−	−	−	±	±	−	±	−

1. Background cpm (in presence of "irrelevant" first antibody MS1.20A) subtracted from each individual value.
2. Counts recorded in the presence of anti-HLA-class I antibody H92.1 (after background subtraction) taken as the 100% level.
3. Counts recorded in the presence of each Workshop antibody are expressed as a percentage of the 100% level defined above. The values are denoted as follows: 0–10%, −; 11–25%, ±; 26–50%, +; 51–75%, ++; >75%, +++.

Western blotting was attempted with antibodies L3, L5, L8, L12, L13, L19, and L20, all of which showed appreciable levels of binding to several cell types. Lysates were prepared from K562 cells and from the B follicular lymphoma (BLLma in Table 28.2). These were run in SDS–PAGE either unboiled in 0.1% SDS buffer or after boiling in 2% SDS buffer with 5% mercaptoethanol (this procedure separates proteins into their individual polypeptide chains). None of the L series antibodies tested showed any binding to electrotransferred proteins from either cell source although control antibodies PD7.26 (anti-leukocyte common), DA6.147 (anti-HLA-class II α chain), and CR3.43 (anti-HLA-class II β chain), run in parallel in every test, clearly identified bands of the expected molecular weight in B lymphoma-derived material. Only PD7.26 reacted with K562 lysate since this line is negative for HLA-class II antigens.

Discussion

It seems likely that all of the L series antibodies show rather low affinity for their respective target antigens as judged by the prolonged incubation required to demonstrate binding to intact cells and the uniformly negative results of Western blotting.

The protocol which gave positive results in the radioimmunobinding assay was somewhat arbitrary but neither time nor the quantities of antibody available permitted exhaustive investigation of all possible variables. The patterns of reactivity apparent in Table 28.2 should therefore be interpreted with considerable caution. Given the caveat, it seems that L8, L12, and L20 are pan-reactive, L12 giving by far the strongest overall binding; L6 is almost pan-reactive but with a preference for certain T and null cells. L8 shows selectivity for B cells and L9 virtually the reverse pattern. L19 seems particularly reactive with dividing cells since it binds to B-lymphoblastoid cells from a culture in log phase growth but not to those from a static culture. It also gives positive results with several samples of acute leukemia (ALL, AML, and monocytic leukemia) but not with B or T cell CLL. L13 may have a similar reaction pattern but is evidently much weaker. Only one antibody, L20, detects an antigen which is clearly inducible with TPA in one patient's CLL cells.

The sporadic strong positive results given by L3 and L10 against a "background" of almost uniform nonreactivity could either be technical artifacts or very significant findings. Judgment should perhaps be reserved until the present data can be compared with those from other sources.

Summary

Twenty-one Workshop L series antibodies have been tested, in an indirect radioimmunobinding assay, against an extensive panel of cells from human leukocyte lines and from individual cases of lymphoma and leukemia. Prolonged incubation was required before any binding could be demonstrated and attempts to investigate the molecular weight of the target antigens by Western blotting were uniformly unsuccessful. The antibodies therefore appear to be of low affinity. However several distinctive patterns of reactivity with the cell panel could be identified.

Acknowledgments. The authors are grateful to Dr. D.Y. Mason (Oxford), Dr. B. Fleischer, (Ulm) and to their colleagues Dr. Veronica van Heyningen and Mr. D.N. Crichton for generous supplies of monoclonal antibodies used in controls.

References

1. Rosenfeld, C., A. Goutner, C. Choquet, A.M. Venuat, B. Kayibanda, J.L. Pico, and M.F. Greaves. 1977. Phenotypic characteristics of a unique non-T, non-B acute lymphoblastic leukemia cell line. *Nature* **267:**841.
2. Lozzio, C.B., and B.B. Lozzio. 1975. Human chronic myelogenous leukemia cell line with positive Philadelphia chromosome. *Blood* **45:**321.
3. Collins, S.J., R.C. Gallo, and R.E. Gallacher. 1977. Continuous growth and differentiation of human myeloid leukemia cells in suspension culture. *Nature* **270:**347.
4. Sundstrom, C., and K. Nilsson. 1976. Establishment and characterisation of a human histiocytic lymphoma cell line. *Int. J. Cancer* **17:**565.
5. Matsuoka, Y., G.E. Moore, Y. Yagi, and D. Pressman. 1967. Production of free light chains by a hemopoietic cell line derived from a patient with multiple myeloma. *Proc. Soc. Exp. Biol. Med.* **125:**1246.
6. Nilsson, K. 1971. Characteristics of established myeloma and lymphoblastoid cell lines derived from an E myeloma patient. A comparative study. *Int. J. Cancer* **7:**380.
7. Minowada, J., T. Ohunuma, and G.E. Moore. 1973. Rosette-forming human lymphoid cell lines. 1. Establishment and evidence of origin from thymus-derived lymphocytes. *J. Natl. Cancer Inst.* **49:**891.
8. Foley, G.E., H. Lazarus, S. Farber, B.G. Uzman, B.A. Boone, and R.E. McCarthy. 1965. Continuous culture of human lymphocytes from peripheral blood of a child with acute leukemia. *Cancer* **18:**522.
9. Huang, C.C., Y. Hous, L.K. Woods, G.E. Moore, and J. Minowada. 1974. Cytogenetic study of human lymphoid T-cell lines derived from lymphocytic leukemia. *J. Natl. Cancer Inst.* **53:**655.
10. Minowada, J. 1978. Markers of human leukemia–lymphoma cell lines reflect haematopoietic cell differentiation. In: *Human lymphocyte differentiation: Its application to cancer,* INSERM symposium No 8, B. Serrou and C. Rosenfeld, eds. Elsevier/North Holland, Amsterdam, pp. 337–344.
11. Totterman, T.H., K. Nilsson, and C. Sundstrom. 1980. Phorbol ester-induced differentiation of chronic lymphocytic leukemia cells. *Nature* **288:**176.
12. Guy, K., V. van Heyningen, E. Dewar, and C.M. Steel. 1983. Enhanced expression of human la antigens by chronic lymphocytic leukemia cells following treatment with 12-*O*-tetradecanoylphorbol-13-acetate. *Eur. J. Immunol.* **13:**156.
13. Caligaris-Cappio, F., G. Janossy, D. Campana, M. Chilosi, L. Bergin, R. Foa,

D. Delia, M.C. Guibellino, P. Preda, and M. Gobbi. 1984. Lineage relationship of chronic lymphocytic leukemia and hairy cell leukemia: studies with TPA. *Leuk. Res.* **8**:567.

14. Towbin, H., T. Staehelin, and Gordon. J. 1979. Electrophoretic transfer of proteins from polyacrylamide gels to nitrocellulose sheets: procedure and some applications. *Proc. Natl. Acad. Sci. U.S.A.* **76**:4350.
15. Cohen, B.B., D.L. Deane, V. van Heyningen, K. Guy, D. Hutchins, M. Moxley, and C.M. Steel. 1983. Biochemical variations of human Ia-like antigens detected with monoclonal antibodies. *Clin. Exp. Immunol.* **53**:41.
16. Laemmli, U.K. 1970. Cleavage of structural proteins during the assembly of the head of bacteriophage T4. *Nature* **277**:680.

CHAPTER 29

Subclassification of Leukemia Using Monoclonal Antibodies

Glenn R. Pilkington, Grace T.H. Lee, Patricia M. Michael, O. Margaret Garson, Norbert Kraft, Robert C. Atkins, and David G. Jose

Introduction

The rapid proliferation of monoclonal antibodies since 1975 (1) has resulted in vast numbers of antibodies to leukocyte antigens being made available for typing of leukemia. We have collected a panel of monoclonal antibodies selected from those which we have found most specific (2) in the First International Workshop on Leucocyte Differentiation Antigens and complemented these antibodies with others available locally or internationally, but not tested in the First Workshop. The panel of 44 antibodies is listed in Table 29.1. This panel of antibodies has been tested against leukemic cells from patients with childhood acute lymphocytic leukemia (ALL) and myeloid leukemia as well as T-lymphoma cell suspensions from patients with nodal disease. The results obtained have been compared to the normal ranges of reactivity in the tissues being tested and to results from patients in remission.

We have previously found heterogeneity within the Null (CALLA$^-$, non-B/T)-ALL subgroup of patients using a rabbit antiserum to Null-ALL cells (3) and other authors have reported a pre-B (4,5) subgroup of ALL patients. We have also applied serological typing to myeloid leukemia patients (3) and preliminary results have suggested poorer survival of the p50$^+$ group of Null-ALL patients and PHM2$^+$ myeloid patients (3). Therefore the primary aim of this study has been to examine the benefit of serological typing, using our panel, for diagnosis and monitoring of leukemia. However, the secondary aim has been to look for heterogeneity within the established subgroups of ALL (cALL, Null-ALL, and T-ALL) and myeloid leukemia and test for differences in survival or correlation with specific chromosome abnormalities if these exist.

Table 29.1. Antibodies used in study.

Monoclonal antibody	Specificity	Supplier	Ref. no.
Common			
W6/32	HLA-A,B,C	Sera-lab	9
TDR31.1	HLA-DR (Ia)	T. De Kretser	10
BA2/FMC8	p24	Hybritech/Ser-lab	11
BA1	p30	Hybritech	12
PHM1	Common leuko.	N. Kraft	13
CIPAN	Pan human	G. Lee	14
B-cell			
y29/55	Pan-B/pre-B	H. Forster	11
BL13	Germ. center	J. Brochier	15
Tul/BL14	Mantle/mature-B	A. Bernard/J. Brochier	11/15
FMC1	Mature-B	Sera-lab	11
FMC7	Mature-B/B-PLL	Sera-lab	16
PHM14	Mature-B/NPDLL	N. Kraft	17
F29.132	Mature-B/NPDLL	N. Kraft	17
Non-B/T			
J5	CALLA	Coulter	11
PHM6	CALLA	N. Kraft	18
CIMT	Null/T/myeloid (LFA-1)	G. Lee	(this volume)
T-cell			
9.6/T11	E-receptor	DuPont/Coulter	11
T3/UCHT1	Mature-T	A. Bernard	11
10.2/T101	Mature-T/B-CLL	DuPont/Hybritech	11
T4/Leu3	Helper-T	Coulter/Becton Dickinson	11
T8/Leu2	Suppressor-T	Coulter/Becton Dickinson	11
Thymocyte			
OKT10	Common/B-blast	Ortho	19
NA1/34	Cortical/plasma	Sera-lab	11
RPH1	T-ALL assoc.	R. Herrmann	20
Myeloid/monocyte			
Mo1	Mature myeloid	Coulter/Ortho	11
PHM2	AML-M2/M4/M5a	N. Kraft	13
Tu2	Myelo./promono.	A. Bernard	11
CIKM5	Monobl./metamy.		3
Granulocyte			
Tu5	Metamy. → neutr.	A. Bernard	11
FMC13	Promy. → neutr.	Sera-lab	11
Monocyte			
FMC17/Mo2	Monocyte	Sera-lab/A. Bernard	11
VIMD2	Macrophage	W. Knapp	11
PHM3	Mono/macro/thymo.	N. Kraft	13
Platelet/mega.			
AN51/FMC25	F.VIII receptor	A. McMichael/H. Zola	21
J15/H5	Megakaryoblast	A. McMichael	21
Erythroid			
CMRF4/R10	Glycophorin	D. Hart/P. Edwards	(unpublished)/22

Materials and Methods

Patients' Cells

Mononuclear cells enriched for blasts were separated from heparinized blood and marrow by single-step density gradient centrifugation on Ficoll–Hypaque (1.077g/cm^3, 400g × 30 min, 20°C).

Microcytotoxicity Test

Complement-mediated lysis of patients' target cells was measured using a modification of the cytotoxicity test of Amos *et al.* (6) as previously described (7).

Flow Cytofluorometry

The binding of mouse monoclonal antibodies was detected using an F(ab')$_2$ fragment of goat anti–mouse Ig (affinity-purified) FITC (cat. no. 1311-0111, Cappel Laboratories, Cochranville, PA). Samples were prepared as previously described (8).

Indirect immunofluorescence was measured using an Ortho System 50H cell sorter coupled to a 2150 computer (Ortho Instruments, Westwood, MA), or using an Ortho FC200/4800A cytofluorograph coupled to an ND100 multichannel analyzer (Nuclear Data, Chicago, IL). Instruments were standardized with 2-μm fluorescent spheres (cat. no. 9847, Polysciences, Warrington, PA).

Cell suspensions were analyzed by using two-dimensional scatter gating (90° scatter vs. axial light loss). The sensitivity of measurements was optimized by setting the instruments to minimize background binding of conjugate, NS1 supernatant, normal mouse Ig (1/100), and other negative controls such as CMRF4 or R10 (Table 29.1) and to maximize binding of appropriate positive controls (W6/32, CIPAN, TDR31.1, etc.) depending on the cell population being analyzed.

Mouse Monoclonal Antibodies

The monoclonal antibodies used in this study are listed in Table 29.1 along with the suppliers.

Cell Markers

All mononuclear cell fractions were tested for the established markers, E-rosette receptor (E-RFC), both 4°C and 37°C, surface Ig(SIg), myeloperoxidase, and α-naphthyl butyrate esterase.

Chromosome Studies

These were performed on 24-hr bone marrow cultures followed by trypsin G-banding.

Results

Normal Ranges

The normal ranges of reactivity of all monoclonal antibodies in Table 29.1 were established for bone marrow ($n = 22$), peripheral blood ($n = 19$), lymph node ($n = 9$), and juvenile thymus ($n = 7$). The normal ranges for bone marrow and thymus are expressed in Tables 29.2–29.4. Patients were only recorded as positive with a monoclonal antibody when results were >10% above the normal range for the tissue being tested. Results for all remission patients tested (ALL $n = 30$, AML/CML $n = 10$) fell within the normal ranges.

Subgrouping of Common (CALLA$^+$, non-B/T) ALL Patients

Results of surface marker analysis of cells from cALL patients are summarized in Table 29.2 [results for antibodies positive (e.g., CIPAN) or negative (e.g., B cell antibodies) with all specimens and not relevant to the discussion have been omitted]. Cells from all patients expressed CALLA, and the majority expressed HLA-DR (59 patients) and p24 (51

Table 29.2. Serological subgrouping of common ALL (CALLA$^+$, non-B/T) patients by cytofluorometry and cytotoxicity.

Marker/monoclonal antibody	Group 1 ($n = 9$)	Group 2 ($n = 47$)	Group 3 ($n = 8$)	Bone marrow normal range (%)
HLA-DR (Ia)[a]	9/9[b]	42/47	8/8	1–18
P24 (BA2/FMC8)[a]	9/9	35/40	7/8	0–14
Pan B cell (Y29/55)[a]	9/9	0/46	0/8	0–20
Pan T cell (9.6/T11)[a]	0/9	1/47	0/8	0–15
CALLA (J5/BA3/PHM6)[a]	9/9	47/47	8/8	0–13
T/myeloid (CIMT)	1/4	0/12	4/6	0–25
Common myeloid (Mo1)	1[c]/9	0/25	2/7	20–64
Monocyte (FMC17)	0/3	0/21	0/4	0–8
Mono/macrophage (VIMD2)	0/2	0/7	1/4	0–6
Promyelo → neut. (Tu5/FMC13)	0/4	0/30	0/8	12–64
Myeloblast/mono (PHM2)	1/4	0/21	7/8	0–13
Promyelo/promono (Tu2)	0/2	0/15	1/5	0–9
Monoblast/myelo (CIKM5)	1/3	0/17	5/8	0–15
Factor VIIIR (AN51/FMC25)	0/4	0/18	0/7	0–11
Glycophorin (R10/CMRF4)	1[c]/3	0/17	1/7	1–54

[a] 22 patients were tested by complement-mediated cytotoxicity.
[b] Number patients positive (>10% above normal range)/number tested.
[c] Normal erythroblasts?

Table 29.3. Serological subgrouping of null-ALL ($CALLA^-$, non-B/T) patients by cytofluorometry.

Marker/monoclonal antibody	Group 1 (*n* = 5)	Group 2 (*n* = 7)	Bone marrow normal range (%)
HLA-DR (Ia)	5/5 (28–65)[a]	0/7 (0–16)	1–18
p24 (BA2/FMC8)	3/5	2/5	0–14
Pan B cell (Y29/55)	1/3	1/5	0–20
Centrocyte (Tu1)	0/2	1/6	0–19
B-blast (PHM14)	1[b]/2	0/2	0–12
Thymocyte (OKT10)	1[b]/2		0–10
Cort.Thymo. (NA1/34)	1[b]/3	0/4	0–12
Pan T cell (9.6/T11)	0/5	0/7	0–15
T/myeloid (CIMT)	0/4	0/7	0–25
CALLA (J5/PHM6)	0/5 (0–8)	0/7 (0–8)	0–13
Common myeloid (Mo1)	0/4 (0–2)	0/7 (0–12)	20–64
Promyelo → neut. (Tu5/FMC13)	0/4 (0–10)	0/7 (0–9)	12–64
Myeloblast/mono (PHM2)	0/4 (0–14)	0/7 (0–9)	0–13
Monocyte (FMC17)	0/4 (0–8)	0/5 (0–3)	0–8
Monoblast/myelo (CIKM5)	0/5	1/6	0–15
Factor VIIIR (AN51/FMC25)	0/4 (0–4)	3/7 (21–49)	0–11
Platelet/mega (J15/H5)	0/4 (0–11)	0/2 (0–5)	1–13
Glycophorin (R10/CMRF4)	0/4 (0–14)	0/5 (0–12)	1–54

[a] Number patients positive (>10% above normal range)/number tested (range in %).
[b] Peripheral blood.

Table 29.4. Serological subgrouping of T-ALL/lymphoma patients by cytofluorometry.

Marker/monoclonal antibody	T-ALL Group 1 (*n* = 5)	T-ALL Group 2 (*n* = 6)	T-Lymphoma node (*n* = 12)	Normal range thymus (%) (*n* = 7)
HLA-DR (Ia)	0/5[a]	1/6	0/12	0–6
p24 (BA2/FMC8)	0/2	0/5	0/6	31–38
p30 (BA1)		2/4	0/10	0–1
Pan B cell (Y29/55)	0/3	0/5	0/12	0–2
CALLA (J5/PHM6)	1/4	0/6	0/12	2–10
Pan T cell (9.6/T11)	5/5	6/6	12/12	53–93
Mature T cell (T3/UCHT1)	1/4	0/6	7/12	19–46
Pan T cell/B-CLL (10.2/T101)	1/3	3/4	5/9	78–87
Helper T cell (T4/Leu3a)	2/3	2/5	5/12	39–85
Suppressor T cell (T8/Leu2a)	2/3	2/5	2/12	52–82
Common thymocyte (OKT10)	2/3	1/2	5/10	76–90
Cortical thymocyte (NA1-34)	5/5	0/5	1/12	46–86
T-ALL Assoc. (RPH1)	2/3	4/5	3/12	55–86
T cell/Myeloid (CIMT)	1/4	3/4	7/12	82–94
Thymocyte/monocyte (PHM3)	0/2	0/4	2/6	38–74
Common myeloid (Mo1/ OKM1)	0/5	0/5	0/11	2–7
Myeloblast/mono (PHM2)	0/3	0/4	0/8	0–2
Glycophorin (R10/CMRF4)	0/3	0/6	0/9	2–8

[a] Number patients positive (>10% above normal range)/number tested.

patients). Cells from 15 of 23 patients in this group were p30 (BA1) positive also (data not shown).

The patients in Table 29.2 however have been divided into three groups (numbered 1–3) on the basis of reactivity of their cells with the pan B cell antibody Y29/55 (Group 1) or the myeloid-associated antibodies CIMT (LFA-1), PHM2, CIKM5, Mo1, VIMD2, or Tü2 (Group 3). Cells from Group 2 patients expressed HLA-DR (Ia) antigen (42/47 patients) and p24 antigen (35/40 patients) in the majority of cases, characteristic of cALL. Cells from one patient in Group 2 expressed the E-rosette receptor (positive with 9.6 and T11). None of the myeloid, monocyte, platelet, or erythrocyte associated antibodies reacted with cells from patients in Group 2 (Table 29.2).

Survival and Cytogenetics of Y29/55-Positive Subgroups of cALL

Follow-up time for most patients in Group 1 has been insufficient to test for prognostic significance of this serological subgroup, with 5/9 patients having survived to date less than 2 years. However, one patient, whose cells also expressed the myeloid-associated antigen, CIKM5, survived 3 months and one patient has survived to date 5 years with one relapse at 3 years.

Cytogenetic analysis of cells from the patients within Group 1 has shown hyperdiploidy (>50) in four patients, normal karyotypes in two patients, and hypodiploidy with multiple abnormalities in one patient. No karyotype data was available for the longest and shortest surviving patients in this group.

Survival and Cytogenetics of Myeloid Antigen-Positive Subgroup of cALL

As with Group 1, duration of follow-up for patients in Group 3 has not been sufficiently long to determine the prognostic significance of this serological subgroup. To date maximum follow-up period has been 2 years (one patient) with most patients (six) having been followed for 1 year or less.

Cells from five patients within this serological subgroup were hyperdiploid (>50) and one patient had a normal karyotype. Cells from two patients were hypodiploid, one with a Ph′ chromosome 45, XY, 5p+, *t(9;22)(q34;q11)*,-10, *i(17q)* and one with the karyotype 45, X-X, *t(4;5)(q35;q13)*, del(6)(q21).

Subgrouping of Null (CALLA$^-$, Non-B/T) ALL Patients

Table 29.3 summarizes the results of surface marker analysis of cells from the 12 null-ALL patients tested. (As above, results for antibodies positive or negative with all specimens and not relevant to the discussion have been omitted.) The cells from all patients were negative when tested for reactivity with myeloid, monocyte, or erythrocyte associated antibodies (Mol, FMC13, FMC17, or CMRF4 R10) as well as CALLA.

The patients in Table 29.3 have been arbitrarily divided into two groups on the basis of their cells being positive or negative for HLA-DR (Ia) antigen. Cells from Group 1 patients also expressed p24 (3/4 patients) and p30 (BA1, 1/2 patients) (data not shown). Cells from one patient in Group 1 reacted with the B cell-associated antibodies Y29/55 and PHM14 as well as OKT10 and NA1/34. In Group 2, cells from 3/5 patients tested expressed the platelet-associated antigen Factor VIII receptor (AN51/FMC25). Rarely did cells from patients in this group express p24 (1/5 patients), p30 (BA1, 0/4 patients) (data not shown), or the platelet-associated antigen identified by J15/H5 (0/5 patients). Cells from one patient in Group 2 were positive with the B cell antibody Y29/55. Of the three patients whose cells were positive for Factor VIII receptor, one was also positive with the myeloid-associated antibody CIKM5 and one positive with the B cell-associated antibody Tül. Cells from three patients in Group 2 however were positive only for HLA-A,B,C and CIPAN (data not shown) but no other markers tested.

Survival and Cytogenetics of Null-ALL Subgroups

Of the five Group 1 (HLA-DR positive) patients one has died after eight months with four surviving to date between 5 months and 4½ years.

Cells from two patients in this subgroup were diploid, one pseudodiploid 46, XY, 6q−, 11q−, 19p+, one hypertriploid (~70), and one hyperdiploid with multiple abnormalities 48, XX, *t(4;6;11)* (q21; p12; q23), +8, +22.

Survival of Group 2 (HLA-DR negative) patients has been poor with 4/7 patients dying between 1 month and 16 months post diagnosis. Two of the patients whose cells were positive with Factor VIII receptor survived only 1 month and 4 months. The other two patients whose cells expressed only HLA-A,B,C and the antigen recognized by CIPAN died 15 and 16 months post diagnosis. The patient whose cells were Y29/55-positive is the longest survivor (4½ years).

Cytogenetic analysis of the cells from Group 2 patients has shown that in two of the patients whose cells were devoid of markers (except HLA-A,B,C and CIPAN) karyotypes were normal but in two of the patients whose cells expressed Factor VIII receptor multiple abnormalities were present: 47, XX, 10p+, −15, + mar1, + mar 2 in one patient and 45, XY, t(1;9) (1;20) (p36; q22; q42; q13), −7, 17q+.

Subgrouping of T-ALL/T-Lymphoma Patients

When cell suspensions from 11 patients with ALL of the T cell type (T-ALL) and 12 patients with T-lymphoma (nodal disease) were analyzed for surface marker expression the results summarized in Table 29.4 were obtained.

Note that the T-ALL patients have been divided into two groups on the basis of the expression of the cortical thymocyte antigen identified by NA1/34 (or OKT6). Cells from the NA1/34-positive group (1) of patients rarely expressed the mature T cell markers recognized by T3/UCHT1 (1/4 patients) or 10.2/T101 (1/3 patients) but more frequently expressed the antigens identified by OKT10 (2/3 patients) and CIMT (1/4 patients). Conversely cells from the Group 2 patients were frequently positive with the mature T cell marker 10.2/T101 (3/4 patients) and CIMT (3/4 patients). In this respect cells from Group 2 patients were more similar to those of T-lymphoma patients (Table 29.4). RPH-1 was frequently positive with cells from both Group 1 and 2 T-ALL patients but less frequently with cells from T-lymphoma patients (3/12). In addition, cells from most T-lymphoma patients expressed the mature antigen T3/UCHT1 and cells from 5/10 patients were positive with OKT10. CALLA was positive on cells from only 1/10 T-ALL patients tested. B cell myeloid, monocyte, platelet (data not shown), and erythrocyte associated antibodies were not positive with cells from Groups 1 or 2 of the T-ALL patients except HLA-DR which was positive on cells from one T-ALL patient.

Subgrouping of Myeloid Leukemia Patients

Summary results of surface marker analysis of cells from 47 myeloid leukemias are presented in Tables 29.5 and 29.6 (results for antibodies positive or negative with all specimens and not relevant to the discussion have been omitted). HLA-DR (Ia) antigen was present in most cases of AML (M1, 2, 4, and 5; 17/25 patients) (Table 29.5). Cells from AML-M2 were the least frequently positive (4/9 patients). p24 antigen was present on cells from 2/6 AML-M2, 3/5 AML-M4, and one AML-M5a patient. Cells from patients with AML-M1 were negative with all other markers tested except CIMT in 2/6 cases, whereas cells from AML-M2 patients were frequently positive with antibodies PHM2 (6/9 patients) and CIMT (5/7 patients) but rarely with antibodies reacting with mature myeloid (Tu5 or FMC13) or with monocytic cells (CIKM5 or FMC17). Cells from patients with AML-M4 could thus be differentiated from cells from AML-M2 in the expression of PHM2 (2/7 patients positive in M4) and CIMT (0/5 patients positive in M4). Cells from patients with AML-M5 were positive with CIMT (2/2 patients) and CIKM5 (2/2 patients) and cells from the AML-M5a patient were positive in addition with the markers Mo1, PHM2, Tu2, FMC17, VIMD2, and PHM3.

Similarly cells from patients in different phases of chronic myeloid leukemia (CML), including myeloid and lymphoid blast crisis, could be distinguished on the basis of their surface marker expression (Table 29.6). Thus cells from chronic phase were positive with the mature granulocyte

Table 29.5. Serological subgrouping of acute myeloid leukemia patients by cytofluorometry.

	AML		AMML	AMoL		
Marker/monoclonal antibody	M1 (n = 5)	M2 (n = 9)	M4 (n = 7)	M5a	(n = 4)	M5b
HLA-DR (Ia)	4/5[a]	4/9	5/7		4/4	
CD24 (BA2/FMC8)	0/4	2/6	3/5	1/1		0/1
Pan B cell (Y29/55)	0/4	1/8	0/6		0/2	
Pan T cell (9.6/T11)	0/5	0/9	0/6		0/2	
CALLA (J5/PHM6)	0/5	1/9	0/7		0/2	
T/myeloid (CIMT)	2/4	5/7	0/5		2/2	
Common myeloid (Mo1)	0/4	0/7	0/7	1/1		0/1
Myeloblast/mono (PHM2)	0/5	6/9	2/7	1/1		0/1
Promyelo/promono (Tu2)		1/5	0/4	1/1		0/1
Monoblast/myelo (CIKM5)	0/5	0/5	1/7		2/2	
Promyelo → neut. (Tu5/FMC13)	0/5	0/9	0/6		0/2	
Monocyte (FMC17/Mo2)	0/5	1/9	0/7	1/1		0/1
Mono/macro (VIMD2)	0/1	1/8	0/4	1/1		0/1
Thymo/mono/macro (PHM3)	0/5	0/7	0/6	1/1		0/1
Factor VIIIR (AN51/FMC25)	0/4	0/7	0/6		0/2	
Glycophorin (CMRF4/R10)	0/5	0/8	0/6		0/2	

[a] Number patients positive (>10% above normal range)/number tested.

Table 29.6. Serological subgrouping of chronic myeloid leukemia patients by cytofluorometry.

			Blast crisis		
Marker/monoclonal antibody	Chronic phase (n = 6)	Accelerated phase (n = 4)	Myeloid (n = 6)	Lymphoid (n = 4)	Megakaryo. (n = 2)
HLA-DR (Ia)	0/6[a]	0/4	4/6	4/4	0/2
CD24 (BA2/FMC8)	1/3	0/4	2/3	1/4	2/2
Pan B cell (Y29/55)	0/4	0/4	0/5	0/4	1/1
Pan T cell (9.6/T11)	0/5	0/4	0/5	0/4	0/1
CALLA (J5/PHM6)	0/6	0/4	0/6	4/4	0/2
T/myeloid (CIMT)	0/4	3/4	1/5	0/4	0/1
Common myeloid (Mo1)	2[b]/6	1[b]/4	1[b]/6	0/4	0/1
Myeloblast/mono (PHM2)	2/6	1/4	3/6	2/4	0/1
Promyelo/promono (Tu2)	0/5	4/4	1/6	0/2	0/1
Monoblast/myelo (CIKM5)	1/6	1/4	1/5	0/4	0/1
Promyelo → neut. (Tu5/FMC13)	4[b]/5	0/3	1/6	0/4	1[b]/1
Monocyte (FMC17/Mo2)	0/6	0/3	0/6	0/4	0/1
Mono/macro (VIMD2)	0/5		0/5	0/3	
Thymo/mono/macro (PHM3)	0/4	0/3	0/5	0/3	0/1
Factor VIIIR (AN51/FMC25)	0/5	0/3	0/6	0/4	2/2
Glycophorin (CMRF4/R10)	0/3	0/3	0/3	0/4	0/1

[a] Number patients positive (>10% above normal range)/number tested.
[b] Peripheral blood Ficoll–Hypaque fraction.

markers Tu5 and FMC13 (4/5 patients) and cells from accelerated phase were reactive with CIMT (3/4 patients) and Tu2 (4/4 patients). Cells from patients in lymphoid or myeloid blast crisis expressed HLA-DR in most cases (8/10 patients); however, cells from patients in lymphoid blast crisis expressed CALLA (4/4 patients) and PHM2 (2/4 patients) but none of the other myeloid-associated antibodies. In contrast, cells from the patients in myeloid blast crisis were positive with PHM2 (3/6 patients) or CIKM5 (2/5 patients) or Tu2 (1/6 patients) or Tu5/FMC13 (1/6 patients). Two patients in blast crisis with megakaryoblasts present in blood were also tested. Cells from both of these patients were positive for Factor VIII receptor but negative for HLA-DR antigen.

Discussion

In conclusion, our results from typing cells from childhood ALL and from myeloid leukemia patients have indicated heterogeneity within defined subgroups of ALL. The common ALL (cALL) subgroup can be divided into a pre-B group [as previously described (4,5)] using the monoclonal antibody Y29/55 (23) and a group expressing myeloid-associated antigens as well as the larger Group 2 expressing the markers CALLA, p24, p30, and HLA-DR(Ia) antigens. Follow-up time has been insufficient to establish prognostic significance of these subgroups; however, one patient whose cells expressed the myeloid-associated antigen CIKM5 survived only 3 months. Cytogenetic data for these patients indicated in general that karyotypes were similar to the cALL group as a whole (24). However, cells from two of the patients in the myeloid antigen-positive subgroup of cALL had abnormalities which may be found in myeloid leukemias [one patient with a Ph′ chromosome, the other patient a translocation t(4;5)] suggesting a correlation between surface marker expression and karyotype in 2/7 patients.

Typing of the Null(CALLA$^-$, non-B/T)-ALL patients' cells also indicated heterogeneity within this group of patients. Group 1 patients (HLA-DR$^+$) to date have survived longer. Karyotypes for Group 1 and 2 patients together were atypical of the CALLA$^+$ group and more in keeping with the Null-ALL group (24). Two of the three patients in Group 2, whose cells expressed the Factor VIII receptor, and who died after 1 and 4 months, had multiple abnormalities. Thus data for these patients also suggested a link between surface marker expression, survival, and karyotype although survival data were not significant for the two Group 1 patients with abnormalities due to length of follow-up.

Our results for T-ALL and T-lymphoma patients indicated two main subgroups of T-ALL which reflected the maturation of these cells. The more immature cells expressed the cortical thymocyte antigen identified by the antibody NA1/34, but both groups of T-ALL could be distinguished from T-lymphoma on the basis of surface marker expression. The myeloid leukemias tested yielded characteristic patterns of marker expression for each morphological type of myeloid leukemia including AML-M1, M2, M4, M5a, M5b, and CML-chronic and accelerated phases, as well as lymphoid, myeloid, and megakaryoblast crisis (see also this volume, Chapter 20). Moreover our previous data (3) have suggested correlation of expression of the antibody PHM2 (which was not clearly correlated with the morphological types of myeloid leukemia) with patient survival. However, survival and cytogenetic data for the T-ALL, T-lymphoma, and myeloid leukemia patients in this study have not yet been analyzed.

Overall our results have indicated serological subgroups within the cALL, Null-ALL, T-ALL, and myeloid leukemia groups of patients. Initial results indicate these subgroups may be significant. Specific clusters of antibodies reacted with each morphological type of leukemia significantly above the normal ranges in cases of active disease whereas cells from patients in clinical and hematological remission gave results within the normal ranges.

Summary

Samples (blood and bone marrow) from 150 leukemia patients and 50 hematologically normal donors have been tested with a panel of 44 monoclonal antibodies (mAbs) selected from the First International Leucocyte Workshop and other antibodies available locally and internationally. Normal ranges were established for the blood and bone marrow and results compared for patients with definitive clinical and hematological diagnoses, in remission, and relapse. Results indicated serological subgroups within the cALL, Null-ALL, T-ALL, and myeloid leukemia groups of patients and departures from the normal range with specific groups of mAbs in these patients. cALL patients could be divided into three subgroups on the basis of reactivity with the B cell mAb Y29/55 (pre-B), reactivity with the myeloid-associated antibodies (PHM2, CIMT, Tu2, CIKM5, VIMD2, Mo2) and nonreactivity with any of these mAbs. Null-ALL (CALLA$^-$, non-B/T ALL) was subdivided into groups reacting with anti-platelet-associated mAbs AN51/FMC25 (FVII Receptor) and BA2/FMC8 or with TDR31.1 (Ia). Myeloid mAbs gave patterns of reactivity associated with CML, CML in accelerated phase, CML-BC (lymphoid), CML-BC (myeloid), AML-M1, AML-M2, AMML-M4, or AMoL-M5. Reactivity of mAbs with remission patients was identical to control results but changes were obvious in relapsing patients. Preliminary results suggested a possible correlation of one phenotypic subgroup within the common ALL group and karyotype, and a possible correlation of one phenotypic subgroup of Null-ALL with karyotype and survival.

Acknowledgments. This work was supported by grants from the Research Committee, Cancer Institute, Melbourne and the National Health and Medical Research Council of Australia. We are grateful to Dr. H.

Ekert for supplying specimens from childhood ALL patients and survival data on these patients, to J. Quirk and S. Rockman for expert technical assistance, and to G. White for typing the manuscript.

References

1. Kohler, G., and C. Milstein. 1975. Continuous cultures of fused cells secreting antibody of predefined specificity. *Nature* **256:**495.
2. Pilkington, G.R., J. Quirk, T. de Kretser, G.T.H. Lee, W.W. Hancock, N. Kraft, R.C. Atkins, and D.G. Jose. 1984. Analysis of B cell-CALLA and monocyte-granulocyte protocol reagents using immunofluorescence, tissue sections (PAP) and molecular weight determination. In: *Leucocyte typing,* A. Bernard, L. Boumsell, J. Dausset, C. Milstein, and S.F. Schlossman, eds. Springer-Verlag, Berlin, Heidelberg, p. 481–487.
3. Pilkington, G.R., N. Kraft, V. Murdolo, G.T.H. Lee, S. Hunter, R.C. Atkins, and D.G. Jose. 1984. Serological typing of acute leukemia using the monoclonal antibodies PHM1,2,3,6, CIKM5, and the rabbit antisera RARC2a(Ad) and RAALLP50. In: *Leucocyte typing,* A. Bernard, L. Boumsell, J. Dausset, C. Milstein, and S.F. Schlossman, eds. Springer-Verlag, Berlin, Heidelberg, p. 588–595.
4. Vogler, L.B., W.M. Crist, D.E. Bockman, E.R. Pearl, A.R. Lawton, and M.D. Cooper. 1978. Pre-B-cell leukemia. A new phenotype of childhood lymphoblastic leukemia. *New England J. Med.* **298:**872.
5. Brouet, J.C., J.L. Preud'Homme, C. Penit, F. Valensi, P. Rouget, and M. Seligmann. 1979. Acute lymphoblastic leukemia with preB-cell characteristics. *Blood* **54:**269.
6. Amos, D.B., H. Bashir, W. Boyle, M. MacQueen, and A. Tiilikainen. 1969. A simple microcytotoxicity test. *Transplantation.* **7:**220.
7. Pilkington, G.R., G.T.H. Lee, D. O'Keefe, M. Plain, F.C. Wilson, and D.G. Jose. 1980. Classification of childhood acute lymphocytic leukaemia using rabbit antisera to leukaemia cells and lymphoblastoid cell lines. *Aust. J. Exp. Biol. Med. Sci.* **58:**27.
8. Pilkington, G.R., W.W. Hancock, S. Hunter, D.J. Jacobs, R.C. Atkins, and D.G. Jose. 1984. Monoclonal anti-T-cell antibodies react with circulating myeloid leukemia cells and normal tissue macrophages. *Pathology.* **16:**447.
9. Barnstable, C.J., W.F. Bodmer, G. Brown, G. Galfre, C. Milstein, A.F. Williams, and A. Zeigler. 1978. Production of monoclonal antibodies to group A erythrocytes, HLA and other human surface antigens—new tools for genetic analysis. *Cell* **14:**9.
10. De Kretser T.A., M.C. Crumpton, J.G. Bodmer, and W.F. Bodmer. 1982. Demonstration of two distinct light chains in HLA-DR-associated antigens by two-dimensional gel electrophoresis. *Eur. J. Immunol.* **12:**214.
11. Bernard, A., L. Boumsell, and C. Hill. 1984. Joint report of the First International Workshop on Human Leucocyte Differentiation Antigens by the investigators of the participating laboratories. In: *Leucocyte typing,* A. Bernard, L. Boumsell, J. Dausset, C. Milstein, and S.F. Schlossman, eds. Springer-Verlag, Berlin, Heidelberg, p. 9–124.
12. Abramson, C., J. Kersey, and T. Le Bien. 1981. A monoclonal antibody

(BA-1) primarily reactive with cells of human B lymphocyte lineage. *J. Immunol.* **126:**83.
13. Becker, G.J., W.W. Hancock, N. Kraft, H.C. Lanyon, and R.C. Atkins. 1981. Monoclonal antibodies to human macrophage and leukocyte common antigens. *Pathology* **13:**669.
14. De Kretser, T.A., G.T.H. Lee, H.J. Thorne, and D.G. Jose. Monoclonal antibody CI-PANHU defines a pan-human cell-surface antigen unique to higher primates. *J. Immunol. Methods* (Submitted).
15. Brochier, J., D. Schmitt, E. Yonish-Rouach, G. Codier, and J. Viac. 1984. Use of tissue distribution studies to determine the specificity of monoclonal antilymphocyte antibodies. In: Leucocyte typing, A. Bernard, L. Boumsell, J. Dausset, C. Milstein, and S.F. Schlossman, eds. Springer-Verlag, Berlin, Heidelberg, p. 465–469.
16. Zola, H., J.G. Bradley, D.A. Brooks, P.J. Macardle, P.J. McNamara, H.A. Moore, and A. Nikoloutsopoulos. 1984. The human B cell lineage studied with monoclonal antibodies. In: *Leucocyte typing,* A. Bernard, L. Boumsell, J. Dausset, C. Milstein, and S.F. Schlossman, eds. Springer-Verlag, Berlin, Heidelberg, p. 363–371.
17. Barr, I.G., W.W. Hancock, N. Kraft, B.H. Toh, and R.C. Atkins. 1984. PHM14: A novel monoclonal antibody that reacts with both normal and neoplastic human B cells but not B-CLL. *Scand. J. Haematol.* **33:**187.
18. Pilkington, G.R., N. Kraft, G.T.H. Lee, W.W. Hancock, R.C. Atkins, and D.G. Jose. 1983. PHM6 a monoclonal antibody to common acute lymphocytic leukaemia antigen (CALLA): Analysis of human leukemias. *Med. Ped. Oncol.* **11:**200 (abstract).
19. Reinherz, E.L., P.C. Kung, G. Goldstein, R.H. Levey, and S.F. Schlossman. 1980. Discrete stages of human intrathymic differentiation: Analysis of normal thymocytes and leukemic lymphoblasts of T lineage. *Proc. Natl. Acad. Sci. U.S.A.* **77:**1588.
20. Meyer, B.F., V.M. Chugg, R.P. Herrmann, and R.E. Davis. 1983. A unique T-cell monoclonal antibody with potential uses in autologous bone marrow transplantation. *Pathology* **15:**315.
21. McMichael, A.J., N.A. Rust, J.R. Pilch, R. Sochynsky, J. Morton, D.Y. Mason, C. Ruan, G. Tobelem, and J. Caen. 1981. Monoclonal antibody to human platelet glycoprotein I. I. Immunological studies. *Brit. J. Haematol.* **49:**501.
22. Edwards, P.A. 1980. Monoclonal antibodies that bind to the human erythrocyte-membrane glycoproteins glycophorin A and Band 3. *Biochem. Soc. Trans.* **8:**334.
23. Hirt, A., C. Baumgartner, H.K. Forster, P. Imbach, and H.P. Wasner. 1983. Reactivity of acute lymphoblastic leukemia and normal bone marrow cells with the monoclonal anti-B-lymphocyte antibody, anti-Y29/55. *Cancer Res.* **43:**4483.
24. Third International Workshop on Chromosomes in Leukemia (Lund, Sweden) 1981. Chromosomal abnormalities in acute lymphoblastic leukemia: Structural and numerical changes in 234 cases. *Cancer Genet. Cytogenet.* **4:**101.

CHAPTER 30

Immunological Classification of "Unclassifiable" Acute Leukemia

Friedhelm Herrmann, Bernd Dörken, Annette Gatzke, and Wolf Dieter Ludwig

Introduction

Cytochemical and immunological markers have helped to elucidate the ontogeny of hematopoietic cells and have improved the subclassification of acute leukemia (reviewed in Refs. 1–3). Nonetheless, a small fraction (5–10%) of acute leukemias lack expression of standard cytochemical and immunological markers and have no morphological features providing evidence for cell lineage (4). Such cases—usually referred to as "acute unclassifiable leukemia" (AUL)—may reflect leukemic counterparts of the earliest hematopoietic stem cells. The development of monoclonal antibodies (mAbs) which include earlier differentiation stages in their reactivity may be helpful in determining cell type affiliation of such leukemic cells.

In the present study cryopreserved blasts from 49 patients initially considered to have AUL based on morphology, cytochemistry, and immunological marker analysis were reinvestigated using a battery of recently developed mAbs. The data suggest that unclassifiable leukemias are extremely rare. A scheme relating leukemic cells to counterpart primitive hematopoietic progenitor cells is proposed.

Material and Methods

Patients

Of 546 cases with acute leukemia (438 children, 108 adults) diagnosed between April 1982 and May, 1984, 338 were considered to have ALL (301 children, 37 adults) and 159 (100 children, 59 adults) to have AML. In the remaining 49 cases (38 children, 11 adults) the diagnosis of AUL was established by virtue of their morphological appearance on Wright–

Giemsa staining, lack of specific cytochemical reaction pattern using MPO, AP, PAS, and ANAE staining, the absence of reactivity using mAbs towards cALL antigen (J5) (5), T antigens [T11 (6), L17F12 (7), NA134 (8)], and granulo-mono-erythroid antigens [OKM1 (10), VIM-D5 (9), VIE-G4 (11)]. In none of these cases could surface Ig or cytoplasmic μ be detected (data not shown).

Cells

Cryopreserved marrow or blood blasts were available from all 49 patients and were thawed in the presence of DNAs, 100 μg/ml (Worthington Biochemicals, Freehold, NJ), washed twice, and centrifuged in Ficoll–Hypaque resulting in a recovery of more than 85% viable blasts per sample.

Monoclonal Antibodies

mAbs used for the rephenotyping study are listed in Table 30.1. For more detailed information about mAbs selected, see the references given. In addition to the mAbs listed, some other anti-B mAbs [B1 (19), B2 (20), HD28 (this volume, Chapter 46), Y29/55 (21)] and mAbs directed against platelet glycoproteins [AN51, J15 (22)] were tested but did not react with any of the 49 leukemias (data not shown). In all cases expression of cALL and Ia antigens were reassessed using mAbs J5 (5) and L243 (23). Binding of mAbs was detected by FITC-conjugated goat anti–mouse IgG and IgM (Tago, Burlingame, CA) and evaluated using immunofluorescence microscopy. Cytofluorographic analysis (FACS I, Becton Dickinson, Mountain View, CA) was also done in selected samples. Controls were performed using isotype identical irrelevant ascites. TdT estimation was done on methanol-fixed cytospins according to standard procedures (24) using a commercially available test kit (BRL, Bethesda, MA).'

Table 30.1. Monoclonal antibodies selected.

mAb	Immunogen	M.W. (Kd)	Reference
WT1	Thymocytes	40	Tax (12)
4H9	T-ALL	40	Link (13)
BA1	Pre-B ALL line (Nalm-6-MI)	45/55/65	Abramson (14)
BA2	Pre-B All line (Nalm-6-MI)	24	Kersey (16)
VIB-C5	cALL line (Reh-6)	NR[a]	Knapp (16)
HD37	Hairy cell leukemia	NR	Pezzutto (this volume, Chapter 33)
HD6	Hairy cell leukemia	NR	Moldenhauer (this volume, Chapter 7)
MY7	AMML cells	150/160	Griffin (17)
MY9	Myeloid CML-BC cells	70	Griffin (18)

[a] NR: Not reported.

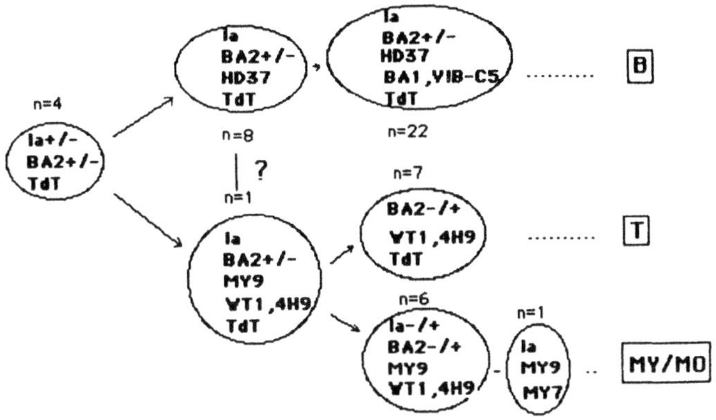

Fig. 30.1. Hypothetical scheme relating leukemic cells to primitive counterpart hematopoietic cells. n = Number of cases expressing each phenotype.

Results

The results are summarized in Tables 30.2–30.5. A schematic overview is given in Fig. 30.1. All but four cases had an identifiable lineage affiliation. Seven cases had a myeloid and 38 a lymphoid phenotype. The ALL cases were composed of eight cases with an early thymic phenotype and 30 cases which were assigned to the presumed early B lineage. One quarter of the latter group was Ia, TdT positive with a variable expression of the BA2-defined p24 antigen. The B restriction of these cases was documented by HD37 expression. The remaining part resembled a more ad-

Table 30.2. "AUL" cases typed as early B lineage-associated ALL.

mAb	No. tested	No. positive	Percent positive (range)
WT1	24	0	0–11
4H9	30	0	0–15
B1	30	0	1–12
BA1	23	18	43–90
BA2	21	13	31–84 (0–5[a])
VIB-C5	30	22	45–90
Ia	30	29	29–99 (0[a])
J5	30	0	0
HD37	30	30	45–87
HD6	21	2	0–3 (48–62[b])
MY7	30	0	0–1
MY9	22	0	0–5
TdT	30	30	16–99

[a] Results of negative cases.
[b] Results of positive cases.

Table 30.3. "AUL" cases typed as early T-ALL.

mAb	No. tested	No. positive	Percent positive (range)
WT1	8	8	71–97
4H9	8	8	72–97
B1	8	0	0–2
BA1	8	0	0–4
BA2	5	2	29–90 (0–1[a])
VIB-C5	4	0	0–2
Ia	8	1	0–3 (49[b])
J5	8	0	0
HD37	5	0	0–2
HD6	4	0	0–1
MY7	8	0	0–3
MY9	8	1	0–2 (69[b])
TdT	8	8	78–98

[a] Results of negative cases.
[b] Results of positive cases.

vanced stage within the pre-B development as manifested by the expression of the BA1- and VIB-C5-defined antigens (Table 30.2). A further seven TdT-positive cases were Ia negative. Based upon their reactivity with WT1 and 4H9—both shown to have a broad spectrum of T-ALL recognition (4,13,25)—these cases were retyped as early T-ALLs. One of the WT1, 4H9, TdT positive cases coexpressed Ia antigens as well as the myelo-monocytic marker MY9. We have no definitive proof that this phenotype represents an early T-ALL. However, there are several other

Table 30.4. "AUL" cases typed as AML.

mAb	No. tested	No. positive	Percent positive (range)
WT1	7	6	53–81 (18[a])
4H9	6	6	57–81
B1	7	0	0–3
BA1	4	0	0–8
BA2	4	2	30–52 (3–7[a])
VIB-C5	5	0	2–8
Ia	7	1	1–9 (65[b])
J5	7	0	0
HD37	6	0	0–2
HD6	4	0	0–1
MY7	7	1	2–6 (52[b])
MY9	7	7	48–89
TdT	7	0	0

[a] Results of negative cases.
[b] Results of positive cases.

Table 30.5. "AUL" cases without lineage affiliation.

mAb	No. tested	No. positive	Percent positive (range)
WT1	4	0	2–4
4H9	4	0	2–5
B1	4	0	0–2
BA1	4	0	0–7
BA2	4	2	38–84 (0[a])
VIB-C5	3	0	1–6
Ia	4	3	43–92 (3[a])
J5	4	0	0
HD37	4	0	0–2
HD6	2	0	0–1
MY7	4	0	0–2
MY9	4	0	0–5
TdT	4	3	79–86 (0[a])

[a] Results of negative cases.

cases with similar phenotype described (25) presenting with mediastinal masses and/or focal AP activity, which led us categorize this case within the T-ALL group (Table 30.3). Six further cases (all children) showed a similar composite WT1, 4H9, MY9 phenotype. In contrast to the former cases they were Ia negative and repeatedly TdT negative. Furthermore, the lack of any clinical or morphological clues for a lymphoid origin suggested that these cases represented a distinct subset of AML. Indeed, this assumption has been supported by the subsequent clinical courses. One additional case considered to have AML showed an Ia, MY9, MY7 positive pattern which may place the leukemic clone in a phenotypical relation to normal cells on the CFU-GM stage (Table 30.4). Four further cases remained without recognizable lineage affiliation (Table 30.5). Exposure of blasts from these four leukemias to inducers of cell differentiation (10^{-6}, 10^{-7}, 10^{-8}, 10^{-9} *M* TPA, 1.1% DMSO, 10^{-5}, 10^{-6}, 10^{-7} *M* retinoic acid) did not result in acquisition of lineage-associated surface antigens (data not shown). However, three of these cases showed high numbers of Ia- and TdT-positive cells and two were BA2 positive, at least suggesting a lymphoid nature.

Discussion

Leukemias without distinct morphological, cytochemical, or immunological markers of a specific cell lineage are often termed "acute unclassifiable leukemias." They may be analogous to hematopoietic stem cells and are therefore sometimes referred to as "stem cell leukemia." The task of further subclassifying such leukemias would be facilitated by finding specific markers reactive with these progenitors. Nevertheless, previous monoclonal reagents have not clearly identified the earliest progenitors of a single lineage. Initial attempts to develop such reagents led to mAbs restricted in their reactivity to progenitor cells, but which were not lineage restricted (e.g., BA2 was shown to react with some leukemias tested irrespective of their lineage, including cases without recognizable lineage affiliation). Recent antibodies have shown more restricted patterns of reactivity. Even so, absolute lineage fidelity is uncommon. WT1 and 4H9, reagents of choice in the routine diagnosis of T-ALL, and MY9 in that of AML, are coexpressed on occasional leukemias. BA1 and VIB-C5, recognizing B lineage cells prior to CALLA expression, are known to react with mature granulocytic cells. Nevertheless, despite their lack of lineage specificity and even the translineage reactivity of some of these mAbs, preliminary evidence from this study suggests that these mAbs are extraordinarily useful for leukemia analysis when the composite phenotypes are taken into consideration rather than single antibody reactions. To give an example: Caution should certainly be exercised in identifying T-ALL on the basis of WT1 and 4H9 reactivity since in this as well as in other studies (25,26) CD7 antibodies have been shown in a few instances to react with AML samples. Nonetheless, the composite phenotype WT1$^+$, 4H9$^+$, TdT$^+$, Ia$^-$ is of proven value in the identification of very early (pre-T11) T-ALL.

The six cases of MY9$^+$, WT1$^+$, 4H9$^+$, BA2$^{-/+}$, MY7$^-$, Ia$^-$, TdT$^-$, leukemias assigned within the AML group were of particular interest. Similar cases have been described elsewhere (25). However, it is noteworthy that in four of these six cases the leukemias were congenital and associated with trisomy 21. We speculate that we are dealing with a distinct subset of AML derived from very early myeloid progenitor cells rather than detecting antigens associated with cell activation or proliferation common to different cell lineages.

Taken together, based on the phenotypic analysis of AUL done in this study, a hypothetical model (Fig. 30.1) of the relationship of leukemic cells to the presumed normal counterpart cells is proposed. The most immature cell type showing a variable expression of Ia, TdT, and BA2 gives rise to a Ia$^+$, BA2$^{+/-}$, MY9$^+$, WT1$^+$, 4H9$^+$, TdT$^+$ common progenitor cell for both the T and the myelo-monocytic lineage. In addition, the recent finding of an Ia$^+$, WT1$^+$ leukemia with rearranged μ heavy-chain gene alleles (Chan and Greaves, personal communication) may suggest this cell type can even be a B cell precursor.

However, further studies are needed to clarify the earliest pathways of hematopoietic differentiation. This would be facilitated by further phenotypic characterization of normal counterpart cells and the identification of other substances which are able to induce differentiation of undifferentiated leukemias along the myeloid or lymphoid axis.

Summary

Of 546 cases of acute leukemia (438 children, 108 adults) 338 were diagnosed as ALL and 159 as AML. The remaining 49 were considered to be marker-less, i.e., MPO$^-$, AP$^-$, PAS$^-$, ANAE$^-$, SmIg$^-$, CALLA$^-$, not reactive with anti–T cell antibodies (T11, L17F12, NA134) as well as negative for myeloid/erythroid-associated antigens (OKM1, VIM-D5, VIE-G4), and therefore typed as unclassifiable leukemia. Cryopreserved leukemic blasts of these patients were retrospectively investigated using an extensive panel of additional monoclonal reagents directed against the T (WT1, 4H9), B (B1, B2, VIB-C5, Y29/55, HD6, HD28, HD37), and granulo-mono-megakaryocytic (MY7, MY9, AN51, J15) lineages. Expression of CALLA, Ia, and TdT was reevaluated.

Seven out of 49 cases were shown to have a myeloid and 38 a lymphoid phenotype. The latter comprised 8 cases with an early thymic phenotype and 30 cases with an affiliation to be presumed early B lineage. Only 4 cases remained unclassifiable even after culturing with inducers of cell differentiation such as TPA, DMSO, and retinoic acid.

The data underscore the diagnostic value of immunophenotyping in acute leukemia.

Acknowledgment. This work was supported by Deutsche Krebshilfe and Deutsche Forschungsgemeinschaft (He 1380/1-1).

References

1. Foon, K.A., R.W. Schroff, and R.P. Gale. 1982. Surface markers on leukemia and lymphoma cells: recent advances. *Blood* **60:**1.
2. Nadler, L.M., J. Ritz, J.D. Griffin, R.F. Todd, E.L. Reinherz, and S.F. Schlossman. 1981. Diagnosis and treatment of human leukemias and lymphomas using monoclonal antibodies. *Progr. Hematol.* **7:**187.
3. Ritz, J., and J.D. Griffin. 1983. Cell surface antigens in acute leukemia. In: *Biological responses in cancer,* E. Mihich, ed. Plenum Publishing, New York, pp. 1–21.
4. Greaves, M.F., R. Bell, J. Amess, and T.A. Lister. 1983. ALL masquerading as AUL. *Leuk. Res.* **7:**735.
5. Ritz, J., J.M. Pesando, J. Notis-Mc Conarty, H. Lazarus, and S.F. Schlossman. 1980. A monoclonal antibody to acute lymphoblastic leukemia antigen. *Nature* **283:**583.
6. van Wauwe, J., J. Goossens, W. de Cock, P. Kung, and G. Goldstein. 1981. Suppression of human T-cell mitogenesis and E-rosette formation by the monoclonal antibody OKT IIa. *Immunology* **44:**865.
7. Engleman, E.G., R. Warnke, R.I. Fox, J. Dilley, C.J. Benike, and R. Levy. 1981. Studies of human T lymphocyte antigen recognized by a monoclonal antibody. *Proc. Natl. Acad. Sci. U.S.A.* **78:**1791.

8. McMichael, A.J., J.R. Pilch, G. Galfre, D.Y. Mason, J.W. Fabre, and C. Milsten. 1979. A human thymocyte antigen defined by a hybrid myeloma monoclonal antibody. *Eur. J. Immunol.* **9:**205.
9. Majdic, O., K. Liszka, D. Lutz, and W. Knapp. 1981. Myeloid differentiation antigen defined by a monoclonal antibody. *Blood* **58:**1127.
10. Breard, J., E.L. Reinherz, P.C. Kung, G. Goldstein, and S.F. Schlossman. 1980. A monoclonal antibody reactive with human peripheral blood monocytes. *J. Immunol.* **124:**1943.
11. Liszka, K., O. Majdic, P. Bettelheim, and W. Knapp. 1983. Glycophorin A expression in malignant hematopoiesis. *Am. J. Hematol.* **15:**219.
12. Tax, W.J.M., N. Tidman, G. Janossy, L. Trejdosiewicz, R. Willems, J. Leeuwenberg, T.J.M. de Witte, P.J.A. Capel, and R.A.P. Koene. 1984. Monoclonal antibody (WTI) directed against a T cell surface glycoprotein: characteristics and immunosuppressive activity. *Clin. Exp. Immunol.* **55:**427.
13. Link, M., R. Warnke, J. Finlay, M. Amylon, R. Miller, J. Dilley, and R. Levy. 1983. A single monoclonal antibody identifies T cell lineage of childhood lymphoid malignancies. *Blood* **62:**722.
14. Abramson, C.S., J.H. Kersey, and T.W. LeBien. 1981. A monoclonal antibody (BAI) reactive with cells of human B lymphocyte lineage. *J. Immunol.* **126:**83.
15. Kersey, J.H., T.W. LeBien, C.S. Abramson, R. Newman, R. Sutherland, and M.F. Greaves. 1981. p24: A human leukemia-associated and lymphohemopoietic progenitor cell surface identified with monoclonal antibody. *J. Exp. Med.* **153:**726.
16. Knapp, W., P. Bettelheim, O. Majdic, K. Liszka, W. Schmidmeier, and D. Lutz. 1984. Diagnostic value of monoclonal antibodies to leukocyte differentiation antigens in lymphoid and nonlymphoid leukemias: In: *Leucocyte typing,* A. Bernard, L. Boumsell, J. Dausset, C. Milstein, and S.F. Schlossman, eds. Springer-Verlag, Berlin, Heidelberg, pp. 564–572.
17. Griffin, J.D., J. Ritz, R. Beveridge, J.M. Lipton, J.F. Daley, and S.F. Schlossman. 1983. Expression of MY7 antigen on myeloid precursor cells. *Int. J. Cell. Clon.* **1:**33.
18. Griffin, J.D., D. Linch, K. Sabbath, P. Larcom, and S.F. Schlossman. 1984. A monoclonal antibody reactive with normal and leukemic human myeloid progenitor cells. *Leuk. Res.* **8:**521.
19. Stashenko, P., L.M. Nadler, R. Hardy, and S.F. Schlossman. 1980. Characterization of a human B lymphocyte-specific antigen. *J. Immunol.* **125:**1678.
20. Nadler, L.M., P. Stashenko, R. Hardy, A. van Agthoven, C. Terhorst, and S.F. Schlossman. 1981. Characterization of a human B-cell specific antigen (B2) distinct from BI. *J. Immunol.* **126:**1941.
21. Forster, H.K., F.G. Gudat, M.F. Girard, R. Albrecht, J. Schmidt, C. Ludwig, and J.P. Obrecht. 1982. Monoclonal antibody against a membrane antigen characterizing leukemic human B lymphocytes. *Cancer Res.* **42:**1927.
22. McMichael, A.J., N.A. Rust, J.R. Pilch, R. Sochynsky, J. Morton, D.Y. Mason, C. Ruan, G. Tobelem, and J. Caen. 1981. Monoclonal antibody to human platelet glycoprotein I.I: Immunological studies. *Brit. J. Haematol.* **49:**501.

23. Lampson, L.A., and R. Levy. 1980. Two forms of Ia molecules on human B cell line. *J. Immunol.* **125:**293.
24. Bollum, F.J. 1979. Terminal deoxynucleotidyl transferase as a hematopoietic cell marker. *Blood* **54:**1203.
25. Vodinelich, L., W. Tax, Y. Bai, S. Pegram, P. Capel, and M.F. Greaves. 1983. A monoclonal antibody (WT1) for detecting leukemias of T-cell precursors (T-ALL). *Blood* **62:**1108.
26. Royston, I., J. Minowada, T.W. LeBien, G. Pavlov, G. Vosika, C. Bloomfield, and R.R. Ellison. 1984. Phenotype of adult acute lymphoblastic leukemia defined by monoclonal antibodies. In: *Leucocyte typing,* A. Bernard, L. Boumsell, J. Dausset, C. Milstein, and S.F. Schlossman, eds. Springer-Verlag, Berlin, Heidelberg, pp. 558–564.

CHAPTER 31

Age Predilection of Distinct Phenotypically Defined Subgroups of Non-T Cell ALL: Studies Using B Cell-Restricted or -Associated Monoclonal Antibodies in 359 Patients

Friedhelm Herrmann, Bernd Dörken, Wolf Dieter Ludwig, and Hansjörg Riehm

Introduction

Only a small fraction (15–20%) of non-T cell acute lymphoblastic leukemia (non-T ALL) cases express the classic markers of the B cell, i.e., cytoplasmic μ and monoclonal membrane Ig (1,2). Nevertheless, phenotype studies using B lineage-restricted monoclonal antibodies (3,4) and *in vitro* differentiation of CALLA-positive blasts to more mature B cells (5,6) as well as genetic probes to show gene rearrangement of Ig heavy and light chains (7) suggest that a substantial number of non-T ALLs may be committed to the B lineage.

In this study, we have investigated the cellular origin of non-T ALL in a large number of both pediatric and adult cases using B lineage-restricted or -associated monoclonal antibodies.

Materials and Methods

Patients

Leukemic cells from 359 out of 428 consecutively admitted, immunologically evaluable cases of acute leukemia (383 children, 45 adults) were classified as being *de novo* non-T ALL. Diagnosis was made in all cases using standard morphological and cytochemical criteria as well as lack of reactivity with anti-T (WT1, 4H9, T11, Na134, L17F12) and anti-myelomonocytic antibodies (MY7, MY9, VIMD5, MOP9, OKM1). In addition 13 cases (9 children, 4 adults) of relapsed non-T ALL were studied.

Cells

Isolated (on Ficoll–Hypaque density gradients) malignant bone marrow or peripheral blood blasts were studied either fresh or on cryopreserved cells thawed in the presence of DNAs, 100 μg/ml (Worthington Biochemicals, Freehold, NJ). The criterion for immunological evaluability was that in all cases malignant cells made up more than 75% of the viable population tested.

Monoclonal Antibodies

Monoclonal antibodies (mAbs) listed in Table 31.1 were used in all 359 cases phenotyped except HD37 which has been tested in 71 patients. For more detailed information about mAbs selected, see references given. In addition, antibodies from the B cell and Leukemia Workshop panel were assessed on blasts from 45 patients. Binding of mAbs was detected by FITC-conjugated goat anti–mouse IgG + IgM (Tago, Burlingame, CA) and evaluated using immunofluorescence microscopy. In selected samples cytofluorometric analysis has been performed using a FACS I (Becton Dickinson, Mountain View, CA). Controls were done using isotype identical irrelevant ascites.

Cytoplasmic μ and Membrane Ig

Enumeration of blasts containing cytoplasmic μ and expressing membrane Ig was performed according to standard procedures as previously described (8).

Results and Discussion

By their differential expression and coexpression of B cell restricted and associated antigens five phenotypic subgroups could be identified which included 352 out of 359 cases with non-T ALL (Table 31.2). Antigen

Table 31.1. Monoclonal antibodies selected.

mAb	Antigen detected	Reference
L243	HLA-DR (Ia)	Lampson (9)
J5	CALLA	Ritz (10)
VIB-C5	B-associated	Knapp (11)
B1	B-restricted	Nadler (3) Stashenko (12)
Y29/55	B-restricted	Forster (13)
HD37	B-restricted	Pezzutto (this volume, Chapter 33)

Table 31.2. Phenotypically defined subgroups of non-T ALL.

Subgroup type	Phenotype	No. of Non-T ALLs
I	Ia	11 (3.1%)
II	Ia,VIB-C5	64 (18.2%)
III	Ia,VIB-C5,CALLA	153 (43.5%)
IV	Ia,VIB-C5,CALLA,B1	108 (30.7%)
V	Ia,VIB-C5,CALLA,B1,Y29/55	16 (4.5%)

density (as estimated by fluorescence intensity) varied from case to case within each subgroup type, suggesting that these subgroups may not represent distinct stages of differentiation, but rather may reflect a continuum of malignant transformation of several possible steps of B cell ontogeny.

Another B-restricted mAb termed HD37 was tested in 71 patients (Table 31.3). HD37 reacted with all non-T ALLs from type II to V. It was of interest to note that one out of five type I leukemias tested was HD37 positive, suggesting that some of the subgroup type expressing Ia alone may be already committed to differentiate towards the B lineage.

We also studied the presence of cytoplasmic μ (tested in 55 patients) and monoclonal membrane Ig expression (tested in all patients) (Table 31.4). Our data revealed that within type I–III, cytoplasmic μ was not detectable. In eight of 25 cases tested within type IV and in one case of the type V group, blasts expressed cytoplasmic μ. Monoclonal membrane Ig was seen in all type V leukemias tested, thus allowing us to suggest that the acquisition of cytoplasmic μ may occur between types IV and V. Phenotypes of seven of 359 patients did not conform with the proposed subgroup types and may represent aberrant antigen expression rather than distinct subgroup types: four patients of those showed blasts expressing Ia, VIB-C5, and B1 and lacked CALLA and Y29/55. Blasts from two patients expressed Ia and CALLA but lacked VIB-C5, B1, and Y29/55. One Ia-negative case expressed only CALLA and VIB-C5.

Those phenotypic subgroups provided a framework for us to study the clinical and biological heterogeneity of non-T ALLs in more detail.

In light of the well-known fact that adults have significantly worse prognosis than children, it was of considerable interest to note that when

Table 31.3. Reactivity of HD37 with subgroup types of non-T ALL.

Subgroup type	No. positive/no. tested
I	1/5
II	20/20
III	20/20
IV	20/20
V	6/6

Table 31.4. Presence of cytoplasmic μ and membrane Ig in different subgroup types of non-T ALL.

Subgroup type	cμ No. positive/no. tested	mIg No. positive/no. tested
I	0/2	0/11
II	0/7	0/64
III	0/15	0/153
IV	8/25	0/108
V	1/6	16/16

cases were divided into three groups by age (A = under 2 years, B = 2–18 years, C = more than 18 years) adult leukemia tended to cluster in subgroup type IV (55.9% of all leukemias tested in group C) (Table 31.5). Leukemia in the age group under 2 years, which has been shown to have a worse prognosis than leukemia in children between 2 and 10 years (14,15), predominated in subgroup II (64.4% of all leukemia tested in group A). The most common leukemia resided within subgroup type III and age group B (53.3% of all leukemias tested in group B). This difference in subgroup frequency between groups A and C may be related to the predominant pre–B cell in the bone marrow at the moment of leukemogenesis. This assumption is supported by recently available data suggesting that subgroup type II cells correspond to the pre–B cell stage predominating in the fetal marrow (16) whereas subgroup type IV cells resemble pre–B cells in the adult marrow (17).

It is too early to evaluate possible differences in prognosis in these groups since most of the treatment protocols of our patients are still open. However, with respect to the known prognostic differences outlined above and the subtype-related age clusters demonstrated, it is reasonable to anticipate that these phenotype groups may bear prognostic significance.

In addition, the observation that 10 out of 13 relapsed leukemias tested (Table 31.6) clustered in subgroups IV and V may lend further support to this supposition. When correlating the blast morphology as assessed by

Table 31.5. Age distribution of subgroup types of non-T ALL.

	No. of patients in each age group		
Subgroup type	<2 years (A)	2–18 years (B)	>18 years (C)
I	2 (3.4%)	7 (2.7%)	2 (5.9%)
II	38 (64.4%)	21 (8.1%)	5 (14.7%)
III	11 (18.7%)	138 (53.3%)	4 (11.8%)
IV	7 (11.8%)	82 (31.7%)	19 (55.9%)
V	1 (1.7%)	11 (4.2%)	4 (11.7%)

Table 31.6. Subgroup types of relapsed leukemias.

Subgroup type	No. positive/no. tested
I	0/13
II	2/13
III	1/13
IV	7/13
V	3/13

the FAB type (18) with the phenotypic subgroups (Table 31.7), 17 out of 26 leukemias of FAB type L2 and 15 out of 16 leukemias of FAB type L3 clustered in subgroup types IV and V, respectively. No apparent clustering within any subtype group was observed when evaluating PAS reactivity (Table 31.8). PAS-positive leukemias were found in all subgroups to a variable extent, except in type V leukemia.

Hematological data (initial platelet counts, Hb values, and peripheral blast counts) are tabulated according to subgroup type in Table 31.9. Again, lowest initial platelet counts and Hb values were seen in type IV. In contrast, highest initial blast counts were found in types I and II. However, the latter data must be interpreted with some caution because of the small number tested within these groups. Nevertheless, they are in line with data from other studies (19,20).

Taken together, the data reported here provide further evidence to support the notion that non-T ALLs are B lineage-derived malignancies. Non-T ALLs are heterogeneous and may be clustered in clinical and biological relevant groups using B cell-restricted or -associated mAbs. This has been further documented when analyzing non-T ALL subgroups with Workshop B cell and leukemia panel antibodies. Results of 45 completely phenotyped cases—representatives from each subgroup type (except type V)—are depicted in Table 31.10, which demonstrates the expression of the listed Workshop antibodies within the described subgroup types.

Table 31.7. FAB-type distribution of subgroup types of non-T ALL.

		FAB		
Subgroup type	*n*	L1	L2	L3
I	7	5	2	0
II	25	21	4	0
III	59	57	2	0
IV	43	26	17	0
V	16	0	1	15
Total	150	109	26	15

Table 31.8. PAS reactivity in non-T ALL subgroup types.

Subgroup type	PAS positive/no. tested
I	1/7
II	7/25
III	34/59
IV	21/43
V	0/16

The detection of relevant subgroups by immunophenotyping shows that this procedure is an important factor in the management of ALL.

Summary

In the present study the expression of B cell-restricted or -associated antigens on leukemic cell samples from 359 patients with non-T ALL was determined using a panel of monoclonal antibodies (mAbs). All cases were unreactive with anti-T and anti-myeloid/monocytic mAbs. According to their reaction pattern, all but seven cases could be clustered in five different phenotypes: An Ia positive only (Type I, n = 11), an Ia,VIB-C5 positive (Type II, n = 64), an Ia,VIB-C5,CALLA positive (Type III, n = 153), an Ia,VIB-C5,CALLA,B1 positive (Type IV, n = 108), and an Ia,VIB-C5,CALLA+/−,B1,Y29/55 positive (Type V, n = 16). Another mAb (HD37) tested on samples from 71 patients reacted with all Type II–V leukemias. When cases were derived into three groups by age (A = <2 years, B = 2–18 years, C = >18 years), our data revealed that Type IV and V were most common in Group C (56% and 12%, respectively, of cases in this group). Type II predominated in the Group A (64%) and Type III in the Group B (53%) cases. Type I showed no age predilection. Samples from relapsed leukemias tended to cluster in Type IV leukemia irrespective of patient's age suggesting that a stage-related phenotyping may be helpful in explaining survival differences.

Table 31.9. Initial platelet count (PC), Hb value (Hb), and peripheral blast count (PBC) in non-T ALL subgroup types.

Subgroup type	PC(<50 × 10^9/liter) No.+/no. tested	Hb(<8g/dl) No.+/no. tested	PBC(>25 × 10^9/liter) No.+/no. tested
I	2/4	0/4	3/4
II	6/15	3/15	7/15
III	32/58	28/58	14/58
IV	32/41	23/41	11/41
V	2/9	1/9	0/9

Table 31.10. Clustering of B cell and leukemia panel workshop antibodies in subgroup types of non-T ALL.

Subgroup type	Expression of Workshop antibodies
I	B1, 50, 51, L8, 12
II	B8, 14, 15, 18, 28, 34, 43, 47, 48, L17
III	B12, 38, 44, L2, 4, 6, 10, 11, 14, 15, 18, 21, 22
IV	B5, 22, 24
V	not tested

Further data are presented to correlate these phenotypes with those obtained from studies using the B and Leukemia Workshop panel, with the expression of classical B markers, with morphology (FAB), cytochemistry, and initial hematological data.

Acknowledgment. This work was supported by Deutsche Krebshilfe and Deutsche Forschungsgemeinschaft (He 1380/1).

References

1. Greaves, M.F., W. Verbi, L.B. Vogler, M.D. Cooper, R. Ellis, G. Ganeshaguru, V. Hoffbrand, G. Janossy, and F.J. Bollum. 1979. Antigenic enzymatic phenotypes of the pre-B subclass of acute lymphoblastic leukemia. *Leuk. Res.* **3**:353.
2. Pullen, D.J., J.M. Falletta, W.M. Crist, L.B. Vogler, B. Dowell, G.B. Humphrey, R. Blackstock, J. van Eys, M.D. Cooper, R.S. Metzgar, and E.F. Meydrech. 1981. Southwest oncology group experience with immunological phenotyping in acute lymphoblastic leukemia of childhood. *Cancer Res.* **41**:4802.
3. Nadler, L.M., P. Stashenko, J. Ritz, R. Hardy, J.M. Pesando, and S.F. Schlossman. 1981. A unique cell surface antigen identifying lymphoid malignancies of B cell origin. *J. Clin. Invest.* **67**:134.
4. Nadler, L.M., K.C. Anderson, G. Marti, M.P. Bates, E.K. Park, J.F. Daley, and S.F. Schlossman. 1983. B4, a human B lymphocyte associated antigen expressed on normal mitogen activated and malignant B lymphocytes. *J. Immunol.* **131**:244.
5. Nadler, L.M., J. Ritz, M.P. Bates, E.K. Park, K.C. Anderson, and S.F. Schlossman. 1982. Induction of human B cell antigens in non-T cell acute lymphoblastic leukemia. *J. Clin. Invest.* **70**:433.
6. Cossman, J., S.L.M. Necker, A. Arnold, and S.J. Korsmeyer. 1982. Induction of differentiation in a case of common acute lymphoblastic leukemia. *New England J. Med.* **307**:1251.
7. Korsmeyer, S.J., A. Arnold, A. Bakhshi, J.V. Ravetch, V. Siebenlist, P.A. Heiter, P.O. Sharrow, T.W. LeBien, J.H. Kersey, D.G. Poplack, P. Leder, and T. Waldmann. 1983. Immunoglobulin gene rearrangement and cell surface antigen expression in acute lymphocytic leukemias of T cell and B cell precursor origins. *J. Clin. Invest.* **71**:301.

8. Herrmann, F., and R. Wirthmüller. 1982. Cell surface marker phenotyping in patients with non-Hodgkin lymphomas of low and intermediate malignancy. *Immunobiol.* **163:**77.
9. Lampson, L.A., and R. Levy. 1980. Two forms of Ia molecules on human B cell line. *J. Immunol.* **125:**293.
10. Ritz, J., J.M. Pesando, J. Notis-McConarty, H. Lazarus, and S.F. Schlossman. 1980. A monoclonal antibody to acute lymphoblastic leukemia antigen. *Nature* **283:**583.
11. Knapp, W., P. Bettelheim, O. Maijdic, K. Liszka, W. Schmidmeier, and D. Lutz. 1984. Diagnostic value of monoclonal antibodies to leukocyte-differentiation antigens in lymphoid and non-lymphoid leukemias. In: *Leucocyte typing,* A. Bernard, L. Boumsell, J. Dausset, C. Milstein, and S.F. Schlossman, eds. Springer-Verlag, Berlin, Heidelberg, pp. 564–572.
12. Stashenko, L.M. Nadler, R. Hardy, and S.F. Schlossman. 1980. Characterization of a human B lymphocyte specific antigen. *J. Immunol.* **125:**1678.
13. Forster, H.K., F.G. Gudat, M.F. Girard, R. Albrecht, J. Schmidt, C. Ludwig, and J.P. Obrecht. 1982. Monoclonal antibody against a membrane antigen characterizing leukemic human B lymphocytes. *Cancer Res.* **42:**1927.
14. Riehm, G., H. Gadner, G. Henze, B. Kornhuber, H.J. Langermann, S. Muller-Weihrich, and G. Schellong. 1983. Acute lymphoblastic leukemia: Treatment results in three BFM studies (1970–1981). In: *Leukemia research: Advances in cell biology and treatment,* S.B. Murphy and J.R. Gilbert, eds. Elsevier, New York, pp. 251–260.
15. Sallan, S.E., S. Hitchcock-Bryan, R. Gelber, J.R. Cassady, E. Frei III, and D.G. Nathan. 1983. Influence of intensive asparaginase in the treatment of childhood non-T cell acute lymphoblastic leukemia. *Cancer Res.* **43:**5601.
16. Rosenthal, P., I.J. Rimm, T. Umiel, J.D. Griffin, S.F. Schlossman, and L.M. Nadler. 1983. Characterization of fetal lymphoid tissue by monoclonal antibodies. *J. Immunol.* **131:**232.
17. Hokland, P., L.M. Nadler, J.D. Griffin, and J. Ritz. 1984. Purification of common acute lymphoblastic leukemia antigen positive cells from normal human bone marrow. *Blood* **64:**662.
18. Bennett, J.M., D. Catovsky, M.T. Daniel, G. Flandrin, D.A.G. Galton, H.R. Gralnick, and C. Sultan. 1976. Proposals of the classification of acute leukemias. *Brit. J. Haematol.* **33:**451.
19. Bowman, W.P. S.L. Melvin, R.J.A. Aur, and A.M. Mauer. 1981. A clinical perspective on cell markers in acute lymphocytic leukemia. *Cancer Res.* **41:**4794.
20. Greaves, M.F., G. Janossy, J. Peto, and H. Kay. 1981. Immunologically defined subclasses of acute lymphoblastic leukemia in children: Their relationship to presentation features and prognosis. *Brit. J. Haematol.* **48:**170.

CHAPTER 32

Heterogeneity of B-CLL Cells Defined by Monoclonal Antibodies

Noelle Genetet, Dominique Bourel, Bernard Grosbois, Genevieve Merdrignac, Michele Marty, Renee Fauchet, Francois Lancelin, Robert Leblay, and Bernard Genetet

Introduction

Leukemic cells from patients with B chronic lymphocytic leukemia (B-CLL) are thought to represent immature B lymphocytes, at a differentiation stage intermediate between pre–B cell and mature B lymphocyte, a stage not normally found in the peripheral blood (1–2). The characterization of B-CLL cells by their expression of cytoplasmic (cIg) or monoclonal surface immunoglobulin (SIg) (3) and by monoclonal antibodies directed against antigens restricted or associated to B cell lineage (2) suggests a large degree of heterogeneity. Whether this phenotypic diversity is related to significant heterogeneity of organ localization, degree of bone marrow involvement, disease course, or response to therapy, is not clear. In an attempt to achieve a better B-CLL cell characterization we analyzed 25 CLLs with a panel of monoclonal antibodies defining both B-restricted and B-associated antigens. In addition we tried to relate phenotypic characteristics to the extent of dissemination as measured by biological and clinical criteria, according to Binet's classification (4).

Immunological Study

Patients

All 25 patients (mean age: 68 $\pm$ 9 years) exhibited increased relative (>50%) and absolute (>$8.10^3/\mu l$) lymphocyte counts at the time of analysis [mean lymphocytosis: 78.36 $\pm$ 11.77% and (23.10 $\pm$ 18.10) $\times$ $10^3/\mu l$, respectively]. The cases were divided among the following stages according to Binet's classification based on easily measurable biological and clinical parameters: stage A (N = 16), no anemia or thrombocytopenia and less than three areas of lymphoid enlargement; stage B (N = 6), no anemia or thrombocytopenia with three or more involved areas; stage C ($N = 3$), anemia (Hb < 100 g/l) and thrombocytopenia (platelets < 100 × 10^9/l) regardless of the number of areas of lymphoid enlargement. Five patients were treated prior to sampling and 20 were previously untreated.

Methods

Ficoll–Hypaque mononuclear cells were used either fresh or cryopreserved in −196°C vapor-phase liquid nitrogen in 7.5% dimethyl sulfoxide (DMSO) until the time of characterization. B lineage of CLL cells was defined by the presence of "Ia-like" molecules using anti-HLA-DR nonpolymorphic monoclonal antibody. The percent of T cells was determined using a pan T monoclonal antibody, T11 (Coultronics). Surface immunoglobulin expression was revealed by staining with goat anti–human (H + L) immunoglobulin, anti-IgM and anti-IgG conjugated with fluorescein isothiocyanate (FITC) (Cappel). Isotypic characterization with heavy- and light-chain-specific reagents was not analyzed in all cases. Phenotypes of B-CLL cells were determined using a panel of monoclonal antibodies (Table 32.1) which included (i) *B cell-specific antibodies:* BL 14 (5) reacting with the majority of human B lymphocytes, even at immature stages; BL 13 (5) directed against an antigen present on B cells which do not circulate under normal conditions, and staining normal B cells located only in the germinal centers of the secondary follicles; Y 29.55 (6) specific for tissue B cells and expressed on a small proportion of normal blood B cells; B 121 (unpublished) produced by immunizing mice with normal peripheral blood lymphocytes (PBLs) and detecting a small subset of circulating B cells (<5%) and some B-CLLs; (ii) *Antibody reacting with*

Table 32.1. Characteristics of the monoclonal B cell-restricted and -associated antibodies.

Antibody	Reactivity	Reference
B cell restricted		
BL 14	The majority of human B lymphocytes even at immature stages	Brochier (5)
BL 13	Normal B cells in the germinal centers of secondary follicles	Brochier (5)
Y 29.55	Tissue B cells and a small proportion of normal blood B cells	Forster (6)
B 121	A small subset of normal PBLs, and some B-CLLs	Unpublished
B cell associated		
BA_2	CALLA antigen (p24)	Kersey (7)
49.9	Mature T cells and B cells of most B-CLLs	Unpublished
102.3	Mature T cells and B cells of some B-CLLs	Unpublished

possible precursors of B lymphocytes: BA_2 (CALLA, p24) (Hybritech) (7); (iii) *Antibodies not restricted to the B cell lineage*—"T1-like" (unpublished): 49.9 produced against PHA-activated T cells and recognizing mature T lymphocytes and B cells of most CLLs; 102.3 made against PBLs and showing a similar reactivity pattern, recognizing mature T cells, some B-CLL cells, and a small subset of circulating peripheral B cells (tissue distribution has not yet been established). Cells were analyzed for surface immunoglobulin by direct immunofluorescence and microscopic examination. Staining by monoclonal antibodies was assessed by indirect immunofluorescence using goat anti–mouse IgG–FITC (I. Pasteur) and cytofluorometric analysis (Ortho 50H).

Results

As shown in Fig. 32.1(a) the majority of cells expressed "Ia-like" antigens and BL 14, whilst T11-positive cells were always less than 25% in all the 25 B-CLLs. Surface immunoglobulin was expressed on 17/25 cases. Surprisingly monoclonal IgG was the predominantly expressed isotype: 8 versus 6 IgM-positive B-CLLs. This could be explained by the lack of sensitivity of direct immunofluorescence for assessment of weak amounts of surface IgM and/or by the presence of cytophilic IgG on B-CLL cells. Nonetheless the coexpression of Ia, BL 14, and SIg defined the phenotype of most B-CLLs. A subgroup was identified, lacking detectable SIg.

Phenotypic characterization of B-CLL cells with monoclonal antibodies defining B cell-restricted or -associated antigens is depicted in Fig. 32.1(b). Out of 25 patients, 8 expressed BL 13, but reactivity was often

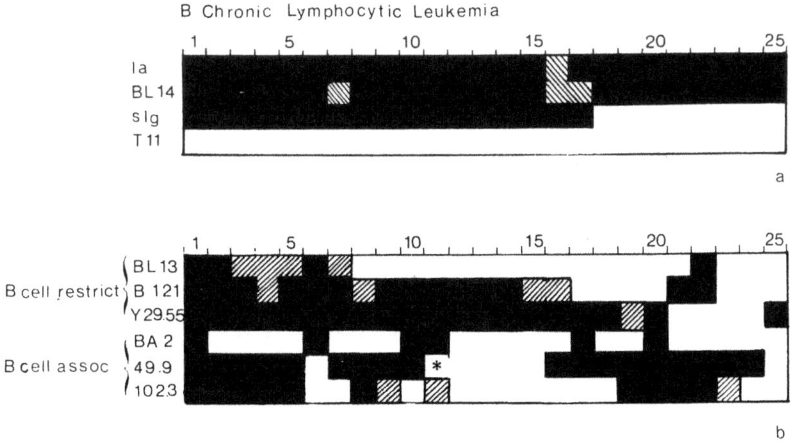

Fig. 32.1. Reactivity patterns of antibodies against 25 CLLs. ■: >50% positive cells; ▨: 25–50% positive cells; □: <25% positive cells; *: not tested.

moderate with a mean percent of positive cells of 49 ± 14%. The majority of B-CLLs were reactive with Y 29.55 (21/25). B 121 showed a similar but more restricted distribution on B-CLL cells (18/25). Both recognized antigens were often simultaneously expressed or lacking. In contrast the CALLA antigen was only present on a small subset (6/25) of B-CLLs. 49.9 was expressed in 18/24 patients, including nearly all the 13 B-CLLs reactive with 102.3.

Discussion

The expression of these B cell-restricted and -associated antigens on leukemia B cells revealed distinct patterns of antigenic expression. Three predominant phenotypes were identified using the B cell-restricted monoclonal antibodies: seven cases were BL 13^+, B 121^+, Y 29.55^+; nine were B 121^+, Y 29.55^+; and five were Y 29.55^+ alone. There was no association between BA_2 expression and one of these phenotypes. Reactivity with 102.3 and/or 49.9 occurred independently of the presence of one or an association of B cell-restricted antigens. None of these antigens seemed to be associated with the degree of maturation of CLL cells, as manifested by surface immunoglobulin expression (Fig. 32.2), except for 102.3: the overwhelming majority of B-CLLs reactive with 102.3 (12/13) expressed surface immunoglobulins, in contrast to one of the 8 SIg-negative B-CLLs. These data in their entirety suggest that the B-CLL cells may be derived from distinct subpopulations of B lymphocytes. Phenotypic diversity of cells within histologically comparable B-CLLs may reflect heterogeneity within the currently defined maturation stages of B cell differentiation.

Whether heterogeneity of organ localization and disease course may be related to this diversity is not yet clear. We did not find any segregation of either surface or no surface immunoglobulin detectable (8) and of heavy-

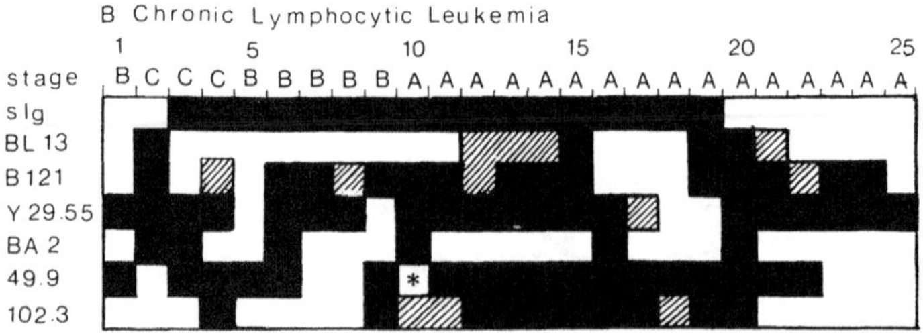

Fig. 32.2. Surface phenotypes of stage A, B, and C B-CLLs with or without detectable surface immunoglobulin. ■: >50% positive cells; ▨: 25–50% positive cells; □: <25% positive cells; *: not tested.

chain isotype (data not shown) (9) with a more advanced clinical stage. The search for correlation between phenotypic characteristics of B-CLL cells and clinical stages revealed two associations with stage A CLLs, the less disseminated stage having the best prognosis (4). BL 13 is expressed in 7/16 stage A CLLs versus 1/9 stage B and C CLLs. 102.3 reacted against relatively more mature B-CLLs, as suggested by the presence of SIg in the majority of these patients (11/12), predominantly within stage A (10/16) in contrast to 1/6 stage B and 1/3 stage C. Additional and longitudinal studies are in progress to provide a better characterization of B-CLL cells and to determine whether antigenic characteristics can indicate which of these patients will develop a progressive or aggressive clinical course.

Acknowledgment. This work was supported by the Association pour la Recherche sur le Cancer.

References

1. Johnstone, A.P. 1982. Chronic lymphocytic leukaemia and its relationship to normal B lymphopoiesis. *Immunology Today* **3**:343.
2. Anderson, K.C., M.P. Bates, B.L. Slaughenhoupt, G.S. Pinkus, S.F. Schlossman, and L.M. Nadler. 1984. Expression of human B cell-associated antigens on leukemias and lymphomas: A model of human B cell differentiation. *Blood* **63**:1424.
3. Han, T., H. Ozer, M. Bloom, K. Sagawa, and J. Minowada. 1982. The presence of monoclonal cytoplasmic immunoglobulins in leukemic B cells from patients with chronic lymphocytic leukemia. *Blood* **59**:435.
4. Binet, J.L., D. Catovsky, P. Chandra, G. Dighiero, E. Montserrat, R. Raik, and A. Sawitsky. 1981. Chronic lymphocytic leukaemia: Proposals for a revised prognostic staging system. *Brit. J. Haematol.* **48**:365.
5. Brochier, J., D. Schmitt, E. Yonish-Rouach, G. Cordier, and J. Viac. 1984. Use of tissue distribution studies to determine the specificity of monoclonal anti-lymphocytes antibodies. In: *Leucocyte typing,* A. Bernard, L. Boumsell, J. Dausset, C. Milstein, and S.F. Schlossman, eds. Springer-Verlag, Berlin, Heidelberg, pp. 465–469.
6. Forster, H.K., F.G. Gudat, M.F. Girard, R. Albrecht, J. Schmidt, C. Ludwig, and J.P. Obrecht. 1982. Monoclonal antibody against a membrane antigen characterizing leukemic human B-lymphocytes. *Cancer Res.* **42**:1927.
7. Kersey, J.H., T.W. Lebien, C.S. Abramson, R. Newman, R. Sutherland, and M. Greaves. 1981. A human hemopoietic progenitor and acute lymphoblastic leukemia-associated cell surface structure identified with monoclonal antibody. *J. Exp. Med.* **153**:726.
8. Hamblin, T., and D. Hough. 1977. Chronic lymphocytic leukemia: Correlation of immunofluorescent characteristics and clinical features. *Brit. J. Haematol.* **36**:359.
9. Ligler, F.S., J.R. Kettman, R. Graham Smith, and E.P. Frenkel. 1983. Immunoglobulin phenotype on B cells correlates with clinical stage of chronic lymphocytic leukemia. *Blood* **62**:256.

CHAPTER 33

HD37 Monoclonal Antibody: A Useful Reagent for Further Characterization of "Non-T, Non-B" Lymphoid Malignancies

Antonio Pezzutto, Bernd Dörken, Alfred Feller, Gerhard Moldenhauer, Reinhard Schwartz, Peter Wernet, Eckhard Thiel, and Werner Hunstein

Introduction

A crucial point for diagnosis and therapy selection in hematological malignancies is the correct characterization of proliferating cells. This is particularly relevant for acute leukemias which require different treatment regimens for myeloid and lymphoid lineage. Moreover a closer characterization within the group of the acute lymphatic leukemias can detect patients with different prognosis and possibly different therapeutic requirements (1,2).

Distinction between cells of myeloid and lymphoid lineage is usually made by cytochemistry (3) but biochemical activities responsible for staining may be absent on immature leukemias. Therefore a number of them may remain "unclassified." The conventional lymphoid markers such as intrinsic surface membrane immunoglobulins and receptors for complement components, for immunoglobulin Fc portions, and for sheep and murine erythrocytes are usually of little value since they recognize mature phenotypes which constitute only a minority (<20%) of acute leukemias (4).

A useful reagent used as a prognostic tool among acute lymphatic leukemias lacking conventional T or B markers has been the common ALL-associated antigen (CALLA) (5) which defines a group of patients with a less aggressive clinical course. The finding that about 20% of patients with $CALLA^+$ non-T ALL express intracytoplasmic μ chains has suggested that a number of patients within this group may in fact have a pre–B cell tumor (6,7). This has been further supported by studies based on gene rearrangements (8,9), and by the demonstration of a monoclonal antibody-defined B cell determinant (B1) on 50% of non-T cell ALL (10). The majority of lymphoid blast crises of chronic myeloid leukemia (CML) also express a B-related phenotype (9,11).

The application of new monoclonal antibodies (mAbs) to the characterization of ALL has provided new insights about the origin of the leukemic cells and also about normal hematopoietic differentiation (12–19). In the present report we describe the generation and characterization of a B cell-specific broadly reacting mAb named HD (Heidelberg) 37 which proves to be a very useful reagent for the characterization of lymphoid malignancies previously reported as non-T, non-B. This antibody is present on the large majority of cells in a series of 51 ALL cases. Moreover the combined use of HD37 and a broadly reacting T cell marker such as Tü14 (WT56 in the First Workshop on mAbs) allowed us to dissect the whole group of lymphoblastic lymphomas into T or B lineage malignancies.

Material and Methods

Hybridoma Production and Screening

Spleen cells from BALB/c mice immunized with cells obtained from a hairy cell leukemia patient were fused with NS1-Ag4/1 myeloma cells according to classical procedure (20). Supernatants of clones were screened by an immunoenzymatic assay on cells plated in Terasaki plates (21) and by standard immunofluorescence methods.

Other mAbs were also used in this study. Anti-I2 (Coulter) (22) defines a nonpolymorphic Ia-like antigen expressed on B cells, monocytes, and activated T cells. Tü14 is a broadly-reacting T cell marker which has been characterized during the First International Workshop and Conference on Human Leucocyte Differentiation Antigens held in Paris (Workshop number WT56). It was clustered together with mAb 3A1 and seems to define the same antigen (17). T6 is a T cell marker which is not present on normal circulating peripheral blood cells; it reacts with about 70% of thymocytes and is found on a number of T-ALLs (23). VIL-A1 (a gift from Dr. Knapp, Vienna) (24) recognizes the CALLA antigen (gp100) which is expressed on leukemic cells in about 80% of non-T ALLs, in the majority of lymphoid blast crises of CML, and in a number of non-Hodgkin's lymphomas. This antigen is also found in a small population of cells in normal bone marrow and fetal liver. HD28 is a B cell-associated antigen found on mature stages of B cell differentiation, but weakly expressed also on T cells and on cells of the myelo-monocytic lineage (25). B1 (Coulter) identifies a unique B cell surface determinant present on all normal B lymphocytes (10). B2 (Coulter) is lacking in early members of the B lineage, being associated with more mature phenotypes (12). BA1 is a B-associated marker which is expressed on a high proportion of non-T ALLs but is also found on neutrophils and non-lymphoid cell lines (13).

Preparation of Target Cells

Human peripheral blood mononuclear cells (PBMC) were isolated on Ficoll–Hypaque (F/H) density gradients (26). Granulocytes were prepared from the cell pellet; further purification was done by Dextran sedimentation followed by NH_4 lysis of erythrocytes. Monocytes were obtained by removing cells adherent to plastic petri dishes. T cells were isolated by rosetting PBMC with neuraminidase-treated sheep red blood cells ($SRBC_n$). Repeated F/H gradients were performed until more than 95% of rosettes were counted in the pellet. In some experiments the E^- preparation was further enriched in B cells by removing contaminating monocytes by passage through a Sephadex G 10 column (27). Bone marrow mononuclear cells were isolated from heparinized marrow aspirates by F/H gradient centrifugation. Tonsil, thymus, and spleen cells were recovered by gentle teasing of surgically removed material and subsequent F/H centrifugation. Leukemic cells were prepared from peripheral blood when leukemic involvement was present. If involved, bone marrow was analyzed. Histopathological diagnoses were determined according to Kiel classification. Clinical criteria together with routine hematological techniques including cytochemistry were used for diagnosis.

In a number of cases cryopreserved cells from patients with acute leukemias were evaluated after thawing and F/H centrifugation to remove dead cells. Tumor cells in every case accounted for more than 70% of cell samples.

Indirect Immunofluorescence

Cell suspensions were analyzed on a Zeiss epiilluminated fluorescence microscope (200 cells were counted) or by flow cytometry (Ortho Diagnostic System) after incubation with mAbs and FITC-conjugated goat anti–mouse IgG/IgM (Tago). Double-marker analysis was performed by further incubation with TRITC-conjugated goat anti–human polyvalent Ig (Nordic).

Immunohistological Study

Cryostat sections of snap-frozen material were stained by using either alkaline phosphatase-conjugated second and third antibodies (goat anti-mouse and rabbit anti-goat, Sigma) or an indirect AP–AAP method (rabbit anti-mouse followed by an alkaline phosphatase–monoclonal anti-alkaline phosphatase immune complex produced in the mouse). Reactions were developed by incubation with naphthol AS-BI phosphate/Fast Red TR salt in the presence of levamisole (28).

Cellular Radioimmunoassay

A cellular radioimmunoassay (CRIA) was performed on viable cells at 10^6/well in microtiter plates. Reactivity with mAbs was measured by binding of ^{125}I-labeled rabbit anti-mouse Ig.

Results

BALB/c mice were immunized with tumor cells isolated from a patient with a hairy cell leukemia. After somatic cell hybridization, several clones reacting with immunizing cells were established. By means of an immunoenzymatic staining assay developed in our laboratory, supernatants were screened against a variety of cell preparations including PBMC, enriched T cells, monocytes, and leukemia/lymphoma cells of B origin. The antibody named HD37 proved to be B cell specific in these preliminary tests and was further characterized by different techniques. It is a IgG1 immunoglobulin; the molecular weight of the corresponding antigen has not yet been characterized.

Reactivity of HD37 with Normal Hematopoietic Cells

The antibody has proven to be B lineage specific by evaluation on human peripheral blood elements including PBMC, E-rosette-enriched T cells,

Table 33.1. Expression of HD37 on normal cells (indirect immunofluorescence).

Cells	Number tested	HD37
Peripheral blood		
Mononuclear cells	11	5, 7 (3–14)[a]
B-enriched cells[b]	2	67 (64–70)
SIg^+ cells[c]	3	89 (82–96)
T cells (E^+)	5	1 (0–2)
Monocytes	3	0
Granulocytes	3	0
Erythrocytes	3	0
Platelets	3	0
Bone marrow[d]	8	
Lymphoid cells		5 (1–8)
Myeloid cells		0
Erythroid cells		0
Thymus	3	1 (0–2)
Tonsil	3	59 (46–72)
Spleen	1	53

[a] Percentage of positive cells: mean (range).
[b] E^- fraction depleted of monocytes by passage through Sephadex G 10 (70% SIg^+ cells).
[c] Double-marker staining of a B-enriched population.
[d] Examination by phase contrast morphology.

E-rosette-negative populations depleted of adherent cells, monocyte/macrophage-enriched adherent cells, granulocytes, erythrocytes, and platelets (Table 33.1). Within the bone marrow only a small fraction of cells with lymphoid morphology as judged by examination by phase contrast was stained. Number of reacting cells as well as results of double-staining experiments are consistent with a B cell-specific reaction. This is confirmed by analysis of thymus, spleen, and tonsil cell suspensions. These data were recorded by indirect immunofluorescence methods and were confirmed by using a highly sensitive cellular radioimmunoassay (CRIA) test (data not shown). With the CRIA procedure also 12 different cell lines (T, B, and non-lymphoid cell lines) reacted with a B-specific pattern (data not shown, manuscript in preparation).

Immunohistological analysis of normal and reactive lymph nodes has shown a typical B-specific reaction.

Reactivity of HD37 with Leukemia/Lymphoma Cells

This study was done by evaluation of mononuclear cell suspensions by indirect immunofluorescence (see Table 33.2). HD37 was reactive with 58 out of 59 cases of non-Hodgkin's lymphomas, the only negative case being a lymphoplasmacytic lymphoma (a tumor which may be considered as intermediate between a nonsecreting lymphoma and the plasmocytoma). The antibody did not show any reaction with T cell malignancies or with cases of acute myeloid or myelo-monocytic leukemias. Among a total of 51 acute lymphoblastic leukemias 43 were classified as HD37$^+$ (see below).

Table 33.2. Reactivity of HD37 with leukemia/lymphoma cells (indirect immunofluorescence).

Cells[a]	HD37
ALL	43/51[b]
B-CLL	25/25
B-PLL	4/4
HCL	10/10
Lymphoplasmacytic lymphoma	19/20
MM	0/6
T-CLL/CTCL	0/10
AML/AMML	0/15

[a] ALL = Acute lymphoblastic leukemia of T and non-T type; CML = chronic myelogenous leukemia; CLL = chronic lymphocytic leukemia; PLL = prolymphocytic leukemia; HCL = hairy cell leukemia; MM = multiple myeloma; CTCL = cutaneous T cell lymphomas, including Sézary syndrome; AML = acute myelocytic leukemia; AMML = acute myelo-monocytic leukemias.
[b] Number of positive cases/number studied.

Table 33.3. Immunohistological evaluation of "blast-type" non-Hodgkin's lymphomas.[a]

	VIL-A1	B1	B2	BA1	HD37
Centroblastic lymphoma	9/13[b]	8/12	3/13	3/12	12/13
Lymphoblastic lymphoma,[c] non-T, non-Burkitt	11/16	9/18	9/17	11/16	18/18
Burkitt's lymphoma	10/10	10/10	0/10	7/8	10/10
Immunoblastic lymphoma	8/12	12/13	10/13	6/13	9/13
Total	33/51	39/53	22/53	27/49	49/54

[a] Lymphomas were classified according to Kiel classification.
[b] Number of positive cases/number studied.
[c] Not reacting with the T cell-specific mAb Tü14.

Immunohistological Studies

Among a total of 103 biopsies from non-Hodgkin's lymphomas (NHL) classified according to the Kiel classification HD37 stained positively 91 cases (data not shown). In Table 33.3 the staining pattern of "blast-type" NHL is shown. These tumors often represent the immature or "blastic" counterpart of low-grade malignant lymphomas; many of them are usually called "diffuse histiocytic lymphomas" in other classification schemes. HD37 clearly shows the broadest reaction, being positive in 49 out of 54 cases of these "high malignant" lymphoid neoplasms. Four of the five negative cases belong to the "immunoblastic" subtype, which is an Ig-secreting tumor whose cells are morphologically and immunologically related to plasmoblasts and plasmacells. On the other hand, HD37 stained all cases of lymphoblastic lymphomas of the non-Burkitt type which were negative with the T-marker Tü14. These tumors can be considered as the histological counterpart of acute leukemias. Even the broadly reacting mAb BA1 was expressed only in a proportion of these lymphomas.

Evaluation of ALL Cases by Monoclonal Antibody Reactivity (Table 33.4)

As markers for the T lineage we used the antibodies OKT6 and Tü14; as B cell markers we evaluated the antibodies HD37, B1, and HD28. Moreover, the broadly reacting, non-lineage-specific antibodies I_2 (anti-Ia) and VIL-A1 (CALLA specificity) were used. Among 51 tested ALL cases six were assigned to the T lineage. Cells from three of these patients were Ia positive, did not express the T6 antigen, and were recognized as early T members because of their positive staining with mAb Tü14. On the other hand, the three cases which lacked Ia antigens did show a reaction with mAb T6.

The remaining 45 cases were considered to have a non-T acute leukemia. Ia antigens were detected on all of the cases. Thirty-nine patients ble 33.4. ALL Phenotypes.

. positive cases	Ia	Vil-A1[a]	HD37	B1	HD28	Tü14	T6		
2	+	+	−	−	−	−	−		?
5	+	−	+	−	−	−	−	B-commited	B lineage
21	+	+	+	−	−	−	−		
11	+	+	+	+	−	−	−	early B	
5	+	+	+	+	+	−	−		
1	+	−	+	+	+	−	−		
1	+	+	−	−	−	+	−		T lineage
2	+	−	−	−	−	+	−		
3	−	−	−	−	−	+	+		

ALLA specificity.

showed also a "CALLA" reactivity. Among these, 23 had previously been considered as having a non-T, non-B, "common type" ALL, because of a lack of cytoplasmic immunoglobulin and the B cell marker B1. Given their positivity with the monoclonal antibody HD37, 21 out of these 23 ALL cases could, however, be reclassified as B-committed ALL. Moreover, positivity for HD37 clearly indicated the B origin of leukemia in 5 additional CALLA⁻ cases, which would otherwise have been considered as "Null" or "unclassified" ALL.

Seventeen patients (37% of our non-T ALL) had a more mature phenotype, as demonstrated by positivity for B1; six of them also reacted with HD28, so that they should be considered as having a rather mature phenotype.

Discussion

In the present study we describe a new B lineage-specific monoclonal antibody HD37. Virtually all B lymphocytes in blood and in lymphoid organs so far evaluated bear the corresponding antigen which is also present on a small lymphoid-like fraction of marrow mononuclear cells. Normal and neoplastic T cells, monocytes, and myeloid cells were consistently negative when challenged with HD37. Besides being lineage specific within the hematological compartment, the antigen could not be demonstrated in a variety of normal and neoplastic tissues; this must however be further investigated. Given its B cell specificity, HD37 is a valuable tool for the immunological characterization of leukemia and lymphoma cells.

Recent introduction of mAbs as diagnostic reagents for hematological malignancies has provided important advances in disease classification; it has also improved our understanding of normal leukocyte differentiation (19). Unfortunately, a number of reagents so far characterized are not lineage specific, a fact that may strongly affect their usefulness in leukemia/lymphoma typing. This is particularly important for the group of acute leukemias and malignant lymphomas which require different therapeutical strategies according to their origin. CML in lymphoid blast crisis, for example, often respond to vincristine and prednisone which is a mild therapy compared to regimens necessary for non-lymphoid blast crisis (29).

Despite the limited variance in morphology acute leukemias constitute a very heterogeneous group of malignancies with respect to enzymatic, cytochemical, and immunological patterns. This probably reflects the complexity of differentiation stages at which, during maturation, the malignant transformation has taken place (19,30). Classification of acute leukemias corresponding to the phenotypes of early progenitor cells may be difficult: while a number of specific markers accumulate along with maturation, immature cells of different lineages often share common determinants. This has been demonstrated for Ia-like antigens (31), CALLA antigens (32), and TdT (33), and stresses the importance of having reliable lineage-specific reagents.

By using conventional markers such as E-rosetting and SIg only 20% of ALL can be classified as T or B. The majority of cases lacking these markers express the CALLA antigen and this has been associated with a good prognosis (4). At present time we know that a large majority of CALLA$^+$ ALL probably are early members of the B lineage. In about 20% of the cases intracytoplasmic μ chains can be demonstrated (pre-B phenotype) (6) and in nearly 50% of cases the B cell-specific mAb B1 binds to the blasts. In accordance with these findings, studies done by Korsmeyer *et al.* (9) have shown that immunoglobulin gene rearrangement may indicate the B origin of nearly 100% of non-T ALL. A negative correlation between expression of antibody 3A1 and heavy-chain Ig gene rearrangement was demonstrated in 37 patients. However, while the maintenance of germ-line Ig genes may be used as a proof that B cell commitment has not yet taken place, heavy-chain rearrangement is not a reliable marker: it may be found even in T cells (34) and in myeloid leukemia (35).

Recently Nadler *et al.* (36) demonstrated the presence of an antigen defined by the mAb B4 on 94% of non-T ALL cases. Since blast cells of all these patients also displayed heavy-chain Ig rearrangement this has been interpreted as a further proof that non-T ALL should indeed be considered as early members of the B lineage. The evaluation of 51 ALL cases in our study confirms these data: we found that 43 out of 45 patients with non-T leukemia express the antigen recognized by our B cell-specific mAb HD37.

The phenotypes observed in our study are similar to those reported by Nadler. In our series however we still found two patients with "uncommitted" ALL: whether this represents a true subgroup of patients or is

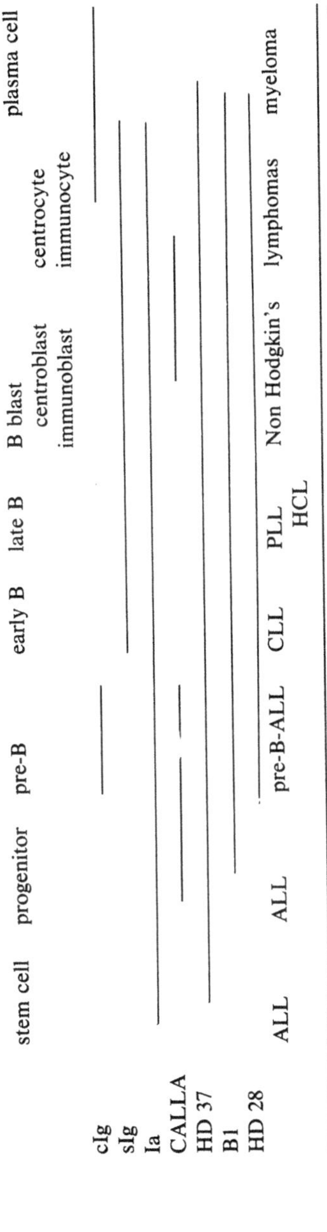

ALL = Acute lymphoblastic leukemia; CLL = Chronic lymphocytic leukemia; PLL = Prolymphocytic leukemia; HCL = Hairy cell leukemia

Fig. 33.1. Hypothetical B cell maturation scheme.

due to artifacts (weak antigen expression on the blasts) remains open. On the other hand, the whole group of non-Burkitt lymphoblastic lymphomas could be divided among T and B tumors. This suggests that slightly different phenotypes may be associated with the older age of these patients.

On the basis of our observations a hypothetical maturation scheme of B cell differentiation is shown in Fig. 33.1.

Currently studies are under way in our laboratory in order to assess whether mAbs HD37 and B4 may recognize the same antigenic determinant. Among the antibodies submitted for evaluation in the B cell Workshop, mAbs B14, B34, and B43 also had a similar reaction pattern.

References

1. Tsukimoto, I., K.Y. Wong, and B.C. Lampkin. 1976. Surface markers as prognostic factors in childhood acute leukemia. *New England J. Med.* **294:**245.
2. Hoelzer, D., E. Thiel, H. Löffler, H. Bodenstein, *et al.* 1984. Intensified therapy in acute lymphoblastic and acute undifferentiated leukemia in adults. *Blood* **64:**38.
3. Bennet, J.M., D. Catovsky, M.T. Daniel, D.A. Flandrin, H.R. Galton, H.D. Gralnik, and C. Sultan. 1976. Proposal for the classification of acute leukemia. *Brit. J. Haematol.* **33:**451.
4. Thiel, E., 1984. Biological and clinical significance of immunological cell markers in leukemia. *Recent Results in Cancer Research* **93:**102–158.
5. Greaves, M.F., G. Brown, N.T. Rapson, and T.A. Lister. 1975. Antisera to acute lymphoblastic leukemia cells. *Clin. Immunol. Immunopathol.* **4:**67.
6. Vogler, L.B., W.M. Crist, D.E. Bockmann, E.R. Pearl, A.R. Lawton, and M.D. Cooper. 1978. Pre-B cell leukemia: a new phenotype of childhood lymphoblastic leukemia. *New England J. Med.* **298:**872.
7. Brouet, J.C., J.L. Preud'homme, C. Penit, F. Valensi, P. Rouget, and M. Seligmann. 1979. Acute lymphoblastic leukemia with pre-B cell characteristics. *Blood* **54:**269.
8. Korsmeyer, S.J., P.A. Hieter, J.V. Ravetch, D.G. Poplack, T.A. Waldmann, and P. Leder. 1981. Developmental hierarchy of immunoglobulin gene rearrangements in human leukemic pre-B cells. *Proc. Natl. Acad. Sci. U.S.A.* **78:**7096.
9. Korsmeyer, S.J., A. Arnold, A. Bakhshi, J.V. Ravetch, V. Siebenlist, P.A. Hieter, S.O. Sharrow, T.W. LeBien, J.H. Kersey, D.G. Poplack, P. Leder, and T.A. Waldmann. 1983. Immunoglobulin gene rearrangement and cell surface antigen expression in acute lymphocytic leukemias of T cell and B cell precursor origin. *J. Clin. Invest.* **71:**301.
10. Nadler, L.M., J. Ritz, R. Hardy, J.M. Pesando, S.F. Schlossman, and P. Stashenko. 1981. A unique cell surface antigen identifying lymphoid malignancies of B cell origin. *J. Clin. Invest.* **67:**134.
11. Bakhshi, A., J. Minowada, A. Arnold, J. Cossman, J.P. Jensen, J. Whang-Peng, T.A. Waldmann, and S.J. Korsmeyer. 1983. Lymphoid blast crisis of chronic myelogenous leukemia represent stages in the development of B-cell precursors. *New England J. Med.* **309:**826.

12. Nadler, L.M., P. Stashenko, R. Hardy, A. van Agthoven, C. Terhorst, and S.F. Schlossman, 1981. Characterization of a human B cell specific antigen (B2) distinct from B1. *J. Immunol.* **126:**1941.
13. Abramson, C.S., J.H. Kersey, and T.W. LeBien. 1981. A monoclonal antibody (BA-1) reactive with cells of human B lymphocyte lineage. *J. Immunol.* **126:**83.
14. Yokochi, T., R.D. Holly, and E.D. Clark. 1982. B lymphoblast antigen (BB-1) expressed on Epstein–Barr activated B cell blasts, B lymphoblast and cell lines and Burkitt's lymphoma. *J. Immunol.* **128:**283.
15. Royston, I., M.R. Omary, and I.S. Trowbridge. 1981. Monoclonal antibodies to a human T cell antigen and Ia-like antigen in the characterization of lymphoid leukemia. *Transplant Proc.* **13:**761.
16. Nadler, L.M., K.C. Anderson, G. Marti, M.P. Bates, E.K. Park, J.F. Baley, and S.F. Schlossman. 1983. B4, a human B lymphocyte associated antigen expressed on normal, mitogen activated, and malignant B lymphocytes. *J. Immunol.* **131:**244.
17. Haynes, B.F., R.S. Metzgar, J.D. Minna, and P.A. Bunn. 1981. Phenotypic characterization of cutaneous T-cell lymphoma. *New England J. Med.* **34:**1319.
18. Koziner, B., D. Gebhard, T. Denny, S. McKenzie, B.D. Clarkson, D.A. Miller, and R.L. Evan. 1982. Analysis of T-cell differentiation antigens in acute lymphatic leukemia using monoclonal antibodies. *Blood* **60:**752.
19. Foon, K.A., R.W. Schroff, and R.P. Gale. 1982. Surface markers on leukemia and lymphoma cells: Recent advances. *Blood* **60:**1.
20. Köhler, G., and C. Milstein. 1975. Continuous cultures of fused cells secreting antibodies of predefined specificity. *Nature* **256:**495.
21. Dörken, B., A. Pezzutto, G. Moldenhauer, R. Schwartz, S. Kresel, and W. Hunstein. An immunoenzymatic staining assay (ISA) for the rapid screening of monoclonal antibodies detecting membrane and cytoplasmic antigens. Manuscript submitted for publication.
22. Nadler, L.M., P. Stashenko, R. Hardy, J.M. Pesando, E.Y. Yunis, and S.F. Schlossman. 1981. Monoclonal antibodies defining serological distinct HLA-DIDR-related Ia-like antigens in man. *Hum. Immunol.* **1:**77.
23. Kung, P.C., G. Goldstein, E.L. Reinherz, and S.F. Schlossman. 1979. Monoclonal antibodies defining distinctive human T cell surface antigens. *Science* **206:**347.
24. Knapp, W., O. Majdic, P. Bettelheim, and K. Liszka. 1982. VIL-A1 a monoclonal antibody reactive with common acute lymphatic leukemia cells. *Leuk. Res.* **6:**137.
25. Schwartz, R., G. Moldenhauer, A. Pezzutto, and B. Dörken. 1984. Characterization of monoclonal antibody HD28 reactive with human B lymphocytes. This conference.
26. Boyum, A. 1968. Isolation of mononuclear cells and granulocytes from human blood. *Scand. J. Clin. Lab. Invest.* (Suppl.) **97:**77.
27. Jerrels, T.R., J.H. Dean, G.L. Richardson, and R.B. Herbermann. 1980. Depletion of monocytes from human peripheral blood mononuclear leukocytes: comparison of the Sephadex G-10 column method with other commonly used techniques. *J. Immunol. Methods* **32:**11.
28. Stein, H., K. Lennert, A. Feller, and D.Y. Mason. 1984. Immunohistological

analysis of human lymphoma. Correlation of histological and immunological categories. *Advances in Cancer Research* **42:**67.
29. Griffin, J.D., R.F. Todd, J. Ritz, L.M. Nadler, G.P. Canellos, D. Rosenthal, M. Gallivan, R.P. Beveridge, H. Weinstein, D. Karp, and S.F. Schlossman. 1983. Differentiation patterns in the blastic phase of chronic myeloid leukemia. *Blood* **61:**85.
30. McCulloch, E.A. 1983. Stem cells in normal and leukemic hemopoiesis (Henry-Stratton Lecture, 1982). *Blood* **62:**1.
31. Bodger, M.P., C.A. Izaguirre, H.A. Blachlock, and A.V. Hoffbrand. 1983. Surface antigenic determinants on human pluripotent and unipotent hematopoietic progenitor cells. *Blood* **61:**1006.
32. Hokland, P., P. Rosenthal, J.D. Griffin, L.M. Nadler, J. Daley, M. Hokland, S.F. Schlossman, and J. Ritz. 1983. Purification and characterization of fetal hematopoietic cells that express the common acute lymphoblastin leukemia antigen (CALLA). *J. Exp. Med.* **157:**114.
33. Jani, P., W. Verbi, M.F. Greaves, D. Bevan, and F.J. Bollum. 1983. Terminal deoxynucleotidyl transferase in acute myeloid leukemia. *Leuk. Res.* **7:**17.
34. Kurosawa, Y., H. von Boehmer, W. Mass, H. Sakano, A. Trauneker, and S. Tonegawa. 1981. Identification of D segments of immunoglobulin heavy-chain genes and their rearrangement in T lymphocytes. *Nature* **209:**565.
35. Rovigatti, U., J. Mirro, G. Kitchingman, G. Dahl, J. Ochs, S. Murphy and S. Stass. 1984. Heavy chain immunoglobulin gene rearrangement in acute nonlymphocytic leukemia. *Blood* **63:**1023.
36. Nadler, L.M., S.J. Korsmeyer, K. Anderson, A.W. Boyd, B. Slaughenhaupt, E. Park, J. Jensen, F. Coral, R.J. Mayer, S.E. Sallan, J. Ritz, and S.F. Schlossman. 1984. B cell origin of non-T cell acute lymphoblastic leukemia. A model for discrete stages of neoplastic and normal pre-B cell differentiation. *J. Clin. Invest.* **74:**332.

CHAPTER 34

Characterization of Burkitt's Lymphoma Cell Lines with Monoclonal Antibodies Using an ELISA Technique

John T. Sandlund, James Kiwanuka, Gerald E. Marti, Walter Goldschmidts, and Ian T. Magrath

Introduction

In screening the leukocyte B cell and leukemia mAb panels against undifferentiated lymphoma (UL) cell lines, we hoped to more precisely characterize the phenotype of UL and to determine whether differences in antigen expression exist among various subgroups of UL, including Burkitt's lymphoma (BL), non-Burkitt's lymphoma (NB), UL of African and American origin, and BL occurring in patients with AIDS. Further, we hoped to utilize the Workshop data to provide insights into the identity of the normal counterpart cell of BL. Finally, we wished to compare two methods used commonly in screening mAB reactivity, namely ELISA and FCM.

Methods

Cell Lines

All cell lines were maintained at 37°C in RPMI 1640 supplemented with 20% fetal calf serum, penicillin, and streptomycin. Cells were subcultured every 3–4 days. Table 34.1 is a summary of some of the known characteristics of the 21 cell lines used in this study. All UL cell lines contained either an 8;14 or 8;22 chromosomal translocation.

ELISA

A solid-phase indirect ELISA method using an avidin–biotin–peroxidase detection system was used (1). 10^5 cells (in mid-log phase) per well were bound to poly-L-lysine (Sigma, St. Louis, MO; 0.75 μg/well)-coated wells

Table 34.1. Characteristics of undifferentiated lymphoma cell lines.

Cell line	Origin[a]	Karyotype	EBV nuclear antigen
Namalwa	AF BL	t8;14	+
AG876	AF BL	t8;14	+
P3HR1	AF BL	t8;14	+
Daudi	AF BL	t8;14	+
Raji	AF BL	t8;14	+
EB3	AF BL	t8;14	+
MC116	AM BL	t8;14	–
CA46	AM BL	t8;14	–
KK124	AM BL	t8;22	+
ST486	AM BL	t8;14	–
LW878	AM BL	t8;14	–
AS283	AM BL (AIDS)	t8;14	+
PA682	AM BL (AIDS)	t8;22	+
JD38	AM UL	t8;14	–
JD39	AM UL	t8;14	–
EW36	AM UL	t8;14	–
JLPC119	AM UL	t8;14	–
DS179	AM UL	t8;14	–
CB22	CB	Diploid	+
LKC	NL	Diploid	+
LHA	NL	Diploid	+

[a] AF BL = African Burkitt's lymphoma; AM BL = American Burkitt's lymphoma; AM UL = American undifferentiated lymphoma/non-Burkitt's type; CB = cord blood line; NL = normal lymphoblastoid cell lines.

(96-well, flat-bottomed microtiter plate; Dynatech, Alexandra, VA) and fixed with glutaraldehyde (0.5%). Monoclonal antibodies were tested at two dilutions, 1 : 125 (to ensure antibody excess), and 1 : 500, in phosphate-buffered saline (PBS) containing 2% bovine serum albumin and 10% human serum. Biotinylated horse anti–mouse immunoglobulin (anti-IgM, heavy and light chain specific, which reacted with all mouse immunoglobulins; Vector, Burlingame, CA) was used at a dilution of 1 : 1000, and developed with an avidin/biotin/peroxidase complex (Vector, Burlingame, CA) using *o*-phenylenediamine (Sigma, St. Louis, MO) at a concentration of 1 mg/ml (200 μl/well) as substrate. After a 1-hr incubation with substrate, reactivity was measured by spectrophotometry (Titertek Multiskan) using a 450-nm filter. After background subtraction, optical densities were interpreted as follows: less than 0.100, negative; 0.100–0.199, borderline; 0.200–0.399, positive, greater than 0.400, strongly positive. Positive (anti-HLA, anti-HLA-DR) and negative (NS1 ascites) controls were included on all plates.

Surface Immunofluorescence (SIF)

Three American BL cell lines (CA46, ST486, and KK124) were selected for analysis by flow cytometry (FCM). Mid-log-phase cells were washed and adjusted to 10^7 cells/ml, and Workshop mAbs were used at a final dilution of 1 : 250 in PBS containing 10% human serum. mAb binding was measured indirectly by incubation with a fluoresceinated goat antibody to mouse immunoglobulin (polyvalent; Coulter, Hialeah, FL) and quantitation of fluorescent intensity was performed with an EPICS IV flow cytometer (Coulter, Hialeah, FL). The mean channel fluorescence was determined for each reagent and compared to appropriate negative controls. Interpretation of reactivity depended upon the number of standard deviations (SD) by which the sample mean differed from the control mean: negative, 0–1 SD; borderline, 1–2 SD; positive, 2–4 SD; strongly positive, greater than 4 SD. Bimodal distributions were uncommon and are not considered in this analysis.

Results and Discussion

Results of the mAb screening by ELISA (mAb dilution, 1 : 125) and FCM are summarized in Tables 34.2 and 34.3. We were able to analyze our data after cluster groups had been defined at the leukocyte Workshop. Of immediate note was the marked variation in the reactivity of individual mAbs within a cluster group. These differences were frequently apparent in both ELISA and FCM data and thus are not simply due to fixation artifact. The most probable explanation of this finding is that some of the epitopes on individual antigens are masked or absent in these cell lines. At a practical level, this finding indicates that false negative results may be obtained if an antigen is screened for by one or a small number of mAbs. Mixtures of mAbs may therefore be preferable for screening, and will have the further advantage that simultaneous binding to different epitopes will increase the intensity of the reactivity. Differentiation between absence or inaccessibility of an epitope should be readily accomplished by immunoprecipitation and SDS–PAGE electrophoresis of cell lysates.

In comparing the results of ELISA and FCM some differences in the reactivity of a given mAb were observed. This is exemplified by reactivity with mAb L6 (detecting CALLA), which gave positive results with all cell lines tested by ELISA but a negative result with one line (KK124) by FCM. Our previous experience with several different anti-CALLA antibodies (unpublished data) was similar in that some cell lines were negative when tested by FCM, but all were positive when tested by ELISA. It is possible that the preparation of cells for ELISA (e.g., glutaraldehyde fixation) exposes an epitope not accessible in viable, non-fixed cells. CALLA is believed to be membrane-associated but not an integral membrane protein (2), and this may be responsible for our observations. Alternatively, ELISA positivity/FCM negativity may result from access to cytoplasmic antigens in the ELISA system (3). For some antigens FCM was positive when ELISA was weakly positive or negative (e.g., p140 in cell line KK124). The possibility that fixation modifies the binding of some

Table 34.2. Reactivity of monoclonal antibodies with cell lines.[a]

	CD19 (p95)[b]					CD20 (p35)[b]			CD21 (p140)[b]				CD22 (p135)[b]				
	B14	B28	B34	B43	L17	B5	B22	B24	B9	B33	B35	B41	B7	B25	B31	B40	B49
Namalwa	−	±	±	++	±	±	±	+	−	±	±	±	±	±	−	±	+
AG876	−	±	±	++	±	±	±	+	−	−	±	±	±	±	±	±	+
P3HR1	−	±	+	++	−	±	±	+	−	−	±	±	+	±	±	−	±
Daudi	+	±	−	++	±	++	±	+	−	−	±	±	++	+	±	±	±
Raji	−	+	+	++	+	±	±	++	−	±	+	+	±	±	+	+	+
EB3	±	±	+	++	−	+	±	+	±	±	±	+	++	+	±	±	+
MC116	±	+	+	++	±	±	+	++	−	±	±	+	+	+	+	+	±
CA46	−	+	+	++	+	−	±	++	−	±	±	±	±	+	+	−	+
KK124	+	±	±	++	−	±	±	+	−	±	±	±	+	+	±	±	±
ST486	±	±	+	++	±	+	±	++	±	±	±	±	++	+	±	±	+
LW878	±	±	±	++	−	±	−	+	±	−	−	±	++	+	−	−	+
AS283	±	±	±	++	+	+	±	++	±	±	±	±	+	+	−	−	+
PA682	±	+	±	+	±	+	±	+	±	±	±	±	+	+	±	±	−
JD38	−	+	+	++	+	±	±	++	−	±	±	+	+	−	±	±	+
JD39	+	−	−	++	±	++	−	±	±	−	−	−	++	+	−	−	+
EW36	−	+	±	++	±	+	±	++	−	±	±	±	+	±	±	±	+
JLPC119	±	−	±	++	+	±	±	+	−	±	−	+	++	−	−	−	+
DS179	+	−	±	+	±	±	±	±	−	−	−	±	−	±	−	±	−
CB23	+	−	−	++	−	++	−	±	+	−	−	−	++	−	−	−	++
LKC	±	±	±	±	±	−	±	+	±	±	−	±	+	±	±	±	−
LHA	±	±	±	+	−	+	−	+	−	−	−	±	+	+	−	−	−
CA46[d]	++	++	++	++	+	±	+	+	−	−	−	ND[e]	+	++	+	−	++
KK124[d]	++	++	++	++	++	++	++	++	++	++	++	++	+	++	+	+	++
ST486[d]	++	++	++	++	++	++	++	++	±	−	−	−	±	+	+	+	−

[a] Tested by ELISA unless otherwise noted.
[b] Clustered.
[c] Non-clustered.
[d] Reactivity tested by flow cytometry.
[e] ND: Not determined.

mAbs to cell surface antigens must be considered as a possible explanation for this phenomenon. Thus, fixation of cells may reveal otherwise undetectable antigens, but may also result in lessened reactivity of some antigens readily observed in viable cells. These considerations must be taken into account when interpreting the binding patterns of mAbs.

Several observations can be made regarding the phenotypic characterization of the cell lines we studied by ELISA. The majority of the lymphoma cell lines reacted with antibodies detecting B cell cluster group proteins p95, p35, p135 (CD19, 20, 22), p45/55/65, p29/34, and p220. In addition, most cell lines reacted with non-clustered B cell antibodies B23, B32, B42, B10, B44, B45, and B52. Similar reactivity was observed with antibodies detecting leukemia panel cluster groups CD10 (CALLA), CD9 (p24), the transferrin receptor, and the non-clustered antibody L7. No obvious phenotypic differences were noted between African BL, American BL, and undifferentiated lymphoma (non-Burkitt's type); nor did BL cell lines derived from AIDS patients differ significantly from other BL lines. We have previously observed a greater expression of C3d and EBV receptors in EBV-positive African lines compared to EBV-negative American lines (4). Although we expected to confirm this difference with antibodies reacting with p140, which binds both C3d and EBV, all four

	CD23 (p45)[b]		All resting B[c]				Some resting B[c]				Not expressed on resting B[c]					α-μ[b]		
1	B19	B39	B6	B8	B30	B36	B21	B46	B16	B27	B13	B20	B23	B32	B37	B2	B3	B42
	±	−	±	±	−	−	−	++	±	−	−	−	++	+	+	−	±	++
	±	−	±	±	−	−	−	+	−	−	−	−	++	+	+	−	±	++
	−	+	−	±	−	−	−	+	−	−	−	−	++	+	±	−	±	++
	±	−	+	+	±	±	±	++	+	+	±	±	++	±	±	±	+	++
	−	±	−	±	−	±	±	++	−	±	−	−	++	++	+	−	±	++
	±	−	+	+	−	±	±	++	+	±	±	±	++	+	+	±	±	++
	−	−	±	±	−	±	±	+	±	+	−	±	++	++	+	±	+	++
	−	−	−	−	−	+	−	++	−	−	−	−	++	++	+	−	±	++
	−	−	+	−	−	−	±	++	+	+	±	±	++	+	±	−	+	++
	±	−	+	+	−	−	±	++	+	±	±	±	++	+	±	−	±	++
	±	−	±	+	−	−	−	++	+	++	±	±	++	+	+	±	±	++
	±	−	+	±	−	−	±	++	+	+	±	±	++	+	±	−	+	++
	+	−	+	+	−	±	±	+	+	+	±	±	+	+	+	±	+	+
	−	−	±	±	−	±	−	+	−	±	−	−	++	+	+	±	±	++
	±	−	++	+	−	−	±	++	+	+	±	±	+	±	−	±	+	++
	−	−	−	±	−	−	−	++	−	++	−	−	++	+	+	−	±	++
	±	−	±	+	−	+	±	++	+	±	±	±	++	+	++	±	+	++
	±	−	±	±	−	−	±	+	+	±	±	±	+	±	±	±	±	+
+	++	±	+	++	−	−	+	++	++	−	+	+	+	±	±	−	+	++
	±	±	±	±	−	−	±	+	±	++	±	±	++	+	±	±	±	−
	±	±	+	+	−	−	±	±	+	+	−	±	++	+	±	±	+	±
+	−	−	±	−	±	++	+	−	−	+	−	−	±	−	−	−	+	++
+	++	++	+	++	−	+	−	−	±	++	−	±	++	−	±	+	+	+
	±	−	±	++	±	+	+	+	++	−	−	±	++	−	±	++	++	++

antibodies in this cluster group gave negative or weak reactivity with all cell lines using the ELISA technique. We confirmed this result with commercially available antibody (anti-CR2, Becton Dickinson, Mountain View, CA). However, one cell line tested by FCM (KK124) was positive with p140 mAbs. This cell line is EBNA positive. Presumably, at least some epitopes on the p140 antigen are adversely affected by fixation. Mixtures of mAbs or the use of mAbs recognizing other epitopes on p140 may resolve this problem.

Since UL cells have a blastic morphology, it is surprising that antibodies in the cluster groups recognizing the p135 antigen (expressed by mantle zone cells but not by germinal center cells) and "some resting B cells" gave good reactivity in the ELISA system with virtually all the UL cell lines. Antibodies B47 and B48 (BA-1-like), and, as already mentioned, mAbs of the CD10 (CALLA) group were also very reactive. BA-1, which reacts predominantly with "early B cells," reacts strongly with mantle zone cells and weakly with germinal center cells (5), while CALLA is expressed strongly on B lymphocyte precursor cells and to a limited extent on cells of the germinal follicle (7). Negative or weak reactivity was observed with antibodies recognizing the p45 protein, present in germinal center cells. It is of note that mAb reactivity with lymphoid follicles has

Table 34.3. Reactivity of monoclonal antibodies with cell lines.[a]

	CD10 (CALLA)[b]						CD9 (p24)[b]					Transferrin receptor[b]				IA[b] (p24, 3	
	L2	L6	L10	L14	L15	L21	L4	L16	L18	L22	B38	L3	L13	L19	L20	L8	
Namalwa	++	++	±	−	−	+	±	−	++	+	−	−	−	±	−	+	+
AG876	+	++	−	−	−	±	±	−	+	±	−	−	−	±	−	+	+
P3HR1	++	++	−	−	−	±	±	−	+	+	−	−	−	±	−	+	+
Daudi	++	++	+	−	−	+	±	−	++	++	−	−	−	++	++	+	+
Raji	++	++	+	±	±	++	+	−	++	++	−	±	++	++	+	++	±
EB3	++	++	+	−	−	±	+	−	+	+	±	±	−	+	±	++	+
MC116	+	++	−	−	−	+	±	−	+	+	−	−	±	+	±	+	+
CA46	++	++	+	±	±	+	+	−	++	++	−	±	++	++	±	++	+
KK124	++	++	−	−	−	−	±	−	−	−	−	±	++	++	++	++	+
ST486	++	++	++	−	−	+	+	−	+	+	−	±	−	+	±	++	+
LW878	++	++	+	−	−	−	+	−	+	+	−	±	−	−	−	++	+
AS283	++	++	±	−	±	+	+	−	++	+	−	±	−	++	+	++	+
PA682	±	+	−	−	−	+	−	−	+	+	−	−	±	++	+	+	+
JD38	++	++	+	±	±	+	+	−	++	+	−	±	++	++	+	++	+
JD39	++	++	++	−	−	±	+	−	+	+	−	+	±	++	+	++	+
EW36	++	++	+	−	±	++	+	−	++	++	−	±	++	++	+	++	+
JLPC119	++	++	+	±	±	+	+	±	++	++	−	−	±	+	+	+	+
DS179	±	+	−	−	−	±	−	−	+	+	−	−	±	+	±	±	+
CB23	++	++	±	−	−	±	++	−	+	+	−	±	−	+	±	++	+
LKC	±	++	−	−	−	±	−	−	++	++	−	−	−	+	+	+	+
LHA	±	+	−	−	−	−	−	−	++	+	−	−	−	+	+	±	+
CA46[e]	+	+	+	±	−	++	+	±	±	ND[f]	−	++	++	++	++	++	+
KK124[e]	−	−	+	−	+	−	−	−	−	−	−	++	++	++	++	++	+
ST486[e]	++	++	++	+	+	+	−	−	−	−	−	+	++	++	++	++	+

[a] Tested by ELISA unless otherwise noted.
[b] Clustered.
[c] Clustered B cell associated, not further specified.
[d] Non-clustered.
[e] Reactivity tested by flow cytometry.
[f] ND: Not determined.

usually been examined in fixed tissue by an ELISA technique (6,7) so that although we cannot assume that p45 is absent, for reasons already discussed (see also FCM results), when interpreting these data, it would appear that a meaningful comparison between our results using ELISA with the information generated from the study of tissue sections by a similar technique can be made. Such a comparison, based particularly on the positive reactions observed, would lead to the conclusion that UL cells have a stronger phenotypic resemblance to mantle zone lymphocytes than to germinal center cells. Thus UL, including BL, is unlikely to arise from a germinal center cell as has previously been postulated (8), but is phenotypically more consistent with an immature B cell somewhere in the differentiation pathway between a pre-B cell and a mantle zone or primary follicle cell. Within this "differentiation window," different tumors may have variable degrees of maturity.

This interpretation of our results is consistent with the possibility that Burkitt's lymphoma has no precise normal counterpart cell, but instead is the neoplastic equivalent of an immature B cell which has been unable to become a resting, virgin B cell because of continued expression of the c-

CD24[b] (p45, 55, 65)				(p220)[b]			Unspecified[d]								Unspecified[d]				
B47	B48	B15[c]	B18[c]	B50	B51	L12	B4	B10	B17	B26	B29	B44	B45	B52	L1	L7	L9	L11	L17
±	+	−	−	++	+	++	−	+	±	+	−	++	+	±	±	+	±	++	±
±	±	−	−	++	±	++	−	+	±	±	±	+	++	±	±	+	±	+	±
−	±	−	−	++	+	+	±	+	±	+	±	+	++	±	±	±	±	+	−
−	++	±	±	++	+	++	+	++	−	±	+	++	++	+	±	++	±	++	±
++	++	−	−	++	++	++	±	±	±	+	±	++	++	++	+	+	+	++	+
+	++	±	±	++	+	++	+	++	+	+	+	++	++	+	+	++	+	++	−
±	+	±	−	++	+	++	±	+	+	+	+	+	+	±	±	+	±	+	±
±	+	−	−	++	+	++	±	+	−	++	+	++	++	+	+	++	+	++	+
−	++	±	−	++	+	++	+	+	−	+	++	++	++	+	±	+	±	++	−
++	++	±	±	++	++	++	+	+	+	+	−	±	++	++	+	++	++	++	±
++	++	±	±	++	++	++	±	+	+	±	+	++	++	+	+	++	+	+	−
++	++	±	±	++	++	++	±	+	++	±	−	++	++	+	+	++	+	++	+
±	±	±	±	++	±	++	+	++	+	+	+	−	++	±	−	+	±	++	±
++	++	−	−	++	++	++	±	+	±	+	+	++	++	+	+	+	+	++	+
++	++	±	−	++	++	++	++	++	+	±	±	++	++	+	+	++	++	++	±
++	++	−	−	++	++	++	−	±	−	+	±	+	++	+	+	+	+	++	±
+	++	±	−	++	++	++	±	++	+	−	±	++	++	+	+	++	+	++	+
−	±	±	±	++	±	++	+	++	+	±	±	±	++	−	−	−	−	+	±
++	++	−	+	++	++	++	++	++	++	−	−	++	++	+	++	++	++	+	−
−	±	−	±	++	±	++	±	+	+	±	±	−	++	−	−	±	−	++	±
±	+	−	−	+	−	++	±	++	+	±	±	+	++	−	−	±	−	±	−
−	−	−	−	+	+	++	++	−	+	−	++	++	++	ND[f]	−	−	−	+	+
++	++	±	−	++	++	++	−	++	++	−	++	++	++	++	−	+	++	±	++
−	−	±	+	−	−	++	++	±	+	++	++	−	±	−	−	−	−	++	++

myc gene—a gene known to be important to cell proliferation, and whose expression is altered by the chromosomal translocations characteristic of UL (9).

Summary

Using an ELISA technique, the B cell/leukemia panel of murine monoclonal antibodies (mAbs) was screened against 18 undifferentiated lymphoma (UL) cell lines (African Burkitt's lymphoma: 6; American Burkitt's lymphoma: 7; undifferentiated lymphoma/non-Burkitt's type: 5), two normal lymphoblastoid cell lines, and one cord blood lymphoblastoid cell line. In addition, three of the American BL cell lines were rescreened with the same mAb panel by flow cytometry (FCM). The majority of the lymphoma cell lines studied reacted with antibodies detecting B cell cluster group proteins p95, p35, p135 (CD19, 20, 22), p45/55/65, p29/34, and p220. In addition, most cell lines reacted with non-clustered B cell antibodies B23, B32, B42, B10, B44, B45, and B52. Similar reactivity was observed with antibodies detecting leukemia panel cluster groups CD10 (CALLA), CD9 (p24), the transferrin receptor, and the non-clustered antibody L7. With this panel we were unable to discern phenotypic differences among the subgroups of cell lines using ELISA. Within each cluster group, marked variation in reactivity with both techniques was seen among the individual mAbs, suggesting that the antigens which they bind in lymphoma cells may vary with regard to the presence or accessibility of different epitopes. Some differences were also seen which related to the technique used. The pattern of reactivity we observed is consistent with the possibility that Burkitt's lymphoma originates from an immature B cell.

References

1. Madri, J.A., and K.W. Barwick. 1983. Use of Avidin–Biotin complex in an ELISA system: A quantitative comparison with two other immunoperoxidase detection systems using keratin antisera. *Lab. Invest.* **48:**98.
2. Newman, R.A., R. Sutherland, and M.F. Greaves. 1981. The biochemical characterization of a cell surface antigen associated with acute lymphoblastic leukemia and lymphocyte precursors. *J. Immunol.* **126:**2024.
3. Wilkinson, J.M., D.L. Wetterskog, J.A. Sogn, and T.J. Kindt. 1984. Cell surface glycoproteins of rabbit lymphocytes: characterization with monoclonal antibodies. *Mol. Immunol.* **21:**95.
4. Freeman, C.B., I.T. Magrath, D. Benjamin, R. Makuch, E.C. Douglass, and M.L. Santaella. 1982. Classification of cell lines derived from undifferentiated lymphomas according to their expression of complement and Epstein–Barr virus receptors: Implications for the relationship between African and American Burkitt's lymphoma. *J. Clin. Immunol. Immunopathol.* **25:**103.
5. LeBien, T., J. Kersey, S. Nakazawa, K. Minato, and J. Minowada. 1982. Analysis of human leukemia/lymphoma cell lines with monoclonal antibodies BA-1, BA-2 and BA-3. *Leuk. Res.* **6:**299.
6. Hofman, F.M., E. Yanagihara, B. Byrne, R. Billing, S. Baird, D. Frisman, and C.R. Taylor. 1983. Analysis of B-cell antigens in normal reactive lymphoid tissue using four B-cell monoclonal antibodies. *Blood* **63:**775.
7. Hoffman-Fezer, G., W. Knapp, and S. Thierfelder. 1982. Anatomical distribution of CALL antigen expressing cells in normal lymphatic tissue and in lymphomas. *Leuk. Res.* **6:**761.
8. Mann, R.B., E.S. Jaffe, R. Braylan, K. Nauba, M.M. Frank, J.L. Ziegler, C.W. Berard. 1976. Non-endemic Burkitt's lymphoma. A B-cell tumor related to germinal centers. *New England J. Med.* **295:**686.
9. Magrath, I.T. 1985. Burkitt's lymphoma as a human tumor model: new concepts in etiology and pathogenesis. *Ped. Hem. Onc. Rev.*, in press.

Part VI. Functional Significance of Human B Cell/Leukemia Antigens

CHAPTER 35

Human B Cell Populations Defined by the B1 and B2 Antigens

Kenneth C. Anderson, Andrew W. Boyd, David C. Fisher, John F. Daley, Stuart F. Schlossman, and Lee M. Nadler

Introduction

B cell-specific and -associated monoclonal antibodies have proven useful both for the study of normal B cell differentiation and for the categorization of B cell tumors. The B cell-specific B1 antigen is a 30-Kd nonglycosylated phosphoprotein detected on the B cell surface prior to the development of cytoplasmic μ chains (μ^+ pre-B cells) which is lost at the secretory stage of B cell differentiations (1–5). B2, a 140-Kd B cell-restricted glycoprotein, has a more limited window of expression since it appears on the cell surface after the cytoplasmic μ^+ pre-B cell stage and is lost earlier than B1 at a time when surface IgD is no longer detectable (3–6). The notion that the differential expression or coexpression of these antigens might define distinct stages of B cell differentiation is supported by the observations that 1) most cells isolated from peripheral blood coexpressed both B1 and B2 and that the B2 antigen was lost prior to B1 when B cells were triggered *in vitro* with pokeweed mitogen (1,4,6); and 2) most B cell chronic lymphocytic leukemias coexpressed B1 and B2 whereas the more differentiated B cell lymphomas, such as large-cell lymphoma and Waldenstrom's cells, were $B1^+B2^-$ (7).

We have utilized B cell-enriched cell suspensions and dual fluorescent cell sorting to isolate homogeneous normal B cell populations identified by their cell surface expression and/or coexpression of the B1 and B2 antigens. Most normal B cells are $B1^+B2^+$ whereas a minor population of cells expressed B1 but lacked B2 ($B1^+B2^-$). Moreover, these purified B cell populations are both phenotypically unique and functionally distinct when tested *in vitro* for proliferation and immunoglobulin secretion. We will also show that the relative proportions of these phenotypic subgroups may differ significantly from normal in B cell disease states.

Materials and Methods

Human B Lymphocytes

Peripheral blood was obtained from healthy donors and normal spleen, tonsil, and lymph node were obtained from operative specimens of patients not known to have any systemic or malignant disease. Lymph nodes were also obtained for phenotypic analysis from patients with the Acquired Immunodeficiency Syndrome (AIDS) (8) or AIDS-related complex (ARC) (9), diseases characterized by activated B cells and polyclonal hypergammaglobulinemia (10). Ficoll–Hypaque gradients (11) were used to prepare mononuclear cell fractions from each tissue. Mononuclear cell fractions were further enriched for B cells by E-rosetting and adherence to deplete T cells and macrophages, respectively. Cells were either used fresh or aliquoted and cryopreserved in 50% fetal calf serum at −196°C in the vapor phase of liquid nitrogen. Cells could be recovered at high (>80%) viability for later use without demonstrable change in their cell surface phenotype or functional behavior in the assays described below.

Staining of Mononuclear Populations with Anti-B1 and Anti-B2 Monoclonal Antibodies

Directly biotin-conjugated anti-B2 (Coulter Immunology, Hialeah, FL) and directly fluoresceinated (f/p 6.8) anti-B1 (Coulter Immunology) were ultracentrifuged at 40,000 × *g* for 20 min to remove aggregates immediately prior to staining. Ficoll–Hypaque mononuclear cells were first incubated for 20 min at 4°C with directly fluoresceinated anti-B1 (1 : 100) and directly biotin-conjugated anti-B2 (1 : 100). After washing, cells were developed with Texas Red-TM (Molecular Probes, Junction City, OR) conjugated to avidin (Calbiochem, La Jolla, CA) to label free biotin sites. The percentage of cells expressing only green emitting dye ($B1^+B2^-$), red emitting dye alone ($B1^-B2^+$), or coexpressing both dye fluorochromes ($B1^+B2^+$) was determined by two methods: first, by analysis of at least 10,000 viable cells using a dual-laser cell sorter (EPICS V, Coulter Electronics, Hialeah, FL); and second, by counting at least 200 cells under the fluorescent microscope (Zeiss, West Germany).

Appropriate controls for specificity of fluorescein and Texas Red staining included the staining of cells with an unreactive monoclonal antibody as a negative control as well as incubation with either anti-B1 or anti-B2 monoclonal antibodies as positive controls. All three groups were developed with fluorescein-conjugated goat anti–mouse Ig (FITC) (Coulter Immunology). Staining was also performed with an unreactive directly fluoresceinated goat anti–mouse immunoglobulin (Coulter Immunology) as a negative control and with biotin-conjugated anti-B1 and anti-B2 monoclonal antibodies developed with Texas Red (Molecular Probes) conjugated to avidin (Calbiochem) as positive controls. Staining with Texas Red–avidin alone was performed as a negative control. A further control was to label with a variety of biotin-conjugated, irrelevant monoclonal antibodies developed with Texas Red–avidin.

Cell Sorting of the B Cell Populations

Our strategy for dual staining permitted the assessment of cell viability, calibration of the cell sorter, and highly accurate quantitation of the proportion of cells expressing the $B1^+B2^+$, $B1^+B2^-$, and $B1^-B2^+$ cell surface phenotypes. Routinely 50–100 × 10^6 dual-stained unfractionated or B cell-enriched lymphoid cells were utilized as starting B cell populations for the sorting of phenotypic B cell subpopulations using the dual-laser cell sorter (EPICS V, Coulter Electronics). Light scatter and fluorescent signals (both red and green) were passed through log amplifiers, processed, and integrated in a Multiple Data Acquisition and Display Unit (MDADS, Coulter Electronics) to generate dual fluorescent contour displays (Fig. 35.1).

Analysis of DNA Content

Analysis of DNA content was performed by incubating 1 × 10^6 cells/ml at 37°C for 1 hr in the presence of 5 μg/ml of Hoechst 33342 stain (12). This allowed us to recover viable cells after analysis, wash twice, and culture the cells *in vitro*.

Phenotypic Analysis of B Cell Subsets

Cytocentrifuge preparations of the $B1^+B2^+$ and $B1^+B2^-$ subpopulations were incubated for 30 min at room temperature with peroxidase-conjugated rabbit immunoglobulins to human IgM (μ chains), IgG (γ chains), and IgD (δ chains) (Dakopatts, Copenhagen) diluted 1 : 20 in PBS as previously described (13). After washing, cytocentrifuge preparations were incubated with 3,3′-diaminobenzidine (DAB) (0.6 μg/ml, Sigma, St. Louis, MO) and hydrogen peroxide (0.01%) for 8 min at room temperature. After washing, cells were counterstained with hematoxylin and mounted. The plasma cell-associated monoclonal antibodies anti-PCA-1 and PC-1 were directly conjugated to horseradish peroxidase (14) and utilized to stain the $B1^+B2^+$ and $B1^+B2^-$ subsets as previously described (15).

A second approach to phenotypic characterization of the $B1^+B2^-$ singly stained subset involved staining splenocytes with directly biotinylated

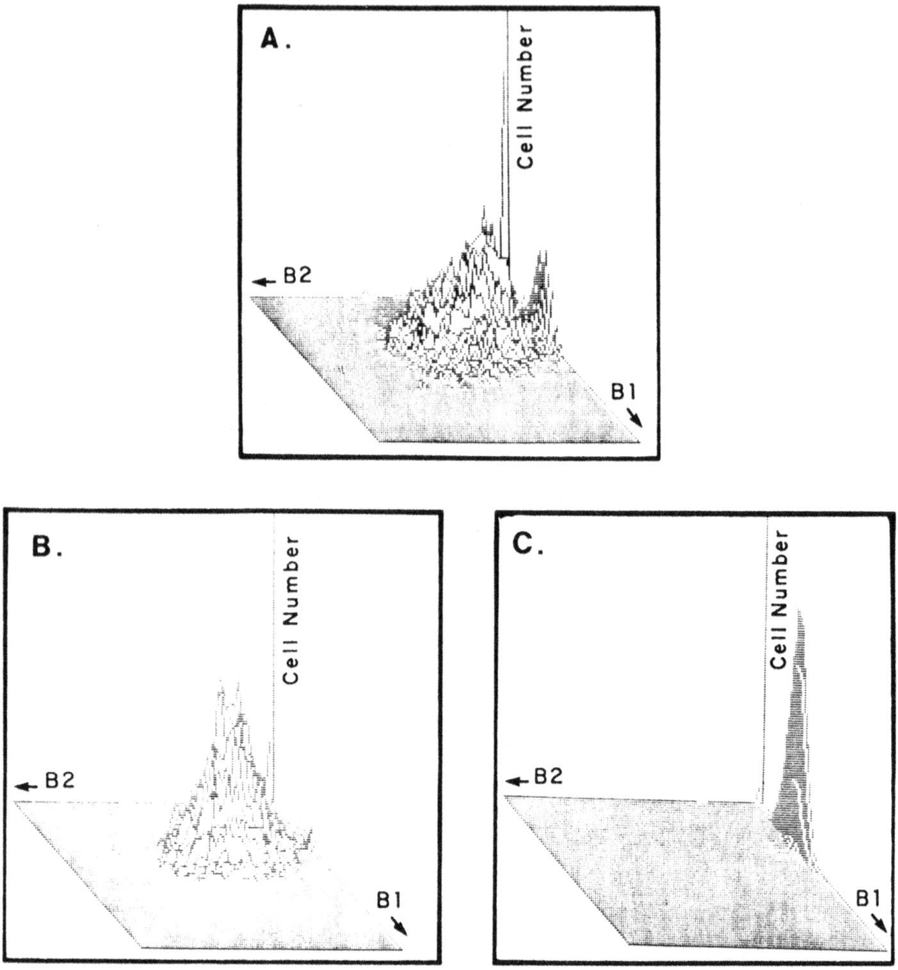

Fig. 35.1. Splenic mononuclear cells enriched for B cells by the depletion of T cells and monocytes were incubated with fluorescein-conjugated anti-B1 and biotin-conjugated anti-B2, developed with Texas Red–avidin, and then analyzed by dual-laser flow cytometry for the presence of green fluorescence alone (B1), red fluorescence alone (B2), and simultaneous green and red fluorescence (B1 and B2). Two distinct subpopulations of B cells are detectable in (A): the largest subgroup (between *x*- and *y*-axes) expresses both B1 and B2 antigens ($B1^+B2^+$) and a smaller subset of cells (*x*-axis) bears B1 only ($B1^+B2^-$). A few cells (*y*-axis) expressed only B2 ($B1^+B2^+$). The fourth population (origin) bears neither B1 nor B2 antigen ($B1^-B2^-$). Reanalysis of the sorted populations indicated that the $B1^+B2^+$ (B) and $B1^+B2^-$ (C) were phenotypically homogeneous.

anti-B1 and directly fluoresceinated anti-B2. Subsequent cell sorting to isolate those cells which express only the biotinylated antibody developed with Texas Red, the $B1^+B2^-$ subset, permitted their further characterization by staining with directly fluoresceinated (16) monoclonal reagents directed against IgM, IgD, IgG, PCA-1 (17), and PC-1 (18).

Preliminary evidence that the B2 antigen can be modulated from the cell surface permitted further phenotypic characterization of the $B1^+B2^+$ subpopulation in a similar manner. Anti-B2 was wholly removed from the cell surface of purified $B1^+B2^+$ cells by modulation at 37°C overnight in the presence of anti-B2 (1 : 100). Since the B1, IgM, IgD, IgG, PCA-1, and PC-1 antigens do not modulate under these conditions, reanalysis of the $B1^+B2^+$ subset with B2 modulated from its surface was done, utilizing directly fluoresceinated anti-IgG, IgM, IgD, PCA-1, and PC-1.

Preparation of Helper T Cell Populations

Unfractionated syngeneic mononuclear cells were first incubated with saturating concentrations of the anti-B1 and anti-B2, anti-Mo1 and anti-Mo2 (19), and anti-T8 (20) monoclonal antibodies for 30 min at 4°C. A second incubation with absorbed rabbit complement (90 min at 37°C) depleted B cells, macrophages and null cells, and T8-positive suppressor/cytotoxic T cells, respectively. The resulting populations were greater than 90% helper T cells ($T4^+$) (20) with few residual B cells as evidenced by 1) lack of anti-B1 or anti-B2 reactivity and 2) inability to produce immunoglobulin under the stimulus of PWM.

Stimulation of B Cell Populations with Rabbit Anti–Human Ig (Anti-μ) Conjugated Beads and/or Phytohemagglutinin-Stimulated Leukocyte-Conditioned Medium (PHA–LCM)

Purified subpopulations of B cells were resuspended at 5×10^5 cells/ml in RPMI/10% fetal calf serum and 100-μl aliquots dispensed in 96-well round-bottomed tissue culture plates (Costar, Cambridge, MA). Cells were cultured in the presence of either rabbit anti–μ chain-conjugated polyacrylamide beads (30 μg/ml, w/v) (Immunobead, Biorad, Richmond, CA), PHA–LCM (10% v/v), or anti-μ beads plus PHA–LCM to a final culture volume of 200 μl/well. Anti-Ig is known to induce B cell proliferation without differentiation (21–25). The PHA–LCM used in this study was produced according to the method of Aye *et al.* (26). To measure proliferation, some cultures were pulsed at 48 hr with 0.2 μCi/well (1 Ci = 3.7×10^{10} becquerels) of tritiated thymidine ([^{3}H]thymidine), harvested 15 hr later onto glass filters, and counted on a liquid scintillation counter (Packard Tri-Carb #4530, Downers Grove, IL).

Results

Dual-Laser Flow Cytometric Analysis of Splenocytes for Expression and/or Coexpression of the B1 and B2 Antigens Identifies Phenotypically Distinct Populations of B Lymphocytes

Mononuclear cells from seven spleens, either unfractionated or enriched for B cells, were fluorochrome labeled by incubating with fluorescein-conjugated anti-B1 and biotin-conjugated anti-B2 developed with Texas Red–avidin. Analysis by dual-laser flow cytometry for the presence of green fluorescence alone (B1), red fluorescence alone (B2), and simultaneous green and red fluorescence (B1 and B2) revealed two populations [Fig. 35.1(A)]: the largest subgroup of cells coexpress both B1 and B2 antigens ($B1^+B2^+$) and a smaller subset of cells bear B1 only ($B1^+B2^-$). A very small percentage of cells appear to stain with B2 and lack B1 ($B1^-B2^+$). A fourth population which bears neither the B1 nor B2 antigens ($B1^-B2^-$) is probably not B cell specific (<5% expressed SIg or PCA-1). These populations could be clearly identified by dual-laser flow cytometric analysis in all seven individuals studied.

Morphologic, Phenotypic, and Cell Cycle Characterization of the B Cell Populations

The populations of cells shown in Fig. 35.1(A) were isolated by dual-laser flow cytometric sorting for further characterization. Reanalysis of the sorted populations indicated that the major $B1^+B2^+$ population [Fig. 35.1(B)] was significantly depleted for $B1^+B2^-$ (x-axis), $B1^-B2^+$ (y-axis), or $B1^-B2^-$ (origin) contaminated cells. Moreover, the $B1^+B2^-$ cells [Fig. 35.1(C)] appeared to be phenotypically homogeneous. In contrast, we have not been able to isolate significant numbers of $B1^-B2^+$ cells utilizing current enrichment and dual-laser cell sorter techniques.

Wright–Giemsa staining of cytocentrifuge preparations of the $B1^+B2^+$ and $B1^+B2^-$ subpopulations indicated that both were composed of ⅔–¾ small and ¼–⅓ large lymphoid cells and were morphologically indistinguishable (Table 35.1). Moreover, both populations had 35–53 mean peak channels of forward light scatter (comparable to peripheral blood lymphocytes) and 30% larger cells (channels 67–148, similar to peripheral blood monocytes). Further phenotypic analysis with either peroxidase- or fluorescein-conjugated antibodies to IgM, IgG, IgD, and the plasma cell-associated and -restricted antigens PCA-1 and PC-1 did, however, reveal differences. The majority of $B1^+B2^+$ cells expressed surface IgM and IgD (90%), only 10% were weakly anti-IgG reactive, and few, if any, (<5%) $B1^+B2^+$ cells expressed PCA-1 or PC-1. In contrast, most $B1^+B2^-$ cells (70%) weakly expressed IgM and only 10% weakly expressed IgD, but

able 35.1. Morphologic, phenotypic, and cell cycle characterization of the B1$^+$B2$^+$ and 1$^+$B2$^-$ subpopulations.

	Morphology[a]		% of cells reactive with monoclonal antibody[b] (intensity of antigen expression[c])				Cell cycle[d]	
	Small	Large	IgM	IgD	IgG	PCA-1 or PC-1	% in G_0/G_1	% in S-G_2/M
1$^+$B2$^+$	⅔	⅓	90(++)	90(+)	10(+/−)	5(+/−)	96	4
1$^+$B2$^-$	¾	¼	70(+)	10(+/−)	20(+)	30(+/++)	99	1

The lymphoid morphology was confirmed on Wright–Giemsa cytospin preparations. The relative size of B1$^+$B2$^+$ nd B1$^+$B2$^-$ cells was comparable with mean peak channels of forward light scatter (35–53) and 30% larger cells hannels 67–148). The small and larger cells are of comparable size to peripheral blood lymphocytes and monocytes, spectively.

Both directly fluoresceinated and peroxidase-labeled monoclonal reagents were utilized. The percentage of cells active with each monoclonal was determined by counting at least 200 cells under the fluorescent microscope and ght microscope, respectively.

Analysis of DNA content was performed by incubating 1×10^6 cells/ml at 37°C for 1 hr in the presence of 5 μg/ml of oechst 33342 stain (12).

The intensity of reactivity was qualitatively assessed as weak (+/−), moderate (+), or strong (++) by counting at ast 200 cells under the fluorescent or light (peroxidase-labeled reagents) microscope.

approximately 20% of the B1$^+$B2$^-$ cells expressed IgG and, most importantly, 30% bear PCA-1 or PC-1. To date, the latter antigens have been demonstrated on plasma cell tumors and not on normal B cells, supporting the view that splenic B1$^+$B2$^-$ cells may be more differentiated than the B1$^+$B2$^+$ population. Analysis with Hoechst DNA vital staining of the B1$^+$B2$^+$ and B1$^+$B2$^-$ cells revealed that the majority (96–99%) of both populations were resting (G0/G1) and only a minority (1–4%) were in cycle (S-G2/M).

B Cell Subpopulations Can Be Demonstrated in All Lymphoid Tissues

The overall distribution of these populations was similar in normal lymph node, tonsil, and peripheral blood; however, within each lymphoid tissue examined, the proportion of cells contained in each population varied (Table 35.2). The majority of cells coexpressed B1 and B2 and only a minor population expressed B1 and lacked B2. A small population of cells appeared to express B2 and lack B1 (B1$^-$B2$^+$) but after purification for B cells the percentage of cells which stained with biotinylated anti-B2 Texas Red–avidin but not with directly fluoresceinated anti-B1 (B2$^+$B1$^-$) was not significantly greater than the staining noted with an irrelevant biotinylated antibody. B1$^+$B2$^-$ cells were also depleted somewhat by purification steps but they, in contrast, could not be depleted below 5–10% of the total cells at a time when very few B1$^-$B2$^-$ cells remained. Thus, while selective losses of B1$^-$B2$^+$ cells cannot be excluded, most losses in the single stained fractions during purification can be accounted for by depletion of nonspecifically stained contaminating cells.

Table 35.2. Phenotypic distribution of B cell subpopulations in lymphoid tissues.

Tissue	No of tests	% of cells in each phenotypic subgroup[a] B1+B2+	B1+B2−	B1−B2+	B1−B2−
Spleen					
Unfractionated[b]	7	52 ± 10	13 ± 7	4 ± 2	31 ± 14
E-rosette and adherence depleted[c]	4	78 ± 5	10 ± 4	3 ± 1	9 ± 2
Lymph node					
Unfractionated	4	37 ± 17	6 ± 2	6 ± 2	51 ± 17
E-rosette and adherence depleted	2	82 ± 2	6 ± 3	3 ± 1	10 ± 5
Tonsil					
Unfractionated	3	55 ± 2	12 ± 7	6 ± 2	27 ± 5
E-rosette and adherence depleted	2	86 ± 3	7 ± 1	4 ± 1	4 ± 1
Peripheral blood					
E-rosette and adherence depleted	4	41 ± 6	12 ± 2	3 ± 2	45 ± 6

[a] The percentage of cells expressing either fluorochrome alone or coexpressing both dyes was determined by two methods: First, by analysis of at least 10,000 viable cells using a dual-laser cell sorter (EPICS V, Coulter Electronics, Hialeah, FL); and second, by counting at least 200 cells under the fluorescent microscope (Zeiss, West Germany). Results obtained by microscopy confirmed the results obtained by cytofluorographic analysis.
[b] Unfractionated cells represent mononuclear cells obtained from each lymphoid tissue by Ficoll–Hypaque density sedimentation.
[c] Ficoll–Hypaque mononuclear cells were enriched for B cells by E-rosetting and adherence to deplete T cells and macrophages, respectively.

Phenotypically Distinct B Cell Populations Have Unique Responses to Triggers of Proliferation and Ig Secretion

Unsorted B cell-enriched populations had comparable responses to anti-μ before and after incubation with anti-B1 and/or anti-B2, confirming that staining with either or both reagents did not alter their response to anti-μ. The $B1^+B2^+$ and $B1^+B2^-$ cells were then isolated by dual-laser cell sorting. Spontaneous proliferation, as measured by overnight incorporation of [^{3}H]thymidine (^{3}H-Tdr), was minimal and comparable in both the $B1^+B2^+$ (374 ± 84 cpm) and $B1^+B2^-$ (416 ± 80 cpm) populations. The two subsets were tested for their response to anti-μ beads, PHA–LCM, and a mixture of the two (Table 35.3). Significant proliferation of the $B1^+B2^+$ cells (S.I. = 4)* was noted to anti-μ beads whereas the $B1^+B2^-$ cells showed little if any response (S.I. = 0.8). Both the $B1^+B2^+$ (S.I. = 1.9) and the $B1^+B2^-$ (S.I. = 1.5) groups had a minimal but reproducible response to PHA–LCM alone. The combination of PHA–LCM and anti-μ

* S.I. = Stimulation index.

Table 35.3. Proliferation of B cell subpopulations in response to anti-μ bound to beads.[a]

	B cell test population[b]			
	$B1^+B2^+$ spleen cells		$B1^+B2^-$ spleen cells	
Stimulus	^{3}H-Tdr uptake (cpm)	S.I.[c]	^{3}H-Tdr uptake (cpm)	S.I.
None	1506 ± 134	—	2256 ± 196	—
Anti-μ	5947 ± 136[d]	4.0	1745 ± 219	0.8
PHA–LCM	2910 ± 87[e]	1.9	3374 ± 138[e]	1.5
Anti-μ + PHA–LCM	17106 ± 490[d]	11.5	3991 ± 213[e]	1.8

[a] Populations of cells were cultured at 3×10^4/well in the presence of either anti-μ conjugated to polyacrylamide beads (1 : 60 dilution) or PHA–LCM (10% v/v) or both. After 2 days the cultures were pulsed with 0.2 cCi/well of [^{3}H]thymidine and the cultures continued for 15 hr. Cells were then harvested onto glass filters and counts per minute enumerated.
[b] Spleen cells were enriched for B cells by depletion of E-rosetting cells and adherent cells. B cell-enriched starting populations contained approximately 80% B1 positive. The prefractionated cells were stained with fluoresceinated anti-B1 antibody and biotin-conjugated anti-B2 antibody, then developed with Texas Red–avidin. The cells were sorted which demonstrated both red (b2) and green (B1) fluorescence or green fluorescence alone (B1) to yield highly purified $B1^+B2^+$ and $B1^+B2^-$ subpopulations, respectively (see Materials and Methods).
[c] Stimulation index = ^{3}H-Tdr uptake of sample/^{3}H-Tdr uptake of medial control.
[d] Significantly greater than control ($p < 0.01$).
[e] Significantly greater than control ($p < 0.05$).

resulted in significant augmentation in proliferation of the $B1^+B2^+$ subset (S.I. = 11.5) whereas the incorporation of ^{3}H-Tdr by $B1^+B2^-$ cells was not greater than that induced by PHA–LCM alone (S.I. = 1.8). These observations suggest that triggers of B cell proliferation including anti-μ beads and PHA–LCM result in different patterns of response by the $B1^+B2^+$ and $B1^+B2^-$ subsets.

Neither the $B1^+B2^+$ nor the $B1^+B2^-$ populations produced Ig spontaneously when cultured overnight in media. To assay for their ability to produce Ig in a pokeweed mitogen (PWM)-driven system, $T4^+$ cells with and without PWM were added to each subset. As can be seen in Table 35.4, unfractionated splenic mononuclear cells, $B1^+B2^+$ cells, $B1^+B2^-$ cells, and $T4^+$ cells did not produce significant amounts of Ig when cultured in media alone. The addition of PWM to $T4^+$ cells also resulted in no Ig production, confirming that depletion of functional B cells was virtually complete. $T4^+$ cells, PWM, or both were then added to purified $B1^+B2^+$ and $B1^+B2^-$ subpopulations. $B1^+B2^+$ cells produced quantities of IgG equivalent to that noted for unfractionated splenic mononuclear cells only when both PWM and $T4^+$ cells were added, but not when either mitogen or accessory cells were added alone. In contrast, the $B1^+B2^-$ cells produced comparable quantities of Ig in response to $T4^+$ cells alone and did not augment their Ig production when PWM was added. Thus, the B1-alone population can be triggered by T4 cells in the absence of PWM to secrete Ig whereas the $B1^+B2^+$ subset requires both T4 cells and PWM to similarly respond.

Table 35.4. Pokeweed mitogen-driven IgG production.

	IgG (ng)[a]	
	Media	PWM[b]
Unfractionated splenic mononuclear cells	<1	>250
$B1^+B2^+$ cells	<1	<1
$B1^+B2^-$ cells	<1	<1
$T4^+$ cells[c]	<1	<1
$B1^+B2^+$ + $T4^+$ cells	<1	>250
$B1^+B2^-$ + $T4^+$ cells	>250	>250

[a] Nanograms of IgG quantitated by radioimmunoassay at day 7 after culture with either PWM or media. Each culture contained 5×10^4 test cells.
[b] PWM concentration 1 : 300.
[c] Splenic mononuclear cells were incubated for 30 min at 4°C with anti-B1 and anti-B2, anti-Mol and anti-Mo2, and anti-T8 monoclonal antibodies. After washing, a second incubation for 90 min at 37°C with absorbed rabbit complement resulted in complete depletion of all B cells, monocytes and null cells, and suppressor T cells, respectively.

Alterations in B Cell Subpopulations Can Be Demonstrated in Disease States

To determine whether these normal B cell populations are present in disease states, we compared hyperplastic lymph nodes to those from patients with AIDS or AIDS-related complex (ARC). Enriched B cell separations were obtained from mononuclear cells by E-rosette depletion of T cells and adherence to remove monocytes. A single patient with AIDS and Kaposi's sarcoma, seven patients with ARC, and seven patients with enlarged but apparently normal nodes were studied (Table 35.5). The percentage of fluorochrome-positive cells coexpressing B1 and B2 ($B1^+B2^+$) in ARC patients (64 ± 6) was less ($p < 0.01$) than that for patients with hyerplastic nodes (87 ± 4). Conversely, the percentage of stained cells bearing B1 only ($B1^+B2^-$) was significantly greater ($p < 0.01$) in ARC patients (31 ± 7) than in those with reactive nodes (10 ± 2). In the single patient with AIDS and Kaposi's sarcoma, the $B1^+B2^-$ cells were in fact the predominant subpopulation (55% $B1^+B2^-$ versus 34% $B1^+B2^+$), a reversal of the normal patterns. Thus, the proportional change of $B1^+B2^+$ relative to $B1^+B2^-$ cells, i.e., 10 : 1 in normal nodes (Tables 35.3 and 35.5) to 2 : 1 or even a reversed ratio in AIDS/ARC cases (Table 35.5), clearly defines this major shift. The relative number of cells bearing the $B1^+B2^-$ phenotype, a subset of normal B cells, may therefore provide a marker for disturbances of normal B cell function.

Table 35.5. B cell subsets in disease states: percentage of fluorochrome-labeled mononuclear cells[a] expressing the B1 and/or B2 antigen.

Diagnosis	Number of tests	$B1^+B2^+$ ($X \pm$ S.E.M.)	$B1^+B2^-$ ($X \pm$ S.E.M.)	$B1^-B2^+$ ($X \pm$ S.E.M.)
AIDS[b]	1	34	55	11
ARC[c]	7	64 ± 6	31 ± 7	4 ± 3
Normal nodes[d]	7	87 ± 4[e]	10 ± 2[f]	2 ± 1[g]

[a] Ficoll–Hypaque mononuclear cells were T cell- and monocyte-depleted by E-rosetting and adherence, respectively prior to phenotypic staining and analysis.
[b] AIDS patient with Kaposi's sarcoma.
[c] AIDS-related complex consisting of persistent generalized lymphadenopathy among homosexuals or those who do not meet the formal CDC definition of AIDS, i.e., no biopsy-proven Kaposi's sarcoma in persons over 60 years old and/or biopsy- or culture-proven *pneumocystis carinii* pneumonia or life-threatening opportunistic infections in young previously healthy persons with no underlying cause of immune deficiency (9).
[d] Includes patients in whom biopsy revealed histologically normal lymphadenopathy and who had no evidence of systemic disease.
[e] Significantly greater than ARC ($p < 0.01$) as determined by student's t test.
[f] Significantly less than ARC ($p < 0.01$).
[g] Not significantly different from ARC.

Discussion

In the present report, we have utilized dual-fluorochrome staining and flow cytometric analysis to identify two B cell populations in normal peripheral blood, lymph node, spleen, and tonsil. The majority of B cells identified coexpressed both B1 and B2 ($B1^+B2^+$), whereas a minority (5–15%) expressed only B1 ($B1^+B2^-$). Although a small number of cells appear to exhibit the $B1^-B2^+$ phenotype, we have been unable to purify or further analyze significant numbers of these cells. Sufficient numbers of $B1^+B2^+$ and $B1^+B2^-$ cells, however, could be obtained, and it was demonstrated by both light microscopy and by cytofluorographic light scatter that both populations were composed of ⅓ large and ⅔ small cells. Wright–Giemsa staining confirmed that these populations were wholly lymphoid and morphologically indistinguishable. Although *in situ* studies of lymphoid tissues may be more sensitive and detect slight variations in the percentage of each B cell population, our studies utilizing single-cell suspensions do suggest that the $B1^+B2^+$ and $B1^+B2^-$ subsets can be identified in normal peripheral blood, lymph node, spleen, and tonsil.

Hoechst vital dye staining indicated that both $B1^+B2^+$ and $B1^+B2^-$ populations were noncycling. Moreover, neither population demonstrated spontaneous thymidine uptake when cultured in media. In contrast, these populations were phenotypically distinct. The moderate to strong expression of IgM and IgD coupled with weak expression of IgG on $B1^+B2^+$ cells suggests that they may be less mature than $B1^+B2^-$ cells, which strongly expressed IgG and weakly expressed IgM and IgD. The presence of the plasma cell-associated antigens PCA-1 and PC-1 on up to ⅓ of $B1^+B2^-$ cells and their absence from $B1^+B2^+$ cells further supports this view.

The distinction between the $B1^+B2^+$ and $B1^+B2^-$ populations was confirmed by their differential responses to mitogen, T cells, and PHA–LCM. The $B1^+B2^+$ subset proliferated to anti-μ whereas the $B1^+B2^-$ did not. In addition, the $B1^+B2^+$ cells required both pokeweed mitogen and T cells to produce Ig whereas the $B1^+B2^-$ cells responded to T cells alone. These observations provide additional evidence that the expression and/or coexpression of the B1 and B2 antigens define B cell populations which are not only phenotypically but also functionally distinct.

Considerable evidence now supports the notion that "resting" B cells must be "activated" either by immunoglobulin cross-linking or by mitogen, and proliferate before differentiating to produce Ig (28–30). T cell-derived lymphokines induce proliferation (B cell growth factor, BCGF) (31–35) and maturation (B cell differentiation factor, BCDF) (34,36–37) of "activated" B cells. Kehrl and Fauci have reported that "resting" B cells are small lymphocytes which lack an activation antigen 4F2 in contrast to "activated" B cells, which are larger and express 4F2 (38). In contrast, no similar cell size differences were noted between our $B1^+B2^+$ and $B1^+B2^-$ cells. Nevertheless, based upon the ability of the $B1^+B2^+$ and $B1^+B2^-$ populations to proliferate and differentiate to known triggers *in vitro*, it can be postulated that the $B1^+B2^+$ and $B1^+B2^-$ cells may represent stages in a continuum between "resting" and more "differentiated" B cells. The $B1^+B2^+$ cells are not cycling (>95% G0/G1) and neither spontaneously proliferate to PHA–LCM nor produce immunoglobulin in response to T4 cells alone. It would appear that these cells lack the ability to respond to both BCGF and BCDF, and can be induced by anti-μ to respond to BCGF. Moreover, they can be triggered by PWM and $T4^+$ cells to respond to BCDF and produce immunoglobulin. In addition, Hoechst vital staining revealed that $B1^+B2^+$ cells stimulated by anti-μ and T cell-derived growth factors began to cycle (>25% S-G2/M). These data provide evidence that the $B1^+B2^+$ cells cannot respond to BCGF or BCDF and are probably "resting." The $B1^+B2^-$ cells are also not spontaneously proliferating (99% G0/G1) or producing immunoglobulin. In contrast to the $B1^+B2^+$ cells, they cannot be driven to proliferate by anti-μ (S-G2/M <5%). The ability of the $B1^+B2^-$ cells to secrete Ig in response to $T4^+$ cells in the absence of mitogens suggests that they may be further along in the differentiation pathway and already have the capacity to respond to BCDF. Thus, they do not require mitogens, as do $B1^+B2^+$ cells, to manifest this response. The expression of plasma cell-associated antigens PCA-1 and PC-1 on 30% of $B1^+B2^-$ cells also supports this view.

Confirmation of the *in vivo* significance of the $B1^+B2^+$ and $B1^+B2^-$ B cell populations was obtained from the observation that shifts in the relative proportion of $B1^+B2^+$ and $B1^+B2^-$ cells could be noted in disease states. In lymph nodes from patients with AIDS and ARC, who have elevated numbers of cells spontaneously secreting immunoglobulins and decreased B cell proliferative responses to T cell-independent B cell mitogens (21–23), there was a marked shift to predominantly $B1^+B2^-$ B cells. This observation supports the view that the $B1^+B2^-$ phenotype may identify a more differentiated B cell phenotype and further suggests that analysis of the B cells may yield additional parameters which reflect disease activity (i.e., the number of $B1^+B2^-$ B cells).

Summary

Distinct phenotypic and functional populations of human B lymphocytes can be identified by their expression and/or coexpression of the B cell-restricted antigens B1 and B2. Dual-fluorochrome staining and flow cytometric cell sorting permitted the isolation of the $B1^+B2^+$ and $B1^+B2^-$ cells to homogeneity. Virtually all $B1^+B2^+$ cells expressed IgM and IgD but lacked IgG and the plasma cell antigens PCA-1 and PC-1 whereas the $B1^+B2^-$ cells more frequently expressed IgG, PCA-1, and PC-1. Both populations were noncycling and were composed of similar percentages of small and large cells. The $B1^+B2^+$ cells proliferate to anti-μ, or to anti-μ + PHA–LCM but not to PHA–LCM alone. They require both T cells and PWM to produce immunoglobulin. In contrast, $B1^+B2^-$ cells do not significantly proliferate to anti-μ, PHA–LCM, or anti-μ and PHA–LCM. They produce Ig in response to T cells alone without PWM. These phenotypic and functional observations provide preliminary evidence that these populations are distinct and that the $B1^+B2^+$ cell may be a "resting" B cell whereas the $B1^+B2^-$ cell appears to be more "differentiated." The present studies further suggest that they will also be helpful in characterizing B cells in some human disease states.

Acknowledgments. The authors would like to thank Mr. Herbert Levine and Mr. David Leslie for technical assistance in flow cytometry. We would also like to thank Ms. Bonnie Frisard for outstanding secretarial assistance during the preparation of this manuscript. This work was supported in part by National Institutes of Health grants CA25369, CA19589, and CA34183. K.C. Anderson is the recipient of a fellowship from the Medical Foundation, Boston, MA. A.W. Boyd is the recipient of a Neil Hamilton Fairley Fellowship of the National Health and Medical Research Council, Australia.

References

1. Stashenko, P., L.M. Nadler, R. Hardy, and S.F. Schlossman. 1980. Characterization of a human B lymphocyte specific antigen. *J. Immunol.* **125:**1678.
2. Nadler, L.M., P. Stashenko, J. Ritz, R. Hardy, J.M. Pesando, and S.F. Schlossman. 1981. A unique cell surface antigen identifying lymphoid malignancies of B cell origin. *J. Clin. Invest.* **67:**134.
3. Oettgen, H.C., P.J. Bayard, W. van Ewijk, L.M. Nadler, and C.P. Terhorst. 1983. Further biochemical studies of the human B-cell differentiation antigens B1 and B2. *Hybridoma* **2:**17.
4. Stashenko, P., L.M. Nadler, R. Hardy, and S.F. Schlossman. 1981. Expression of cell surface markers after human B lymphocyte activation. *Proc. Natl. Acad. Sci. U.S.A.* **78:**3848.
5. Nadler, L.M., S.J. Korsmeyer, K.C. Anderson, A.W. Boyd, B. Slaughenhoupt, E. Park, J. Jensen, F. Coral, R.J. Mayer, S.E. Sallan, J. Ritz, and S.F. Schlossman. 1984. The B cell origin of non-T cell acute lymphoblastic leukemia: A model for discrete stages of neoplastic and normal pre-B cell differentiation. *J. Clin. Invest.* **74:**332.
6. Nadler, L.M., P. Stashenko, R. Hardy, A. van Agthoven, C. Terhorst, and S.F. Schlossman. 1981. Characterization of a human B cell specific antigen (B2) distinct from B1. *J. Immunol.* **126:**1941.
7. Anderson, K.C., M.P. Bates, B.S. Slaughenhoupt, G.S. Pinkus, S.F. Schlossman, and L.M. Nadler. 1984. Expression of human B cell associated antigens on leukemias and lymphomas: A model of human B cell differentiation. *Blood* **63:**1424.
8. Gottlieb, M.S., J.E. Groopman, W.M. Weinstein, J.L. Fahey, and R. Detels. 1983. UCLA Conference: The acquired immunodeficiency syndrome. *Ann. Int. Med.* **99:**208.
9. Persistent, generalized lymphadenapathy among homosexual males. 1982. *M.M.W.R.* **31:**249.
10. Lane, H.C., H. Masur, L.C. Edgar, G. Whalen, A.H. Rook, and A.S. Fauci. 1983. Abnormalities of B cell activation and immunoregulation in patients with the Acquired Immunodeficiency Syndrome. *New England J. Med.* **309:**453.
11. Boyum, A. Isolation of mononuclear cells and granulocytes from human blood. 1968. *Scand. J. Clin. Lab. Invest.* (Suppl) **97:**77.
12. Kruth, H.S. 1982. Flow cytometry: Rapid biochemical analysis of single cells. *Anal. Biochem.* **125:**225.
13. Weir, E.E., T.G. Pretlow, A. Pitts, and E.E. Williams. 1974. A more sensitive and specific technique for the localization of cellular antigen by the enzyme–antibody conjugate method. *J. Histochem. Cytochem.* **22:**1135.
14. NaKane, P.K., and A. Kawaoi. 1974. Peroxidase-labeled antibody. A new method of conjugation. *J. Histochem. Cytochem.* **22:**1084.
15. Warnke, R.A., K.C. Gatter, B. Folini, P. Hildreth, R.E. Woolston, K. Pulford, J.L. Cordell, B. Cohen, C. DeWolf-Peeters, and D.Y. Mason. 1983. Diagnosis of human lymphoma with monoclonal anti-leukocyte antibodies. *New England J. Med.* **309:**1275.
16. Johnson, G.D., E.J. Holborow, and J. Darling. 1978. Immunofluorescence

and immunoenzyme techniques. In: *Handbook of experimental immunology,* D.M. Weir, ed. Blackwell, London, p. 1511.
17. Anderson, K.C., E.K. Park, M.P. Bates, R.C.F. Leonard, S.F. Schlossman, and L.M. Nadler. 1983. Antigens on human plasma cells identified by monoclonal antibodies. *J. Immunol.* **130:**1132.
18. Anderson, K.C., M.P. Bates, B.L. Slaughengoupt, S.F. Schlossman, and L.M. Nadler. 1984. A monoclonal antibody with reactivity restricted to normal and neoplastic plasma cells. *J. Immunol.* **132:**3172.
19. Todd, R.F., L.M. Nadler, and S.F. Schlossman. 1981. Antigens on human monocytes identified by monoclonal antibodies. *J. Immunol.* **126:**1435.
20. Reinherz, E.L., and S.F. Schlossman. 1980. The differentiation and function of human T lymphocytes. *Cell* **19:**821.
21. Sieckmann, D.G. 1980. The use of anti-immunoglobulin to induce a signal for cell division in B lymphocytes via their membrane IgM and IgD. *Immunol. Rev.* **52:**181.
22. Fothergill, J., R. Wistar, J.N., Woody, and D.C. Parker. 1982. A mitogen for human B cells: Anti-Ig coupled to polyacrylamide beads activates blood mononuclear cells independently of T cells. *J. Immunol.* **128:**1945.
23. Yoshizaki, K., T. Nakagawa, T. Kaieda, A. Muraguchi, Y. Yamamura, and T. Kishimoto. 1982. Induction of proliferation and Ig production in human B leukemic cells by anti-immunoglobulins and T cell factors. *J. Immunol.* **128:**1296.
24. DeFranco, A.L., E.S. Raveche, R. Asofsky, and W.E. Paul. 1982. Frequency of B lymphocytes responsive to anti-immunoglobulin. *J. Exp. Med.* **155:**1523.
25. Muraguchi, A., J.L. Butler, J.H. Kehrl, and A.S. Fauci. 1983. Differential sensitivity of human B cell subsets to activation signals delivered by anti-μ antibody and proliferative signals delivered by monoclonal B cell growth factor. *J. Exp. Med.* **157:**530.
26. Aye, M.T., R. Nihoy, J.E. Till, and J.E. McCulloch. 1974. Studies of leukemic cell populations in culture. *Blood* **44:**205.
27. Zuraw, B.L., M. Nonaka, C. O'Hair, and D.H. Katz. 1981. Human IgE antibody synthesis *in vitro:* stimulation of IgE responses by pokeweed mitogen and selective inhibition of such responses by human suppressive factor of allergy (SFA). *J. Immunol.* **127:**1169.
28. Dutton, R.W. 1975. Separate signals for the initiation of proliferation and differentiation in the B cell response to antigen. *Transplant Rev.* **23:**66.
29. Fauci, A.S., H.C. Lane, and D.J. Volkman. 1983. Activation and regulation of human immune responses: Implications in normal and disease states. *Ann. Int. Med.* **99:**61.
30. Falkoff, R.J.M., L.P. Zhu, and A.S. Fauci. 1982. Separate signals for human B cell proliferation and differentiation in response to *Staphylococcus aureus.* Evidence for a two signal model of B cell activation. *J. Immunol.* **129:**97.
31. Maizel, A.L., J.W. Morgan, S.R. Mehta, N.M. Kanttah, J.M. Bator, and C.G. Sahasrabuddhe. 1983. Long term growth of human B cells and their use in a microassay for BCGF. *Proc. Natl. Acad. Sci. U.S.A.* **80:**5047.
32. Ford, R.J., S. Mehta, D. Franzini, R. Montogno, L.B. Lochman, and A.L. Maizel. 1981. Soluble factor activation of human B lymphocytes. *Nature* **294:**261.

33. Muraguchi, A., and A.S. Fauci. 1982. Proliferative responses of normal human B lymphocytes. Development of an assay for human B cell growth factor (BCGF). *J. Immunol.* **129:**1104.
34. Muraguchi, A., T. Kashahara, J.J. Oppenheim, and A.S. Fauci. 1982. B cell growth factor and T cell growth factor produced by mitogen stimulated normal human peripheral blood T lymphocytes are distinct molecules. *J. Immunol.* **129:**2486.
35. Yoshizaki, K., T. Nakagawa, K. Fukunaga, T. Kaieda, Y. Yamamura, and T. Kishimoto. 1983. Characterization of human B cell growth factor (BCGF) from cloned T cells or mitogen stimulated T cells. *J. Immunol.* **130:**1241.
36. Okada, M., N. Yoshimura, T. Kaieda, Y. Yamamura, and T. Kishimoto. 1981. Establishment and characterization of human T hybrid cells secreting immunoregulatory molecules. *Proc. Natl. Acad. Sci. U.S.A.* **78:**7717.
37. Elkins, K., and J.C. Cambier. 1983. Constitutive production of a factor supporting B lymphocyte differentiation by a T cell hybridoma. *J. Immunol.* **130:**1247.
38. Kehrl, J.H., and A.S. Fauci. 1983. Identification, purification and characterization of antigen-activated and antigen-specific human B lymphocytes. *J. Exp. Med.* **157:**1692.

CHAPTER 36

Phenotypic Changes Occurring during *in vitro* Activation of Human Splenic B Lymphocytes

Andrew W. Boyd, Arnold S. Freedman,
Kenneth C. Anderson, David C. Fisher, Jack C. Horowitz,
John F. Daley, Stuart F. Schlossman, and Lee M. Nadler

Introduction

The present study had two motivations: first to develop a system in which normal B cell differentiation could be studied *in vitro*. Second, to find out which antigens were the hallmarks of individual events in B cell development, with the hope that likely candidates for receptors could be identified. We show that highly purified splenocytes can be activated by anti-Ig antibody *in vitro* (1–2) and the phenotypic changes induced by activation defined. When T cell help was provided, these cells could be driven to differentiate (produce Ig) and a further characteristic set of phenotypic changes were shown to occur. Finally, an example is given of how this system can be used to study the function of one of the receptors (IL-2 receptor) induced by activation.

Materials and Methods

Human Spleen Cells

Spleen cells were obtained from operative specimens of normal individuals. After dissociation into single-cell suspensions, the mononuclear cells were isolated on Ficoll–Hypaque gradients. The cells were further fractionated by E-rosetting into B cell-enriched (E^-) and T cell-enriched (E^+) fractions. Both were cryopreserved in liquid nitrogen.

Table 36.1. Monoclonal antibodies used to characterize *in vitro* cultured B cells.

Antibody	Specificity	Reference
B1	Pan-B cell	3
B2	Resting B cells	3
B4	Pan-B cell	3
B5	Activated B cells	This study
BB1	Activated B cells	4
B-LAST 1	Activated B cells	5 and this study
PC 1	Subset of plasma cells	3
PCA-1	Plasma cells, activated T cells	3
anti-IL2	Activated B and T cells	6 and this study
T4	Helper/inducer T cells	3
T8	Cytotoxic/suppressor T cells	3
T10	Plasma cells, T cells	3
T11	Pan-T cell	3
T12	Pan-T cell, activated B cells	3 and this study
Mo1	Monocytes, granulocytes, NK cells	3
Mo2	Monocytes	3

Antibodies

The panel of monoclonal antibodies used in this study is shown in Table 36.1. In addition, an affinity-purified rabbit antiserum against $F(ab')_2$ fragments of human IgG was used in this study.

Interleukin-2 (IL-2)

Cloned IL-2 was the gift of Biogen, Boston, MA.

Anti-Ig Antibody Beads

Two batches of anti-Ig antibody-conjugated beads were used. Both were prepared by coupling $F(ab')_2$ fragments of the rabbit antibody described above to Affigel 702 beads (Bio-Rad, Richmond, CA). One batch induced maximal mitogenesis (when compared to beads and T cell supernatant together) when used alone, and was used in bulk cultures of B cells. A second batch induced only 30–40% of maximal response alone and was used to study augmentation by T cell factors.

Fluorescent Staining

All staining (except for Ig) was performed in 10% pooled AB serum. Cells were mixed with appropriate dilutions of monoclonal antibody and held on ice for 45 min. The cells were washed and stained with fluoresceinated anti-mouse Ig (Tago Inc., Burlingame, CA) for 20 min at 4°C. Direct stains with fluoresceinated reagents or with biotin conjugates which were developed with Texas Red–avidin (Coulter Immunology, Hialeah, FL), were incubated with the cells for 45 min. In all cases the cells were washed twice and prepared for analysis. Samples were analyzed or sorted on an EPICS V cell sorter (Coulter Electronics, Hialeah, FL).

B Cell Cultures

Bulk Cultures

B cells were purified from E^- splenocytes by lysis with anti-T4, anti-T8, anti-Mo1, anti-Mo2, and rabbit complement, and cultured at 10^6/ml in RPMI 1640/10% FCS/2 m*M* glutamine/1 m*M* pyruvate. The cells were activated by adding anti-Ig beads to the cultures. For cultures continued beyond 4 days, E^+ cells (10^5/ml) and anti-$T11_2$ and anti-$T11_3$ antibodies were included to provide activated T cells (7).

Microcultures

Microcultures were established in the same medium, at 10–50,000/well in a final culture volume of 200 μl in 96-well microtiter trays (Costar, Cambridge, MA).

Thymidine Uptake

Tritiated thymidine uptake was used as an index of mitogenesis. Microcultures were pulsed with 0.2 μCi/well of [^{3}H]thymidine (Amersham, Eastbourne, England) and harvested 15 hours later. Dried glass filters were counted on a Packard Tri-Carb scintillation counter (Packard, Downers Grove, IL).

Radioimmunoassay

Radioimmunoassay was performed as previously described (3).

Radiolabeling of Cells

Cell surface iodination and biosynthetic labeling were performed as previously described (8). The cells were solubilized in 0.5% Triton X-100 in 50 m*M* Tris, 0.4 *M* NaCl, 2 m*M* PMSF, 2 m*M* EDTA, pH 8. Immunoprecipitations of Ig were performed by preclearing twice with Sansorbin (Calbiochem, La Jolla, CA) and finally by absorbing Ig to anti–human Ig bound to Pansorbin (Calbiochem, La Jolla, CA). Other samples were precleared twice on Pansorbin, once on protein A–Sepharose, and once on an irrelevant monoclonal linked to Sepharose. Specific complexes were absorbed out by specific antibody linked to Sepharose 4B. All precipitates were washed 4 times in lysis buffer, and analyzed by SDS–PAGE.

Results and Discussion

Characterization of the B Cell Responding to Anti-Ig Antibody

Preliminary experiments on spleen E^- cells demonstrated an intense mitogenic response to anti-Ig beads. The kinetics of this response are depicted in Fig. 36.1. The strong response to beads alone was augmented by the addition of T cell supernatants, as expected from the studies of others (9–12). However, the increase in cell numbers was small, the maximum cell number being typically only 1.5 times the number of cells at the start of the culture. one possible explanation for this discrepancy was that the response was generated by a subpopulation of splenic B cells. To investigate this possibility spleen E^- cells were stained with anti-B1–FITC and anti-B2–biotin/avidin–Texas Red and analyzed cytofluorimetrically. Two distinct populations were identified, one bearing both antigens ($B1^+B2^+$) and the other only expressing B1 antigen. However, the major population ($B1^+B2^+$) was further subdivided into low B2 and high B2 antigen-bearing populations. The two peaks of differing B2 intensity differ in individual spleens as shown in the three examples depicted in Fig. 36.2(A)–(C). The second case [Fig. 36.2(B)] was the most representative of the normal pattern. Analyses of the same spleens dual-stained with both anti-B1 and anti-B2 are shown in Fig. 36.2(D)–(F). Dual-fluorescent histograms clearly demonstrate the three subpopulations: B1 alone, low B2, and high B2.

Having identified these populations we attempted to sort them cleanly, excluding cells in the regions of overlap. The results of sorting the population shown in Fig. 36.2(E) are depicted in Fig. 36.2(G)–(I), which show B1

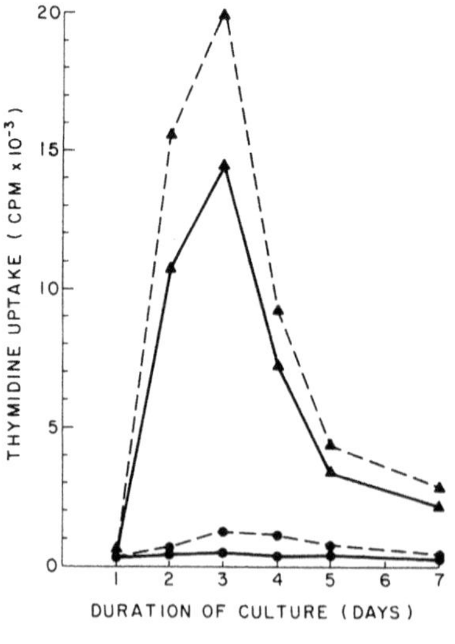

Fig. 36.1. Thymidine uptake by cultured B cells. The responses to medium alone (—●—●—), T cell factors (TCM) alone (-●—●-), bead alone (—▲—▲—), and a combination of beads and TCM (-▲---▲-) are depicted.

Table 36.2. The response to anti-immunoglobulin of B cell subpopulations expressing different levels of B2 antigen.

	Anti-Ig	T cell supernatant	Thymidine uptake by spleen cell subpopulations (cpm)[b]			
			$B1^+B2^-$	$B1^+$lowB2	$B1^+B2^+$	$B1^+$highB2
I[a]	–	–	NT[c]	1142 ± 56	NT	1223 ± 221
	–	+	NT	1749 ± 23	NT	2084 ± 177
	+	–	NT	1128 ± 49	NT	14,333 ± 508
	+	+	NT	3731 ± 78	NT	32,582 ± 1171
II[a]	–	–	1168 ± 63	858 ± 48	735 ± 59	419 ± 28
	–	+	2843 ± 71	521 ± 43	1118 ± 95	775 ± 49
	+	–	1076 ± 55	3669 ± 134	8929 ± 344	16,637 ± 385
	+	+	3247 ± 103	6293 ± 421	12,417 ± 435	22,052 ± 891
III[a]	–	–	1255 ± 96	898 ± 83	908 ± 109	1506 ± 129
	–	+	1754 ± 219	1519 ± 97	942 ± 37	2137 ± 108
	+	–	2037 ± 139	2641 ± 249	5947 ± 153	7354 ± 209
	+	+	3991 ± 231	4876 ± 132	17,106 ± 491	18,410 ± 409

[a] Spleen cells were stained with anti-B1–FITC and anti-B2–biotin–Texas Red–avidin as described and sorted into the groups indicated.
[b] Cells were cultured at 50,000/well in 96-well trays in the presence of anti-Ig antibody (anti-Ig) or supernatant containing T cell-derived factor (T cell supernatant). Thymidine uptake was measured by adding [^{3}H]thymidine after 2 days and continuing the cultures for a further 15 hr.
[c] NT: Not tested.

alone, low B2, and high B2, respectively. These populations were tested for their response to anti-Ig antibody. As shown in Table 36.2, only high-B2 cells showed a vigorous response to anti-Ig. In experiment 2 a more restrictive sort was used to further exclude cross contamination. In this case the low B2 cells showed virtually no response. We concluded that the cell responding to anti-Ig was characterized by the expression of high levels of B2 antigen. This population was about 40% of the total number of splenic B cells.

Phenotypic Changes of Splenic B Cells after Activation

Cell sorting did not provide the number of cells required for re-phenotyping activated cells. For this reason antibody- (anti-T4, anti-T8, anti-Mo1, anti-Mo2) and complement-mediated lysis were used to prepare large numbers of cells. During the first four days a near maximal response was generated with anti-Ig alone. However, for longer-term cultures T cells (10% of the total B cells) plus anti-$T11_2$ and anti-$T11_3$ antibodies to activate the T cells were added. These conditions induced differentiation and also guaranteed optimal survival of the activated cells. Prior to analysis, T cells and monocytes were removed by antibody and complement lysis (as above). Following this procedure residual T cells and monocytes were no greater than 5%. In all cases, the cultures were incubated with 10% AB serum for 1 hr to compete off the anti-Ig beads before staining with monoclonal antibodies.

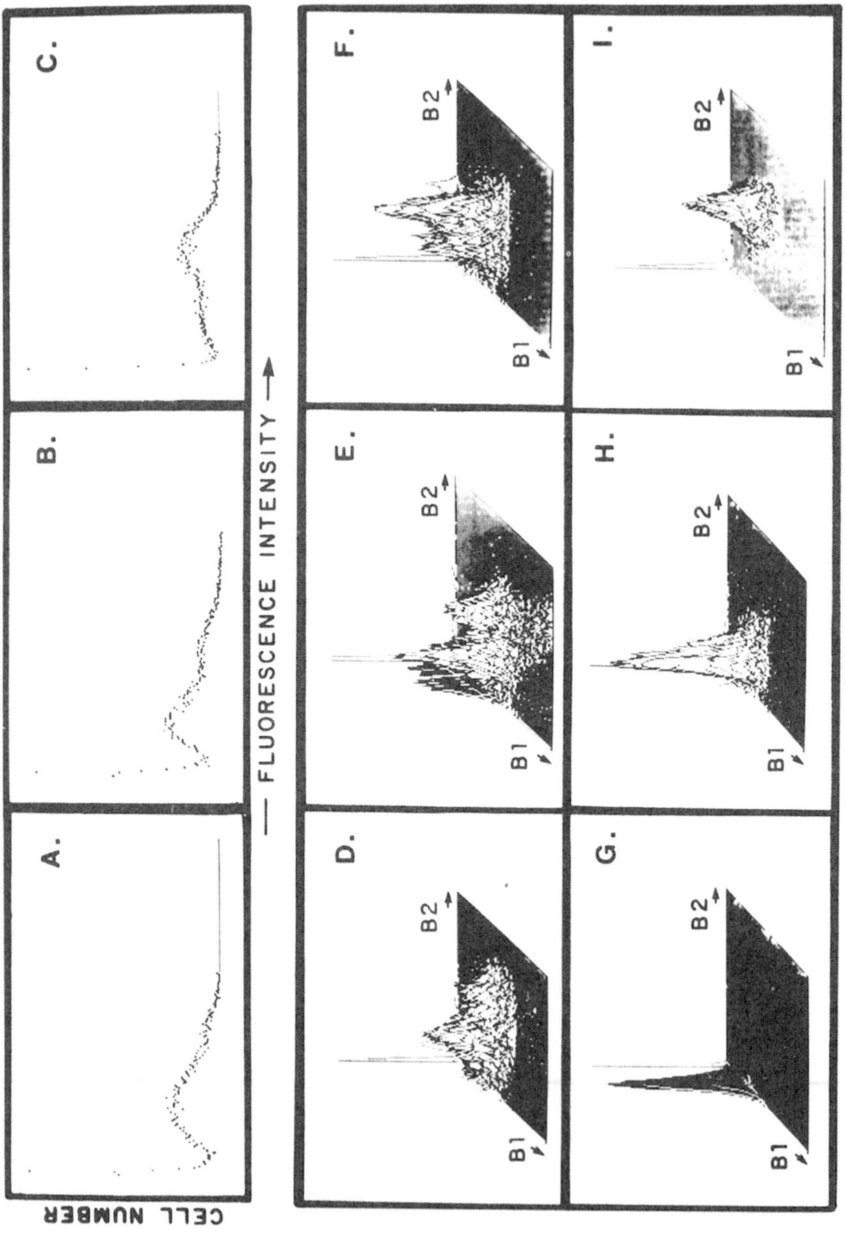
A.
B.
C.
D.
E.
F.
G.
H.
I.
B1
B2
FLUORESCENCE INTENSITY
CELL NUMBER

Table 36.3. Phenotypic changes during *in vitro* culture of splenic B cells.[a]

	Percent positive after *t* days of *in vitro* culture[b]						
	$t = 0$	$t = 1$	$t = 2$	$t = 3$	$t = 4$	$t = 7$	$t = 10$
Group I:							
B1	85	96	94	90	92	57	45
B4	54	83	86	91	94	69	50
Ia	82	89	95	84	NT[c]	76	45
Group II:							
B2	74	29	14	16	9	6	1
Group III:							
B5	1	14	33	76	61	14	2
IL-2R	1	21	37	49	NT	34	29
B-LAST 1	1	3	12	31	NT	7	NT
T12	1	11	25	51	47	36	13
BB1	5	5	29	56	39	15	10
Group IV:							
T10	1	3	11	18	NT	46	77
PCA-1	1	3	4	6	12	27	46
Group V:							
PC 1	0	1	1	1	1	1	3
T4/T8	1	1	2	3	2	1	0
T11	1	4	3	2	NT	3	0
Mo1	1	1	2	1	2	2	1

[a] Cells were cultured at 1.5×10^6/ml in the presence of anti-Ig antibody-conjugated beads. In some cases at early time points (day 1–3) and in all cases where cultures were maintained beyond 3 days, T cells and T cell supernatants were also present.
[b] Cells were stained and analyzed as described in Materials and Methods. A cutoff between positivity and negativity was established using the control antibody.
[c] NT: Not tested.

Phenotypic analyses at varying times after activation are shown in Table 36.3. Five groups were identified by their differing reactivity patterns. The first group, the pan B cell antigens (B1.B4, and Ia), are expressed strongly on resting cells and become still stronger during the proliferative phase. However, during the differentiation phase the level of expression of all these antigens declined.

The B2 antigen stood alone in its characteristic pattern of change. As shown above, B2 antigen is strongly expressed on the resting B cells which responded to anti-Ig, but its expression declined dramatically after activation. By 4 days few cells expressed B2 antigen and those which did, expressed it weakly.

◁ **Fig. 36.2.** Panels A–C depict the intensity of staining by anti-B2 biotin-developed Texas Red–avidin on three spleens. Panels D–F show three-dimensional histograms of B1 and B2 antigen intensity for the same three spleens stained with anti-B2 (as above) and with an anti-B1–FITC. Panels G–I show different subsets of B cells sorted from the spleen population shown in panel.

The third group of antigens were not expressed on resting cells but appeared within 48 hr after activation. The first antigens to appear were B5 and IL-2R, which were detectable within 24 hr. The second group, B-LAST 1, BB1, and T12 were detected from 24–48 hr after activation. All five showed peak expression at 3–4 days and then progressively declined in intensity.

The fourth group of markers appear during the differentiation phase. T10 was detectable on a small percentage of cells as early as 4 days. By contrast, PCA-1 was not detectable until day 7 post-activation. In each case the level of expression increased progressively. At day 10 both were expressed on the majority of the cells.

The final group; Mo1, T11, PC 1, T4, and T8 were negative at all stages, confirming the purity of the B cell preparation. PC 1 is a plasma cell marker but it was not expressed under the culture conditions used, perhaps indicating that it is a late plasma cell marker.

The Expression of IL-2 Receptors on Activated B Cells

The *de novo* expression of certain antigens after activation raised the possibility that some may be growth factor receptors. Expression of IL-2 receptors after activation, which has also been reported by others (13–15), tends to support this notion if the detected antigens could be shown to be true IL-2 receptors which were functional in this context. Most studies have been subject to the criticism that the effects of IL-2 on B cells could be indirectly mediated by T cells. However the B cell purification techniques developed in this study eliminated this problem. Moreover, unlike the other markers a definitive study was possible due to the availability of cloned IL-2.

In Table 36.4 the expression of IL-2R is shown to be increased whatever the mode of B cell activation. To prove that the molecule being

Table 36.4. Activated B cells express a cell surface determinant detected by two anti-IL-2 receptor monoclonal antibodies.

	% of positive cells[a]					
	B1	IL-2R	B-LAST 1	T3	T11	Mo1
Day 0:	86	0	0	0	0	0
Day 3:						
Anti-Ig	96	71	79	0	1	1
Protein A	94	49	47	0	0	1
EBV	87	31	65	1	0	1

[a] Cells were treated with monoclonal antibody for 45 min at 4°C. After washing, the cells were stained with fluoresceinated goat anti–mouse Ig antibody. Cytofluorographic analysis was performed; results are shown as the percentage above background defined by an irrelevant mouse monoclonal antibody.

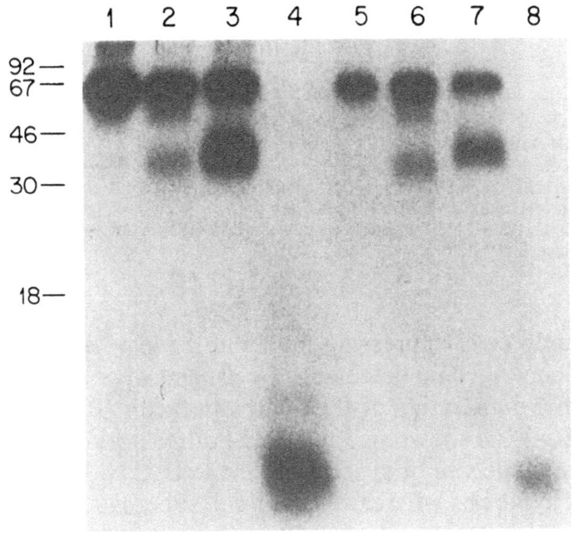

Fig. 36.3. 15% SDS–PAGE of ^{125}I-labeled IL-2 receptor. Tracks 1–4 are from T cell blasts: 1. control—no digestion, 2. staphylococcal V8 protease, 3. chymotrypsin, 4. papain. Tracks 5–8 are from B cell blasts: 5. control—no digestion, 6. staphylococcal V8 protease, 7. chymotrypsin, 8. papain.

detected on anti-Ig activated B cells was identical to that on activated T cells immunoprecipitation studies were performed. Highly purified B cell blasts (day 3 after activation) were prepared by double antibody and complement lysis as described above. Highly purified T cells were prepared from E^+ spleen cells by lysis with anti-B1, anti-PCA-1, anti-Mo1, anti-Mo2, and complement. Both cell populations were surface-labeled and precipitated with anti-IL-2R antibody, the precipitates being analyzed by SDS–PAGE. In both cases a single 60–65-Kd band was detected. To further prove the identity of the receptors on the two cell lineages, limited peptide maps were prepared (Fig. 36.3). An identical pattern of digestion is evident with both B and T cell IL-2 receptors.

The Functional Role of IL-2 Receptors on B Cells

Having shown that the IL-2R molecule on B cells was identical to that on T cells we needed to show that it was functional. This was achieved by a sequential purification beginning with E-rosette negative (E^-) spleen cells (80–90% B cells). These cells were lysed with antibody cocktail plus complement (Methods) after which T cell contamination was 1% or less. This population was cultured with anti-Ig alone for 3 days. The resulting population contained less than 0.1% T cells. To further exclude the T cells, this population of B cell blasts was stained with B1–FITC and

Table 36.5. The effect of IL-2 on B cell blasts.

		[^{3}H]Thymidine uptake (cpm)[a]		
		IL-2	BCGF	IL-2 + BCGF
Resting B cells	220 ± 19	279 ± 10	318 ± 23	594 ± 241
Day-3 blast cells	1165 ± 109	6142 ± 315	8498 ± 445	11206 ± 493

[a] Resting B cells prepared by antibody complement lysis and B cell blasts prepared by sequential purification of activated B cells (see text) were cultured (20,000/well) with IL-2, BCGF, or both. After 24 hr the cells were pulsed with [^{3}H]thymidine and the cultures continued for a further 15 hr.

sorted for only cells expressing moderate to high levels of B1 antigen. These cells contained no detectable T cells and were cultured with cloned IL-2 or with a preparation of B cell growth factor (BCGF) containing no detectable IL-2. As shown in Table 36.5 both factors induced mitogenesis and displayed a degree of synergism. While IL-2 was mitogenic, it failed to induce Ig synthesis whereas crude T cell supernatant did so (Fig. 36.4).

The IL-2 and BCGF Receptors Are Separate Entities

Two lines of evidence were developed which showed that the receptor for BCGF differed from the IL-2 receptor. First, B cells purified as above were cultured with IL-2, BCGF, and anti-IL-2R antibody or combinations of the three. As shown in Table 36.6 the effect of IL-2 was abrogated by the anti-IL-2R antibody but the response to BCGF was unaffected.

The second line of evidence was based on the kinetics of the response to the two factors. The results of such a study are shown in Fig. 36.5;

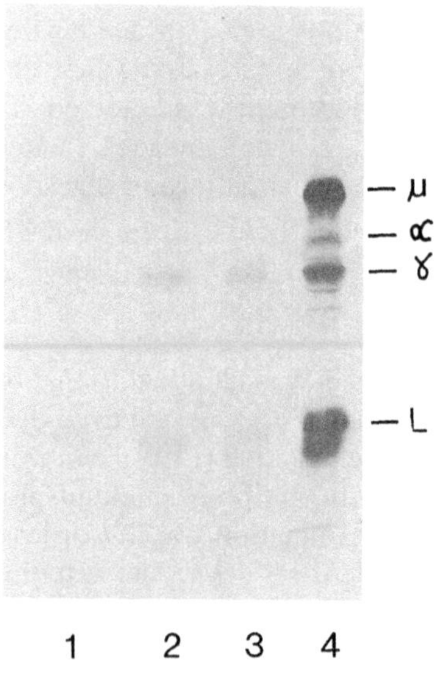

Fig. 36.4. 10% SDS–PAGE of immunoglobulin precipitated from the supernatants of B blasts incubated with: 1. medium alone, 2. cloned IL-2 (Biogen), 3. purified IL-2 (the gift of Dr. K. Smith, Dartmouth, NH), and 4. crude supernatant of activated T cells.

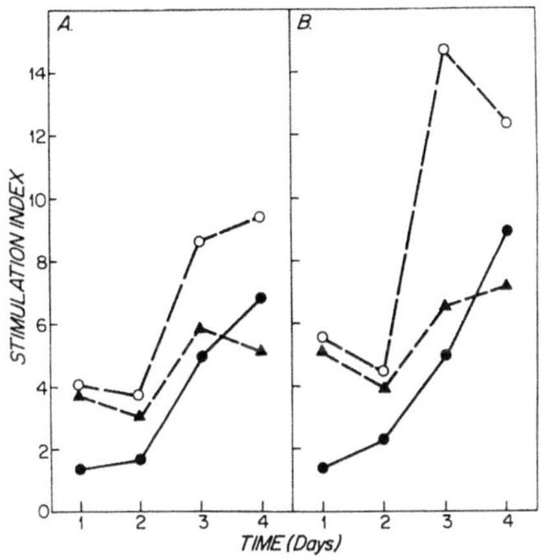

Fig. 36.5. Effect of IL-2 on BCGF on splenic B cells. (A) Purified resting B cells were cultured with IL-2 (-●—●-), BCGF (-▲—▲-), or IL-2 and BCGF (-○—○-), plus anti-Ig beads. Thymidine uptake was assessed by adding thymidine at the times indicated and continuing the cultures for a further 15 hr. The stimulation index was calculated as the number of counts divided by the number of counts incorporated by cells incubated with beads alone. (B) Purified B cells were cultured with anti-Ig beads for the times indicated. The B blast cells were stained with anti-B1–FITC and sorted; the most positive 80% of B1 staining cells were cultured with IL-2 (-●—●-), BCGF (-▲—▲-), or both (-○--○-). In this case thymidine uptake was measured 48 hr later.

BCGF clearly acts very early, if not immediately after activation, whereas IL-2 does not act for 48 hr. After that time the effect of IL-2 becomes similar to or greater than that of BCGF.

Summary

Activation by anti-Ig antibody was studied in culture systems which allowed us to maintain B cells throughout both the proliferative (day 0–4) and differentiative phases (day 5–10). The precursor of anti-Ig activation

Table 36.6. Anti-IL-2R antibody blocks the IL-2 effect but does not block BCGF.

	[^{3}H]Thymidine uptake (cpm)[a]			
		IL-2	BCGF	IL-2 + BCGF
B cell blasts	3541 ± 274	8181 ± 508	8346 ± 433	14630 ± 824
+ Anti-IL-2R	3232 ± 156	3540 ± 304	8773 ± 462	8712 ± 505

[a] Activated B cell blasts were purified as described in the text. As before, they were cultured with IL-2, BCGF, or both in the presence or absence of anti-IL-2R antibody. After 24 hr the cultures were pulsed with [^{3}H]thymidine and continued for a further 15 hr.

was characterized as a B cell expressing high levels of B2 antigen which represented about 40% of spleen B cells. The most dramatic changes during the proliferative phase were loss of the B2 antigen and the appearance of activation antigens including B5, BB1, B-LAST 1, T12, and the IL-2 receptor. During the differentiative phase the major antigenic changes were the appearance of T10 and PCA-1 and the disappearance of both the activation antigens and the pan B cell antigens (B1, B4, and Ia).

Having identified certain antigens which were candidates for growth factor receptors, we analyzed one of these, the IL-2 receptor, in greater detail. This choice was motivated by the availability of cloned IL-2 and B cell purification techniques which allowed us to test this factor on normal B cells which were not contaminated with T cells. The receptor on activated B cells was shown to be structurally identical to that on activated T cells by its pattern of reactivity with monoclonal antibodies, and its protein structure as determined by molecular weight and limited peptide mapping. Finally, the IL-2 receptor on B cells was shown to be functional as it induced mitogenesis comparable to that induced by BCGF in B cell blasts. However, it failed to induce significant differentiation (Ig synthesis). This was distinct from the response to BCGF, in that it was blocked by anti-IL-2R antibodies whereas BCGF was not. Moreover, the kinetics differed—BCGF acted immediately after activation but IL-2 did not exert an effect for about 48 hr.

In summary, the culture systems developed have allowed us to study the phenotypic changes which follow B cell activation, thus identifying a group of antigens which are candidates for the receptors of important growth factors. The studies of the IL-2 receptor outlined in this report provide an example of how we hope to be able to exploit these techniques to study regulation of B cell differentiation.

Acknowledgments. The authors wish to thank Ms. Bonnie Frisard for her excellent typing of this manuscript. We also wish to thank David Leslie, John Daley, Mary Kornacki, and Herb Levine for assistance with cytofluorographic analysis and cell sorting. A.W. Boyd is the recipient of a Neil Hamilton Fairley Fellowship of the National Health and Medical Research Council, Australia. K.C. Anderson is a recipient of a Fellowship of the Medical Foundation, Boston, MA. A.S. Freedman is the recipient of a Damon Runyon-Walter Winchell Cancer Fund Fellowship (DRG-041).

References

1. Fothergill, J.J., R. Wistar, J.N. Woody, and D.C. Parker. 1982. A mitogen for human B cells: Anti-Ig coupled to polyacrylamide beads activates blood mononuclear cells independently of T cells. *J. Immunol.* **128:**1945.
2. Muraguchi, A., J.L. Butler, J.M. Kehrl, and A.S. Fauci. 1983. Differential sensitivity of human B cell subsets to activation signals delivered by anti-mu

antibody and proliferative signals delivered by monoclonal B cell growth factor. *J. Exp. Med.* **157**:530.
3. Anderson, K.C., A.W. Boyd, D.C. Fisher, B.L. Slaughenhoupt, J. Groopman, C. O'Hara, J. Daley, S.F. Schlossman, and L.M. Nadler. 1985. Distinct subsets of human B cells defined by the expression of B1 and B2 antigens. *J. Immunol.* **134**:820.
4. Yokochi, T., R.D. Holly, and E.A. Clark. 1982. B lymphocyte antigen (BB-1) expressed on Epstein Barr virus-activated B cell blasts, B lymphoblastoid cell lines, and Burkitt's lymphomas. *J. Immunol.* **128**:823.
5. Thorley-Lawson, D.A., R.T. Schooley, A.K. Bhan, and L.M. Nadler. 1982. Epstein–Barr virus superinduces a new human B cell differentiation antigen (B-LAST 1) expressed on transformed lymphoblasts. *Cell* **30**:415.
6. Fox, D.A., R.E. Hussey, K.A. Fitzgerald, A. Bensussan, J.F. Daley, S.F. Schlossman, and E.L. Reinherz. 1985. Activation of human thymocytes via the 50kd T11 sheep erythrocyte binding protein induces the expression of interleukin 2 receptors on both T3-positive and T3-negative populations. *J. Immunol.* **134**:330.
7. Meuer, S.C., R.E. Hussey, M. Fabbi, D. Fox, O. Acuto, K.A. Fitzgerald, J.C. Hodgdon, J.P. Protentis, S.F. Schlossman, and E.L. Reinherz. 1984. An alternative pathway of T cell activation: A functional role for the 50kd T11 sheep erythrocyte receptor protein. *Cell* **36**:897.
8. Boyd, A.W., J.W. Goding, and J.W. Schrader. 1981. Regulation of growth and differentiation of a murine B cell hybridoma. I. Lipopolysaccharide induced different actions. *J. Immunol.* **126**:246.
9. Maizel, A.C., Sahasrabuddhe, S. Mehta, J. Morgan, L. Lachman, and R. Ford. 1982. Biochemical separation of a human B cell mitogenic factor. *Proc. Natl. Acad. Sci. U.S.A.* **79**:5998.
10. Howard, M., J. Farrar, M. Hilfiker, B. Johnson, K. Takatsu, T. Hamoaka, and W.E. Paul. 1982. Identification of a B cell growth factor distinct from interleukin 2. *J. Exp. Med.* **15**:914.
11. DeFranco, A.L., E.S. Raveche, R. Asofsky, and W.E. Paul. 1982. Frequency of B lymphocytes response to anti-immunoglobulin. *J. Exp. Med.* **155**:1523.
12. Kehrl, J.H., A Muraguchi, and A.S. Fauci. 1984. Human B cell activation and cell cycle progression stimulation with anti-mu and *Staphylococcus aureus* Cowan strain I. *Eur. J. Immunol.* **14**:115.
13. Korsmeyer, S.J., W.C. Greene, J. Cossman, S.M. Hsu, J.P. Jensen, L.M. Neckers, S.L. Marshall, A. Bakshi, J.M. Depper, W.J. Leonard, E.S. Jaffe, and T.A. Waldmann. 1983. Rearrangement and expression of immunoglobulin genes and expression of Tac antigen in hairy cell leukemia. *Proc. Natl. Acad. Sci. U.S.A.* **80**:4522.
14. Malek, T.R., R. Robb, and E. Shevach. 1983. Identification and initial characterization of a rat monoclonal antibody reactive with the murine interleukin 2 receptor–ligand complex. *Proc. Natl. Acad. Sci. U.S.A.* **80**:5694.
15. Tsudo, M., T. Uchiyama, and H. Uchino. 1984. Expression of the Tac antigen on normal human B cells. *J. Exp. Med.* **160**:612.

CHAPTER 37

Changes with *in vitro* Activation of the B Cell Panel Antigens

Arnold S. Freedman, Andrew W. Boyd, David C. Fisher, Stuart F. Schlossman, and Lee M. Nadler

Introduction

With exposure to specific antigenic or mitogenic stimuli, B cells proliferate and subsequently differentiate into antibody-secreting cells. The initial events which accompany activation include an increase in cell size, augmented DNA and RNA synthesis, and changes in cell surface structures. The expression of surface immunoglobulin (SIg) and Ia, the classical surface markers of B cells, has been extensively studied during *in vitro* stimulation. Upon activation, resting B cells which express IgM and/or IgD will lose SIgD, and then no longer express membrane Ig or switch to the expression of other isotypes (1). Similarly, resting B cells continue to express Ia after activation until the plasma cell stage (2). In our laboratory, utilizing a panel of monoclonal antibodies directed against B cell differentiation antigens, we have noted changes in the cell surface phenotype of resting B cells with activation. Two B cell-restricted antigens, B1 and B4, have been observed to be expressed on resting and activated B cells, and lost prior to the terminal stages of differentiation (3,4). The B2 antigen, present on resting cells is lost during the early stages of activation (4,5). We and others have described additional antigens which are not expressed on resting B cells but only appear with activation. These activation antigens include B-LAST 1 (6), B5 (7), interleukin-2 receptor (IL-2R) (8,9), BB1 (10), 4F2 (11), and the transferrin receptor (11).

In the present study, we have examined the Second International Workshop B cell panel antibodies for their pattern of reactivity with splenic B cells prior to and following *in vitro* activation. We have been able to divide these antigens into four subgroups by their expression on unstimulated B cells and on cells activated with three B cell mitogens including anti-Ig antibody (anti-Ig), protein A, and Epstein–Barr virus (EBV). In the results to be presented, we will show that the B cell panel antibodies display distinct reactivities with activation and that clusters of antigens appear to have identical patterns of expression.

Materials and Methods

Human Spleen Cells

Single-cell suspensions were prepared from operative specimens from patients with no known prior illness or intercurrent infection. Splenic mononuclear cells were obtained by Ficoll–Hypaque density gradient centrifugation of the single-cell suspensions. The mononuclear cells were enriched for B cells by rosetting with 5% sheep erythrocytes (E). Cells were cryopreserved in the vapor phase of liquid nitrogen (−196°C) and thawed immediately prior to use.

Monoclonal Antibodies

In addition to the Second International Workshop B cell panel antibodies (B12 was not studied due to insufficient quantity of antibody), a panel of monoclonal antibodies directed against T, B, and myeloid differentiation antigens were used in this study. The T cell antibodies were directed against T3 (pan T), T4 (helper/inducer), and T8 (cytotoxic suppressor) (12). The B cell antibodies used were directed against B1 (13) which is expressed on all mature B cells except plasma cells, and B5 which defines a 67-Kd protein expressed exclusively on activated B cells. The monocyte specific antibody anti-Mo2, as well as anti-Mo1 (14) which in addition to monocytes detects NK cells and myeloid cells, were used to enumerate monocytes.

Indirect Immunofluorescence

Viable cells were harvested from cultures and analyzed for antigen expression by indirect immunofluorescence and flow cytometric analysis as previously described (3). In brief, 0.5–1 × 10^6 viable cells were treated with 100 μl of 1 : 100 dilution of the B cell panel antibodies (other antibodies used at 1 : 250), incubated at 4°C for 30 min, then washed three times. The cells were then treated with a 1 : 50 dilution of goat anti–mouse IgM and goat anti–mouse IgG conjugated with fluorescein isothiocyanate (Coulter Immunology, Hialeah, FL), incubated at 4°C for 30 min, then washed three times, and analyzed on an EPICS C cell sorter (Coulter Electronics, Hialeah, FL). A significant change in antigen expression was determined to be a 50% or greater increase or decrease in the percentage of resting cells expressing antigen.

In vitro Stimulation

For all stimulations, cells were cultured at 1.5 × 10^6 cell/ml in RPMI 1640 with 10% fetal calf serum, 2 m*M* glutamine, and 1 m*M* pyruvate in tissue culture flasks (Corning, No 25100, Corning, NY). For the 3-day cultures, splenic E^- were used, and for the 6-day cultures, whole splenic mononuclear cells were used. To the day-6 stimulation cultures, anti-$T11_2$ and anti-$T11_3$ ascites were added (1 : 300 final concentration) to facilitate T cell activation (15) as it has been previously observed that B cell viability beyond 3 days was poor without activated T cells. Prior to phenotypic analysis of the day-6 cultures, T cells and monocytes were removed with antibody and complement lysis with anti-T4, anti-T8, anti-Mo1, and anti-Mo2. This yielded a similar number of B cells to that of the E^- day-3 cultures. The cells were cultured with the following mitogens: 1) Anti-Ig: Affinity-purified rabbit anti–human Ig antibody was coupled to Affigel 702 polyacrylamide beads (Bio-Rad, Richmond, CA). Specificity was assessed by coupling bovine serum albumin, anti-B1, or anti-B2 to Affigel beads. These preparations lacked any stimulatory activity; 2) Protein A: protein A (Sigma Co, St. Louis, MO) was used at a final concentration of 10 μg/ml; 3) EBV: Supernatant from the EBV-producing marmoset cell line B955 was used at a 1 : 4 dilution.

Results

In vitro Activation

Human splenic B cells were activated *in vitro* with three T cell-independent B cell mitogens. At a concentration of 1.5×10^6 cell/ml, splenic E^- cells (for 3-day cultures) or whole splenic mononuclear cells (for 6-day cultures) were cultured with anti-Ig coupled to beads, protein A, or EBV for either 3 or 6 days. Viable cells (viability of 60–80% were harvested and examined by indirect immunofluorescence and flow cytometric analysis. Because of the significant number of T cells and monocytes present in the day-6 cultures, these cells were removed by antibody and complement lysis, using anti-T4, anti-T8, and anti-Mo1, and anti-Mo2. Therefore, the day-3 and day-6 cultures had less than 10% T cells and monocytes remaining.

Assessment of Activation

Cells stimulated *in vitro* were determined to be activated by three parameters: 1) increase in cell size, 2) [^{3}H]thymidine incorporation, and 3) expression of a B cell activation antigen. Prior to activation, greater than 90% of the splenic E^- population clustered between peak channel 35–53 as measured by forward-angle scatter. These cells are approximately the size of peripheral blood lymphocytes. However, after stimulation the mean peak channel (67–148) corresponded to that of peripheral blood monocytes, suggesting that these cells were larger, consistent with previous characterization of activated B cells (11). These cultured cells were also observed to be activated by [^{3}H]thymidine incorporation. B cells

Table 37.1. Expression of B cell panel antibodies on resting and activated B cells.[a]

	Mitogen						
	Anti-Ig			Protein A		EBV	
Antibody	Day 0	Day 3	Day 6	Day 3	Day 6	Day 3	Day 6
B1	+++	+++	+++	+++	++	+++	++
B2	0	0	0	0	0	0	0
B3	0	0	0	0	0	0	0
B4	+++	+	+	+++	++	+++	++
B5	++	++	+++	++	++	++	++
B6	++	+	+/0	++	++	++	++
B7	++	+	+	++	+	+/0	0
B8	++	++	+++	++	++	+++	++
B9	+++	+	0	++	+/0	+	+
B10	+++	++	+++	+++	+++	+++	+++
B11	0	0	0	0	0	0	0
B12	QNS[b]						
B13	0	0	0	0	0	0	0
B14	++	++	++	+++	++	+++	+++
B15	++	+	+	++	++	++	+
B16	0	++	0	+	0	+	0
B17	+++	++	+++	+++	+++	+++	+++
B18	0	+	0	0	0	0	0
B19	0	0	0	+	0	0	0
B20	0	0	0	0	0	0	0
B21	0	0	0	0	0	0	0
B22	+++	+++	+++	+++	++	+++	+++
B23	0	++	+++	++	++	++	++
B24	++	++	++	+++	++	++	+++
B25	++	+	+	+	+/0	+	+
B26	0	0	0	0	0	0	0
B27	++	++	++	++	++	++	++
B28	++	++	++	+++	+++	+++	++

cultured with anti-Ig had a stimulation index (S.I.) of 5–10 on day 3 with a gradual decrease to near background level by day 7. In the presence of protein A, a S.I. of 10–15 was observed at day 3, again returning to background by day 7. Finally in the presence of EBV, the S.I. was 10–20 after 6 days in culture. The final parameter assessing activation was the expression of the B cell-restricted activation antigen, B5. Greater than 50% of cells cultured with all three mitogens were observed to express this antigen by day 3. The level of B5 antigen expression decreased by day 6 to approximately 15% of the cells examined.

Antigens Not Expressed by Resting or Activated Cells

Utilizing the B cell panel of monoclonal antibodies, 10 antigens were found not to be expressed on unstimulated splenic E^- cells nor did they

Table 37.1. *(Continued)*

	Mitogen						
	Anti-Ig			Protein A		EBV	
Anbibody	Day 0	Day 3	Day 6	Day 3	Day 6	Day 3	Day 6
B29	0/+	+++	++	0/+	0/+	0/+	0/+
B30	+++	+	0	+++	++	+++	++
B31	++	+	0	+	0	+	0
B32	0	0	0	0	0	0	0
B33	++	+	0	++	0/+	++	0/+
B34	++	++	++	+++	++	+++	+++
B35	++	0	0	+	0	+	+/0
B36	+++	++	+++	+++	+++	+++	++
B37	0	++	+	0	0	++	+
B38	0	0	0	0	0	0	0
B39	0	0	0	0	0	0	0
B40	++	+	0	++	+	+	0
B41	+++	+/0	0	++	+/0	+	+/0
B42	+	+	+	+	+	+	+
B43	++	++	++	++	++	+++	++
B44	+++	++	+++	+++	++	+++	++
B45	+++	+++	+++	+++	++	+++	++
B46	+++	++	+++	++	+++	+++	++
B47	+++	++	+++	++	++	+++	++
B48	++	++	++	++	++	+++	++
B49	+++	++	++	++	+	++	+
B50	+++	++	+++	+++	+++	+++	+++
B51	+++	++	+++	+++	++	+++	++
B52	+++	++	++/+	+++	++	+++	+++

[a] Degree of positivity was assessed by flow cytometry. 0, less than 15% of cells positive; +, 16–35% of cells positive; ++, 36–60% of cells positive; +++, greater than 60% of cells positive.
[b] QNS, quantity not sufficient.

appear with stimulation utilizing any of the three B cell mitogens. These included antigens B2, B3, B11, B13, B20, B21, B26, B32, B38, and B39 (Table 37.1).

Antigens Expressed on Both Resting and Activated Cells

Within the B cell panel, 22 antigens were expressed on resting B cells, and did not change significantly with *in vitro* activation. Within this group were four clusters of antigens, as determined by their molecular weight and pattern of reactivity on normal and malignant cells. The first group of antigens were identified as the B cell-restricted antigen p35 [B1-like as developed in our laboratory (13)] exemplified by B5, B22, and B24 and were all expressed on 60–75% of unstimulated splenic E^- cells (Fig. 37.1). With activation, using all three mitogens, the intensity of antigen expres-

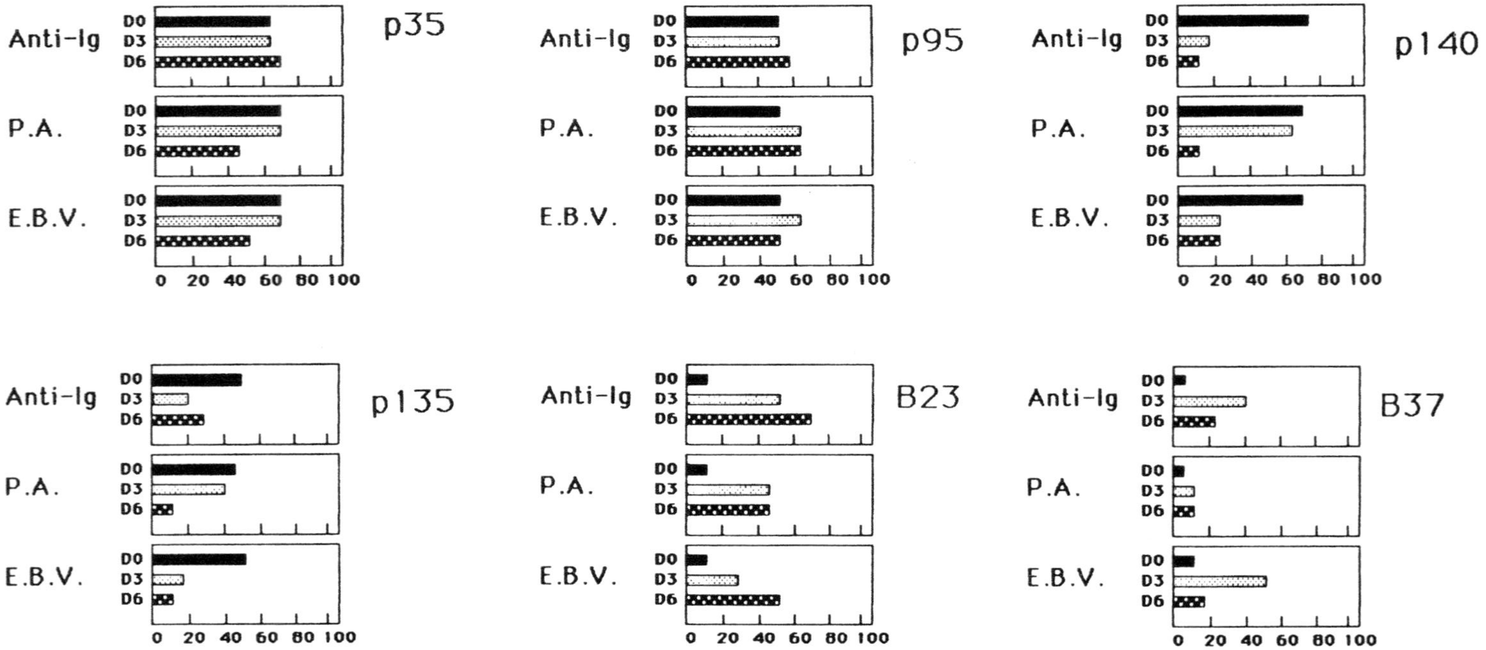

Fig. 37.1. Graphs represent the percent of splenic E^{-} cells expressing antigen (p35, p95, p140, p135, B23, and B37) (horizontal axis) on day 0 (resting cells) or after activation with anti-Ig, protein A, or EBV for 3 and 6 days.

sion increased slightly by day 3 but returned to the pre-stimulation level by day 6 (Fig. 37.2). The percentage of positive cells also minimally decreased by day 6. The second cluster of antigens were identified as the B cell-restricted antigen p95 [B4-like as developed in our laboratory (3)] included B14, B28, B34, and B43. These antigens behaved similarly to those of the p35 cluster, except the intensity of expression and the number of cells expressing the antigens were less (approximately 50% of unstimulated cells expressed p95) (Fig. 37.1). The third cluster of antigens, identified as p45,55,65 [BA-1 like, as developed by Abramson *et al.* (16)], included B47 and B48. These antigens were expressed on approxi-

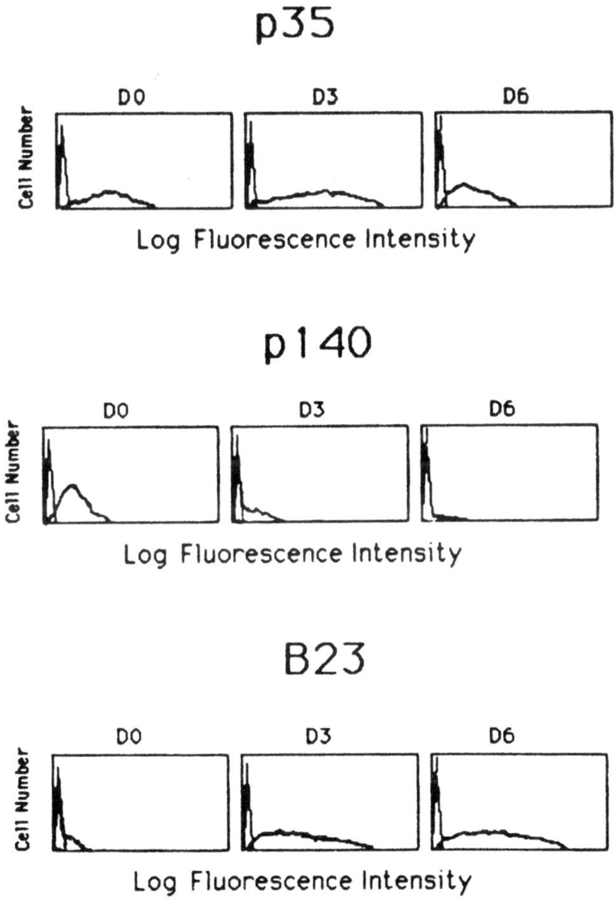

Fig. 37.2 EPICS C histograms of reactivity of anti-p35, anti-p140, and anti-B23 with resting splenic E^- cells (D0) and after stimulation with anti-Ig for 3 and 6 days. Isotype identical control ascites is shown at far left of each histogram.

mately 60% of splenic E$^-$ cells either prior to or after activation with all three mitogens, with no change in intensity. Similarly, the final cluster, identified as the B cell-associated antigen p220 (17), including B50 and B51, did not change with activation from resting cells. Other antigens in the B cell panel which could not be clustered together, and which did not change their expression with activation included B1 (p29,34) which is an Ia-like antigen (18), B8, B10, B15, B17, B27, B36, B42, B44, B45, and B46.

Antigens Expressed on Resting B Cells and Decreased with Activation

Two clusters of B cell-restricted antigens, p140 [B2-like as developed in our laboratory (19)] and p135 (HD6-like as developed by Dorkin *et al.* this volume, Chapter 7), were observed to decrease in antigen expression with activation. The p140 antigens included B9, B33, B35, and B41. These antigens were well expressed on unstimulated splenic E$^-$ cells (approximately 60% of cells expressed p140) (Fig. 37.1). With anti-Ig stimulation, antigen expression decreased to near background level by day 3 (Fig. 37.2). With protein A and EBV, the decrease in antigen expression was more gradual (about 25% of cells still expressed p140 after a 6-day culture with EBV). The p135 cluster included B7, B25, B31, B40, and B49. A somewhat more variable pattern in antigen expression was observed with the three mitogens (Fig. 37.1). Both EBV and protein A had a more profound effect on the decrease of antigen expression, but this effect was less than that seen with the p140 antigen. Four antigens—B4, B6, B30, and B52—decreased only in response to anti-Ig stimulation, and no effect was observed with EBV or protein A.

Antigens Not Expressed on Resting B Cells But Expressed on Activated Cells

Of the 52 antigens in the B cell panel, six antigens were not expressed on unstimulated splenic E$^-$ cells (less than 10% of cells positive) and were detected only after activation (Fig. 37.1). Two antigens, B23 and B16, were induced by all three mitogens. B23 was present on cells cultured for 3 and 6 days (Fig. 37.2). In contrast, B16 was weakly expressed on cells in culture for 3 days, only to return to background level on cells cultured for 6 days. Three antigens were expressed on cells cultured with only one of the three mitogens: B18 and B29 only appeared on cells cultured with anti-Ig, while B19 appeared only on cells stimulated with EBV. One other antigen B37 was induced on splenic E$^-$ cells with both anti-Ig and EBV, but not observed on cells cultured with protein A (Fig. 37.1).

Discussion

In the present report, we have examined the expression of the B cell panel antigens on splenic B cells, prior to and following activation with three B cell mitogens: anti-Ig, protein A, and EBV. After 3 and 6 days in culture, activated B cells demonstrated increased cell size, an increase in [^{3}H]thymidine incorporation, and the expression of the B5 antigen, a 67-Kd protein expressed on activated B cells. These activated B cells were then examined for the expression of 51 B cell-restricted and -associated antigens by indirect immunofluorescence and flow cytometric analysis. Four groups of antigens were identified by their pattern of expression: 1) antigens not expressed by resting or activated B cells, 2) antigens expressed on both resting and activated cells, 3) antigens expressed on resting cells whose expression decreased with activation, and 4) antigens not expressed on resting cells but expressed on activated cells. This study has allowed the identification of antigens, which, by their expression in relation to activation, may lead to an understanding of the proliferation and differentiation of human B cells.

Within the group of antigens expressed on both resting and activated B cells were four clusters of antigens, as determined by their molecular weights and patterns of reactivity with normal and malignant cells. These included p35 (B1-like), p95 (B4-like), p45,55,65 (BA-1-like), and p220 (common leukocyte antigen-like). Also within this group was an Ia-like antigen, I-2, and ten unclustered antigens. These findings were consistent with previous *in vitro* stimulation data where p35, p95, and Ia continued to be expressed with activation and were lost at the terminal stages of B cell differentiation. Immunoperoxidase studies of normal lymph nodes have similarly confirmed our findings, with p35, p95, and Ia present in the mantle zone and germinal center (20,21). The p45,55,65 (BA-1-like) antigen has been reported to be present only in the mantle zones, which is not consistent with our findings with the BA-1-like cluster (21). We have demonstrated that B cell-derived leukemias and lymphomas serve as a model of normal B cell differentiation by the expression of B cell-restricted and -associated antigens (22). Therefore, the expression of p35, p95, p45,55,65, and Ia on malignancies of B cell lineage including non-T cell acute lymphoblastic leukemias, chronic lymphocytic leukemias (CLL), and non-Hodgkin's lymphomas, but not myelomas, supports the notion that these antigens are lost only at the terminal stages of differentiation.

The antigens which were expressed on resting B cells and lost with activation included two clusters of antigens including the p140 (B2-like) and p135 (HD6-like) antigens. It has been previously demonstrated that the p140 antigen is lost early in activation, along with SIgD. Similarly, immunoperoxidase studies of normal lymph nodes have shown that p140

(20) and p135 are present in the mantle zone but only weakly detected in the germinal center. The expression of p140 on B cell CLLs, nodular and diffuse poorly differentiated lymphocytic lymphomas, and some diffuse histiocytic lymphomas supports the notion that this antigen is present on resting B cells and is lost with activation. While it has been found that p140 is both the C3d (23) and EBV receptor (24; this volume, Chapter 44), its precise role in B cell function is unknown, although one would postulate that it may be involved in the early stages of activation. It was also found in this group that four antigens decreased only with stimulation with anti-Ig. This suggests that the mechanisms involved in activation by these three mitogens may be different.

The final group consisted of six antigens which were only expressed on activated B cells. Similar to the observation made for the previous group of antigens, two of these activation antigens were expressed on cells stimulated with all three mitogens, and the remaining four were heterogenous in their expression. This again suggests that activation of B cells by the three mitogens used involves different mechanisms. It has been previously demonstrated that B cells, when activated, respond to growth and differentiation factors, presumably through the induction of specific receptors for these molecules. Two previously described activation antigens, IL-2R and 5E9, have been shown to be receptors for IL-2 and transferrin, respectively. One would postulate that these activation antigens within the B cell panel are excellent candidates for receptors for growth and differentiation factors, or are involved in cell–cell interactions.

The present study has permitted us to analyze antigenic changes in human B cells following activation with three B cell mitogens. The identification and characterization of antibodies which recognize B cell antigens expressed prior to and at different stages of B cell activation may provide the means to study the function of these antigens.

Summary

Human splenic B lymphocytes were examined with the antibodies of the Second International Workshop B cell panel before and after *in vitro* activation with three B cell mitogens. These antibodies detected four groups of antigens based on the pattern of expression on resting and activated cells: 1) antigens not expressed on resting or activated cells, 2) antigens expressed on resting and activated cells, 3) antigens present on resting cells whose expression diminished with activation, and 4) antigens only detected on activated cells. The relationship between antigen expression and the stages of B cell activation may be useful in the study of B cell proliferation and differentiation.

Acknowledgments. The authors would like to thank Bonnie Frisard for preparation of the manuscript. We also appreciate the technical assistance of Mary Kornacki, David Leslie, and Herb Levine. We also thank Dr. Ellis Reinherz for anti–T cell antibodies, and Dr. Robert Todd for anti-monocyte antibodies. A.W. Boyd is the recipient of a Neil Hamilton Fairley Fellowship of the National Health and Medical Research Council of Australia. A.S. Freedman is the recipient of a Damon Runyon-Walter Winchell Cancer Fund Fellowship (DRG-041).

References

1. Goding, J.W., D.W. Scott, and J.E. Layton. 1977. Genetics, cellular expression and function of IgD and IgM receptors. *Immunol. Rev.* **37:**152.
2. Halper, J.S., S.M. Fu, C.Y. Winchester, and H.G. Kunkel. 1978. Patterns of expression of human Ia-like antigens during the terminal stage of B cell development. *J. Immunol.* **120:**1480.
3. Nadler, L.M., K.C. Anderson, G. Marti, M. Bates, E. Park, J.F. Daley, and S.F. Schlossman. 1983. B4, a human B lymphocyte-associated antigen expressed on normal, mitogen-activated, and malignant B lymphocytes. *J. Immunol.* **131:**244.
4. Boyd, A.W., K.C. Anderson, A.S. Freedman, D.C. Fisher, B. Slaughenhoupt, S.F. Schlossman, and L.M. Nadler, 1985. Studies of *in vitro* activation and differentiation of human B lymphocytes. I. Phenotypic and functional characterization of the B cell population responding to anti-Ig antibody *J. Immunol.* **134:**1516.
5. Stashenko, P., L.M. Nadler, R. Hardy, and S.F. Schlossman. 1981. Expression of cell surface markers after human B cell activation. *Proc. Natl. Acad. Sci. U.S.A.* **78:**3848.
6. Thorley-Lawson, D.A., R.T. Schooley, A.K. Bhan, and L.M. Nadler. 1982. Epstein–Barr virus superinduces a new human B cell differentiation antigen (B-LAST 1) expressed on transformed lymphoblasts. *Cell* **30:**415.
7. Freedman, A.S., A.W. Boyd, K.C. Anderson, D.C. Fisher, S.F. Schlossman, and L.M. Nadler. 1985. B5, a new B cell restricted activation antigen. *J. Immunol.* **134:**2228.
8. Tsudo, M., T. Uchiyama, and H. Uchino. 1984. Expression of TAC antigen on activated normal human B cells. *J. Exp. Med.* **160:**612.
9. Boyd, A.W., D.C. Fisher, D. Fox, S.F. Schlossman, and L.M. Nadler. 1985. Structural and functional characterization of IL-2 receptors on activated B cells. *J. Immunol.* **134:**2387.
10. Yokochi, T., R.D. Holly, and E.A. Clark. 1982. B lymphoblast antigen (BB-1) expressed on Epstein–Barr virus-activated B cell blasts, B lymphoblastoid cell lines, and Burkitt's lymphomas. *J. Immunol.* **128:**823.
11. Kerhl, J.H., A. Muraguchi, and A.S. Fauci. 1984. Differential expression of cell activation markers after stimulation of resting B lymphocytes. *J. Immunol.* **132:**2857.
12. Reinherz, E.L., and S.F. Schlossman. 1980. The differentiation and functions of human T lymphocytes: a review. *Cell* **19:**821.

13. Stashenko, P., L.M. Nadler, R. Hardy, and S.F. Schlossman. 1980. Characterization of a human B lymphocyte-specific antigen. *J. Immunol.* **125:**1678.
14. Todd, R.F. III, L.M. Nadler, and S.F. Schlossman. 1981. Antigens on human monocytes identified by monoclonal antibodies. *J. Immunol.* **126:**1435.
15. Meuer, S.C., R.E. Hussey, M. Fabbi, D. Fox, O. Acuto, K.A. Fitzgerald, J.C. Hodgdon, J.P. Protentis, S.F. Schlossman, and E.L. Reinherz. 1984. An alternate pathway of T cell activation: A functional role for the 550Kd T11 sheep erythrocyte receptor protein. *Cell* **36:**897.
16. Abramson, C.S., J.H. Kersey, and T.W. LeBien. 1981. A monoclonal antibody (BA-1) reactive with cells of human B lymphocyte lineage. *J. Immunol.* **126:**83.
17. Omary, M.B., I.S. Trowbridge, and H.A. Battifora. 1980. Human homologue of murine T200 glycoprotein. *J. Exp. Med.* **152:**842.
18. Nadler, L.M., P. Stashenko, R. Hardy, J.M. Pesando, E.J. Yunis, and S.F. Schlossman. 1981. Monoclonal antibodies defining serologically distinct HLA-D/DR related Ia-like antigen in man. *Hum. Immunol.* **1:**77.
19. Nadler, L.M., P. Stashenko, R. Hardy, A. van Agthoven, C. Terhorst, and S.F. Schlossman. 1981. Characterization of a human B cell-specific antigen (B2) distinct from B1. *J. Immunol.* **125:**1941.
20. Bhan, A.K., L.M. Nadler, P. Stashenko, and S.F. Schlossman. 1981. Stages of B cell differentiation in human lymphoid tissues. *J. Exp. Med.* **154:**737.
21. Hsu, S., and E.S. Jaffe. 1984. Phenotypic expression of B lymphocytes. 1. Identification with monoclonal antibodies in normal lymphoid tissues. *Am. J. Pathol.* **114:**387.
22. Anderson, K.C., M.P. Bates, B.L. Slaughenhoupt, G.S. Pinkus, S.F. Schlossman, and L.M. Nadler. 1984. Expression of human B cell-associated antigens on leukemias and lymphomas: A model of human B cell differentiation. *Blood* **63:**1424.
23. Iida, K., L.M. Nadler, and V. Nussenzweig. 1983. Identification of the membrane receptor for the complement fragment C3d by means of a monoclonal antibody. *J. Exp. Med.* **158:**1021.
24. Fingeroth, J.D., J.J. Weis, T.F. Tedder, J.L. Strominger, P.A. Biro, and D.T. Fearon. 1984. Epstein–Barr virus receptor of human B lymphocytes is the C3d receptor CR2. *Proc. Natl. Acad. Sci. U.S.A.* **81:**4510.

CHAPTER 38

Activation of Human B Cells with Monoclonal Antibody to the Bp35 Cell Surface Polypeptide

Edward A. Clark, Geraldine Shu, and Jeffrey A. Ledbetter

Introduction

Resting human B lymphocytes can be triggered to proliferate with high concentrations of antisera to surface immunoglobulin (Ig) or with low doses of anti-Ig in the presence of T cell-derived B cell-stimulating factors (BSF) (1–3). It is thought that after an initial activation signal, e.g., by antigen or anti-Ig, receptors for BSF and other growth and differentiation factors are expressed, and that in the presence of these factors B cells divide and mature into antibody-forming cells (4–6). The receptor for the T cell growth factor interleukin-2 (IL-2) has been defined (7), but analogous structures on B cells have not yet been described. Nor is it clear how binding of the Ig receptor, which has such a small intracytoplasmic tail (8), leads to B cell activation.

Recently a number of B cell-specific cell surface polypeptides have been identified which presumably could play some role in B cell differentiation; up-to-date summaries of these molecules are described in Chapters 1 and 12 of this volume. One 35,000-dalton polypeptide Bp35 first defined by the monoclonal antibody (mAb) B1 (9) has several properties suggesting it may play a role in B cell activation: it is a B cell-specific phosphoprotein (10,11) and is expressed at higher densities on preactivated germinal center B cells than on dense resting B cells (12,13). Most importantly, mAbs to Bp35 alone trigger resting B cells to proliferate (14). Here we describe the characteristics of this activation and outline two possible models for the function of Bp35.

Materials and Methods

Monoclonal Antibodies

The monoclonal antibodies used in this study have been described (11,13,15). They include the 1F5 mAb specific for Bp35, the HB10a mAb specific for HLA-DR, the 2C3 and 4B8 mAbs specific for Ig μ chain, the BB-2 mAb specific for the p76 B cell-associated antigen (11), and the 3AC5 mAb specific for the p220 common leukocyte antigen (11). Fluorescein-conjugated mAb to the C3d receptor (B2) was purchased from Coulter Labs (16). Purified mAbs were conjugated with fluorescein-5-isothiocyanate or R-phycoerythrin (PE) as described (13). mAbs were conjugated to CNBr-activated Sepharose 4B beads (Pharmacia) at a ratio of 10 mg/ml of Sepharose. Fab fragments were prepared by papain digestion followed by Sephacryl S200 column chromatography to remove undigested mAb or Fc fragments.

Lymphoid Cells

Tonsillar lymphocytes were followed by gently teasing tissue as described (14). Lymphocytes from dense Percoll fractions (55/60 or pellet) were used for activation studies. T cells were depleted using AET-treated sheep erythrocytes to rosette out cells bearing the E-rosette receptor. The proliferation of lymphoid cells was measured as described (14) after exposure of cells for 1–3 days with 2–5 μg/ml 1F5 antibody alone or in the presence of 25–50% concentration of supernatants of 2-day-old mixed leukocyte cultures (AS, allogeneic supernatant).

Flow Cytometry

Flow cytometry was performed using a FACS IV cell sorter as described (11,13). In brief, 5×10^5 cells were incubated for 30 min on ice with fluorescein or PE-conjugated mAb, washed twice, and passed through gauze just prior to analysis.

Results

In our previous studies we have found that tonsils contain several major subpopulations of B cells (13,17) including a buoyant preactivated Bp35^{bright}IgM^{dull} population which proliferates in the presence of AS and a dense resting Bp35^{dull}IgM^{bright} population that does not proliferate in AS but is triggered by monoclonal anti-IgM on beads. This second dense population is also stimulated to proliferate by anti-Bp35 in the presence or absence of T cells (14). Proliferation induced by anti-Bp35 or by anti-Ig on beads are similar in several ways (Table 38.1): Both appear to require able 38.1. Characteristics of anti-Bp35-induced B cell proliferation.

Characteristic	Cell treatment			Mean proliferation ± S.E.[c]
	Anti-Bp35[a]	%AS	Other[b]	
ugmentation with AS	—	—	—	295 ± 31
	—	50	—	696 ± 40
	2	—	—	3048 ± 212
	2	50	—	41,284 ± 2093
ab blockade/ cross-linking required	—	—	—	223 ± 77
	—	33	—	424 ± 30
	2	—	—	4352 ± 86
	2	33	—	20,433 ± 1722
	—	—	Fab anti-Bp32	213 ± 21
	2	—	Fab anti-Bp32	194 ± 16
c domain-dependent anti-Ig inhibition	—	—	—	246 ± 4
	—	25	—	486 ± 38
	5	25	—	19,288 ± 867
	5	25	Anti-μ	1979 ± 149
	5	25	Fab anti-μ	15,426 ± 1242
sotype independence of anti-Ig inhibition	—	—	—	331 ± 28
	—	25	—	1273 ± 34
	5	25	—	15,557 ± 310
	5	25	Anti-μ(IgG1)	4080 ± 119
	5	25	Anti-μ(IgG2a)	2171 ± 223
	5	25	Anti-p76(IgG2a)	15,011 ± 477

Dose in μg/ml of IgG2a (1F5) anti-Bp35 antibody added for 1×10^6 dense tonsillar lymphocytes/ml.
Mean proliferation of quadruplicate samples 3 days after activation of 2×10^5 cells/200 μl sample.
Additions of Fab or anti-μ monoclonal antibodies at 20 μg and 10 μg/10^6 cells/ml, respectively.

cross-linking; both are augmented by allogeneic T cell factors; and both are inhibited by an Fc domain-dependent mechanism. However, several findings clearly distinguish these triggering mechanisms. Anti-Bp35 mAb does not have to be coupled to beads to trigger B cells, and unlike free (uncoupled) anti-Ig mAb, does not inhibit proliferation even at high doses. These differences cannot be attributed to IgG subclass differences (Table 38.1). Second, the proliferation induced by anti-Bp35 and anti-Ig together is much more than the proliferation induced by either alone (Fig. 38.1). The simultaneous triggering of surface Ig and Bp35 is additive and may be synergistic.

The proliferation induced by whole anti-Bp35 antibodies is blocked by Fab fragments of anti-Bp35 (Table 38.1). Although these Fab fragments alone do not trigger B cell proliferation, like whole antibody, they can augment proliferation induced by anti-Ig on beads (14). Once again these results suggest that Bp35 and surface Ig may function synergistically in B cell activation.

One explanation for the augmentation of anti-Bp35-induced proliferation by allogeneic T cell factors is that anti-Bp35 induces the expression of receptors for BSF which then induce B cell proliferation (6). We could not test this possibility since BSF receptors have not yet been characterized.

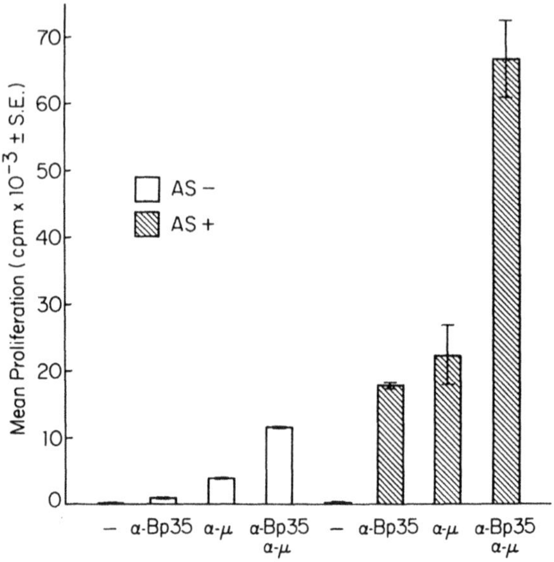

Fig. 38.1. B cell proliferation induced by free monoclonal anti-Bp35 and monoclonal anti-μ on beads is additive. When 5-μg/ml anti-Bp35 mAb was added together with 10-μg/ml anti-μ on beads to tonsillar B cells, proliferation of tonsillar lymphocytes on day 3 was much greater than either inducer alone. This was true in the presence (■) or absence (□) of 25% allogeneic supernatant.

However, we were able to measure the effect of anti-Bp35 on the expression of other membrane antigens. As illustrated in Fig. 38.2, anti-Bp35 induced increased expression of C3d receptors and p76 molecules within 16 hr and induced increased levels of HLA-DR molecules 64 to 96 hr after activation. Both dense and buoyant preactivated tonsillar B cells express similar high levels of HLA-DR, whereas circulating B cells in general have lower levels of HLA-DR molecules (13,17). Thus, it is possible that the dense tonsillar B cells are not primary resting B cells but are secondary B cells which have already been exposed to antigen and thus express higher levels of DR (4,17). This would explain why anti-Bp35 does not induce as rapid an increase in DR expression as seen in other induction systems (19,20).

Figure 38.2 also shows the forward-angle scatter profiles of dense tonsillar B cells 16 and 64 hr after anti-Bp35 activation. The cell size of control cells and of anti-Bp35 plus AS-treated cells were similar at 16 hr; by 48–64 hr 15–20% of the anti-Bp35 treated cells were enlarged suggesting that a subpopulation of B cells had been activated to divide. Similar results were observed with cells exposed to anti-Bp35 alone, while cells treated with AS only had no discernible effect on scatter profile or cell surface phenotype.

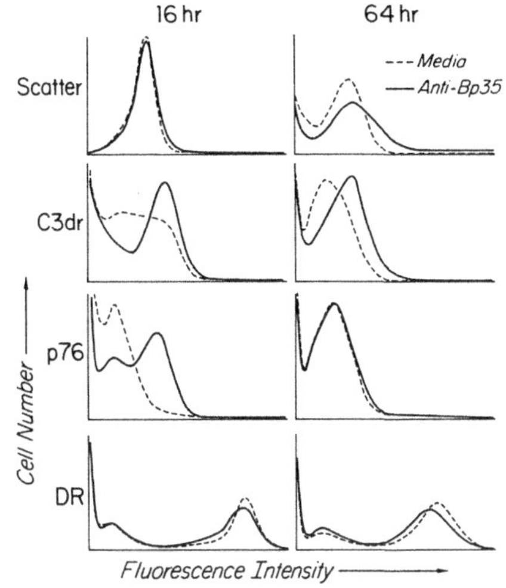

Fig. 38.2. Changes in cell surface phenotype of dense tonsillar lymphocytes after exposure to media (---) or anti-Bp35 with 25% allogeneic supernatant (———). Sixteen hours or 64 hours after induction, scatter profile and fluorescence histogram staining patterns for C3dr, p76, and HLA-DR antigens were determined. Media or AS only controls gave identical staining patterns. Anti-Bp35-induced cells showed a rapid increase in p76 and C3dr and a later increase in HLA-DR.

Discussion

The characteristics of anti-Bp35-induced B cell proliferation are similar to, yet distinct from, anti-Ig-mediated triggering: both appear to require cross-linking; both are augmented by allogeneic factors; and both are inhibited by monoclonal antibody to IgM via an Fc domain-dependent mechanism (2,15). However, anti-Bp35 mAb does not have to be attached to beads to function, and unlike free anti-μ mAb, does not inhibit proliferation even at high doses. Furthermore, the proliferation induced by anti-Bp35 and anti-μ on beads is additive (Fig. 38.1). This is true even when optimal doses of anti-μ or anti-Bp35 are used.

Further evidence that anti-Bp35- and anti-Ig-mediated triggering are in some way interconnected comes from studies using Fab fragments. Anti-Bp35 Fab fragments do not induce proliferation but do block the proliferation induced by whole antibody (14). However, when added together with anti-μ on beads, anti-Bp35 Fab fragments augment proliferation. In other words, monovalent binding of Bp35 in some way promotes B cell proliferation induced by anti-Ig.

To better understand the possible function of Bp35, it is important to consider the mechanisms of B cell activation via membrane Ig. Immunoglobulin is an unusual receptor in two respects: first, not only can it bind its specific ligand, but also after the appropriate signal it is changed into another form for secretion (8). Second, membrane Ig has a very short intracytoplasmic tail of three hydrophilic amino acids which are not phosphorylated. How then can membrane Ig after binding antigen transmit a cytoplasmic signal with such a small tail? Simple internalization seems unlikely since anti-Ig on large beads induces proliferation (2), but a mechanism involving redistribution of surface Ig, e.g., by capping, may be necessary (2,18).

Figure 38.3 presents two related models for how Bp35 may function in B cell activation. According to the first model, membrane Ig can 1) bind antigen and 2) turn into a secreted form, but because of the evolutionary constraints of having to be both a receptor and secreted product, it cannot 3) transmit its own intracytoplasmic signal. The Bp35 molecule would serve as a "bridge" into the cytoplasm for a positive proliferative signal

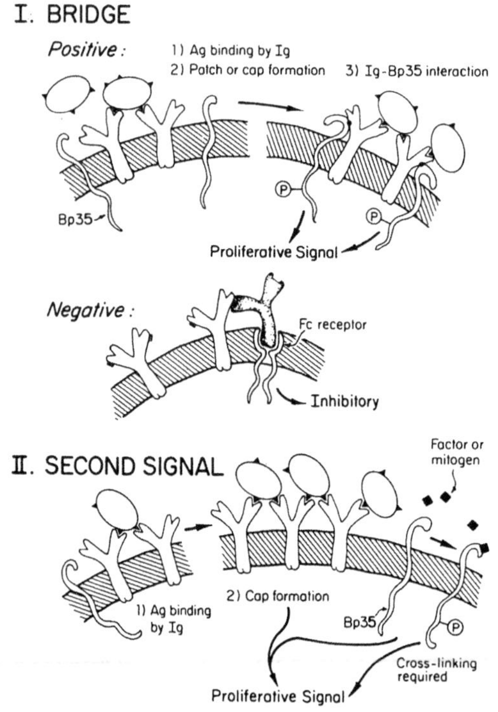

Fig. 38.3. Models for possible role of Bp35 in B cell activation. The *Bridge* model proposed that Bp35 serves as a means for antigen binding surface Ig to transmit a proliferative signal to the cytoplasm. Anti-idiotype bound to surface Ig, in contrast, would transmit a negative signal via the Fc receptor. The alternative *Second Signal* model does not require a specific Ig–Bp35 interaction but rather a "second signal" to Bp35 via a mediator from an accessory cell (e.g., T cell or macrophage) or via a mitogen.

with Ig when it redistributes in patches or caps. Cross-linking of this bridge directly would induce proliferation independently of antigen. The Fab fragment binding of Bp35 may facilitate appropriate Ig–Bp35 interactions or lower the signal threshold in some way. Bp35 is a phosphoprotein and thus could function via a phosphokinase-dependent pathway (19). Similarly, once surface Ig is bound by anti-idiotype another bridging mechanism via Fc receptors would lead to a negative or inhibitory signal (2,21). While this is an attractive explanation of many of our results, no physical association between Ig and Bp35 has been demonstrated yet.

In the second model, instead of a direct Ig–Bp35 interaction, the Bp35 molecule would bind a mediator from an accessory cell or a mitogen and this would lead to B cell proliferation. This pathway could augment or facilitate antigen-induced B cell activation. No close association between Bp35 and membrane Ig would be necessary. With the appropriate cross-linking of Bp35 by factor or mitogen, proliferation would be induced directly, while a monovalent binding would require a second signal. These two models are now in the process of being tested.

Acknowledgments. This work was supported in part by grants CA-34199 and AI-20432 of the National Institutes of Health and by Genetic Systems Corporation.

References

1. Sell, S., and P.G.H. Gell. 1965. Studies on rabbit lymphocytes *in vitro* I. Stimulation of blast transformation with an anti-allotype serum. *J. Exp. Med.* **122:**423.
2. Parker, D.C. 1980. Induction and suppression of polyclonal antibody responses by anti-Ig reagents and antigen-nonspecific helper factors. *Immunol. Rev.* **52:**115.
3. Chiorazzi, N., S.M. Fu, and H.G. Kunkel. 1980. Stimulation of human B lymphocytes by antibodies to IgM and IgG: Functional evidence for the expression of IgG on B-lymphocyte surface membranes. *Clin. Immunol. Immunopathol.* **15:**301.
4. Sieckmann, D.G., R. Asofsky, D.E. Mosier, I.M. Zitron, and W.E. Paul. 1978. Activation of mouse lymphocytes by anti-immunoglobulin. Parameters of the proliferative response. *J. Exp. Med.* **148:**1628.
5. Kehrl, J.H., A. Muraguchi, J.L. Butler, R.J.M. Falkoff, and A.S. Fauci. 1984. Human B cell activation, proliferation and differentiation. *Immunol. Rev.* **78:**75.
6. Moller, G., ed. 1984. B cell growth and differentiation factors. *Immunol. Rev.* **78:**7–224.
7. Leonard, W.J., J.M. Depper, T. Uchiyama, K.A. Smith, T.A. Waldmann, and W. Greene. 1982. A monoclonal antibody that appears to recognize the receptor for human T-cell growth factor; partial characterization of the receptor. *Nature* **300:**267.

8. Early, P., J. Rogers, M. Davis, K. Calame, M. Bond, R. Wall, and L. Hood. 1980. Two mRNAs can be produced from a single immunoglobulin μ gene by alternative RNA processing pathways. *Cell* **20:**313.
9. Stashenko, P., L.M. Nadler, R. Hardy, and S.F. Schlossman. 1980. Characterization of a human B lymphocyte specific antigen. *J. Immunol.* **125:**1678.
10. Oettgen, H.C., P.J. Bayard, W. Van Ewijk, L.M. Nadler, and C.P. Terhorst. 1983. Further biochemical studies of the human B-cell differentiation antigens B1 and B2. *Hybridome* **2:**17.
11. Clark, E.A., J.A. Ledbetter, R.C. Holly, G. Shu, P.A. Dindorf, and J.C. Sturge. 1985. Definition of cell surface polypeptides on human B cells. *Human Immunol.*, in press.
12. Bhan, A.K., L.M. Nadler, P. Stashenko, R.T. McCluskey, and S.F. Schlossman. 1981. Stages of B cell differentiation in human lymphoid tissue. *J. Exp. Med.* **154:**737.
13. Ledbetter, J.A., and E.A. Clark. 1985. Human B cell subsets defined by quantitative two color immunofluorescence. *Human Immunol.*, in press.
14. Clark, E.A., G. Shu, and J.A. Ledbetter. 1985. Role of Bp32 cell surface polypeptide in human B cell activation. *Proc. Nat. Acad. Sci. USA* **82:**1766–1770.
15. Clark, E.A., and T. Yokochi. 1984. Human B cell and B cell blast-associated surface molecules defined with monoclonal antibodies. In: *Leucocyte typing*, A. Bernard, L. Boumsell, J. Dausset, C. Milstein, and S.F. Schlossman, eds. Springer-Verlag, Berlin, Heidelberg, pp. 339–346.
16. Iida, K., L. Nadler, and V. Nussenzweig. 1983. Identification of the membrane receptor for the complement fragment C3d by means of a monoclonal antibody. *J. Exp. Med.* **158:**1021.
17. Ledbetter, J.A., P.J. Martin, and E.A. Clark. 1985. Mantle zone and germinal center B cells respond to different activation signals. In *Immune Regulation* Eds M. Feldmann and N.A. Mitchison (Humana, Clifton, New Jersey) pp. 197–206.
18. Braun, J., and E.R. Unanue. 1980. B lymphocyte biology studied with anti-Ig antibodies. *Immunol. Rev.* **52:**3.
19. Monroe, J.G., J.E. Niedel, and J.C. Cambier. 1984. B cell activation IV. Induction of cell membrane depolarization and hyper I-A expression by phorbol diesters suggests a role for protein kinase in murine B lymphocyte activation. *J. Immunol.* **132:**1472.
20. Clark, E.A., and G. Shu. 1985. A comparison of induction of human B cell proliferation through surface Bp35 polypeptides or immunoglobulin receptors. Submitted.
21. Phillips, N.E., and D.C. Parker. 1984. Cross-linking of B lymphocyte Fcγ receptors and membrane immunoglobulin inhibits anti-immunoglobulin-induced blastogenesis. *J. Immunol.* **132:**627.

CHAPTER 39

B Lymphocyte Surface Antigens Involved in the Regulation of Immunoglobulin Secretion

Josée Golay, Frances Rawle, and Peter Beverley

Introduction

Human B lymphocyte differentiation is a multistep process in which a number of B cell surface antigens must be involved (1). Several B cell surface molecules have been shown to play a role at various stages of B cell differentiation in the mouse (2–7), but little information is yet available on the role of such molecules in immunoregulation in man. We have investigated the role of human B cell surface antigens using relatively simple models of B cell differentiation. In the first, monoclonal human B chronic lymphocytic leukemia (B-CLL) or prolymphocytic leukemia cells (PLL) are induced to secrete immunoglobulin (Ig) *in vitro* by normal activated T cells (8,9). In the second, the B-lymphoblastoid cell line CESS is induced to secrete IgG in response to mixed lymphocyte reaction (MLR) supernatant (10). In both systems we have examined the effect of adding the Workshop panel of monoclonal antibodies (mAbs), and have found that several B lymphocyte surface antigens appear to play a role in human B cell activation and differentiation to immunoglobulin secretion.

Materials and Methods

Cells

Peripheral blood mononuclear cells (PBMC) were isolated by Ficoll–Hypaque (F/H) gradient centrifugation from the blood of patients with B-CLL and of a newly diagnosed patient with PLL, and cryopreserved. T cells were obtained from frozen samples of PBMC from normal volunteers, after rosetting with AET–SRBC and separation by two sequential F/H gradient centrifugation steps.

B Cell Assays

All cell cultures were set up in a total volume of 0.2 ml in Titertek 96-well plates (Flow Laboratories) in RPMI 1640 (Gibco, Grand Island, NY) supplemented with 10% fetal calf serum (FCS) (Gibco), sodium pyruvate, glutamine, and penicillin/streptomycin (50 U/ml, Flow Laboratories). 10^5 B-CLL or PLL cells were cultured with 2×10^5 purified T cells and 10 μl/ml pokeweed mitogen (PWM) (Gibco) in round-bottomed plates. The supernatant was collected after 5 to 7 days and the concentration of Ig measured by an ELISA technique. 2×10^3 CESS cells were cultured in the presence or absence of 10% MLR supernatant for 5 days in flat-bottomed wells and the concentration of Ig determined by ELISA. Monoclonal antibodies were added to both assays at a 1/250 final dilution.

MLR Supernatant

PBMC from several donors were pooled and cultured at a final concentration of $1–2 \times 10^6$ cells/ml in RPMI 1640 supplemented with 10% FCS in tissue culture flasks (Falcon Plastics Co., Oxnard, CA). The supernatant was stored at −20°C.

Enzyme-Linked Immunosorbent Assays (ELISA)

Sheep anti–human IgM (SaHIgM) (Seward Laboratory, London) diluted 1/200 or rabbit anti–human IgG (RaHIgG) (Dakopatts, a/s, Denmark) diluted 1/500 were adsorbed onto flat-bottomed 96-well plates (Titertek) for 1 hr at 37°C. After washing with PBSA-Tween, serial dilutions of culture supernatant and of Ig standards were added and incubated for 1–2 hr. After further washing, plates were incubated with third-layer RaHIgM– or RaHIgG–peroxidase conjugates (Dakopatts) diluted 1/400 and 1/1000, respectively. The last wash was followed by addition of the substrate 0.004% *o*-phenylenediamine (Sigma) with 0.018% H_2O_2 (Fisons, England) in phosphate/citrate buffer, pH 5.0. The reaction was stopped with 50 μl of 1*M* sulfuric acid and the optical density read on a Titertek Multiscan. The Ig concentration in the supernatants was calculated from the log/linear regression curve obtained from the standards.

Indirect Immunofluorescence

2×10^5 cells were stained in flexiplates for 30 min at 4°C with 50 μl of mAbs diluted 1/250 in Hepes buffered minimal essential medium (MEM) with 5% FCS and 0.002% azide. After washing, the cells were incubated with 50 μl of goat anti–mouse Ig–FITC conjugate diluted 1/25 (Sigma) for 30 min at 4°C. Analysis was performed by flow cytometry or fluorescence microscopy.

Proliferation Assays

10^5 HD-PLL cells were cultured in triplicate round-bottomed wells. 1 μCi [^{3}H]thymidine was added 18 hr before the end of culture and the cells harvested in a semiautomated harvester (Titertek).

Results and Discussion

In order to develop an assay for B cell differentiation, PBMC from B-CLL and PLL patients were screened for increased Ig production in the presence of purified normal allogeneic T cells and PWM or PHA. Cells from 2 of 10 patients with B-CLL were found to secrete Ig in response to activated T cells. Patient MB cells secreted IgM and patient GP cells secreted IgG as shown in Fig. 39.1. PBMC from the PLL patient (HD-PLL) showed an increase of up to 50-fold in IgM secretion after five days in culture with T cells and PWM (Fig. 39.1). The Ig concentration was low or background level in the control cultures of unstimulated cells or T cells alone. In addition we used the cell line CESS in a modification of the assay described by Muragushi *et al.* (10) as detailed in Materials and Methods. A 5–10-fold enhancement of IgG secretion can be detected after incubation with 10% MLR supernatant for 5 days (Fig. 39.2).

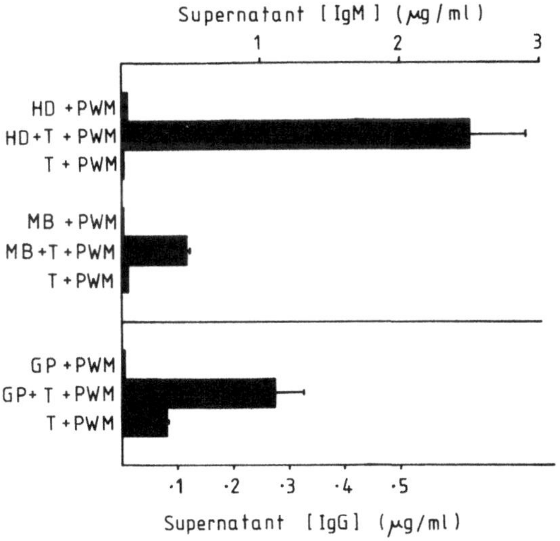

Fig. 39.1. Stimulation by T cells of Ig secretion by B leukemia cells. 10^5 PBMC from leukemia patients were incubated with 2×10^5 normal T cells in the presence of PWM for 5 days in the case of HD-PLL cells and 7 days in the case of MB- and GP-CLL cells. The concentration of immunoglobulin in the supernatants was measured by an ELISA technique as described in Materials and Methods.

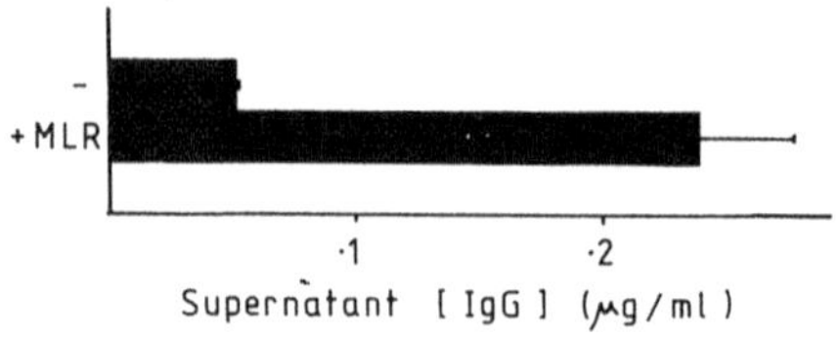

Fig. 39.2. Stimulation of CESS line by MLR supernatant. 2×10^3 CESS cells were incubated in the presence and absence of 10% MLR supernatant for 5 days. IgG levels in supernatant were measured by ELISA.

The effect on the two systems described above of adding the Workshop antibodies was assessed. The results are shown in Table 39.1. Both B and leukemia panels of antibodies were screened on the T-dependent B leukemia assays, whereas only the B cell panel was tested on CESS. A number of antibodies were found to be inhibitory in one or more assays and one antibody was stimulatory. Only the results with mAbs that gave a significant and reproducible effect in any one of the assays, together with the results obtained with mAbs of the same specificity, are presented in Table 39.1. The data have been arranged according to the specificity attributed to the mAbs by the Workshop. Values showing an inhibition of greater than 30% have been underlined in Table 39.1 since such a level of inhibition is considered to be significant when seen in two experiments on a given cell type. In some cases, however, only a single experiment was carried out and this is indicated in the table, as are those cases when only one out of two experiments showed inhibition but the mean inhibition was still more than 30%. It is readily apparent from Table 39.1 that the majority of the inhibitory antibodies are grouped in only three of the clusters defined by the Workshop, namely, the Bp95, Bp35, and Bp135 clusters. Two or more mAbs in each of these clusters are inhibitory for one or more assays. In addition, four inhibitory antibodies and one stimulatory one are found in the non-clustered B cell-specific (B6, B8, B30) or B cell-associated (B38, B17) groups of antibodies. B6 and B30 appear to inhibit all four assays, B8 only the IgG-secreting cells, and B38 only the IgM-secreting cells, although the GP-CLL may also be weakly inhibited. B17 stimulates IgM production by the HD-PLL cells but has an inhibitory effect on the MB-CLL cells.

The correlation between the functional effect of the mAbs and their antigen specificity is not absolute. At least one antibody in each of the Bp95, Bp35, and Bp135 clusters is not inhibitory in any one of the assays and the pattern of inhibition by mAbs in different assays is not always the same for antibodies of the same specificity. There are a number of possible explanations for these apparent discrepancies. Antibody affinity may be important in determining whether the antibody will affect Ig secretion if inhibition is due to competition between a soluble or cell surface molecule and the mAb for a receptor on the B cell surface. Due to the limited

Table 39.1. Effects of mAbs on Ig secretion by B leukemia cells and CESS cell line.[a]

mAb specificity	mAb	HD-PLL IgM	MB-CLL IgM	GP-CLL IgG	CESS IgG
Bp 95	B14	108	105	119	38[b]
	B28	52	57	44	100
	B34	112	98	114	53
	B43	68	50[c]	69	31
	L17	23	43	13	ND[d]
Bp 35	B5	13	31	23	60[b]
	B22	93	92	86	109
	B24	45	57	69[c]	54[c]
Bp 135	B7	72	77	59	33
	B31	22	9	30	63
	B40	120	83	128	119
	B49	36	15	48	58
	B25	106	95	75	112
non-clustered	B6	27	25	37	55[c]
	B8	79	91	60	20
	B30	16	14	53	33
	B38	56	63	75	93
	B17	404	56	104	100

[a] The effects of the mAbs diluted 1/250 on the assays described in Figs. 39.1 and 39.2 were measured. The percentages of the control levels, i.e., in the absence of mAb, are given. The control Ig levels are shown in Figs. 39.1 and 39.2. Data for HD-PLL and MB-CLL are taken from a representative experiment whereas the data for GP-CLL and CESS are the mean of two experiments.
[b] Only one experiment was carried out.
[c] Inhibition was seen in only one out of two experiments.
[d] ND: Not determined.

amount available, antibodies were only tested at a single dilution (1/250) and titration experiments are clearly required. Class and subclass differences may be important in determining the function of mAbs, as has recently been shown in studies of the stimulatory capacity of anti-T3 antibodies of different IgG subclasses (11). This may be particularly relevant to the anti-Bp35 group of antibodies of which only two, B5 and B24, both IgG2a, inhibit Ig secretion, whereas the third, B22, does not and is an IgG2b. Finally the functional effects of different antibodies may be dependent on the epitope specificity. In particular the Workshop data have suggested that the epitope recognized by B25 is different from the epitope(s) seen by the other anti-Bp135 antibodies. This might be an explanation for the lack of inhibition by B25 compared to most of the other anti-Bp135 antibodies.

Table 39.2 shows the staining pattern of the B cells with the relevant mAbs, expressed as a percentage of positive cells. These results correlate well with the clusters defined by the Workshop. Values for CESS may be

Table 39.2. Staining pattern of mAbs on B cells (%).[a]

mAb	HD-PLL	MB-CLL	GP-CLL	CESS
B14	1	49	46	35
B28	3	65	59	64
B34	1	48	40	15
B43	5	58	45	65
L17	5	48	32	ND
B5	53	68	82	93
B22	68	69	82	73
B24	69	58	79	74
B7	11	0	7	70
B31	0	1	4	56
B40	0	2	11	60
B49	3	0	9	40
B25	3	0	11	45
B6	12	15	19	83
B8	1	30	24	33
B30	8	38	13	10
B38	0	1	1	0
B17	45	79	81	61

[a] The percentage of cells stained with the mAbs after subtraction of the negative control values (≤2%) are shown.

considered approximate since they were obtained by microscopy and have not yet been confirmed by flow cytometry. It is worth noting that the "pan B" Bp 95 mAbs are only very weakly expressed on the PLL cells whereas they are strongly positive on the CLL cells and on CESS. There may, therefore, be a small subpopulation of normal B cells which are Bp95 negative and are the normal counterpart of PLL B cells. Alternatively, expression of Bp95 may be altered in some malignant cells. Despite this weak staining, three of the anti-Bp95 antibodies are inhibitory for HD-PLL, suggesting that expression of the antigen may increase during HD-PLL activation. A similar phenomenon could also explain the fact that anti-Bp135 antibodies inhibit Ig secretion by the B leukemia cells but stain them weakly, if at all. The observation that CESS, which may represent a more activated stage of B cell differentiation, expresses Bp135 strongly supports this hypothesis.

Inhibition of B cell maturation was further studied by looking at the time course of the inhibitory effect of most of the mAbs affecting IgM secretion by HD-PLL cells [Fig. 39.3(a) and (b)]. mAbs were added at the beginning of culture or at day 2 or 4. The results demonstrate that inhibition is time dependent, suggesting that the antigens recognized by these antibodies are indeed involved in *in vitro* B cell activation or differentiation, and that the decreased Ig levels are not an artifactual effect of the mAbs in the ELISA. However, there are no clear indications as to the

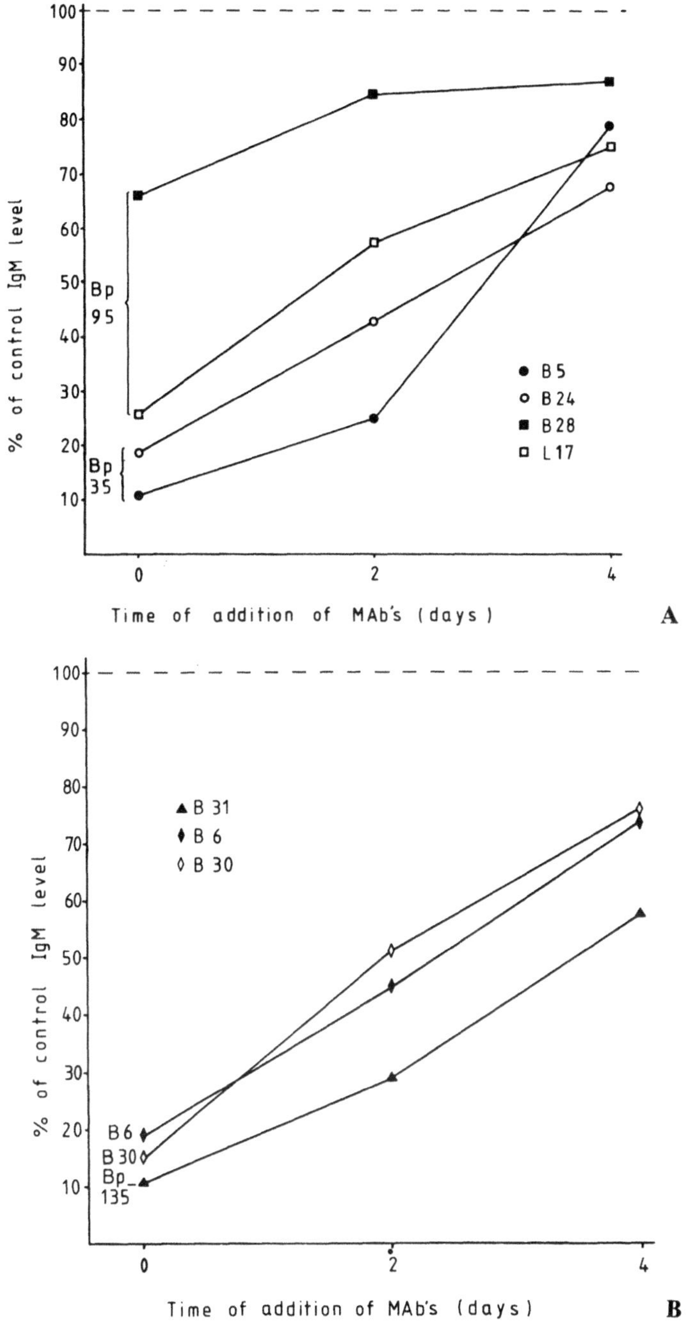

Fig. 39.3. Time course of effects of mAbs on IgM secretion by PLL. The mAbs were added at day 0, 2, and 4 of culture of HD-PLL cells with T cells and PWM, and the IgM concentrations measured at day 7. (a) The percentage of the control Ig concentration obtained with mAbs B5, B24, B28, and L17; (b) the results with mAbs B31, B6, and B30.

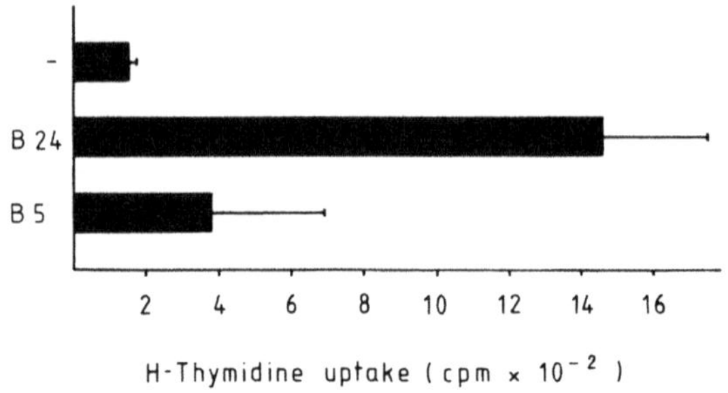

Fig. 39.4. Effect of anti-Bp35 mAbs on [^{3}H]thymidine uptake by HD-PLL. 10^5 HD-PLL cells were incubated in the presence or absence of anti-Bp35 antibodies and [^{3}H]thymidine uptake was measured after 4 days. Data represent the mean of triplicate wells.

stage of differentiation at which the mAbs act, since the time course with all antibodies tested was approximately linear, inhibition being greatest when the mAbs were added at day 0.

Finally, the effect of these mAbs on B cell activation was studied by seeing whether they can induce the HD-PLL cells to proliferate. Interesting results were found with antibodies in the Bp35 group. B24 was found to induce a 5- to 10-fold increase in [^{3}H]thymidine uptake by HD-PLL cells, whereas B5 showed little, if any, effect (Fig. 39.4). B22 was not tested. The difference in the activities of B5 and B24 may be due to the requirement for a higher antibody concentration for induction of proliferation than for inhibition of IgM secretion, or to differences in the epitope specificities of these two antibodies. The observation that B24 has an effect on both proliferation and Ig secretion suggests that the Bp35 molecule is involved in an early stage of B cell differentiation or activation, maybe in a way similar to Lyb2 in the mouse (12,13). Alternatively, Bp35 may interact with, and transmit signals from, more than one B cell surface receptor during activation and differentiation. Anti-Bp35 antibodies would therefore have a pleiotropic effect.

Conclusion

Although still preliminary, the results presented above are the first evidence of the functional role of at least three well-defined B cell-specific surface antigens. These molecules are clearly different from the antigen recognized by the antibody 4F2, which is expressed on activated T and B cells and on monocytes and may be involved in immune regulation (14). Further work will be required to characterize the stages of differentiation at which these molecules are involved and their precise function. Clearly the mAbs used here will prove valuable tools in such studies.

Summary

We have used two systems to study the function of human B lymphocyte surface antigens in the regulation of Ig secretion. In the first, monoclonal B leukemia cells were induced to secrete Ig by activated T cells and, in the second, the B-lymphoblastoid cell line CESS was stimulated to produce IgG by MLR supernatant. A number of antibodies were found to inhibit these assays and most were clustered in the Bp95, Bp35, and Bp135 groups of mAbs, suggesting a functional role for these antigens in B cell maturation to Ig secretion. In addition, some antibodies in the non-clustered group were found to be inhibitory and one had a differential effect on the two IgM-secreting cell types. Finally, B24 induced proliferation of PLL cells, suggesting that Bp35 is involved at least at the proliferative or activation stage of B cell differentiation.

Acknowledgments. The CESS cell line was a kind gift from Dr. T. Kishimoto.

References

1. Melchers, F., and J. Anderson. 1984. B cell activation: three steps and their variations. *Cell* **37:**715.
2. Huber, B.T. 1982. B cell differentiation antigens as probes for functional B cell subsets. *Immunol. Rev.* **64:**57.
3. Kemp, J.D., J.W. Rohrer, and B.T. Huber. 1982. Lyb3: a B cell surface antigen associated with triggering secretory differentiation. *Immunol. Rev.* **69:**127.
4. Yakura, H., F.W. Shen, E. Bourcet, and E.A. Boyse. 1982. Evidence that Lyb-2 is critical to specific activation of B cells before they become responsive to T cells and other signals. *J. Exp. Med.* **155:**1309.
5. Subbarao, B., and D.E. Mosier. 1982. Lyb antigens and their role in B lymphocyte activation. *Immunol. Rev.* **69:**81.
6. Takatsu, K., and T. Hamaoka. 1982. DBA/2 Ha mice as a model of an X-linked immunodeficiency which is defective in the expression of TRF-acceptor site(s) on B lymphocytes. *Immunol. Rev.* **64:**25.
7. Takatsu, K., Y. Sano, S. Tomita, N. Hashimoto, and T. Hamaoka. 1981. Antibody against T cell replacing factor acceptor sites augments *in vitro* primary IgM response to suboptimal doses of heterologous erythrocytes. *Nature* **292:**360.
8. Saiki, O., T. Kishimoto, T. Kuritani, A. Muragushi, and Y. Yamamura. 1980. *In vitro* induction of IgM secretion and switching to IgG production in human B leukaemia cells with the help of T cells. *J. Immunol.* **124:**2609.
9. Fu, S.M., N. Chiorazzi, H.G. Kunkel, J.P. Halper, and S.R. Harris. 1978. Induction of *in vitro* differentiation and immunoglobulin synthesis of human leukaemic B lymphocytes. *J. Exp. Med.* **148:**1570.
10. Muragushi, A., T. Kishimoto, Y. Miki, T. Kuritani, T. Kaeida, K. Yoshisaki, and Y. Yamamura. 1981. T cell replacing factor (TRF) induced IgG secretion

in a human lymphoblastoid cell line and demonstration of acceptors for TRF. *J. Immunol.* **127**:412.
11. Tax, W.J.M., H.W. Willems, P.P.M. Reekers, P.J.A. Capel, and R.A.P. Koene. 1983. Polymorphism in the mitogenic effect of IgG1 monoclonal antibodies against T3 antigen on human T cells. *Nature* **304**:445.
12. Subbarao, B., and D.E. Mosier. 1983. Induction of B lymphocyte proliferation by monoclonal anti Lyb2 antibody. *J. Immunol.* **130**:2033.
13. Subbarao, B., and D.E. Mosier. 1984. Activation of B lymphocytes by monovalent anti Lyb2 antibodies. *J. Exp. Med.* **159**:1796.
14. Gerrard, T.L., C.H. Jurgensen, and A.S. Fauci. 1984. Modulation of human B cell responses by a monoclonal antibody to an activation antigen 4F2. *Clin. Exp. Immunol.* **57**:155.

CHAPTER 40

Inhibition of Immunoglobulin Secretion, But Not Immunoglobulin Synthesis, by a Monoclonal Antibody

Stefania Pittaluga, Jeffrey Cossman, Jane B. Trepel, and Leonard M. Neckers

Immunoglobulin (Ig) secretion is a phenomenon composed of several distinct processes: Ig mRNA accumulation, with, in the case of IgM, a shift from the message coding for membrane Ig to that coding for secretory Ig; translation of that message and assembly of the Ig molecule in the cytoplasm; and, finally secretion of the stored Ig (1–7). We have been studying the signals necessary for the triggering of each of these events in an attempt to define the regulation of normal and malignant B cell differentiation.

We and others have recently determined that CLL cells can be induced to secrete Ig by treatment with phorbol diester (8–9). This induction involves the stimulation of the intracellular events described above. In its action phorbol diester mimics both mitogen (antigen) and T cell-derived soluble factors which act in sequence to trigger Ig secretion (10,11, and Trepel *et al.*, unpublished results). In this study we report that a monoclonal antibody, B42, can inhibit Ig secretion by blocking the ability of phorbol diester and T cell soluble factors to stimulate the last, but not the initial, events in the secretory process.

Methods

Culture Conditions

Cells were cultured at 1–2 × 10^6/ml in RPMI containing 10% fetal calf serum and antibiotics. Monoclonal antibodies from the B/L panel of the Second International Workshop on Human Leukocyte Differentiation Antigens were added as dilutions of culture supernatants as described in the text.

Flow Cytometry

Reactivity of monoclonal antibodies with cells was analyzed by indirect immunofluorescence on a FACS II as previously described (12).

Cell Proliferation Analysis

Cells were pulsed with 1 μCi [^{3}H]methylthymidine for 4 hr and then lysed, filtered, and counted by liquid scintillation spectrometry as previously described (12).

ELISA

The concentration of IgM in culture fluid was determined by ELISA as previously reported (8).

Cell Stimulation Conditions

CLL cells were treated with 10^{-8} *M* TPA as described elsewhere (8). CLL cells and normal B cells [purified as previously reported (13)] were stimulated with Cowan *Staphylococcus aureus* I (CSA; final dilution 1 : 100,000) and 10% T cell supernatant as previously described (13). T cell supernatant was obtained from mitogen-activated T cell cultures as reported elsewhere (13).

Results and Discussion

In an initial screening of the entire monoclonal antibody B/L panel (72 antibodies), five mAbs were determined to inhibit IgM secretion by TPA-treated CLL cells (Table 40.1). Of these, one mAb, B42, inhibited nearly all IgM secretion when added at the start of the 5-day culture period. This antibody was chosen for further study.

Table 40.1. Effect of mAbs on TPA-induced IgM secretion by CLL cells.[a]

mAb	IgM	% Inhibition
—	1.134	—
B42	0.077	94
B3	0.469	59
B2	0.545	52
B4	0.550	52
B47	0.555	51

[a] CLL cells (2 $\times 10^6$/ml) were treated with TPA for 5 days. 20 μl of the B/L panel monoclonal antibodies were added at initiation of culture. IgM (μg/ml) was measured in the culture supernatant on day 5. The antibodies shown were the five most effective in inhibiting secretion.

Table 40.2. Effect of B42 on IgM secretion by normal and malignant B cells.[a]

	Normal B cells +CSA/T sup	CLL cells +TPA	CLL cells +CSA/T sup	Burkitt's cells
−B42	0.220	1.1	0.498	1.000
+B42	0.015	0.077	0.100	0.200

[a] Normal peripheral blood B cells (1 $\times 10^6$/ml were treated with CSA and T cell supernatant for 5 days; CLL cells were treated similarly or received TPA for 5 days; and Burkitt's lymphoma cells (spontaneously secreting) were treated throughout the 5-day culture period with a 1 : 4000 dilution of B42. IgM, expressed as μg/ml, was determined in culture supernatant on day 5 of culture.

Experiments were undertaken to determine the ability of B42 to inhibit Ig secretion by various B cell types. The antibody was effective against mitogen-stimulated normal B cells, TPA-or mitogen-stimulated CLL B cells, and spontaneously secreting Burkitt's lymphoma cells (Table 40.2). By flow cytometric analysis, B42 reacted with all of the above types, but not with normal T cells (Table 40.3).

The effective concentration of B42 was determined using TPA-treated CLL cells incubated for 5 days with the antibody. The antibody was found to be effective even at a final dilution of 1 : 4000 (Table 40.4) and this concentration was used in the remaining experiments. To determine whether B42 interfered with the ELISA measurements, 200 ng of human IgM standard was incubated with either 20 μl of control culture medium or 20 μl of B42 (1 : 100 final dilution) for 30 min at room temperature. Then both samples were applied to an anti-IgM-coated microtiter plate and IgM concentrations determined by ELISA. Preincubation with B42 did not result in a significantly lower apparent IgM concentration (Fig. 40.1).

Since proliferation and Ig secretion have been shown to be two separate processes (13), we tested the ability of B42 to inhibit mitogen-induced B cell proliferation. B42 had no effect on mitogen-induced B cell proliferation, while it totally blocked IgM secretion induced by the same treatment conditions (Table 40.5).

The process of Ig secretion has been dissected into several steps, as shown by its calcium requirements and sensitivity to various stimulating

Table 40.3. Flow cytometry of B42.[a]

Peripheral blood lymphocytes	+/−
Normal peripheral blood B cells	+
CLL cells	+
Burkitt's cells	+
Normal T cells	−

[a] Cells were reacted with B42 followed by FITC–goat anti–mouse IgG/IgM and B42 staining was detected by flow cytometry. +/−5–50%, +>50%, −0%.

Table 40.4. Dose-response of B42-induced inhibition of IgM secretion.[a]

Treatment	IgM	% Inhibition
TPA	0.906	—
TPA + B42 (1:100)	0.267	71
TPA + B42 (1:200)	0.190	80
TPA + B42 (1:400)	0.165	82
TPA + B42 (1:1000)	0.122	87
TPA + B42 (1:2000)	0.132	86
TPA + B42 (1:4000)	0.095	90

[a] CLL cells (2×10^6/ml) were cultured for 5 days with TPA and various dilutions of B42. IgM is expressed as μg/ml.

agents (10,11,14,15; this system will be described in detail elsewhere by Trepel *et al.*). Briefly, in mitogen-stimulated B cell Ig secretion, mitogen is required for Ig mRNA accumulation and its intracytoplasmic translation, but addition of T cell-derived soluble factors (replaceable by exogenous phospholipase C) are necessary to obtain Ig secretion. The phorbol diester, TPA, subserves the dual role of mitogen and T cell lymphokine in its ability to independently induce Ig secretion in B cells and CLL cells. In the next series of experiments we attempted to determine if B42 blocked secretion by affecting one or more of these defined steps.

When TPA was added to CLL cells a slight increase in IgM secretion was observed within two days (Fig. 40.2), although secretory (2.4 kilobase) IgM mRNA is maximally induced within 24 hr of TPA addition (8,16). The addition of B42 to CLL cells 2 days after exposure to TPA completely blocked subsequent IgM secretion (Fig. 40.2). These results demonstrate that B42 can inhibit secretion even when added after secretion has already begun to occur, suggesting a site of action directed at the end of the cascade leading to secretion.

To test this hypothesis further, we used two signals to stimulate Ig secretion by CLL cells—sequential addition of mitogen (CSA) followed by addition of T cell soluble factors. Addition of CSA alone has minimal effects on IgM secretion, but induces IgM mRNA and its translation into

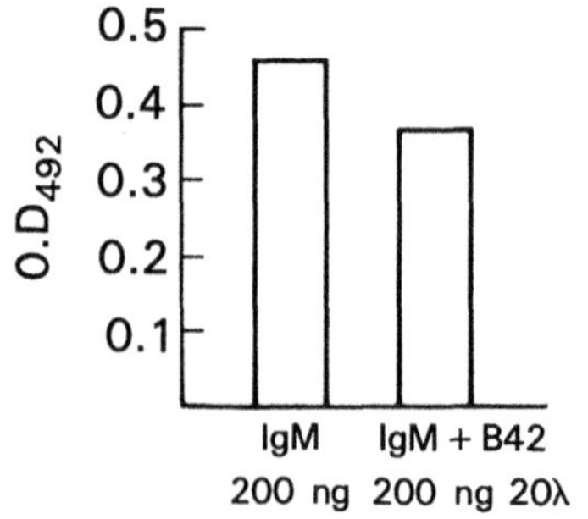

Fig. 40.1. B42 does not prevent recognition of human IgM in the ELISA system. B42 did not block the binding of human IgM standard to anti–human IgM-coated ELISA plates. Concentrations are expressed as fractional O.D. units at a wavelength of 492 nm.

Table 40.5. Effect of B42 on mitogen-induced thymidine incorporation by B cells.[a]

Treatment	^{3}H-Tdr
No mitogen	259
CSA/T sup	2085
CSA/T sup + B42	2186

[a] CSA and T cell supernatant were added to normal B cells (2×10^5/ml) on day 0 and thymidine incorporation determined on day 5. Results are expressed as cpm/well.

cytoplasmic IgM. We have found that secretion occurs once T cell soluble factors are added to these primed cells (10,11). CLL cells were treated with CSA for 3 days and then T cell soluble factors were added for 2 additional days. B42 was added either at the beginning of the 5-day period (with CSA) or with the T cell soluble factors (for the last 2 days) (Fig. 40.3). IgM secretion was measured on the 5th day. B42 completely blocked secretion of IgM when added on the third day along with the T cell soluble factors. This suggests that the antibody is either directly inhibiting the ability of T cell soluble factors to induce secretion or is inhibiting a process set in motion by interaction of the T cell soluble

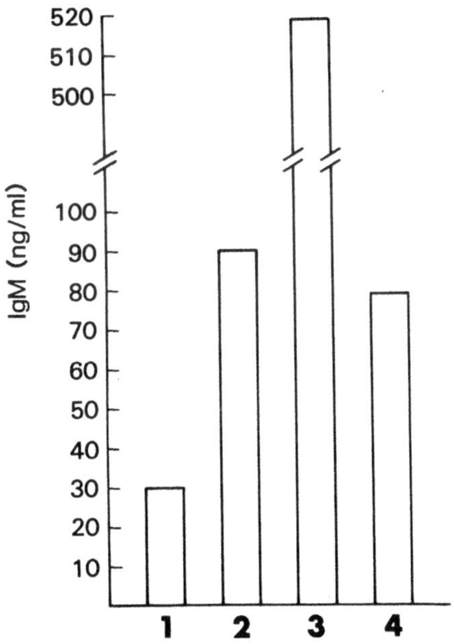

Fig. 40.2. B42 blocks TPA-induced IgM secretion by CLL cells immediately upon addition to cultures. CLL cells were cultured at 1×10^6/ml with TPA for 5 days and B42 was added after the 2nd day. IgM, expressed as μl/ml, was measured on day 2 (bars 1 and 2), or day 5 (bars 3 and 4).

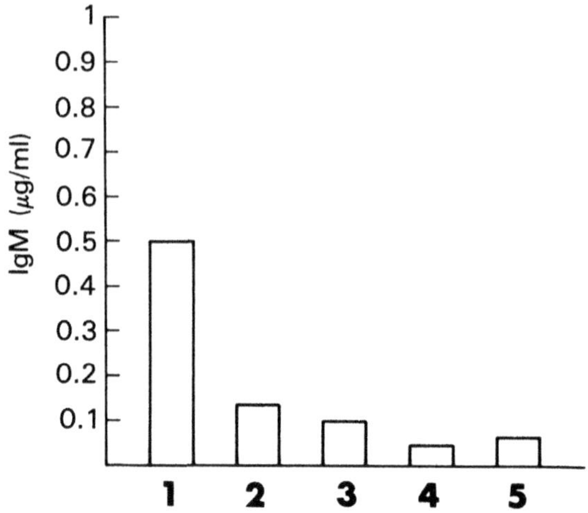

Fig. 40.3. B42 inhibits IgM secretion when added with T cell supernatant. CLL cells (1×10^6/ml) were cultured with CSA for 3 days and then treated as follows: T cell supernatant for 2 days (bar 1), CSA alone (bar 2), CSA for 3 days followed by T supernatant together with B42 for 2 days (bar 3), CSA with B42 for 3 days followed by T supernatant for 2 days (bar 4), or CSA for 3 days, T supernatant for 2 days with B42 added at both points (bar 5). IgM, expressed as μg/ml, was determined on day 5.

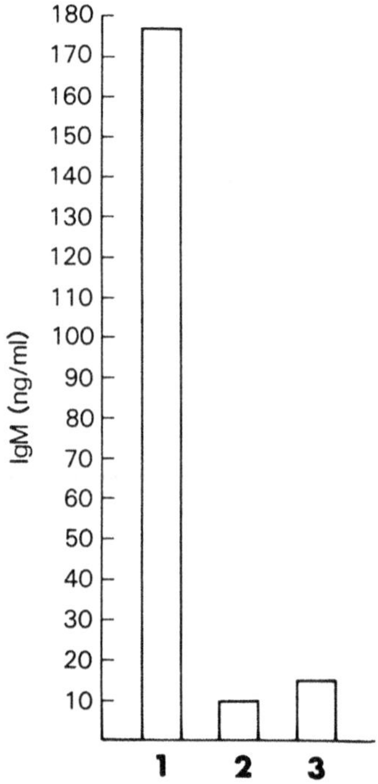

Fig. 40.4. B42 blocks IgM secretion by CSA-pretreated normal B cells when the antibody is added together with the T cell soluble factors for 1 day. B cells (2×10^6/ml) were treated with CSA for 4 days and T cell supernatant was then added for 1 additional day (bar 1). B42 was added with CSA (bar 2) or with the T cell supernatant (bar 3). IgM is expressed as ng/ml.

factors with the cell membrane. The latter hypothesis is supported by the fact that TPA-induced Ig secretion is also inhibited by B42 and TPA has no known surface receptors but binds directly to intracellular protein kinase C (17–20).

Similar results were obtained when normal B cells were studied. Sequential stimulation of target cells by CSA (4 days) followed by T cell soluble factors (1 day) was found to induce secretion of IgM and this was inhibitable by B42 when it was added either on day 1 or on day 5 (Fig. 40.4).

Further evidence that B42 functions near the end of the secretory cascade was sought by examining the effect of B42 on phospholipase C-induced Ig secretion of CSA-pretreated CLL cells (Fig. 40.5). Phospholipase C mimics the effects of TPA by hydrolyzing phosphatidylinositol into diacylglycerol, the physiological analogue of TPA that activates protein kinase C (17–21). Exogenously added phospholipase C caused rapid secretion of intracellular IgM by CSA-treated CLL cells and when B42 was added with phospholipase C for 18–24 hr, IgM secretion was inhibited (Fig. 40.5). This again implicates a site of action of B42 in the late stages of IgM secretion.

The ability of the antibody to block phospholipase C and TPA-induced Ig secretion suggests that B42 may recognize a surface protein normally phosphorylated by protein kinase C during the secretory process. Since B42 so effectively blocks secretion, one may speculate that phosphorylation may lead to conformational changes or intramembrane movement of this protein without which Ig secretion cannot occur.

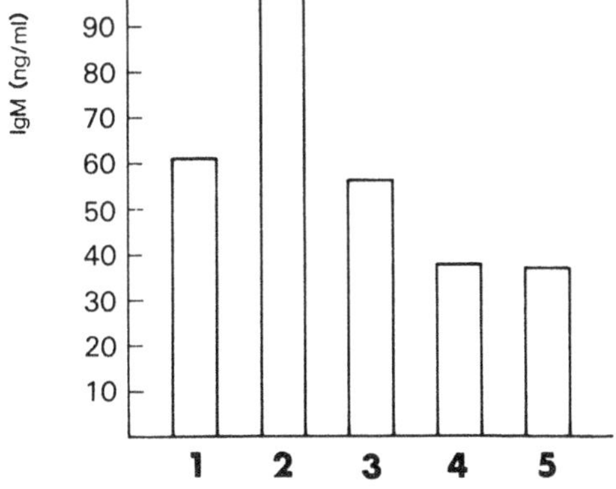

Fig. 40.5. B42 inhibits the ability of exogenous phospholipase C to stimulate IgM secretion in CSA-pretreated CLL cells. CLL cells (2×10^6/ml) were treated with CSA for 4 days. Phospholipase C (0.5 units/ml) was added on the 4th day and IgM concentration of culture supernatant determined on day 5. B42 was added with phospholipase C (bar 3), with CSA (bar 4), or with both (bar 5). IgM is expressed as ng/ml.

Summary

In a continuing study on the regulation of immunoglobulin secretion in both normal and malignant B cells we have screened the B cell leukemia monoclonal antibody B/L panel for antibodies that inhibit phorbol diester and mitogen-induced IgM secretion by chronic lymphocytic leukemia (CLL) and normal B cells. One of the antibodies, B42, completely blocked secretion even when added at a final dilution of 1:4000. B42 bound to CLL cells, normal B cells, and Burkitt's lymphoma cells, and inhibited Ig secretion by these cell types without any apparent effects on Ig accumulation in the cytoplasm. Thus, B42 may recognize a surface structure necessary for actual mechanical functioning of the secretory process.

References

1. Koshland, M.E. 1983. Molecular aspects of B cell differentiation. *J. Immunol.* **131:**i–ix.
2. Warner, N.L., A.W. Harris, I.F.C. McKenzie, D. De Luca, and G. Gutman. 1975. Lymphocyte differentiation as analyzed by the expression of defined cell surface markers. In: *Membrane receptors of lymphocytes,* M. Seligman J.L. Preud'homme, and F.M. Kourilsky, eds. American Elsevier, New York, p. 203–216.
3. Kim, K.J., C. Kanellopoulos-Langevin, G. Chaouat, and R. Asofsky. 1981. Differential effect of mitogens and antigen nonspecific T cell factors on the expression of surface immunoglobulin (sIg) and Ia antigens on B lymphoid cell lines. In: *B lymphocytes in the immune response: Functional, developmental and interactive properties,* N. Klinman, D. Mosier, I. Scher, and E. Vitetta, eds. Elsevier/North Holland, New York, p. 507–514.
4. Mestecky, J., J. Winchester, T. Hoffman, and H.G. Kunkel. 1977. Parallel synthesis of immunoglobulins and J chain in pokeweed mitogen-stimulated normal cells and in lymphoblastoid cell lines. *J. Exp. Med.* **145:**760.
5. Roth, R., A., E.L. Mather, and M.E. Koshland. 1979. Intracellular events in the differentiation of B lymphocytes to pentamer IgM synthesis. In: *Cells of immunoglobulin synthesis,* H. Vogel and P. Pernis, eds. Academic Press, New York, p. 141.
6. Parker, D.C., D.C. Wadsworth, and G.B. Schneider. 1980. Activation of mouse B lymphocytes by anti-immunoglobulin is an indicative signal leading to immunoglobulin secretion. *J. Exp. Med.* **152:**138.
7. Melchers, F., J. Anderson, W. Lernhardt, and M.H. Schreier. 1980. H-2 unrestricted polyclonal maturation without replication of small B cells induced by antigen-activated T cell help factors. *Eur. J. Immunol.* **10:**679.
8. Cossman, J., L.M. Neckers, R.M. Braziel, J.B. Trepel, S.J. Korsmeyer, and A. Bakhshi. 1984. *In vitro* enhancement of immunoglobulin gene expression in chronic lymphocytic leukemia. *J. Clin Invest.* **73:**587.
9. Gordon, J., H. Mellstedt, P. Aman, P. Biberfeld, and G. Klein. 1984. Phenotypic modulation of chronic lymphocytic leukemia cells by phorbol ester:

induction of IgM secretion and changes in the expression of B cell-associated surface antigens. *J. Immunol.* **132:**541.

10. Lipford, E.H., J.B. Trepel, J. Cossman, and L.M. Neckers. 1984. Separate membrane receptors mediate immunoglobulin messenger RNA induction and secretion of translated product. *J. Cell. Biochem.* Suppl **8A:**296.
11. Trepel, J.B., R.C. McGlennen, J. Cossman, E.H. Lipford, and L.M. Neckers. 1984. Mechanism of phorbol ester-induced immunoglobulin secretion in leukemic B cells. *J. Cell. Biochem.* Suppl **8A:**286.
12. Neckers, L.M., and J. Cossman. 1983. Transferrin receptor induction in mitogen-stimulated human T lymphocytes is required for DNA synthesis and cell division and is regulated by interleukin 2. *Proc. Natl. Acad. Sci. U.S.A.* **80:**3494.
13. Neckers, L.M., S.P. James, and G. Yenokida. Role of transferrin receptors in B cell activation. *J. Immunol.* 135:2437–2441, 1984.
14. McGlennen, R.C., J.B. Trepel, J. Cossman, E.S. Jaffe, and L.M. Neckers. 1984. Calcium regulation of immunoglobulin secretion. *J. Cell. Biochem.* Suppl. **8A:**285.
15. Saiki, O., and P. Ralph. 1981. Induction of human immunoglobulin secretion. *J. Immunol.* **127:**1044.
16. Neckers, L.M., J.B. Trepel, E.H. Lipford III, S.P. James, and G. Yenokida. Transferrin receptor regulates B cell proliferation but not differentiation. *J. Cell. Biochem.* 27:377–389, 1985.
17. Nishizuka, Y. 1984. The role of protein kinase C in cell surface signal transduction and tumor promotion. *Nature* **308:**693.
18. Niedel, J.E., L.J. Kuhn, and G.R. Vandenbark. 1983. Phorbol diester receptor copurifies with protein kinase C. *Proc. Natl. Acad. Sci. U.S.A.* **80:**36.
19. Castagna M., Y. Takai, K. Kaibuchi, K. Sano, U. Kikkaw, and Y. Nishizuka. 1982. Direct activation of calcium-activated, phospholipid dependent protein kinase by tumor-promoting phorbol esters. *J. Biol. Chem.* **257:**7847.
20. Kikkawa, U., Y. Takai, Y. Tanaka, R. Miyaka, and Y. Nishizuka. 1983. Protein kinase C as a possible receptor protein of tumor promoting phorbol ester. *J. Biol. Chem.* **258:**11442.
21. Michell, R.H. 1982. Inositol lipid metabolism in dividing and differentiating cells. *Cell Calcium* **3:**429.

CHAPTER 41

Monoclonal Antibody AB1 Identifies a Human B Cell Activation Antigen and Inhibits Growth Factor-Dependent Human B Cell Proliferation

Lawrence K.L. Jung, Shu Man Fu, John Morgan, and Abby L. Maizel

In both human and animal systems, various growth factors have been shown to be involved in B cell proliferation and differentiation (reviewed in Ref. 1). B cell growth factor (BCGF), also named B cell stimulatory factor (BSF), has been intensively studied. It has also been postulated that resting B cells are required to be activated to express BSF receptors during the early stage of activation prior to becoming responsive to BSF. Although many B cell activation antigens (2–5) have been described in man, none of them have been shown to be related to the initiation process of B cell proliferation.

A series of monoclonal antibodies have been produced against activated human B cells in our laboratory. Several were found to inhibit B cell proliferation in a BSF-dependent system. One of them was shown to stain specifically activated B cells. This antibody has been named AB1. In the present report, the staining characteristics of this monoclonal antibody and its effect on B cell proliferation are described.

Materials and Methods

Cell Preparation

Tonsillar and peripheral blood mononuclear cells were prepared by procedures described previously (6). Tonsillar B cells and peripheral blood B cells were isolated by two cycles of sheep red cell (SRBC) rosette depletion and adherence to plastics was used to deplete monocytes. The tonsillar B cell preparations contained greater than 90–95% mIg$^+$ cells and those of peripheral blood 60–80%. SRBC-rosetting lymphocytes were used as T cells.

Monoclonal Antibody (mAb) Production

CF1 females were immunized with non-SRBC-rosetting cells from peripheral blood lymphocytes which had been previously activated with pokeweed mitogen (PWM) for 3 days. The spleen cells were fused with SP2/0 myeloma cells and hybrids selected on HAT selection medium. Hybridoma supernatants were screened for their inhibition activity in the B cell proliferation assay as described below and binding activity to activated B cells. The desired hybridomas were cloned on soft agar. Details of these procedures have been previously described (7).

Lymphocyte Proliferation Assays

10^5 lymphocytes were stimulated in microtiter wells with phytohemagglutinin (PHA-P; Difco, Detroit, MI), PWM, concanavalin A (Con A; GIBCO, Grand Island, NY), and formalinized *Staphylococcus aureus* (a gift of Dr. S. Pahwa) at optimal concentrations for 3 days at 37°C in a humidified atmosphere containing 5% CO_2. 0.5 μCi of tritiated thymidine was added for the last 6–8 hr of incubation. The amount of radioactive thymidine incorporated was measured after the cells were harvested onto glass filter disks with a Beckman scintillation counter (Beckman, Boston, MA). Triplicate samples were counted. Variations of less than 10% were found.

B cell proliferation assay was performed using affinity-purified rabbit anti–human IgM as a first stimulant at a final concentration of 5–50 μg/ml. Condition medium containing B cell stimulatory factor (CM-BSF), produced at previously described (8), was added to 10^5 B cells per microtiter well at a final concentration of 10% at the initiation of culture. Supernatants of hybridomas to be screened for inhibitory activity were also added at the initiation of culture at a final concentration of 25%. After 3 days of culture, the cells were labeled with tritiated thymidine and radioactivity counted as described earlier.

In certain experiments, partially purified BSF prepared according to Maizel *et al.* (9) was used. The preparation contained no IL-2 or B cell differentiation factor activities.

Results

Selective Staining of Activated B Cells by Monoclonal Antibody AB1

Monoclonal antibody AB1 was identified to be an IgG1 antibody. It precipitated a 58-Kd polypeptide from activated B cells. It stained 10–60% of activated B cells which were induced by low dose of anti-IgM (5–10 μg/ml) and 10% CM-BSF in six experiments involving four tonsillar B cell preparations and two peripheral blood B cell samples. In the majority of the experiments more than 30% of the B cells were stained by AB1. AB1 did not stain three samples of activated T cells, three resting T cell samples, six resting B cell preparation, and six samples of separated granulocytes and monocytes. In two bone marrow samples, it failed to stain an appreciable number of cells. In addition, it did not react with cells from seven patients with B cell chronic lymphocytic leukemia, two patients with T cell chronic lymphocytic leukemia, three patients with acute lymphocytic leukemia, two patients with acute myelogenous leukemia, and one patient with mono-myelocytic leukemia. It did not stain four cell lines of T cell leukemia origin, six cell lines of non-lymphoid origin, and over twenty EBV-transformed B cell lines. These data are summarized in Table 41.1.

The expression of antigen AB1 did not depend on the presence of CM-BSF. In three separate experiments, anti-IgM antibodies alone induced the expression of this antigen by 40–60% of the tonsillar B cells within 24 hours.

Kinetics of Antigen AB1 Expression

The kinetics of antigen AB1 was studied. A significant number of B cells expressed this antigen after 3 hours of incubation. Although the staining

Table 41.1. Selective expression of AB1 antigen on activated B cells.

Cell types	%
Normal	
Activated B (4 tonsils and 2 PBLs)[a]	10–60
Activated T (3)	<1
B (6)	<1
T (3)	<1
Monocytes (6)	<1
Granulocytes (6)	<1
Bone marrow (2)	<1
Leukemias	
B-CLL (7)	<1
T-CLL (2)	<1
ALL (3)	<1
AML (2)	<1
AMML (1)	<1
Cell lines	
B lymphoblastoid (20)	<1
T leukemia (4)	<1
Non-lymphoid (6)	<1

[a] The number of individuals studied.

was not bright, this was the earliest time point of our experiment. By 12 hours, bright stained cells were easily detectable. In this experiment, 60% of the activated B cells were positive by 38 hours.

Inhibition of CM-BSF-Dependent Proliferation of B Cells by AB1

The supernatant of hybridoma AB1 consistently inhibited B cell proliferation induced by low-dose anti-IgM and conditioned medium containing BSF. Realizing that hybridoma supernatants often contain nonspecific inhibitory activities, purified AB1 antibodies from mouse ascites fluid were prepared by goat anti–mouse Ig affinity chromatography. The isolated antibody has been demonstrated to have similar staining characteristics as the supernatant. For the majority of the experiments to be presented, isolated antibodies were used.

Tonsillar B cells were stimulated with 10 μg/ml of anti-IgM and conditioned media for 3 days. AB1 or a control antibody of IgG1 subclass were added to the cultures. Without the addition of the antibodies, the control cultures showed thymidine uptake of 14,474 cpm (Table 41.2). With increasing concentrations of AB1, B cell proliferation was inhibited. This was evident at 10 μg/ml. 60% inhibition was achieved at 25 μg/ml and greater than 90% inhibition was seen at 50 μg/ml. The control antibody showed little or no inhibition.

This inhibition was specific for B cell proliferation induced by anti-IgM and CM-BSF. AB1 had no inhibitory effect on T cell mitogenesis and growth factor-independent B cell proliferation. Three T cell mitogens, PHA-P, Con A, and PWM, were used (Table 41.3). 25 μg/ml of the antibody AB1 was shown to inhibit more than 60% of B cell proliferation. The same dose did not significantly inhibit T cell proliferation induced by the three mitogens studied. For example, Con A activation resulted in a thymidine uptake of 58,019 cpm. In the presence of AB1, 53,237 cpm were detected.

Table 41.2. Effect of AB1 on B cell proliferation induced by anti-IgM and T cell conditioned medium.[a]

Monoclonal antibody	Amount of antibody (μg/ml) 0	0.5	5	10	25	50
	cpm					
Control Ab[b]	14,747	16,137	12,679	16,968	16,262	12,171
AB1	—	15,563	11,080	9224	5596	1153

[a] 10^6/ml tonsillar B cells were stimulated with 10 μg/ml anti-IgM and 10% CM-BSF for 3 days. Monoclonal antibodies purified on goat anti–mouse Ig affinity columns were added at the initiation of the culture at the indicated concentrations.

[b] IgG1 monoclonal antibody.

Table 41.3. Lack of inhibitory effect of AB1 on PBL proliferation induced by mitogens.[a]

Mitogen	Antibody	Amount of antibody added (μg/ml) 0	25
		cpm	
PHA-P	Control Ab	45,392	38,641
(10 μg/ml)	AB1	38,468	29,147
Con A	Control Ab	58,019	53,237
(5 μg/ml)	AB1	57,915	53,095
PWM	Control Ab	5027	5901
(1%)	AB1	4972	5192

[a] 10^6/ml PBL were stimulated with mitogens at the stated concentration for 3 days in microtiter wells along with monoclonal antibodies. The cultures were labeled with ^{3}H-Tdr for the last 6–12 hr.

It is known that anti-IgM alone can induce B cell proliferation if high doses are used (6). With increasing concentrations of anti-IgM, significant proliferation of B cells was induced in the present study (Table 41.4). Monoclonal antibody AB1 did not inhibit such proliferation. The B cell-specific mitogen, formalinized *Staphylococcus aureus,* activates B cells to proliferate without the need for growth factors. Again, proliferation of B cells induced by this mitogen was not inhibited by AB1 (Table 41.4).

A number of growth factors are known to influence B cell proliferation; these include interleukin-1, interleukin-2, and B cell stimulation factors.

Table 41.4. Lack of inhibition by AB1 on growth factor-independent B cell proliferation.

	cpm Control	AB1
A. Anti-IgM-induced B cell proliferation[a]		
Concentration of anti-IgM (μg/ml)		
0	806	—
10	1029	1900
25	4659	4550
50	7207	7558
B. Formalinized *S. aureus*-induced B cell proliferation[b]		
Experiment 1	31,486	39,330
Experiment 2	66,851	57,919

[a] 10^6/ml B cells were cultured in the presence of monoclonal antibodies at various concentrations of rabbit anti–human IgM for 3 days. 0.5 μCi ^{3}H-Tdr was added to microtiter wells for the last 6–12 hr of culture.
[b] 10^6/ml B cells were cultured in the presence of monoclonal antibodies with 0.005% v/v formalinized *Staphylococcus aureus* for 3 days. ^{3}H-Tdr was added for the last 6–12 hr of culture.

Table 41.5. Effect of AB1 on B cell proliferation in response to partially purified BSF.[a]

Blocking agent	Dilutions of BSF 1/16	1/32	1/64	1/128
Medium	21,710	8815	3381	1811
25% AB1 Sup.	1,000	833	540	925
25 μg/ml AB1	1,239	359	196	283

[a] Tonsillar B cells were stimulated with 10 μg/ml anti-IgM and BSF at the designated concentration. AB1 antibody was added at the initiation of culture.

CM-BSF used in the above experiments contained many of them. A partially purified BSF preparation was obtained. It was shown to lack IL-2 and B cell differentiation factor activities, to contain ten bands on SDS–PAGE, and to be able to support the growth of B cells in long-term cultures. AB1 in supernatant and isolated IgG forms inhibited B cell proliferation induced by anti-IgM antibodies and this purified BSF (Table 41.5).

Discussion

Immunofluorescence studies indicates that AB1 selectively stains activated B cells. This antigen has been tentatively identified as a 58-Kd polypeptide. From the cell distribution studies of AB1 antigen and its molecular weight, antigen AB1 is distinct from those B cell activation antigens previously described (2–5). It is of interest to note that AB1 did not stain activated T cells. This is in contrast to the Tac antigen which is expressed by both activated T and B cells (8,10).

AB1 inhibited specifically the proliferation of B cells which were cultured in CM-BSF after they were initially stimulated by low doses of anti-IgM antibodies. Although further work is needed, the findings that AB1 blocked B cell proliferation induced by a low dose of anti-IgM antibodies and a partially purified BSF preparation and that anti-IgM alone induced B cells to express this antigen suggest that the AB1 antigen is closely related to the BSF receptor. Further investigation is needed to document this fact.

Summary

Monoclonal antibody AB1, raised against PWM-stimulated B cells, stained activated B cells specifically. The molecule identified by AB1 appeared on B cells 3 to 4 hours after activation. Anti-IgM antibodies by themselves induced the expression of this antigen. The molecular weight was tentatively determined to be 58,000 daltons. AB1 failed to react with resting T cells, B cells, monocytes, granulocytes, various hematopoietic cell lines, and leukemic cells. AB1 was shown to inhibit anti-IgM induced, growth factor-dependent B cell proliferation. AB1 did not inhibit B cell proliferation induced by high-dose anti-IgM and formalinized *Staphylococcus aureus*. It was also not inhibitory for T cell mitogenesis. These findings suggest that antigen AB-1 may be closely related to the receptor of the B cell stimulatory factor.

Acknowledgment. This work was supported in part by a Public Health Service Grant CA-34546 from the National Institutes of Health.

References

1. Moller, G. 1984. B cell growth and differentiation factors. *Immunol. Rev.* **78:** whole volume.
2. Yokochi, T., R.D. Holly, and E.A. Clark. 1982. B lymphoblast antigen (BB-1) expressed on Epstein–Barr virus-activated B cell blasts, B lymphoblastoid cell lines and Burkitt's lymphoma. *J. Immunol.* **128:**823.
3. Thorley-Lawson, D.A., R.T. Schooley, A.K. Bhan, and L.M. Nadler. 1982. Epstein–Barr virus superinduces a new human B cell differentiation antigen (B-LAST1) expressed on transformed lymphoblasts. Cell **30:**415.
4. Slovin, S.F., D.M. Frisman, C.D. Tsoukas, I. Royston, S.M. Baird, S.B. Wormsley, D.A. Caron, and J.H. Vaughn. 1982. Membrane antigen on Epstein–Barr virus infected human B cells recognized by a monoclonal antibody. *Proc. Natl. Acad. Sci. U.S.A.* **79:**2649.
5. Posnett, D.N., C.Y. Wang, N. Chiorazzi, M.K. Crow, and H.G. Kunkel. 1984. An antigen characteristic of hairy cell leukemia cells is expressed on certain activated B cells. *J. Immunol.* **133:**1635.
6. Chiorazzi, N., S.M. Fu, and H.G. Kunkel. 1980. Stimulation of human B lymphocytes by antibodies to IgM and IgG. Functional evidence for the expression of IgG on B lymphocyte surface membrane. *Clin. Immunol. Immunopathol.* **15:**301.
7. Yen, S.H., F. Gaskin, and S.M. Fu. 1983. Neurofibrillary tangles in senile dementia of the Alzheimer type share an antigenic determinant with intermediate filaments of the vimentin class. *Am. J. Pathol.* **113:**373.
8. Jung, L.K.L., T Hara, and S.M. Fu. 1984. Detection and function studies of p60-65 (Tac antigen) on activated human B cells. *J. Exp. Med.* **160:**1597.
9. Maizel, A., C. Sahasrabuddhe, S. Mehta, J. Morgan, L. Lachman, and R. Ford. 1982. Biochemical separation of a human B cell mitogenic factor. *Proc. Natl. Acad. Sci. U.S.A.* **79:**5998.
10. Leonard, W.J., J.M. Depper, T. Uchiyama, K.A. Smith, T.A. Waldmann, and W.C. Greene, 1982. A monoclonal antibody that appears to recognize the receptor for human T cell growth factor: partial characterization of the receptor. *Nature* **300:**267.

CHAPTER 42

Detection and Functional Studies of IL-2 Receptors on Activated Human B Cells

Lawrence K.L. Jung, Toshiro Hara, and Shu Man Fu

Initial description of the IL-2 receptor identifiable by monoclonal antibody anti-Tac suggested that such receptors were specific for activated T cells (1). However, using monoclonal antibody AT1 developed in our laboratory, and which exhibits similar specificity to anti-Tac, we have been able to demonstrate the presence of this antigen on activated B cells. Furthermore, after appropriate activation, B cells have been found to respond to IL-2 by proliferation.

Materials and Methods

Cell Preparation

T cells and B cells were separated from tonsils as previously described (2).

B Cell Activation

Affinity-purified $F(ab')_2$ rabbit anti-IgM antibodies were prepared as described (2). Conditioned medium containing BSF was purchased from Electro-nucleonics (Silver Springs, MD) or prepared in our laboratory. In the latter case, pooled mononuclear cells were stimulated with PHA-P (Difco, Detroit, MI) and 1% PWM (Gibco, Grand Island, NY) for 48 hr. 30–80% $(NH_4)_2SO_4$ precipitate of the supernatant was dialyzed against 0.1 *M* phosphate buffer, pH 7.4, and fractionated on a DE52 column with a 0–0.3 *M* NaCl gradient. Fractions containing BSF activity were pooled. This preparation, which also contained IL-2 and B cell differentiation factors, will be called CM-BSF. 10% CM-BSF and 25 $\mu g/\mu l$ anti-IgM were routinely used for B cell activation; 0.005% v/v formalinized *Staphylococcus aureus* and 20% v/v of B95-8 supernatant were also used to activate B cells. 1% PWM was used to activate B cells in PBLs.

Monoclonal Antibody Production

Six- to eight-week-old BC_3F_1 females were immunized with PHA-activated T cells. Their spleen cells were fused with SP2/0 myeloma cells with PEG 1000 and the fusion products were grown in HAT selection medium, and supernatants were screened for their binding activity and blocking of T cell proliferation. The desired hybridomas were cloned on soft agar. Details of these procedures have been described previously (3).

Other Assays

Lymphocyte proliferation assays and reverse plaque assays were performed as described (4). Cell iodination, [^{35}S]methionine labeling, immunoprecipitation, and autoradiographs were performed as described (5,6).

Results

Monoclonal AT-1 Identified p60–65 (IL-2 Receptor)

Two monoclonal antibodies against activated human T cells were found to precipitate a p60–65 molecule from activated T cells. AT-1 was studied in detail and was found to immunoprecipitate a similar p60–65 molecule to that immunoprecipitated by anti-Tac (kind gift of Drs. T. Waldman and T. Uchiyama). These results are shown in Fig. 42.1 (lanes 1 and 2). After

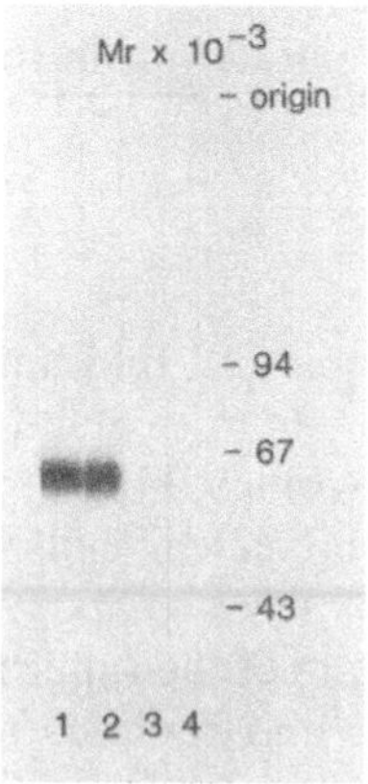

Fig. 42.1. Immunoprecipitation of p60–65 by AT-1 and anti-Tac. ^{125}I-labeled polypeptides were precipitated by AT-1 (lane 1) and by anti-Tac (lane 2). No precipitated peptides were detected by either AT-1 (lane 3) or anti-Tac (lane 4) after the cell lysate was cleared with anti-Tac.

clearance with anti-Tac, neither antibody precipitated any molecule from the cell lysate (Fig. 42.1, lanes 3 and 4). The reciprocal experiment also showed similar results. Thus, both AT-1 and anti-Tac bind to the same p60–65 molecule, which has been shown to be the putative IL-2 receptor (7).

Expression of p60–65 on Activated B Cells

B cells activated by anti-IgM and CM-BSF were found to bind AT-1. This was demonstrated after extensive care was taken to deplete T cells from B cell preparations. To rule out passive adsorption of shed IL-2 receptors present in the CM-BSF, biosynthesis was done with [^{35}S]methionine. As shown in Fig. 42.2, a p60–65 molecule was immunoprecipitated by both T and B cell blasts labeled with [^{35}S]methionine using AT-1.

The expression of IL-2 receptors on B blasts was inducible by various mitogens (Table 42.1). Using two-color fluorochrome immunofluorescence microscopic analysis, anti-IgM with CM-BSF was found to induce approximately 45% of B cells to express the IL-2 receptor at day 3. *Staphylococcus aureus* also induced significant expression of the IL-2 receptor by this time. However, EBV-induced IL-2 receptor expression was delayed and maximal expression was found at day 5. Isolated activated B cells from pokeweed mitogen-stimulated cultures of mononuclear cells from both blood and tonsils were also shown to be stained by AT-1.

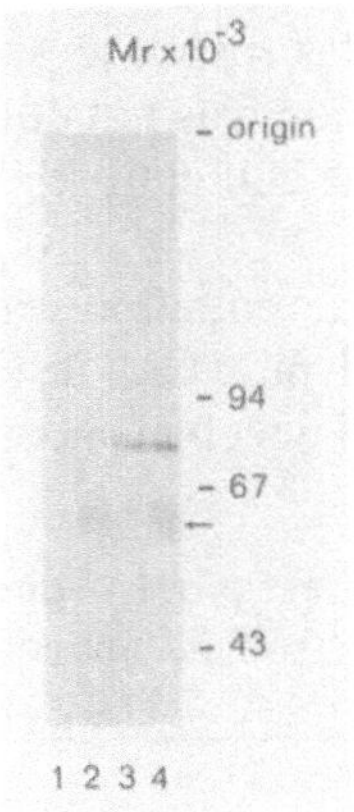

Fig. 42.2. p60–65 molecules immunoprecipitated from activated T and B lymphocytes. 1 × 10^7 activated T lymphocytes and 2 × 10^7 activated B blasts were labeled with [^{35}S]methionine. Lysates were precleared with Sepharose 4B coupled with goat anti-mouse Ig, which had been preincubated with control antibody. The treated lysates were immunoprecipitated with control antibody and AT-1. No proteins were precipitated by the control antibody (lane 1 for T blasts and lane 3 for B blasts). p60–65 molecules were brought down by AT-1 from lysates of T blasts (lane 2) and B blasts (lane 4).

Table 42.1. Expression of IL-2 receptors on B cells activated by various mitogens.

		Time of examination after activation	
Cell source	Stimulation	Day 3	Day 5
Tonsil B cells	Anti-IgM (25 μg/ml) +10% CM-BSF	30.5[a]	
		37.7	
		32.9	
		42.8	
		35.4	
		58.1	
	S. aureus (0.005% v/v)	48.5	59.7
		21.8	35.0
		23.1	
	EBV	16.7	49.5
		1.5	40.7
		2.3	38.2

[a] Percentage of mIgM$^+$ cells stained positive for IL-2 receptors by indirect immunofluorescence microscopy.

Effect of Recombinant IL-2 (rIL-2) on Activated B Cells

Functional significance of IL-2 receptors on B cells was studied using recombinant IL-2 (rIL-2). Resting B cells did not proliferate in the presence of anti-IgM and rIL-2 (Table 42.2). However, after B cells were first activated with anti-IgM and CM-BSF, rIL-2 induced proliferation of these activated cells (Table 42.3). This response was dependent on the concentration of rIL-2. The proliferation was not due to contaminating T cells as no T cells were detected in the cell cultures before and after stimulation with rIL-2.

Effect of rIL-2 on B cell differentiation was also studied. B cells were activated with either anti-IgM plus CM-BSF or *Staphylococcus aureus* for three days. When CM-BSF which also contained B cell differentiation

Table 42.2. Effect of rIL-2 on B cell proliferation: rIL-2 did not act on resting B cells.[a]

	cpm	
Culture condition	Exp. 1	Exp. 2
Medium	846	143
Anti-IgM + 10% CM-BSF	11,117	3898
rIL-2 (250 u/ml)	376	394
Anti-IgM +rIL-2	706	1324

[a] 10^6/ml tonsillar B cells were activated with rabbit anti-IgM (25 μg/ml) plus either CM-BSF or rIL-2 for 3 days. 0.5 μCi ^{3}H-Tdr was added to the microtiter wells for the last 6–12 hr.

Table 42.3. Effect of recombinant IL-2 on activated B cells.[a]

Experiment 1	Agents added	
Time	Medium	rIL-2 (500 units/ml)
24 hr	8988	18,329
48 hr	1221	10,503
60 hr	1332	8,561
Experiment 2		
Time	[rIL-2](μ/ml)	cpm
24 hr	0	2752
	100	4287
	200	6702
	500	5605
	10% CM-BSF	20,690

[a] Tonsillar B cells were activated with rabbit anti-IgM (25 μg/ml) with 10% CM-BSF for 3 days. Cells were then washed 3 times and cell concentration adjusted to 10^6/ml. 10^5 cells/well were cultured with rIL-2 added at the indicated concentrations and cells harvested at the indicated times. No T cells were detectable at the beginning and the end of the incubation period.

factor(s) was added to the activated cells, differentiation into antibody-secretive plaque-forming cells (PFC) was seen. On the other hand, rIL-2 did not increase PFC when added to these cells (Table 42.4). Attempts to demonstrate synergy between low doses of CM-BSF and rIL-2 gave varied results. In the majority of the experiments, no synergy was detected.

Table 42.4. Effect of recombinant IL-2 on differentiation of activated B cells.

		PFC[a]/10^6 cultured cells	
BSF (%)	[rIL-2](u/ml)	A	B
—	—	100	40
10	—	1640	1740
5	—	1170	1500
1	—	780	480
—	100	60	40
—	50	90	60
—	25	120	20
5	100	2160	1200
5	50	1730	1480
5	25	1870	1340

[a] Tonsillar B cells were first activated with either anti-IgM and CM-BSF (A) or *Staphylococcus aureus* (B) for 3 days. After thorough washings, the activated B cells were cultured with either CM-BSF and/or rIL-2. Three days later, plaque-forming cells (PFC) were determined.

Discussion

A monoclonal antibody AT-1 produced against activated T cells identified a p60–65 molecule which is similar to the Tac antigen described by Uchiyama *et al.* (1). Clearance studies confirmed that AT-1 and anti-Tac bind to identical molecules on the T cell surface. This antigen has been shown to represent the IL-2 receptor (7).

Using two-color fluorochrome analysis, IL-2 receptors were identified on $mIgM^+$ cells activated by various mitogens. Passive adsorption of this receptor by B cells was ruled out as activated B cells were shown to synthesize this protein. Thus, the IL-2 receptors are made by activated human B cells. These results are in agreement with those of other investigators (8,9).

The functional significance of IL-2 receptor expression by activated B cells was studied using recombinant IL-2. While rIL-2 had no effect on the proliferation of resting B cells, it was shown to augment the proliferation of activated B cells. Similar findings have been reported in the murine system (10,11). On the other hand, rIL-2 appeared to have no effect on the differentiation of activated B cells.

It is intriguing that B cells were responsive to various proliferation signals at various times after activation: BSF at 2–4 hr in the early G_1 phase, IL-1 at 16–18 hr in the late G_1 phase (12), and then IL-2 at 3 days. The existence of this cascade of growth factors involved in B cell activation is of considerable interest and underscores the complexities of this process.

Summary

A monoclonal antibody AT-1 produced against activated T cells was shown to have properties similar to anti-Tac. Both monoclonal antibodies immunoprecipitated identical molecules on activated T cells. Activated B cells were found to express IL-2 receptors as detected by AT-1. Biosynthesis of this molecule by B cells after activation was documented. B cell mitogens such as anti-IgM, *Staphylococcus aureus*, Epstein–Barr virus (EBV), and pokeweed mitogen (PWM) could activate B cells to express this antigen although its expression varied with different mitogens. While IL-2 did not act to induce proliferation of resting B cells in conjunction with anti-IgM, it induced activated B cells to proliferate. IL-2 did not induce activated B cells to differentiate into antibody-secreting cells. These findings indicate that IL-2 may play certain roles in B cell proliferation.

Acknowledgment. This work was supported in part by a Public Health Service grant CA-34546 from the National Institutes of Health.

References

1. Uchiyama, T., S. Broder, and T.A. Waldmann. 1981. A monoclonal antibody (anti-Tac) reactive with activated and functional mature human T cells. I. Production of anti-Tac monoclonal antibody and distribution of Tac (+) cells. *J. Immunol.* **126:**1393.
2. Chiorrazi, N., S.M. Fu, and H.G. Kunkel. 1980. Stimulation of human B lymphocytes by antibodies to IgM and IgG: Functional evidence for the expression of IgG on B-lymphocyte surface membrane. *Clin. Immunol. Immunopathol.* **15:**301.
3. Yen, S.H., F. Gaskin, and S.M. Fu. 1983. Neurofibrillary tangles in senile dementia of the Alzheimer type share an antigenic determinant with intermediate filaments of the vimentin class. *Am. J. Pathol.* **113:**373.
4. Mayer, L., S.M. Fu, and H.G. Kunkel. 1982. Human T cell hybridomas secreting factors for IgA: Specific help, polyclonal B cell activation and B cell proliferation. *J. Exp. Med.* **156:**1860.
5. McCune, J.M., S.M. Fu, and H.G. Kunkel. 1981. J chain biosynthesis in pre-B cells and other possible precursor B cells. *J. Exp. Med.* **154:**138.
6. Wang, C.Y., A. Al-Katib, C.L. Lane, B. Koziner, and S.M. Fu. 1983. Induction of HLA-DC/DS (Leu 10) antigen expression by human precursor B cell lines. *J. Exp. Med.* **158:**1757.
7. Leonard, W.J., J.M. Depper, T. Uchiyama, K.A. Smith, T.A. Waldmann, and W.C. Greene. 1982. A monoclonal antibody that appears to recognize the receptor for human T cell growth factor; partial characterization of the receptor. *Nature* **300:**267.
8. Muraguchi, A., J.H. Kehrl, D.L. Longo, D.J. Volkman, and A.S. Fauci. 1984. Tac expression on activated B cells and on a human T cell leukemia virus-infected B cell line. *Fed. Proc.* **43:**1676.
9. Tsudo, M., T. Uchiyama, and H. Uchino. 1984. Expression of Tac antigen on activated normal human B cells. *J. Exp. Med.* **160:**612.
10. Zubler, R.H., C. Mingarie, L. Moretta, and H.R. MacDonald. 1984. Effect of interleukin-2 on murine and human B cells. *Fed. Proc.* **43:**1591.
11. Mond, J.J., C. Thompson, F.D. Finkelman, M. Schaefer, and R.J. Robb, 1984. Immuno-affinity purified interleukin-2 (IL-2) induces proliferation of large but not small B cells. *Fed. Proc.* **43:**1591.
12. Howard, M., and W. Paul, 1983. Regulation of B cell growth and differentiation by soluble factors. *Annu. Rev. Immunol.* **1:**307.

CHAPTER 43

The C3d Receptor Identified by the HB-5 Monoclonal Antibody: Expression and Role as a Receptor for Epstein–Barr Virus

Thomas F. Tedder, Loran T. Clement, Janis J. Weis, Douglas T. Fearon, and Max D. Cooper

A number of monoclonal antibodies which react with human B cell antigens have been reported from several laboratories. Some of these antibodies are specific for the B lineage and some serve as useful markers of B cell maturation; however, the function of the majority of the surface molecules identified remains unknown. In this report we have summarized the characterization of a recently produced monoclonal antibody, HB-5. This antibody identifies a B cell-specific membrane antigen that is expressed during a discrete stage of differentiation (1). The HB-5 antibody has allowed the identification of this antigen as the C3d receptor (CR2)* of human B lymphocytes which mediates the binding of immune complexes that bear the C3b, iC3b, C3d,g, or C3d fragments of C3 (2). Through the identification of this receptor it has also been shown that CR2 binds the Epstein–Barr virus (EBV) (3). Results presented here demonstrate that CR2 serves as the functional receptor which binds EBV, resulting in B cell infection.

Materials and Methods

Antibody and C3d,g Production

The HB-5 antibody (γ2a,κ) producing hybridoma was generated by the fusion of the AG8.653 myeloma with spleen cells from a mouse immunized with the human B cell line, SB (1). The YZ-1 monoclonal antibody (γ1,κ) reacts with the 250,000 M.W. human C3b receptor (CR1) (prepared by Dr. Rick Jack). $F(ab')_2$ antibody fragments were generated by pepsin diges-

* CR2: Complement receptor type 2 that has primary specificity for iC3b and C3d.

tion and purified by column chromatography. Indirect immunofluorescence staining and fluorochrome-conjugated secondary antibodies were prepared as described previously (1). The mitogenic anti–human μ heavy chain monoclonal antibody, SA.DA4.4 (γ1,κ) was used as described elsewhere (4).

C3d,g was prepared from outdated human plasma by saturated ammonium sulfate precipitation with subsequent anion exchange and gel filtration chromatography in the presence of protease inhibitors (5). Purity was determined by sodium dodecyl sulfate—polyacrylamide gel electrophoresis (SDS–PAGE) analysis and antigenically by antibodies against C3 fragments.

Cells and Cultures

Blood mononuclear cells (MNC) and B cell-enriched (E^-) MNC—nonadherent MNC remaining after the removal of sheep erythrocyte-rosetting (E^+) cells—were prepared as described (1). Assays for B cell proliferation and differentiation were as described (6). EBV was used as culture supernatant fluid from the B95-8 marmoset cell line with the 50% transformation dose being greater than 1×10^4 (7).

Results

Cell Lineage Specificity of the HB-5 Monoclonal Antibody

The reactivity of the HB-5 antibody with hematopoietic cells was tested by indirect immunofluorescence analysis. Flow cytometry analysis indicated that HB-5 antibodies were only reactive with blood cells contained within an E^- MNC fraction and did not react with E^+ MNC, plastic-adherent cells, neutrophils, and erythrocytes. In studies using two-color indirect immunofluorescence microscopy, the HB-5 antibody was only reactive with cells bearing surface immunoglobulin (Ig) and was not reactive with cells bearing T cell, NK cell, and myelo-monocytic cell markers (1). HB-5 reacted with the majority of adult blood B cells bearing IgM, IgD, or IgA. Binding studies using radiolabeled HB-5 antibody also demonstrated that monocytes do not express the HB-5 antigen (8).

HB-5 reacted on average with 67% of spleen MNC, 61% of tonsil MNC, and 3% of bone marrow MNC. The HB-5 antibody was reactive exclusively with B cells among cells from these tissues and was not expressed by thymocytes as determined by two-color immunofluorescence. Equivalent numbers of HB-5 binding sites were present on the majority of B cells from adult blood, spleen, and tonsil as determined by quantitative flow cytometry analysis of indirect immunofluorescence staining. However, a subpopulation of B cells (25–28%) from spleen expressed higher levels of HB-5 antigen.

The HB-5 antibody was also restricted in expression among lymphoid and myeloid malignancies and cell lines to those of B origin. HB-5 was not reactive with the monocytic cell line U 937, the erythroid line K562, or with three of four T cell lines. Interestingly, the MOLT-4 T cell line was HB-5$^+$. HB-5 was not expressed by two pre-B cell lines but was expressed by four of six B cell lines. Similarly, most chronic B cell leukemias and lymphomas were HB-5$^+$. In contrast, tumor cells from patients with pre-B or B cell acute lymphocytic leukemias rarely expressed the HB-5 antigen. Cells from patients with American type Burkitt's lymphoma, or hairy cell, T cell, null cell, and myeloid leukemias were HB-5$^-$.

Biochemical Characterization of the HB-5 Antigen

The HB-5 antibody immunoprecipitated a single cell surface molecule from ^{125}I-labeled blood B cells of 145,000 M.W. nonreduced and 150,000 under reducing conditions as determined by SDS–PAGE (1). A molecule of similar molecular weight was precipitated from the MOLT-4 T cell line (1) and from the B cell lines Raji (1,2) and SB (1,3). HB-5 did not immunoprecipitate any molecule from normal T cells, monocytes, or neutrophils or from HB-5$^-$ cell lines.

Expression of HB-5 Antigen during B Cell Differentiation

Pre-B cells did not express the HB-5 antigen at levels detectable by indirect immunofluorescence analysis (1). There was also a lag in the expression of this antigen by newly generated B cells since most (>98%) of the IgM$^+$ B cells from fetal liver and bone marrow (12–15 weeks gestational age) were HB-5$^-$ (Fig. 43.1). However, some (25%) B cells from fetal spleen expressed this antigen suggesting that a more mature subpopulation of B cells resides in fetal spleen. Since the entire process of B cell differentiation is continuously recapitulated in the bone marrow throughout life, the finding that only half of the IgM$^+$ B cells from adult bone marrow were HB-5$^+$ supports the notion that recently generated B cells are HB-5$^-$. In contrast, the majority (76%) of IgM$^+$ B cells from newborn

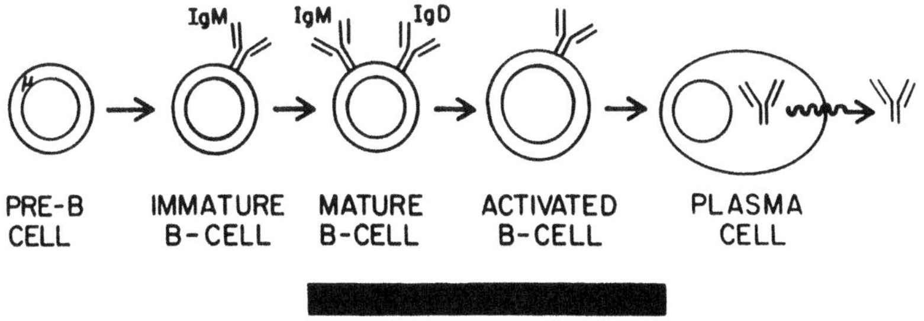

Fig. 43.1. Expression of CR2 (■) with B cell development.

blood were HB-5$^+$ and virtually all adult blood B cells (95%) expressed this antigen. Newborn and adult B cells from blood expressed similar amounts of HB-5 antigen as determined by quantitative flow cytometry analysis. Similarly, most IgM$^+$ B cells from adult lymph node and tonsil (>96%) were HB-5$^+$ while a smaller proportion of IgM$^+$ cells (75%) from adult spleen expressed detectable HB-5 antigen. The subpopulation of preactivated B cells from adult blood which differentiate into antibody-producing plasma cells in response to pokeweed mitogen-induced T cell factors were HB-5$^+$ since the removal of all HB-5$^+$ cells by fluorescence-activated cell sorting removed these plasma cell precursors. However, plasma cells identified by morphology and presence of cytoplasmic Ig rarely expressed detectable HB-5 antigen (12%).

Inhibition of C3 Receptor Function by HB-5

Rosette formation between Raji cells and erythrocytes (E) bearing C3b, iC3b, or C3d was inhibited by HB-5 (2). Raji cells were cultured for 24 hr with HB-5 antibody or with a control IgG2a monoclonal antibody. Following this treatment the percent of rosette formation by Raji cells decreased from 38%, 98%, and 93% for EC3b, EiC3b, and EC3d, respectively, to 8%, 75%, and 81%. Additional treatment of the HB-5-cultured Raji cells with secondary goat anti–mouse Ig antibodies further decreased the percentage of rosette formation to 5%, 26%, and 39% for EC3b, EiC3b, and EC3d, respectively. Treatment of Raji cells with a control monoclonal antibody with or without secondary anti–mouse Ig antibodies did not significantly alter the C3 receptor function of these cells. Similarly, treatment of blood MNC with HB-5 for 30 min did not inhibit their ability to form rosettes with EC3d. However, the further treatment of cells previously treated with 0.5, 1, 2, 4, or 8 μg/ml of HB-5 with anti–mouse Ig decreased their ability to form rosettes with EC3d by 40%, 27%, 60%, 57%, and 76%, respectively.

CR2 Binds to Erythrocyte Intermediates Bearing C3 Fragments

Protein A-bearing *Staphylococcus aureus* with absorbed HB-5 antibody was reacted with detergent-solubilized Raji cell extract or E$^-$ MNC extract under conditions leading to uptake of a single 145,000 M.W. membrane protein. Since HB-5 antibodies did not react with the ligand binding site of CR2, these CR2-bearing particles were shown to bind EiC3b, EC3d, and, to a lesser extent, EC3b. *S. aureus* pretreated with an irrelevant IgG2a monoclonal antibody was unable to absorb CR2 from detergent extracts and was also unable to form rosettes with E-bearing complement intermediates. Therefore, HB-5 binds the molecule which serves as a B cell receptor for C3d, iC3b, and C3b.

Coexpression of EBV Receptors and CR2 on Cell Lines

The levels of binding of fluorescein isothiocyanate (FITC)-labeled EBV were compared with HB-5 antibody binding by the B cell lines SB, Raji, and JY and the T cell line MOLT-4. The median fluorescence intensity of FITC–EBV binding or of HB-5 antibody binding by indirect immunofluorescence was quantified by flow cytometry. In addition, HB-5 binding was quantified using ^{125}I-labeled HB-5 F(ab′)$_2$ antibody fragments. Results from these studies indicated that the rank order of EBV binding was identical to that for HB-5 antibody binding with the B cell lines (SB, Raji, and JY), expressing an average of 24,000–36,000 HB-5 antibody binding sites per cell, and with MOLT-4, expressing an average of 8,000 sites per cell (3).

HB-5 Inhibits EBV Binding to B Cell Lines

Treatment of the B cell lines and the T cell line with HB-5 antibody or a control monoclonal antibody failed to inhibit the binding of FITC-conjugated EBV as determined by flow cytometry analysis (3), as did treatment with monoclonal antibodies to surface molecules other than CR2. However, treatment of these cell lines with HB-5 antibody and secondary anti–mouse Ig antibodies completely inhibited EBV binding, indicating that CR2 was closely associated with the EBV receptor, but that HB-5 was not directed against the virus binding site.

Immunoabsorbed CR2 Binds EBV

HB-5 antibody-treated protein A-bearing *S. aureus* particles were incubated with detergent lysates of SB cells to specifically absorb CR2. These particles bound threefold more ^{125}I-labeled EBV than did control particles bearing irrelevant monoclonal antibodies treated with the lysate of SB cells or HB-5-coated particles treated with the lysate of a CR2$^-$ cell line (3). Therefore, the CR2 molecule was able to specifically absorb EBV particles.

Modulation of CR2 by HB-5 Antibody

B cell-enriched MNC were treated with C3d,g, HB-5 antibody, or HB-5 antibody cross-linked with anti–mouse γ2a heavy chain antibodies for 3 hr at 37°C. Following these treatments, the cells were restained for CR2 with HB-5 antibody and FITC-conjugated anti–mouse Ig antibodies (Fig. 43.2). The pretreatment of cells with C3d,g or HB-5 antibody alone had no effect on the fluorescence intensity of HB-5 reactivity as determined by flow cytometry. In contrast, pretreatment with cross-linked HB-5 antibody modulated CR2 from the cell surface such that no HB-5 antigen was

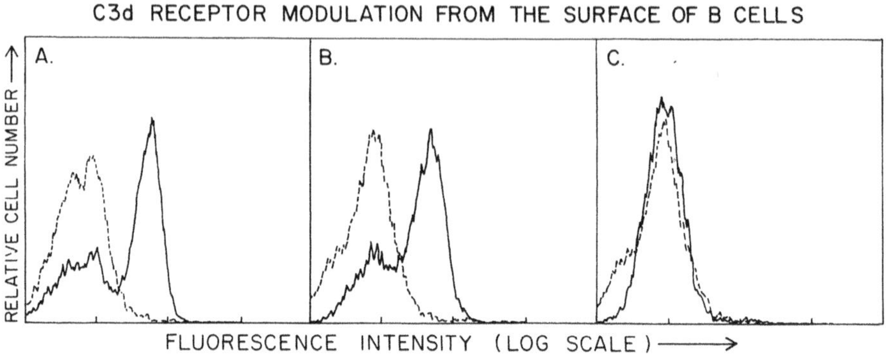

Fig. 43.2. Cross-linking of CR2 results in modulation of this molecule from the cell surface. E^- MNC (60% IgM^+) were cultured for 3 hr at 37°C either alone (A), with HB-5 antibodies (B), or following treatment with HB-5 antibodies which were subsequently cross-linked with FITC-conjugated anti–mouse γ2a antibodies (C). The cells were washed, reacted with HB-5 antibodies, and stained with FITC-conjugated anti–mouse Ig antibodies. The fluorescence intensity of HB-5 staining was analyzed by flow cytometry (solid lines). Dashed lines represent background staining of the cells treated with the FITC-conjugated anti–mouse Ig developing reagent without prior HB-5 treatment.

detectable. The modulation of surface CR2 with HB-5 antibody did not affect the immunofluorescence staining pattern for other B cell surface antigens such as IgM or IgD as determined by two-color indirect immunofluorescence microscopy analysis.

HB-5 Antibody Blocks EBV-Induced Proliferation and Differentiation

E^- MNC were treated with C3d,g, HB-5 alone, or with HB-5 or other monoclonal antibodies (YZ-1, HB-2, and B1) cross-linked with secondary anti–mouse Ig antibodies for 3 hr at 37°C. Following this pretreatment, the cells were cultured in the presence of EBV for 3 hr at 37°C, then extensively washed to remove unbound virus. These cells were cultured for 6 days and assayed for proliferation by [^{3}H]thymidine uptake or were cultured for 7 days and B cell differentiation assessed by antibody secretion. Following such modulation, EBV-induced polyclonal B cell proliferation and Ig production was inhibited by 83% and 90%, respectively. In contrast, modulation of other surface molecules (HB-2, B1, and the C3b receptor) or pretreatment of B cells with C3d,g or HB-5 antibody alone minimally inhibited EBV infection (0–37%).

C3d,g or HB-5 Antibody Did Not Induce B Cells to Proliferate or Differentiate

$F(ab')_2$ fragments of HB-5 antibodies, HB-5 antibodies coupled to beads or purified C3d,g were cultured with E^- cells or fluorescence-activated cell sorter-purified IgM^+ cells for 3 days and proliferation assessed by [^{3}H]thymidine uptake. Culturing B cells in the presence of these reagents at concentrations between 1–500 μg/ml did not result in proliferation above that obtained in control cultures. Similarly, culturing B cells with the above reagents in conjunction with a mitogenic anti-μ monoclonal antibody (4) did not enhance or suppress the resultant B cell proliferation. However, the possibility remained that C3d,g or HB-5 was not the correct or sufficient signal to have resulted in proliferation, but may have induced the B cells to express receptors for growth factors. To examine this, small IgM^+ B cells were isolated by fluorescence-activated cell sorting as described (9) and cultured in the presence of mitomycin C-treated (blocks cell proliferation) E^+ cells and pokeweed mitogen. C3d,g, HB-5 antibodies, and HB-5 antibodies coupled to beads were added to the cultures at various concentrations. Resultant B cell proliferation was assessed at 3 days. The addition of the above reagents did not result in proliferation above that obtained in control cultures. However, anti-μ antibodies (10 μg/ml) in the presence of PWM significantly increased the level of proliferation above that seen with anti-μ alone. Therefore, C3d,g or HB-5 do not appear to activate B cells to proliferate or induce them to express receptors for T cell-derived factors. Also, the addition of HB-5 antibodies to cultures of nonadherent MNC did not induce B cells to produce Ig or affect pokeweed mitogen-induced plasma cell formation.

Comparison of HB-5 with the Anti-B2 Antibody

The anti-B2 and HB-5 antibodies appear to react with the same 145,000 M.W. CR2 molecule (9,10). However, HB-5 antibodies did not inhibit the antigen-binding abilities of anti-B2 antibodies (1). SB cells were incubated with HB-5 antibody at a concentration (500 μg/ml) 100 times that which gave optimal staining. After 20 min, anti-B2 antibody was added to the reaction mixture. The cells were washed and reacted with FITC-conjugated antibodies specific for the anti-B2 heavy-chain isotype. The fluorescence intensity of anti-B2 staining was equivalent to that achieved by anti-B2 without prior treatment of the cells with HB-5 antibodies, as determined by flow cytometry analysis. Similarly, the staining intensity achieved by HB-5 antibodies was not diminished by the prior treatment of the cells with anti-B2 antibodies. Therefore, HB-5 and anti-B2 antibodies recognize distinct antigenic sites on the CR2 molecule.

Discussion

Based on the reactivity of the HB-5 antibody, this 145,000 M.W. antigen appeared to be a B cell-specific determinant. Similarly, CR2 and the EBV receptor are known to be B cell-specific receptors and to be closely associated with one another (11,12), and the EBV receptor has a molecular weight of 145,000 (13). Therefore, through the isolation of this B cell-specific molecule it has been shown that the HB-5 antigen, CR2, and the EBV receptor are the same molecule. The ability of HB-5 antibodies to inhibit CR2 function only in the presence of second antibody indicates that HB-5 bound to an epitope on the receptor that was distinct from the ligand binding site. In addition, HB-5 antibodies alone did not block EBV binding, and C3d,g binding did not block HB-5 binding or the binding of EBV to this molecule. Therefore, there appear to be at least three distinct epitopes present on this molecule; one binds C3d,g, one binds EBV, and one serves as the HB-5 antigen. CR2 has also been identified by another laboratory using a different monoclonal antibody, anti-B2 (9,10). This antibody identifies an antigenic site distinct from that of HB-5 since neither of these antibodies inhibited the antigen-binding ability of the other antibody.

CR2 expression is a relatively late event in B lymphocyte maturation as determined by indirect immunofluorescence analysis. However, EBV-transformed cell lines of the pre–B cell phenotype has been described (14), suggesting that some pre–B cells may express CR2 at levels below our ability to detect them. Nonetheless, pre–B cells and immature B cells are more difficult to transform than newborn and adult B cells from blood, suggesting they have lower receptor numbers (C.F. Webb and M.D. Cooper, unpublished observations). Newborn cord blood B cells respond extremely well to EBV and can be more easily transformed into lymphoblastoid cell lines than adult B cells (15,16). However, newborn and adult B cells from blood expressed similar levels of HB-5 antigen, suggesting they have similar numbers of EBV receptor sites. Therefore, the preferential response of newborn B cells to EBV may be attributable to suppression of transformation by adult T lymphocytes (16).

EBV is able to induce B lymphocytes to grow indefinitely in a test tube through transformation, a process that may be related to virus-associated cases of Burkitt's lymphoma. The mechanisms behind the induction of proliferation and the transformation process remain unknown. We have tried to assess the role which CR2 may play in polyclonal B cell proliferation. Binding of C3d,g by this receptor did not lead to B cell proliferation, nor did cross-linking the receptor with HB-5 antibody or HB-5 antibody-coated beads. Similar treatments also failed to induce Ig production or plasma cell formation. These findings are in agreement with studies demonstrating that inactivation of EBV by exposure to ultraviolet light totally removed the ability of the virus to induce polyclonal B cell activation but did not antigenically alter the virus since subsequent B cell activation by live virus was blocked (15). Therefore, live virus must penetrate the cell, indicating that stimulation is not a surface event. Thus, CR2 appears to be the functionally relevant receptor for EBV on B cells, but the normal role for this membrane protein in B cell function remains unidentified.

Summary

The HB-5 antibody identifies a single cell surface molecule of 145,000 M.W. that is the receptor for C3d (CR2). Among cells from hematopoietic and lymphoid tissues, CR2 is expressed exclusively by mature B cells but not by plasma cells. CR2 is capable of binding EBV and the HB-5 antibody inhibits uptake of EBV by B cell lines. CR2 was modulated from the surface of normal B cells by cross-linking with HB-5 and a secondary goat anti–mouse antibody. Following such modulation, EBV-induced polyclonal B cell proliferation and Ig production was significantly inhibited. In contrast, modulation of other cell surface molecules minimally inhibited EBV infection. C3d,g or HB-5 antibody did not induce resting B cells to proliferate or alter anti-μ antibody-induced proliferation. Also, HB-5 did not induce B cells to produce Ig or affect pokeweed mitogen-induced plasma cell formation. Thus, CR2 appears to be the functionally relevant receptor for EBV on B cells, but the normal role for this membrane protein in B cell function remains unknown.

Acknowledgments. This work was supported by grants CA-16673, CA-13148, AI-17917, AI-07722, and AI-06838 from the National Institutes of Health; 1-608, March of Dimes Birth Defects Foundation; and 4Mo1-RR-32, DRR/NIH.

References

1. Tedder, T.F., L.T. Clement, and M.D. Cooper. 1984. Expression of C3d receptors during human B cell differentiation: Immunofluorescence analysis with the HB-5 monoclonal antibody. *J. Immunol.* **133:**678.
2. Weis, J.J., T.F. Tedder, and D.T. Fearon. 1984. Identification of a 145,000 Mr membrane protein as the C3d receptor (CR2) of human B lymphocytes. *Proc. Natl. Acad. Sci. U.S.A.* **81:**881.
3. Fingeroth, J.D., J.J. Weis, T.F. Tedder, J.L. Strominger, P.A. Biro, and D.T. Fearon. 1984. Epstein–Barr virus receptor of human B lymphocytes is the C3d receptor (CR2). *Proc. Natl. Acad. Sci. U.S.A.* **81:**4510
4. Maruyama, S., H. Kubagawa, and M.D. Cooper. 1984. Effects of monoclonal anti-μ antibodies on human B cells. *Fed. Proc.* **43:**1676A.
5. Vick, D.P., and D.T. Fearon. 1984. Human neutrophil receptor capable of binding C3d,g. *Fed. Proc.* **43:**1665.
6. Clement, L.T., M.K. Dagg, and G.L. Gartland. 1984. Small, resting B cells

can be induced to proliferate by direct signals from activated helper T cells. *J. Immunol.* **132:**740.

7. Miller, G., and M. Lipman. 1973. Comparison of the yield of infectious virus from clones of human and simian lymphoblastoid lines transformed by Epstein–Barr virus. *J. Exp. Med.* **138:**1398.
8. Vargas, I., T. Gaither, C. Hammer, J. O'Shea, and M. Frank. 1984. Interaction of C3d with human monocytes. *Fed. Proc.* **43:**1665A.
9. Iida, K., L. Nadler, and V. Nussenzweig. 1983. Identification of the membrane receptor for the complement fragment C3d by means of a monoclonal antibody. *J. Exp. Med.* **158:**1021.
10. Nadler, L.M., P. Stashenko, R. Hardy, A. van Agthoven, C. Terhorst, and S.F. Schlossman. 1981. Characterization of a human B cell-specific antigen (B2) distinct from B1. *J. Immunol.* **126:**1941.
11. Lambris, J.D., and G.D. Ross. 1982. Assay of membrane complement receptors (CR1 and CR2) with C3b- and C3d-coated fluorescent microspheres. *J. Immunol.* **128:**186.
12. Jondal, M., G. Klein, M.B.A. Oldstone, V. Bokish, and E. Yefenof. 1976. Surface markers on human B and T lymphocytes. VIII. Association between complement and Epstein–Barr virus receptors on human lymphoid cells. *Scand. J. Immunol.* **5:**401.
13. Simmons, J.G., L.M. Hutt-Fletcher, E. Fowler, and R.J. Feighny. 1983. Studies of the Epstein–Barr virus receptor found on Raji cells. I. Extraction of receptor and preparation of anti-receptor antibodies. *J. Immunol.* **130:**1303.
14. Fu, S.M., J.N. Hurley, J.M. McCune, H.G. Kunkel, and R.A. Good. 1980. Pre–B cells and other possible precursor lymphoid cell lines derived from patients with X-linked agammaglobulinemia. *J. Exp. Med.* **152:**1519.
15. Bird, A.G., and S. Britton. 1979. A new approach to the study of human B lymphocyte function using an indirect plaque assay and a direct B cell activator. *Immunol. Rev.* **45:**41.
16. Thorley-Lawson, D.A., L. Chess, and J.L. Strominger. 1977. Suppression of *in vitro* Epstein–Barr virus infection. A new role for adult human T lymphocytes. *J. Exp. Med.* **146:**495.

CHAPTER 44

The B Cell-Restricted Glycoprotein (B2) Is the Receptor for Epstein–Barr Virus

Lee M. Nadler, Andrew W. Boyd, Edward Park, Kenneth C. Anderson, David Fisher, Bruce Slaughenhoupt, David A. Thorley-Lawson, and Stuart F. Schlossman

Introduction

Human B lymphocytes proliferate, secrete immunoglobulin (Ig), and transform into lymphoblastoid cells when infected with Epstein–Barr virus (EBV). Although EBV binding is restricted to B lymphocytes, the structure and normal physiological function of the EBV receptor is presently unknown. Previous studies have suggested that the EBV receptor may be associated with a receptor for the C3d component of complement (1–10). A significant unresolved controversy still exists as to whether these receptors are expressed on a single molecule (1–10). In previous studies, we have developed and extensively characterized a monoclonal antibody (anti-B2) which defines a 140-Kd B cell-restricted glycoprotein (B2) (11,12). The B2 antigen is expressed on the cell surface following the cytoplasmic μ pre–B cell stage and is lost during the mid-stages of B cell differentiation (13–15), suggesting a possible role in B cell activation. Recent studies have demonstrated that the B2 molecule (detected by anti-B2 and anti-HB5) binds human C3d (16,17) and the question arose as to whether the B2 molecule also expressed the EBV receptor. In this report, we show that anti-B2 specifically blocks the binding of EBV. Moreover, we show that EBV can no longer induce B cells to proliferate and secrete Ig when the B2 molecule is selectively depleted from the cell surface, supporting the notion that a receptor on the B2 molecule is essential for EBV infectivity.

Materials and Methods

Normal Human B Cells

Human tonsil or spleen was obtained from operative specimens of patients not known to have any systemic or malignant diseases. After gentle dissociation, single-cell suspensions were layered onto Ficoll–Hypaque density gradients and mononuclear cell fractions prepared from each tissue. Splenic mononuclear cells were isolated by Ficoll–Hypaque density sedimentation (20) and B cells were isolated by E-rosetting (21) and then by complement-mediated lysis to further deplete T cells and monocytes with anti-T13, anti-T11, anti-Mo1, and anti-Mo2. These splenic B cells were greater than 95% B cells as assessed by indirect immunofluorescence with anti-B1 or anti-B4.

Epstein–Barr Virus

EBV was obtained from culture supernatants from the B95-8 lymphoblastoid marmoset line.

Culture Conditions and Proliferation Assays

Purified B cells were resuspended at 5×10^5 cells/ml in RPMI 1640 containing 10% fetal calf serum and 100-μl aliquots were dispensed in 96-well round-bottomed tissue culture plates (Costar, Cambridge, MA). For EBV-induced proliferative assays, splenic B cells were incubated with 50 μg/ml of anti-J13, anti-B2, or anti-B4, at 4°C for 30 min. Cells were washed three times and then cultured for 7 days with either media, EBV, or anti-μ immunobeads (1 : 200, Biorad, NY) and T cell-conditioned media (PHA–LCM). After 6 days of culture, cells were pulsed with 0.2 μCi of tritiated thymidine and harvested on a PHD cell harvester (Cambridge Technology, Inc.). Counts ± SD were determined on a Packard Tri-Carb 4530 Liquid Scintillation Counter.

Radioimmunoassay for IgG

The solid-phase RIA for IgG was a minor modification of previously described techniques. In brief, the assay was performed in flexible polyvinyl chloride microtiter plates containing 96 U-bottomed wells (Falcon 3911, Becton-Dickinson, Oxnard, CA). The wells were coated with 100 μl of 1 : 20 dilution of the immunoglobulin fraction of rabbit anti–human IgG antiserum (Dakopatts, A/S, Denmark) overnight at room temperature. The plates were washed and residual binding sites blocked by incubating with 200 μl of 10% BSA in PBS for 3 hr at room temperature. The plates were washed 3 times with PBS containing 1% sodium azide. Standardcurves were prepared using doubling dilutions of a reference batch of normal human serum (NHS). Aliquots (100 μl) of standard or test culture supernatants were added to individual wells in triplicate and allowed to stand overnight at room temperature. The plates were then washed three times with 1% BSA in PBS diluent, and 100 μl of I^{125}-labeled (50–100 $\times$ 10^3 cpm/well) affinity-purified rabbit anti–human IgG antiserum (Calbiochem, La Jolla, CA) were added to each well. After 4- to 6-hr incubation, the plates were washed 15 times. The individual wells were cut apart and counted in an automated gamma counter (Riagamma 1274, KLB Instruments, Gaithersburg, MD).

Anti-B2 Induced Antigenic Modulation

Anti-B2, anti-B1, anti-B4, and anti-57F were conjugated to biotin as previously described (28). Tonsillar B cells were prepared and stained with biotinylated antibody and Texas Red–avidin. Staining of premodulated cells were evaluated and the staining patterns were identical for cells directly stained with anti-B2–biotin–Texas Red–avidin or indirect immunofluorescence with anti-B2 antibody and fluoresceinated goat anti–mouse Ig antibody (Coulter Immunology, Hialeah, FL). Tonsillar B cells were cultured for 24 hr at 37°C with saturating concentrations of anti-B2, anti-B1, anti-B4, or anti-57F. After culture, cells were washed and stained with goat anti–mouse Ig–FITC to assess the modulated antigen. Cells were also stained with directly biotinylated antibodies and Texas red–avidin to assess the expression of the other antigens post-modulation. Identical data could be demonstrated for B cells isolated from tonsil (n = 10) and spleen (n = 12).

Results and Discussion

Anti-B2 Inhibits the Binding of EBV

To determine whether EBV binds to the B2 molecule, a series of experiments were undertaken to investigate whether anti-B2 could inhibit the binding and infectivity of EBV. Purified splenic B cells were isolated and incubated with saturating concentrations of purified anti-B2, anti-J13 (an isotype identical nonreactive antibody) (18), and anti-B4 (a monoclonal antibody defining another B cell-restricted antigen) (19). Antibody-treated B cells were washed and then cultured for 7 days in media alone or in media containing EBV. After 6 days, cells were pulsed with tritiated thymidine (^{3}H-Tdr) and proliferation was measured on day 7. As seen in Table 44.1, preincubation with anti-B2 inhibited the EBV-induced proliferative response by approximately 60% whereas preincubation with anti-J13 or anti-B4 produced no detectable suppression. The proliferative re-

Table 44.1. Blocking of EBV-induced proliferation by anti-B2.

Culture condition	Proliferative response (cpm ± SD)
Media	170 ± 41
Anti-J13	250 ± 39
Anti-B2	460 ± 82
Anti-B4	330 ± 61
EBV 1:4	11,119 ± 286
EBV 1:4 and anti-J13	12,059 ± 382
EBV 1:4 and anti-B2	4444 ± 127
EBV 1:4 and anti-B4	10,979 ± 293
Anti-μ and PHA–LCM	2773 ± 159
Anti-μ and PHA–LCM and anti-J13	2954 ± 231
Anti-μ and PHA–LCM and anti-B2	2823 ± 124
Anti-μ and PHA–LCM and anti-B4	2839 ± 209

sponse of splenic B cells to the stimulus of anti-immunoglobulin and T cell-conditioned media was unaffected by anti-B2, anti-J13, or anti-B4 (Table 44.1). These data suggest that anti-B2 partially blocked the infection of B cells by EBV.

With the observed suppression of EBV-induced proliferation by anti-B2, we then attempted to determine whether anti-B2 could directly block EBV binding. Raji, a Burkitt's lymphoma cell line which strongly expresses the B4, Ia, and B2 antigens, was incubated at 0°C for 30 min with saturating concentrations of anti-B4, anti-Ia, anti-J13, or anti-B2. After washing, cells were then incubated with EBV for 1 hr at 37°C and the binding of EBV was assessed by indirect immunofluorescence utilizing a specific rabbit anti-EBV antibody (22,23) which was developed with a fluoresceinated goat anti–rabbit Ig antibody. As seen in Fig. 44.1, anti-B4, anti-Ia (24), and anti-J13 did not block EBV binding to Raji cells whereas anti-B2 specifically blocked binding of EBV by approximately 90%. This observation is the first demonstration of direct blocking of EBV binding by a single monoclonal antibody. Moreover, the use of a highly specific anti-EBV antibody allows us to conclude that we are measuring virion binding rather than the binding of contaminants present in the virus preparation.

Anti-B2 Induced Modulation

The blocking of both infectivity and binding of EBV by anti-B2 provided the impetus to determine whether the removal of the B2 molecule from the cell surface would abrogate infectivity. Previous studies have demonstrated that some monoclonal antibodies when bound to antigen induce either internalization or shedding of the antigen–antibody complex (antigenic modulation) (25,26). The addition of anti-B2 antibody (dilutions

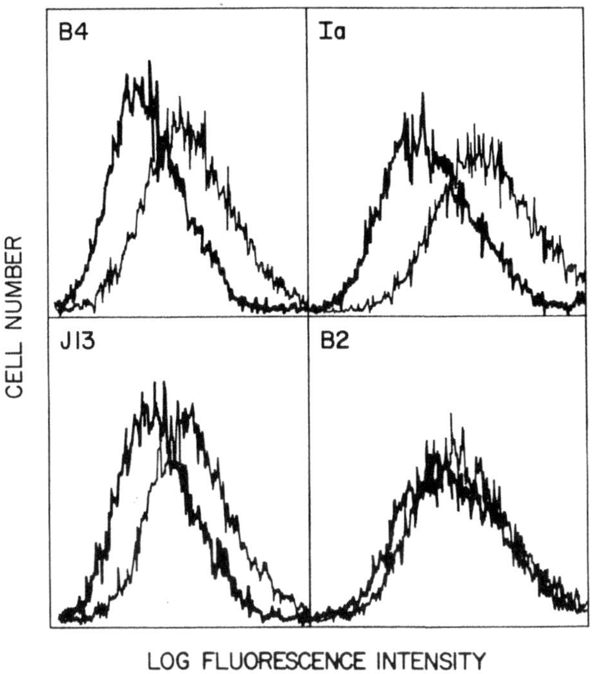

Fig. 44.1. Anti-B2 blocks binding to EBV. Burkitt's lymphoma Raji cells were incubated with 50 μg/ml of either anti-B4, anti-Ia (24), anti-J13, or anti-B2 for 30 min at 4°C. After three washes, EBV was incubated at 37°C for 1 hr with Raji cells coated with anti-B4, anti-Ia, anti-J13, or anti-B2. Binding of EBV was assessed by incubation with a 1 : 30 dilution of a highly specific rabbit anti-EBV serum (22,23), followed by two washings. Cells were then developed with a 1 : 20 dilution of affinity-purified goat anti–rabbit Ig fluoresceinated antibody (Cappel), washed twice, and then fluorescent binding was determined by flow cytometric analysis with log amplification.

ranging from 1 : 100 to 1 : 1,000,000) to B cells for 24 hr at 37°C induced loss of the B2 antigen from the cell surface. As shown, tonsillar B cells were cultured for 24 hr in the presence of anti-B2, anti-B1, anti-B4, or anti-57F (directed against the C3b receptor) (27) (Fig. 44.2). Modulation with anti-B2 lead to the loss of greater than 95% of detectable cell surface B2 whereas the expression of B1, B4, or 57F was unchanged (Fig. 44.2). In addition, anti-B2 modulation did not effect the expression of IgM or Ia (data not shown). Moreover, B cells cultured with monoclonal antibodies anti-B1, anti-B4, or anti-57F for 24 hr showed no modulation of B2, B1, B4, or 57F antigens. Culturing B cells for 24 hr with either anti-J13, anti-IgM, anti-Ia, or Texas Red–avidin produced no change in cell surface antigen expression for the panel of monoclonal antibodies tested. Thus, incubation with anti-B2 allowed for the selective removal of B2 from the cell surface.

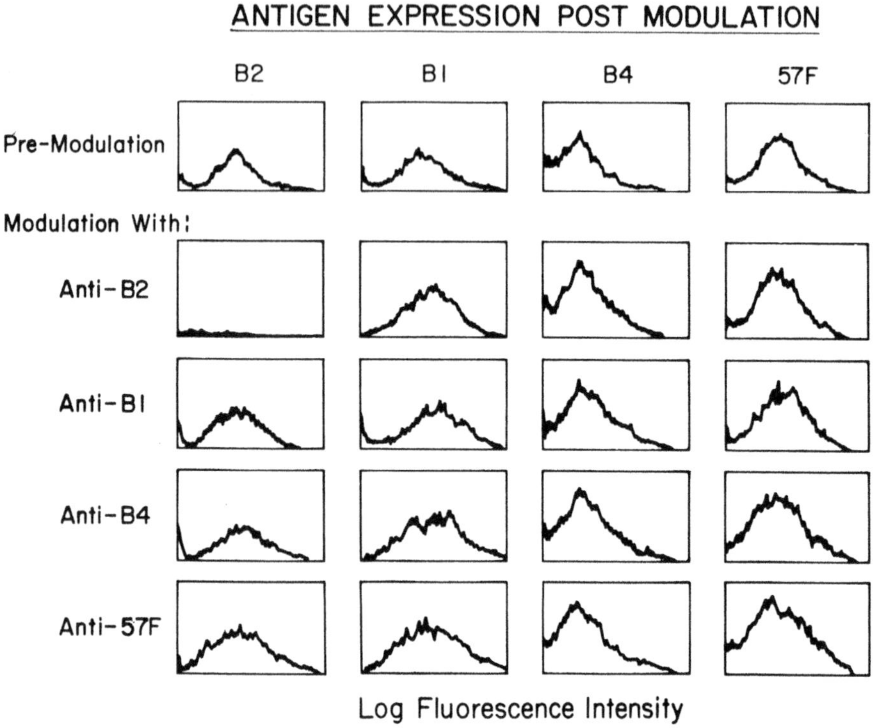

Fig. 44.2. Anti-B2-induced antigenic modulation. Tonsillar B cells were cultured for 24 hr at 37°C with saturating concentrations of anti-B2, anti-B1, anti-B4, or anti-57F. After culture, cells were washed and stained with goat anti–mouse Ig–FITC to assess the modulated antigen. Cells were also stained with directly biotinylated antibodies and Texas Red–avidin to assess the expression of the other antigens post-modulation. Identical data could be demonstrated for B cells isolated from tonsil (n = 10) and spleen (n = 12).

Anti-B2-Induced Modulation and Loss of EBV Infectivity

To determine whether anti-B2-modulated cells could be infected with EBV, the following experiments were undertaken. Purified B cells were treated for 24 hr at 37°C with either anti-B2 or anti-J13 (Table 44.2). Cells incubated overnight with anti-B2 were stained with FITC-conjugated goat anti–mouse Ig to detect residual B2 antigen. In each experiment, greater than 90% of cell surface B2 was lost. Treatment with anti-J13, in contrast, did not induce any change in B2 expression. At the end of the 24-hr culture, anti-J13- or anti-B2-treated cells were washed and re-cultured for an additional 7 days in the presence of either a) media, b) anti-B2, c) EBV, or d) anti-B2 and EBV. On the sixth day, cells were pulsed with ^{3}H-Tdr to assess proliferation and on the seventh day supernatants were harvested to measure IgG synthesis. As seen in Table 44.2, cells modulated with

Table 44.2. Anti-B2-modulated B cells lose the ability to proliferate and synthesize immunoglobulin in response to EBV infection.

	Initial 1-day modulation with:	
	Control antibody	Anti-B2
A. Proliferation		
7-day culture with:	cpm ± SD	cpm ± SD
Media	389 ± 62	318 ± 43
Anti-B2	407 ± 33	376 ± 21
EBV	8118 ± 146	3336 ± 301
Anti-B2 + EBV	3407 ± 211	1155 ± 134
14-day culture with:		
Media	371 ± 21	316 ± 33
Anti-B2	349 ± 19	378 ± 24
EBV	13,135 ± 405	2112 ± 197
Anti-B2 + EBV	2917 ± 116	735 ± 54
B. Immunoglobulin synthesis		
7-day culture with:	ng IgG/ml	ng IgG/ml
Media	7 ± 2	20 ± 4
Anti-B2	16 ± 3	21 ± 5
EBV	829 ± 37	288 ± 11
Anti-B2 + EBV	162 ± 18	86 ± 9

either anti-J13 or anti-B2 and then re-cultured for either 7 or 14 days with media or anti-B2 demonstrated no evidence of proliferation or IgG synthesis. The addition of EBV to cells treated with control antibody (J13) led to a dramatic increase in cellular proliferation and Ig synthesis. In contrast, cells incubated with anti-B2 demonstrated an approximate 60% reduction of proliferation and Ig synthesis. Even greater inhibition of proliferation and Ig synthesis was seen with cells treated with anti-B2 and then cultured with additional anti-B2 and EBV for 7 days. In this case, the inhibition of proliferation was 86% on day 7 and 95% by day 14. The inhibition of Ig synthesis was approximately 90% on day 7. These experiments suggest that the selective removal of B2 from cell surface by antigenic modulation is accompanied by a loss of infectivity by EBV. Additional support for this conclusion is provided by recent preliminary experiments which demonstrate that the re-expression of B2 following B2 modulation is associated with the return of EBV infectivity. Moreover, the re-expression of B2 in culture may partially explain our inability to completely block the infectivity of EBV with a single treatment of cells with anti-B2.

The present data suggest that EBV binds to the B2 molecule. Moreover, the suppression of EBV-induced proliferation and Ig synthesis by either anti-B2 blocking or anti-B2-induced modulation strongly suggests that, in addition to binding of EBV, an intact B2 molecule is required for infectivity. More importantly, the data indicate that the anti-B2 binding site is very close to, if not identical to, the EBV binding site. In contrast, our earlier studies have demonstrated that C3d binds to the B2 molecule but that anti-B2 did not block C3d binding. These observations are consistent with the notion that the B2 molecule has at least two binding sites—one for EBV and a second for C3d. The structural identification of the EBV receptor will not facilitate the study of EBV binding, ontogenic expression, and mechanisms of EBV-induced B cell triggering. Conversely, the elucidation of the process of EBV infection may yield important insights into the physiologic role of a molecule which selectively binds both EBV and C3d.

Acknowledgments. We would like to thank Drs. Kyoko Iida and Victor Nussenzweig for the use of the 57F antibody. We would also like to thank Ms. Bonnie Frisard for excellent secretarial assistance. This work was supported by NIH Grants CA25369 and CA31893. A.W. Boyd is a recipient of a Neil Hamilton Fairley Fellowship of the National Health and Medical Research Council, Australia, and D.A. Thorley-Lawson is a recipient of a Research Career Development Award from NAIAD #00549.

References

1. Yefenof, E., G. Klein, M. Jondal, and M.B.A. Oldstone. 1976. Surface markers on human B and T lymphocytes. IX. Two-color immunofluorescence studies on the association between EBV receptors and complement receptors on the surface of lymphoid cell lines. *Int. J. Cancer* **17:**693.
2. Jondal, M., G. Klein, M.B.A. Oldstone, V. Borkish, and E. Yefenoff. 1976. Surface markers on human B and T lymphocytes. VIII. Association between complement and Epstein–Barr virus receptors on human lymphoid cells. *Scand. J. Immunol.* **5:**401.
3. Yefenof, E., and G. Klein. 1977. Membrane receptor stripping confirms the association between EBV receptors and complement receptors on the surface of human B lymphoma lines. *Int. J. Cancer* **20:**347.
4. Yefenof, E., G. Klein, and K. Kvarnung. 1977. Relationship between complement activation, complement binding, and EBV adsorption by human hematopoietic cell lines. *Cell. Immunol.* **31:**225.
5. Yefenof, E., T. Bakacs, L. Einhorm, I. Ernberg, and G. Klein. 1978. Epstein–Barr virus (EBV) receptors, complement receptors, and EBV infectivity of different lymphocyte fractions of human peripheral blood. I. Complement receptor distribution and complement binding by separated lymphocyte subpopulations. *Cell. Immunol.* **35:**34.
6. Einhorn, L., M. Steinitz, E. Yefenof, I. Ernberg, and G. Klein. 1978. Epstein–Barr virus (EBV) receptors, complement receptors, and EBV infectibility of different lymphocyte fractions of human peripheral blood. II. Epstein Barr virus studies. *Cell. Immunol.* **35:**43.
7. Klein, G., E. Yefenoff, K. Falk, and A. Westman. 1978. Relationship between

Epstein–Barr Virus (EBV). Production and the loss of the EBV receptor/ complement receptor complex in a series of sublines derived from the same original Burkitt's lymphoma. *Int. J. Cancer* **21:**552.
8. Magrath, I., C. Freeman, M. Santaella, J. Gadek, M. Frank, R. Spiegel, and L. Novikovs. 1981. Induction of complement receptor expression in cell lines derived from human undifferentiated lymphomas. II. Characterization of the induced complement receptors and demonstration of the simultaneous induction of EBV receptor. *J. Immunol.* **127:**1039.
9. Wells, A., N. Koide, H. Stein, J. Gerdes, and G. Klein. 1983. The Epstein–Barr virus receptor is distinct from the C3 receptor. *J. Gen. Virol.* **64:**449.
10. Hutt-Fletcher, L.M., E. Fowler, J.D. Lambris, R.J. Feighny, J.G. Simmons, and G.D. Ross. 1983. Studies of the Epstein–Barr virus receptor found in Raji cells. II. A comparison of lymphocyte binding sites for Epstein Barr virus and C3d. *J. Immunol.* **130:**1309.
11. Nadler, L.M., *et al.* 1981. Characterization of a human B cell antigen (B2) distinct from B1. *J. Immunol.* **126:**1941.
12. Oettgen, H.C., P.J. Bayard, W. van Ewijk, L.M. Nadler, and C.P. Terhorst. 1983. Further biochemical studies of the human B cell differentiation antigens B1 and B2. *Hybridoma* **2:**17.
13. Stashenko, P., L.M. Nadler, R. Hardy, and S.F. Schlossman. 1981. Expression of cell surface markers after human B lymphocyte activation *Proc. Natl. Acad. Sci. U.S.A.* **78:**3848.
14. Bhan, A.K., L.M. Nadler, P. Stashenko, and S.F. Schlossman. 1981. Stages of B cell differentiation in human lymphoid tissue. *J. Exp. Med.* **154:**737.
15. Nadler, L.M., K.C. Anderson, M. Bates, E. Park, B. Slaughenhoupt, and S.F. Schlossman. 1984. Human B cell associated antigens: Expression on normal and malignant B lymphocyte. In: *First international congress of human leukocyte antigens,* J. Dausset, C. Milstein, and S.F. Schlossman, eds. pp. 354–362. Springer-Verlag, Berlin, Heidelberg, New York, Tokyo.
16. Iida, K., L.M. Nadler, and V. Nussenzweig. 1983. Identification of the membrane receptor for the complement fragment by means of a monoclonal antibody. *J. Exp. Med.* **158:**1021.
17. Weiss, J.J., T.F. Tedder, and D.T. Fearon. 1984. Identification of a 145,000 Mr membrane protein as the C3d receptor (CR2) on human B lymphocytes. *Proc. Natl. Acad. Sci. U.S.A.* **81:**881.
18. Ritz, J., and J.D. Griffin. 1982. Cell surface antigens in acute leukemia. In: *Biological responses in cancer: progress towards potential applications,* Vol 1, E. Michich, ed. Plenum Press, New York, pp. 2–21.
19. Nadler, L.M., K.C. Anderson, G. Marti, M. Bates, E. Park, J.F. Daley, and S.F. Schlossman. 1983. B4, a human B lymphocyte association antigen expressed on normal, mitogen activated, and malignant B lymphocytes. *J. Immunol.* **131:**244.
20. Boyum, A. 1968. Isolation of mononuclear cells and granulocytes from human blood. *Scand. J. Clin. Lab. Invest.* **21:**51.
21. Mendes, N.F., M.C.A. Tolnai, P.A. Silveira, R.B. Gilbertson, and R.S. Metzgar. 1973. Technical aspects of rosette tests used to detect human complement receptor (B) and sheep erythrocyte-binding (T) lymphocytes. *J. Immunol.* **111:**860.

22. Thorley-Lawson, D.A. 1979. Characterization of cross-reacting antigens on the Epstein–Barr virus envelope and plasma membrane of producer cells. *Cell* **16**:33.
23. Thorley-Lawson, D.A., and C.M. Edson. 1979. Polypeptides of the Epstein–Barr virus membrane antigen complex. *J. Virol.* **32**:458.
24. Nadler, L.M., P. Stashenko, R. Hardy, J.M. Pesando, E.J. Yunis, and S.F. Schlossman. 1981. Monoclonal antibodies defining serologically distinct HLA-D/DR related Ia-like antigens in man. *Hum. Immunol.* **1**:77.
25. Ritz, J., J.M. Pesando, J. Notis-McConarty, and S.F. Schlossman. 1980. A monoclonal antibody to human acute lymphoblastic leukemia antigen. *J. Immunol.* **125**:1506.
26. Reinherz, E.L., S. Meuer, K.A. Fitzgerald, R.E. Hussey, H. Levine, and S.F. Schlossman. 1982. Antigen recognition by human T lymphocytes is linked to surface expression of the T3 molecular complex. *Cell* **30**:735.
27. Iida, K., R. Mornaghi, and V. Nussenzweig. 1982. Complement receptor (CR_1) deficiency in erythrocytes from patients with systemic lupus erythematosus. *J. Exp. Med.* **155**:1427.
28. Bayer, E., and M. Wilcheck. 1978. The avidin biotin complex as a tool in molecular biology. *Trends Biochem. Sci.* **3**:N257.
29. Anderson, K.C., A.W. Boyd, D. Fisher, B. Slaughenhoupt, J.F. Daley, S.F. Schlossman, and L.M. Nadler. 1985. Isolation and functional analysis of human B cell populations: I. Characterization of the B1+B2+ and B1+B2− subset. *J. Immunol.* **134**:820.

CHAPTER 45

Monoclonal Antibody-Defined Cell Surface Molecules Regulate Lymphocyte Activation

Donald R. Howard, Allen C. Eaves, and Fumio Takei

Introduction

Molecular mechanisms underlying T and B cell activation are poorly defined. Evidence has accumulated implicating cell–cell interactions as well as soluble factors in the initiation of proliferative responses to antigens and mitogens. Clearly, an appreciation of those events governing normal lymphopoiesis may lead to a greater understanding of processes involved in neoplastic transformation. Lymphocyte activation requires at least two signals. For B cells the first signal appears to be cross-linking of surface immunoglobulin, followed by a second signal transmitted via T cell-derived B cell growth factor (BCGF) (1,2). T cells may be activated by antigen in the context of an antigen-presenting cell (APC) or mitogen, and induced to express receptors for T cell growth factor (TCGF). This T helper cell-derived soluble factor is then capable of supporting the proliferation of activated T cells (3,4).

Monoclonal antibodies which modulate cellular growth regulatory processes are rare. In order to define those cell surface molecules which are important in lymphocyte activation, we have generated monoclonal antibodies to lymphocyte surface determinants and tested these for the ability to inhibit lymphocyte proliferation as measured by lipopolysaccharide (LPS) and anti-μ stimulation of B cells, and phytohemagglutinin (PHA) and the mixed lymphocyte response (MLR) as indicators of T cell activation (5,6). In this report we describe two distinctly different antibodies, each capable of inhibiting lymphocyte responses to LPS, anti-μ, and the MLR. These findings suggest the molecules defined by these antibodies play an important role in the regulation of the immune response.

Materials and Methods

Monoclonal Antibodies

(C3H × BALB/c) F1 mice were immunized with LPS-stimulated spleen cells from a patient with B cell lymphoma (NB-107) or with the B lymphoma cell line DHL-10 generously provided by Dr. Henry Kaplan, Stanford University (DH-84). Cell fusion was performed as previously described by Kohler and Milstein (7). Two of the monoclonal antibodies obtained, NB-107 and DH-84, were selected for their ability to inhibit lymphocyte activation. Antibodies were used either as hybridoma culture supernates or as purified antibody with identical results. Antibodies were purified from ascitic fluid by ammonium sulfate precipitation, followed by DEAE Affi-Gel Blue column purification (Biorad, Richmond, CA). Fractions containing pure antibody were pooled and used in subsequent experiments.

Monoclonal antibody NHL-30.5 (IgG1) is directed against a myeloid-specific antigen (8) and was used as one of several negative control antibodies. NB-65 is a monoclonal antibody reactive with the transferrin receptor on human cells. Anti–transferrin receptor antibodies have been shown to inhibit cell proliferation and were used as a positive control (9).

Cells and Reagents

Peripheral blood mononuclear cells (PBMC) were obtained from normal volunteers by Ficoll–Hypaque separation according to established procedures (10). Affinity-purified $F(ab')_2$ fraction of goat anti–human IgM was obtained from Cappel Laboratories (Cochranville, PA). The optimal stimulating concentration for goat anti-μ was 50 μg/ml. Lipopolysaccaride (*E. coli,* Sigma, St. Louis, Mo) was utilized at 100 μg/ml; phytohemagglutinin (PHA, Gibco, Grand Island, NY) at 1% final concentration.

Stimulation/Inhibition Assays

Lymphocytes (2×10^5 cells/well) were incubated in wells of flat-bottomed microtiter plates with appropriate mitogen and monoclonal antibody. One-way mixed lymphocyte cultures were performed in microtiter wells using mitomycin C-treated stimulator cells (1.5×10^5 cells/well) and responder cells at the same concentration (11). All cultures were performed in RPMI 1640 media plus penicillin, streptomycin, and 10% fetal calf serum. Cells were cultured for 4 days (anti-μ, LPS, MLR) or 3 days (PHA) at 37°C, 100% humidity, and 5% CO_2, then pulsed with 1 μCi/well of [^{3}H]thymidine (Amersham Int., UK). Cells were harvested four hours later using a multiple automated sample harvester and then counted in a liquid scintillation counter (Beckman LS 7500).

All tests were set up in triplicate and included negative controls (cells + media, cells + test monoclonal antibody), positive controls (cells + mitogen), and test samples (cells + mitogen + antibody). Results are expressed as mean counts per minute (cpm) with standard error of the mean (S.E.M.) of triplicate wells. Percent inhibition was calculated by dividing mean test cpm by positive control cpm after subtracting background counts according to the following equation:

$$\left(1 - \frac{(\text{test}) - (\text{antibody control})}{(\text{positive control}) - (\text{negative control})}\right) \times 100$$

Results expressed in each table are representative of four individual experiments, all giving comparable results.

Results

NB-107 is an IgG1 monoclonal antibody which immunoprecipitates a heterodimer from the cell surface of ^{125}I-labeled B-lymphoma cells. The molecular weight of this dimer under nonreducing conditions is 160,000 and 115,000; under reducing conditions 160,000 and 90,000. Therefore, the antigen defined by NB-107 consists of two noncovalently linked chains. NB-107 reacts with a mean of 82 ± 13% of peripheral blood mononuclear cells (PBMC) when tested by FACS analysis against cells from 27 normal donors.

DH-84 is an IgG2a antibody which immunoprecipitates a heterodimer of approximately 35,000 and 25,000 under both reducing and nonreducing

Table 45.1. Inhibition of anti-μ stimulation[a]

Antibody	Mean cpm ± S.E.M.		% Inhibition
	Control[b]	Test[c]	
Media	1167 ± 192	3248 ± 402	
NB-107	566 ± 173	1456 ± 119	57
NHL-30.5[d]	1547 ± 130	3679 ± 377	0
NB-65[e]	586 ± 92	1735 ± 129	45
Media	232 ± 124	3436 ± 637	
DH-84	197 ± 46	298 ± 54	97

[a] Results are expressed as mean counts per minute (cpm) ± standard error of the mean (S.E.M.) of triplicate wells containing 2×10^5 peripheral blood mononuclear cells (PBMC). Percent inhibition was calculated as described in Materials and Methods.
[b] Cultures without mitogens.
[c] Cultures with mitogens.
[d] NHL-30.5 is a monoclonal antibody directed against a myeloid differentiation antigen (8); one of several monoclonals used as negative controls.
[e] Antibody to transferrin receptor.

Table 45.2. Inhibition of LPS stimulation

Antibody	Mean cpm ± S.E.M. Control	Test	% Inhibition
Media	310 ± 27	4469 ± 839	
NB-107	157 ± 31	1159 ± 111	76
NHL-30.5	387 ± 85	4390 ± 1528	4
NB-65	246 ± 25	997 ± 97	82
Media	1520 ± 339	10,951 ± 1323	
DH-84	222 ± 38	1332 ± 256	88

conditions from ^{125}I-labeled B-lymphoma cells. By FACS analysis, DH-84 reacted with a mean of 14% of PBMC from a total of 20 normal donors. By dual labeling with fluorescein and phycoerythrin, DH-84 was shown to react with the same population of cells as does anti–human HLA-DR (Becton Dickinson). DH-84 thus fulfills accepted criteria for defining non-polymorphic (monomorphic) HLA-class II determinants (12,13).

Monoclonal antibodies NB-107 and DH-84 showed profound inhibition of B cell stimulation by anti-μ and LPS (Tables 45.1 and 45.2). The MLR was similarly inhibited by these antibodies, but no effect was demonstrable on the ability of PHA to stimulate T cells (Tables 45.3 and 45.4). Anti–transferrin receptor antibody showed significant inhibition of T and B cell stimulation with anti-μ, LPS, MLR, and PHA, while the negative control antibody NHL-30.5 did not.

Discussion

The regulatory events underlying lymphocyte activation and proliferation are not well understood. In an attempt to clarify these processes we have screened a series of monoclonal antibodies for their ability to inhibit various *in vitro* B and T lymphocyte functions (5,6). Two of these, NB-107 and DH-84, profoundly inhibit LPS and anti-μ stimulation of B cells, and the T cell response in the mixed lymphocyte reaction. Stimulation of T cells by PHA was not affected.

DH-84 is directed against a common portion of HLA-class II molecules. Class II antigens are now thought to include at least three distinct

Table 45.3. Inhibition of MLR

Antibody	Mean cpm ± S.E.M. Control	Test	% Inhibition
Media	459 ± 204	17,998 ± 797	
NB-107	219 ± 104	3811 ± 826	80
NHL-30.5	679 ± 141	19,810 ± 1941	0
NB-65	343 ± 93	4028 ± 502	79
DH-84	111 ± 20	4748 ± 286	74

Table 45.4. Inhibition of PHA

	Mean cpm ± S.E.M.		
Antibody	Control	Test	% Inhibition
Media	519 ± 62	85,767 ± 1665	
NB-107	164 ± 60	98,442 ± 12,123	0
NHL-30.5	272 ± 48	87,022 ± 3951	0
NB-65	305 ± 82	48,373 ± 3083	44
DH-84	132 ± 20	89,735 ± 5416	0

groups of molecules—DR, DC, and SB (13–15). The function of these molecules is largely unknown, although it is clear through studies of MHC restriction that DR plays an associative role in antigen recognition and is involved in T helper cell activation and in the induction of allogeneic T cell proliferative responses (12,13,16). A monoclonal antibody against DC antigens has been shown to lack any effect on cell proliferation but selectively inhibits the generation of effector T cells mediating specific cytolytic activity (17). The complexity of HLA-DC molecules may be considerably greater than realized (18). These findings suggest that class II DR and DC molecules may have different regulatory functions in the immune system.

Anti-class II polyclonal or monoclonal antibodies have been shown to inhibit a variety of cellular responses in human and mouse systems including: proliferative and plaque-forming responses to Con A and pokeweed mitogen (19–21), primary and secondary MLR (22–24), and antigen-specific lymphoproliferative responses (25–27). To our knowledge the effect of anti-class II monoclonal antibodies on human B cell stimulation by anti-μ and LPS and T cell stimulation by PHA has not been previously reported. Although many of the above reports have designated their monoclonal antibodies as anti-DR, unless cross-blocking and sequential immunoprecipitation studies are performed with known anti-DR and anti-DC antibodies, this is not certain. We prefer the less restrictive designation: anti-class II.

Inhibition of MLR by anti-class II antibodies as shown in this study confirms previous reports (22,24). The complete lack of inhibition of PHA stimulation suggests that activation by this mitogen, which is primarily stimulatory to T cells, does not involve class II molecules. Inhibition of anti-μ and LPS-induced proliferation by anti-class II monoclonal antibodies is of particular importance because it indicates class II molecules may play an important role in B cell stimulation. Furthermore it suggests that the mechanism of action of anti-μ either involves class II molecules or that these molecules are able to transduce a direct growth regulatory signal to B cells. Previous studies have suggested that anti-μ generates a proliferative signal in B cells by cross-linking surface immunoglobulin (2,28). At high concentrations of anti-μ, proliferation apparently occurs in the absence of accessory cells, T cells, or soluble factors (1). In order to inhibit B cell stimulation by anti-μ, anti-DR must either have a direct inhibitory effect on B cells or alternatively anti-μ must act through a mechanism other than simple cross-linking of surface immunoglobulin and one that allows for the participation of DR molecules. Although the present study does not clearly separate inhibitory effects mediated via accessory cells from direct effects of HLA-class II antibodies on B cells, studies are now in progress to distinguish between these alternatives using purified cell populations and growth factors (29).

Monoclonal antibody NB-107 is directed against a cell surface heterodimer of molecular weight 160,000 and 90,000 under reducing conditions, which is present on a proportion of peripheral blood mononuclear cells from all normal individuals. The function of this molecule is not yet known, although its unique structure suggests it may function as a receptor. The importance of the antigen defined by NB-107 in lymphocyte activation is shown by the ability of this antibody to inhibit LPS and anti-μ stimulation of B cells and T cell proliferation in the MLR. Whether the inhibitory effect of this antibody is mediated via a direct effect on lymphocytes or through effects on accessory cells remains to be determined. That NB-107 does not interfere with the uptake of some essential nutrient or substance, analogous to the effect of anti–transferrin receptor antibody, is suggested by the lack of inhibition of the *in vitro* growth of neoplastic cell lines (unpublished observations).

Summary

Despite intense investigation the regulatory events underlying lymphocyte activation and proliferation are poorly understood. In order to characterize cell surface molecules involved in these processes we have generated monoclonal antibodies to human lymphocyte surface determinants and tested these antibodies for the ability to inhibit B cell function, as measured by LPS and anti-μ stimulation, and T cell function, as indicated by the mixed lymphocyte reaction and PHA stimulation. Two antibodies, NB-107 and DH-84, were shown to profoundly inhibit anti-μ, LPS, and MLR responses, but had no effect on PHA stimulation. DH-84 is an antibody directed against HLA-class II antigens. Inhibition of anti-μ stimulation of B cells by DH-84 suggests a hitherto unrecognized role for HLA-class II molecules in B cell activation. The identity of the antigen defined by NB-107 has not yet been determined. However, its unique structure and the ability of NB-107 antibody to inhibit lymphocyte proliferation suggests it plays an important role in the regulation of cell growth, possibly as a cell surface receptor.

Acknowledgments. The authors wish to thank Cam Smith and Trudy Bagan for technical assistance and Judy Waite for preparation of the manuscript. This research is supported by the National Cancer Institute of Canada, the Cancer Control Agency of British Columbia, and the B.C. Cancer Foundation. D.R. Howard is a Terry Fox Training Centre Fellow of the National Cancer Institute of Canada. F. Takei is a Research Scholar of the Medical Research Council of Canada.

Note added in proof: Since submission of this manuscript we have shown that NB-107 reacts with the lymphocyte function associated (LFA-1) molecule (6).

References

1. Kehrl, J.H., A. Muraguchi, J.L. Butler, R.J.M. Falkoff, and A.S. Fauci. 1984. Human B cell activation proliferation and differentiation. *Immunol. Rev.* **78:**75.
2. Howard, M., K. Nakanishi, and W.E. Paul. 1984. B cell growth and differentiation factors. *Immunol. Rev.* **78:**75.
3. Ruscetti, F.W., and R. Gallo. 1981. Human T-lymphocyte growth factor: Regulation of growth and function of T lymphocytes. *Blood* **57:**379.
4. Gillis, S. 1983. Interleukin 2: Biology and biochemistry. *J. Clin. Immunol.* **3:**1.
5. Howard, D.R., A. Eaves, and F. Takei. 1985. A new regulatory role for HLA class II molecules in B lymphocyte activation: Inhibition by monoclonal anticlass II antibodies. Submitted.
6. Howard, D.R., A. Eaves, and F. Takei. 1985. Lymphocyte function associated antigen (LFA-1) is involved in B cell activation. Submitted.
7. Kohler, G., and C. Milstein. 1975. Continuous cultures of fused cells secreting antibody of predefined specificity. *Nature* **256:**495.
8. Askew, D.S., A.C. Eaves, and F. Takei. 1985. NHL-30.5: A monoclonal antibody reactive with an acute myeloid leukemia (AML)-associated antigen. *Leuk. Res.* **9:**135.
9. Trowbridge, I. and F. Lopez. 1982. Monoclonal antibody to transferrin receptor blocks transferrin binding and inhibits human tumor cell growth *in vitro*. *Proc. Natl. Acad. Sci. U.S.A.* **79:**1175.
10. Howard, D.R. 1983. T-antigen does not induce cell mediated immunity in patients with breast cancer. *Cancer* **51:**2053.
11. DeWolf, W.C., J.J. O'Leary, and E.J. Yunis. 1980. Cellular typing. In: *Manual of clinical immunology,* N.R. Rose and H. Friedman, eds. American Society for Microbiology, Washington, D.C., p. 1006–1025.
12. Shackelford, D.A., J.F. Kaufman, A.J. Korman, and J.L. Strominger. 1982. HLA-DR antigens: Structure, separation of subpopulations, gene cloning and function. *Immunol. Rev.* **66:**133.
13. Kaufman, J.F., C. Auffray, A.J. Korman, D.A. Shackelford, and J. Strominger. 1984. The class II molecules of the human and murine major histocompatibility complex. *Cell* **36:**1.
14. Brodsky, F.M. 1984. A matrix approach to human class II histocompatibility antigens: Reactions of four monoclonal antibodies with the products of nine haplotypes. *Immunogenetics* **19:**179.
15. Bohme, J., D. Owerbach, M. Denaro, A. Lernmark, P.A. Peterson, and L.

Rack. 1983. Human class II major histocompatibility antigen α-chains are derived from at least three loci. *Nature* **301:**82.

16. Rogozinski, L., A. Bass, E. Glickman, M.A. Talle, G. Goldstein, J. Wang, L. Chess, and Y. Thomas. 1984. The T4 surface antigen is involved in the induction of helper function. *J. Immunol.* **132:**735.
17. Corte, G., A. Moretta, M.E. Cosulich, D. Ramarli, and A. Bargellesi. 1982. A monoclonal anti-DC1 antibody selectively inhibits the generation of effector T cells mediating specific cytolytic activity. *J. Exp. Med.* **156:**1539.
18. Karr, R.W., C. Alber, S. Goyert, J. Silver, and R. Duquesnoy. 1984. The complexity of HLA-DS molecules. *J. Exp. Med.* **159:**1512.
19. Friedman, S.M., J.M. Breard, R.E. Humphries, J.L. Strominger, S.F. Schlossman, and L. Chess. 1977. Inhibition of proliferative and plaque-forming cell responses by human bone-marrow-derived lymphocytes from peripheral blood by antisera to the p23,30 antigen. 1977. *Proc. Natl. Acad. Sci. U.S.A.* **74:**711.
20. Broder, S., D.L. Mann, and T.A. Waldmann. 1980. Participation of suppressor T cells in the immunosuppressive activity of a heteroantiserum to human Ia-like antigens. *J. Exp. Med.* **151:**257.
21. Mizouchi, T., Yamashita, T. Hamaoka, and K. Morinaki. 1981. The role of Ia antigens in the activation of T cells by Con A: An evidence for the species restriction between T cells and accessory cells. *Cell. Immunol.* **57:**28.
22. Pawelec, G.P., S. Shaw, A. Ziegler, C. Muller, and P. Wernet. 1982. Differential inhibition of HLA-D or SB-directed secondary lymphoproliferative responses with monoclonal antibodies detecting human Ia-like determinants. *J. Immunol.* **129:**1070.
23. Eckels, D.D., J.N. Woody, and R.J. Hartzman. 1981. Monoclonal and xenoantibodies specific for HLA-DR inhibit primary responses to HLA-D but fail to inhibit secondary proliferative (PLT) responses to allogeneic cells. *Hum. Immunol.* **3:**133.
24. Dubreuil, P.C., D.H. Caillol, and F.A. Lemonnier. 1982. Analysis of unexpected inhibitions of T lymphocyte proliferation to soluble antigen, alloantigen and mitogen by unfragmented anti I-A or anti I-E/C monoclonal antibodies. *J. Immunogen.* **9:**11.
25. Triebel, F., V. Missenard-Leblond, M. Couty, D.J. Charron, and P. Debre. 1984. Differential inhibition of human antigen-specific T cell clone proliferative responses by distinct monoclonal anti-HLA-DR antibodies. *J. Immunol.* **132:**1773.
26. Sterkers, G., Y. Henin, J. Kalil, M. Bagot, and J. Levy. 1983. Influence of HLA class I and class II specific monoclonal antibodies on HLA class II restricted lymphoproliferative responses. *J. Immunol.* **131:**2735.
27. Marrack, P. and J.W. Kappler. 1977. Anti-Ia inhibits the activity of B cells but not a T cell-derived helper mediator. *Immunogenetics* **4:**541.
28. Muraguchi, A., J.L. Butler, J.H. Kehrl, and A.S. Fauci. 1983. Differential sensitivity of human B cell subsets to activation signals delivered by anti-μ antibody and proliferative signals delivered by a monoclonal B cell growth factor. *J. Exp. Med.* **157:**530.
29. Maizel, A.L., J.E. Morgan, S.R. Mehta, N.M. Kouttab, J.M. Bator, and C.G. Sahasrabuddhe. 1983. Long-term growth of human B cells and their use in a microassay for B cell growth factor. *Proc. Natl. Acad. Sci. U.S.A.* **80:**5047.

CHAPTER 46

TPA-Induced Modulation of B Cell Differentiation Antigens Defined by Monoclonal Antibodies (HD6, HD28, HD37, HD39)

Reinhard Schwartz, Gerhard Moldenhauer, Bernd Dörken, Antonio Pezzutto, Frank Momburg, and Volker Schirrmacher

Introduction

B cell neoplasms, particularly chronic lymphocytic leukemias (CLL) and non-Hodgkin's lymphomas, can be considered as monoclonal cell populations which are arrested at certain stages of normal B cell differentiation. Efforts to suspend this restriction by treatment with the tumor promoter 12-*O*-tetradecanoylphorbol 13-acetate (TPA) resulted in the stimulation of terminal differentiation-like changes of the leukemic cells (1). This process was assessed by alterations in cytoplasmic and surface immunoglobulin content (2), HLA-class II antigens (3), complement receptor (4), and morphological features (5). Lately, modulations of cell surface structures after TPA stimulation, which were followed up by monoclonal antibodies specific for B cell differentiation antigens, supported the assumption that chronic lymphocytic leukemias can be induced by TPA to further maturation towards the plasma cell stage (6,7). Thus, the monitoring of TPA-induced alterations of differentiation antigens by B cell-specific antibodies could help to i) describe more precisely the reaction pattern of the respective monoclonal antibody, ii) define closer sections along the normal B cell differentiation pathway, and iii) gain information about the genetic program of the various leukemia and lymphoma types.

In an attempt to further clarify these questions we measured the reactions of TPA-treated peripheral blood lymphocytes from several leukemia patients with three of our monoclonal antibodies which seem to be specific for different stages of B cell differentiation.

Materials and Methods

Patients

Peripheral mononuclear blood lymphocytes (PBML) were obtained from 14 patients with B-type chronic lymphocytic leukemia (CLL), 2 patients with prolymphocytic leukemia (PLL), 1 patient with B-type acute lymphoblastic leukemia (ALL), 1 patient with B-lymphoid chronic myelogenous leukemia in blast crisis (CML-BC), and 2 patients with hairy cell leukemia (HCL). Diagnosis of the types of leukemia was based on clinical, cytological, and morphological findings. The histopathological diagnosis was determined according to the Kiel classification (8). In addition, leukemic cells were examined for the presence of various markers [HLA-DR, TDT, CALLA (J5), E-rosettes, T antigens (OKT series), surface and cytoplasmic Ig]. Patients were off cytostatic therapy prior to blood sampling.

Cell Culture and TPA Treatment

Lymphocytes were isolated by Ficoll–Hypaque density gradient centrifugation and cultivated in RPMI 1640 supplemented with 10% fetal calf serum at a cell concentration of 2×10^6/ml in a 5% CO_2 atmosphere at 37°C for further experiments.

TPA (12-*O*-tetradecanoylphorbol 13-acetate, kindly provided by Prof. Hecker, Institute for Biochemistry, German Cancer Center, Heidelberg, FRG) was added at a final concentration of 5×10^{-9} *M* to the cell culture. This concentration was found to be optimal for stimulation experiments. Cell viability was greater than 90% after a cultivation time of 3 days with TPA. After culture the cells were washed 3 times and subjected to the tests.

B Cell-Specific Monoclonal Antibodies

Four of our monoclonal antibodies raised against B leukemias which were used for the detection of differentiation antigens both indirect immunofluorescence and IPSA are characterized in Table 46.1.

Table 46.1. Monoclonal antibodies used in this study.

Designation	Isotype	M.W. (Kd) of target antigen	Specificity
HD28 (B17)[a]	IgG2a	39–52 gp[b]	B cells
HD6 (B25)	IgG1	130, 140 gp	Mature B cells
HD39 (B31)	IgG1	130, 140 gp	Mature B cells
HD37 (B28)	IgG1	Unknown	Pan B

[a] Workshop nomenclature given in brackets.
[b] Glycoprotein.

Indirect Immunofluorescence

Indirect immunofluorescence was performed on cells in suspension. The cells were kept on melting ice for antibody reactions. Second antibody was FITC-conjugated rabbit anti–mouse IgG (whole molecule) from Miles-Yeda Ltd., Israel. Fluorescence intensity distributions were determined with a cytofluorograph (Ortho Diagnostic Systems, Westwood, MA).

Immunoperoxidase Slide Assay (IPSA)

The IPSA was performed as described by Morich *et al.* (9). Cells were attached to poly-L-lysine-coated reaction areas on specially prepared glass slides. Then cells were fixed with 0.05% glutardialdehyde in phosphate-buffered saline (PBS). Prior to incubation with monoclonal antibody, reaction areas were blocked with 0.2% gelatine in PBS. Monoclonal antibodies were followed by rabbit anti–mouse Ig, goat anti–rabbit Ig (Tago, Burlingame, CA), and rabbit PAP complex (Dako, Denmark). Substrate for peroxidase was diaminobenzidine (1 mg/ml) in Tris buffer containing 0.01% H_2O_2. Post-fixing was done with 2% OsO_4 and slides were mounted in glycerol. Membrane staining was evaluated under a light microscope counting 300 to 400 cells. Positive tumor cells exhibited dark brown rings.

Enzyme Immunoassay (ELISA)

Flexible PVC microtiter plates were coated with goat anti–human Ig. After blocking of nonspecific sites with BSA, culture supernatant or cell lysate (100 μl/well) was added and incubated for 1 hr. Plates were washed several times and incubated with peroxidase-conjugated subclass-specific goat anti–human Ig antibody. After another washing cycle the reaction was visualized using *o*-phenylenediamine (OPD).

Radioimmunoassay on Cells (CRIA)

Cells cultured with or without TPA were washed in PBS–2% BSA. 50 μl/well of target cell suspension containing 10^6 viable cells were incubated with 50 μl ^{125} I-labeled heavy-chain isotype-specific goat anti–human Ig for 1 hr. After washing, radioactivity in individual wells was measured.

Incorporation Assay

Cells were cultivated with or without TPA for 48 hr. The following were added to 2×10^5 cells/well (flat-bottomed, Greiner, FRG) indicated times for a 12-hr pulse: for determination of proliferation, 1 μCi of [^{3}H]thymidine; for determination of protein synthesis, 1 μCi of [^{3}H]leucine; and for determination of glycoconjugate synthesis, 2 μCi of [^{3}H]galactose. The cells were harvested by acid precipitation with 0.4 *M* perchloric acid on glass fiber filters using a Skatron multiple cell culture harvester (Flow, FRG). Samples were counted in emulsifier scintillator in a Packard Scintillation counter. All incorporation experiments were carried out in triplicates.

Results

Expression of Differentiation Antigens after TPA Treatment

In vitro cultured peripheral blood lymphocytes of 20 patients with various B-type leukemias were investigated for their expression of differentiation antigens after TPA treatment. Cytofluorograph analysis of these experiments is shown in Table 46.2. Untreated CLL cells were characterized by only weak or no reaction with HD6 and HD39, respectively. A more

Table 46.2. Cytofluorograph analysis of surface antigens of leukemia cells after TPA treatment.

		Positive cells (%)								
		HD6			HD39			HD28		
Patient	Type of leukemia	0[a]	2−[b]	2+[c]	0	2−	2+	0	2−	2+
Ste	CLL	3	2	33	0	1	9	77	78	17
Emm	CLL	0	0	17	0	0	7	65	61	24
Sta	CLL	7	0	29	0	0	16	81	75	4
Lip	CLL	0	0	8	0	0	4	89	77	4
Sei	CLL	1	0	7	0	0	4	57	36	4
Rub	CLL	0	0	26	0	0	17	66	68	6
Pfe	CLL	0	0	11	0	0	2	81	60	19
End	CLL	1	0	9	0	0	2	91	83	9
Aml	CLL	1	0	13	0	0	5	84	52	2
Sce	CLL	7	12	35	ND	ND	ND	65	50	38
Hun	CLL	38	20	61	0	0	16	89	91	95
Koh	CLL	5	49	67	5	12	29	91	99	98
Hap	CLL	0	0	0	0	0	0	88	69	19
Mun	CLL/PLL	31	31	41	7	8	14	97	99	92
Alb	PLL	47	5	75	23	3	42	76	51	43
Lil	PLL	41	24	41	26	18	33	31	32	37
Sch	ALL	28	32	61	10	28	62	16	19	11
Kai	CML-BC	5	16	44	2	14	33	21	22	29
Sca	HCL	36	35	54	29	30	51	30	37	54
Eis	HCL	63	65	50	64	65	49	57	68	51

[a] Control cells without TPA treatment at time 0 of the incubation.
[b] Cells incubated for 2 days without TPA.
[c] Cells incubated for 2 days with TPA.

pronounced reaction with HD6 was seen in two cases diagnosed as CLL, which clinically appeared as lymphoplasmacytic lymphomas (HUN, SCE), and in one case of CLL with a tendency to PLL (Mun). More than 80% of cells from all CLL cases reacted with HD28. The expression of these antigens was not greatly affected by *in vitro* culture of the cells for 3 days. Culture in the presence of TPA for 3 days at a concentration of 5×10^{-9} *M* reversed the reaction pattern. In all cases reaction with HD6/HD39 increased whereas reaction with HD28 decreased. The distribution of fluorescence intensity after 48-hr TPA stimulation is presented for four cases in Fig. 46.1(A). In general, the mean fluorescent intensity on a single cell basis was not altered by TPA treatment. In some cases with high expression of HD28 at time 0, reaction with HD28 decreased only in its intensity per cell [c.f. Fig. 46.1(A)]. The kinetics of antibody reactions

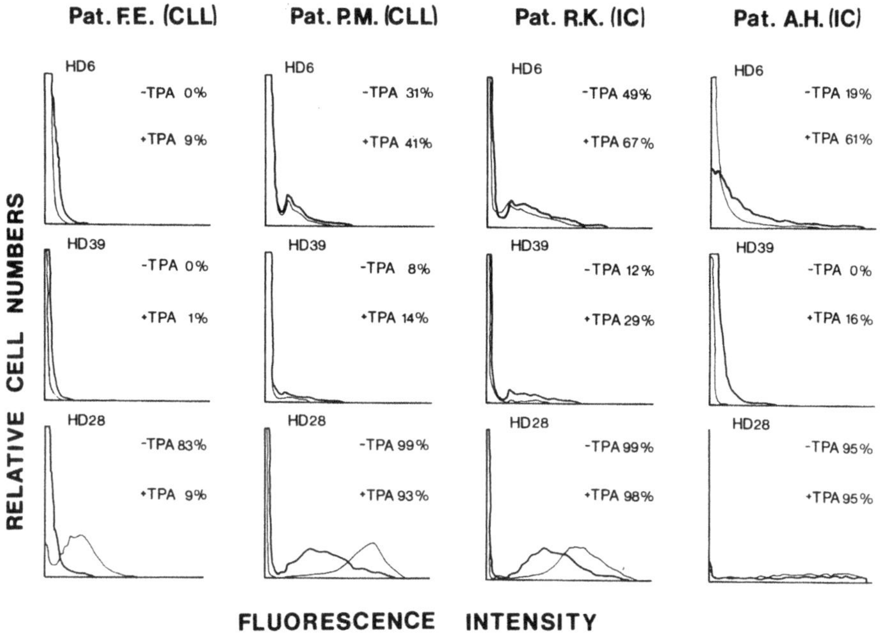

Fig. 46.1. (A) Effect of TPA treatment after 48-hr incubation on expression of HD6, HD39, and HD28 of cultured chronic lymphocytic leukemia (B-CLL) and lymphoplasmacytic lymphoma (IC) cells. Histograms from cytofluorographic analysis. Strong lines represent TPA-treated cells, weak lines untreated cells. Background controls were done by incubating cells with second antibody only (RAM–FITC). 1.5×10^4 cells were analyzed for each histogram; the percentage of cells within a region (channel 80–1000) minus percentage of background controls is given.

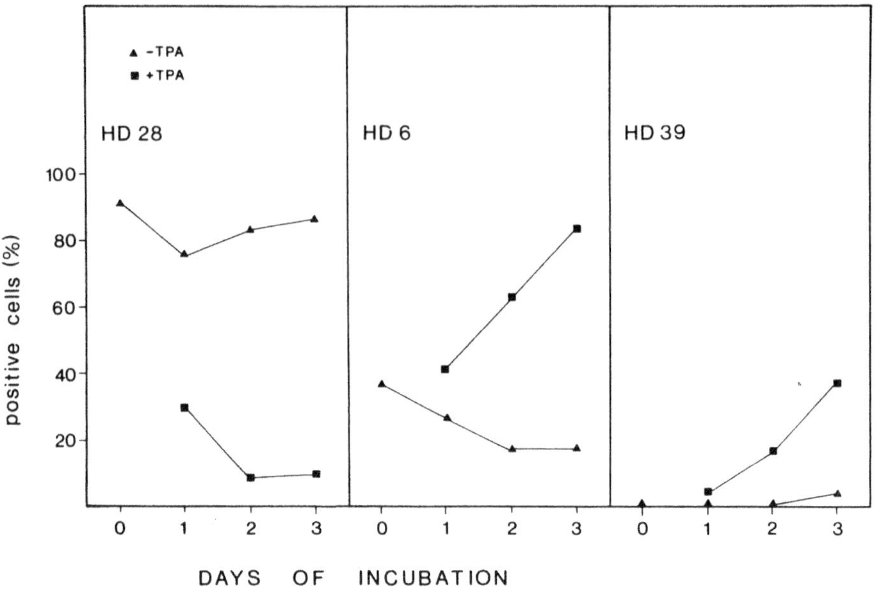

Fig. 46.1. (B) Kinetics of cytofluorograph-analyzed reactions with monoclonal antibodies. Data for expression of HD28 were taken from patient End (CLL); those for expression of HD6/HD39 from patient Hun (CLL/IC).

in a time course of 3 days are shown for three examples in Fig. 41.1(B). Loss of expression of the HD28 antigen was complete after 2 days of TPA treatment in most cases whereas increase of the HD6/HD39 antigen expression still continued after 3 days of TPA treatment. Antibody-specific staining of the cells with the IPSA technique confirmed the results obtained with indirect immunofluorescence. It could also be seen that, for example, HD6/HD39-positive cells had an altered morphology, such as enlargement of cytoplasm, after TPA induction. Sometimes even hairy cell-like structures occurred (Fig. 46.2). In contrast to TPA-induced modulations found in CLL cells, all studied HCL and PLL cases remained virtually unchanged after TPA treatment in their reaction patterns with HD6, HD39, and HD28. Untreated cells of these leukemias showed already positive reactions with the three antibodies. In two single cases of B-ALL and B-lymphoid CML in blast crisis, reaction patterns with HD6, HD39, and HD28 were similar. Both leukemias were characterized by an increased reaction with HD6 and HD39 whereas reaction with HD28 remained constant. For control, we also applied a pan B monoclonal antibody, HD37, in our TPA stimulation experiments. This antibody reacted with 60–90% of cells of all leukemia patients. As expected, TPA did not alter the expression of this antigen in all cases tested.

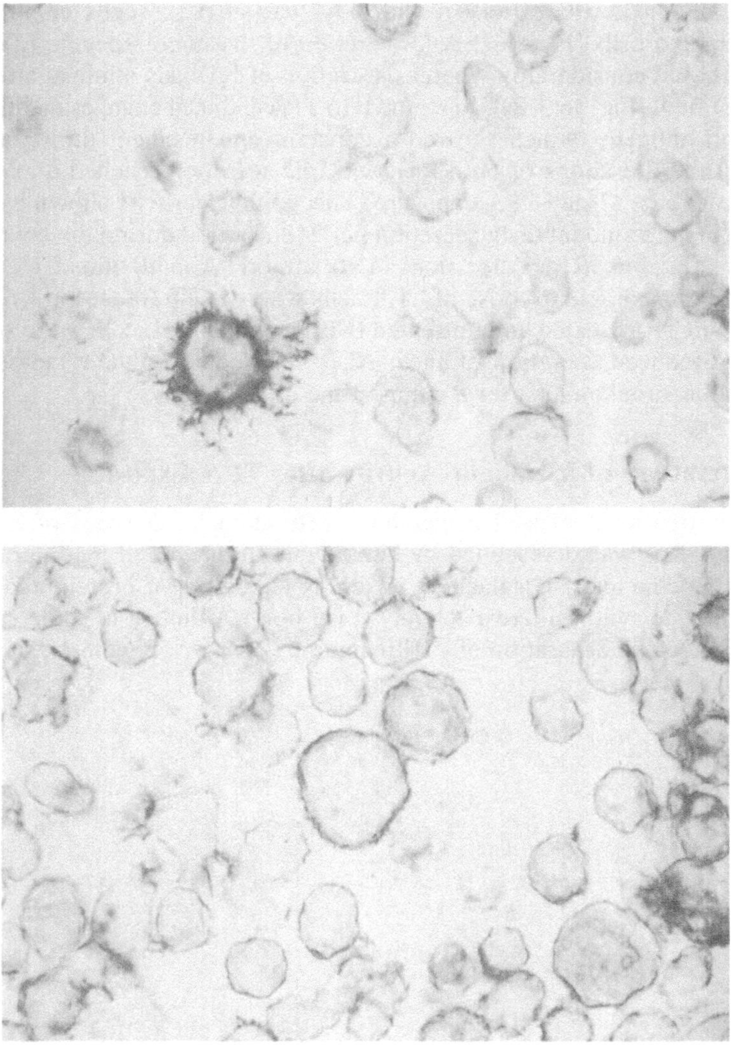

A

B

Fig. 46.2. IPSA-stained B-CLL cells after 48-hr TPA treatment: (A) reaction with HD6. Note the hairy surface structure of the prominent cell; (B) Reaction with HD11, an antibody against the HLA-DR framework. ×1000.

Alterations of Immunoglobulin Expression and Secretion after TPA Exposure

In order to monitor further parameters of differentiation we measured the expression and secretion of immunoglobulin during TPA treatment. As a general feature, a decrease in surface-expressed immunoglobulin, preferentially IgD, was observed with TPA-treated CLL cells compared to untreated cells [Fig. 46.3(A)]. Secretion of the tumor-specific Ig isotype increased considerably whereas secretion of IgD was minimal after TPA treatment [Fig. 46.3(B)]. In contrast to TPA-induced changes in differentiation antigens, which showed a constant enhancement during a 3-day culture, alterations of both IgG, and IgD secretion reached a maximum already after 1-day TPA exposure. This could clearly be shown by determining the amount of Ig secretion per 24-hr period during the course of 6 days [Fig. 46.3(C)]. Alterations in sIg after TPA induction of HCL cells were comparable to those of CLL cells whereas differences in Ig expression in TPA-treated and untreated B-PLL and B-ALL cells were smaller. TPA-induced secretion of Ig in ALL cells could only be measured in significant amounts after a culture time of 7 to 8 days.

Alterations of Metabolic Activity after TPA Exposure

TPA treatment caused a metabolic activation in all types of leukemic cells. This was determined by kinetic measurements of incorporation of [^{3}H]leucine and [^{3}H]galactose, taken as indicators of protein and glycoconjugate synthesis, respectively (Fig. 46.4). Although in some cases an enhanced incorporation of [^{3}H]thymidine was seen (Fig. 46.4), a stimula-

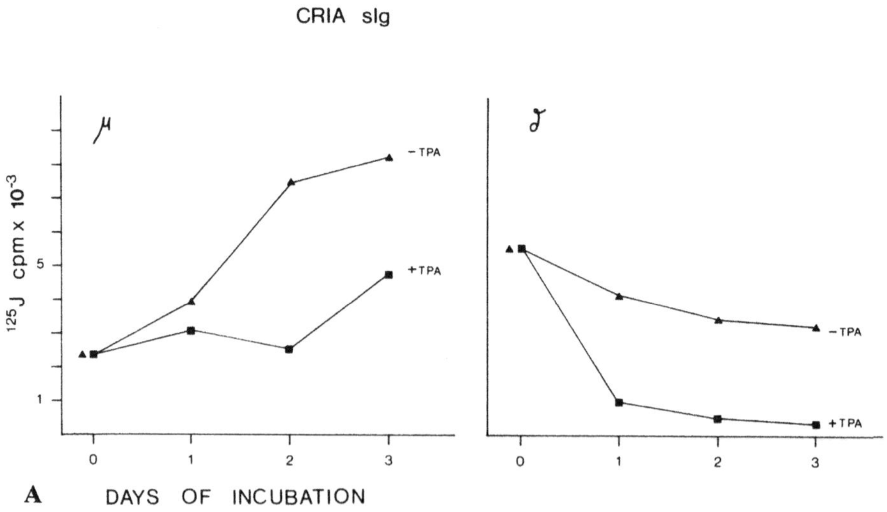

Fig. 46.3. Expression and secretion of immunoglobulin by B-CLL cells after TPA treatment. (A) Surface-expressed Ig was determined in a radioimmunoassay (CRIA) as described in Materials and Methods. (B) Secreted Ig was determined from supernatants of cultured cells by an ELISA technique. (C) Secretion of Ig per 24-hr period was determined by washing cells extensively and culturing them in fresh medium 24 hr prior to determination.

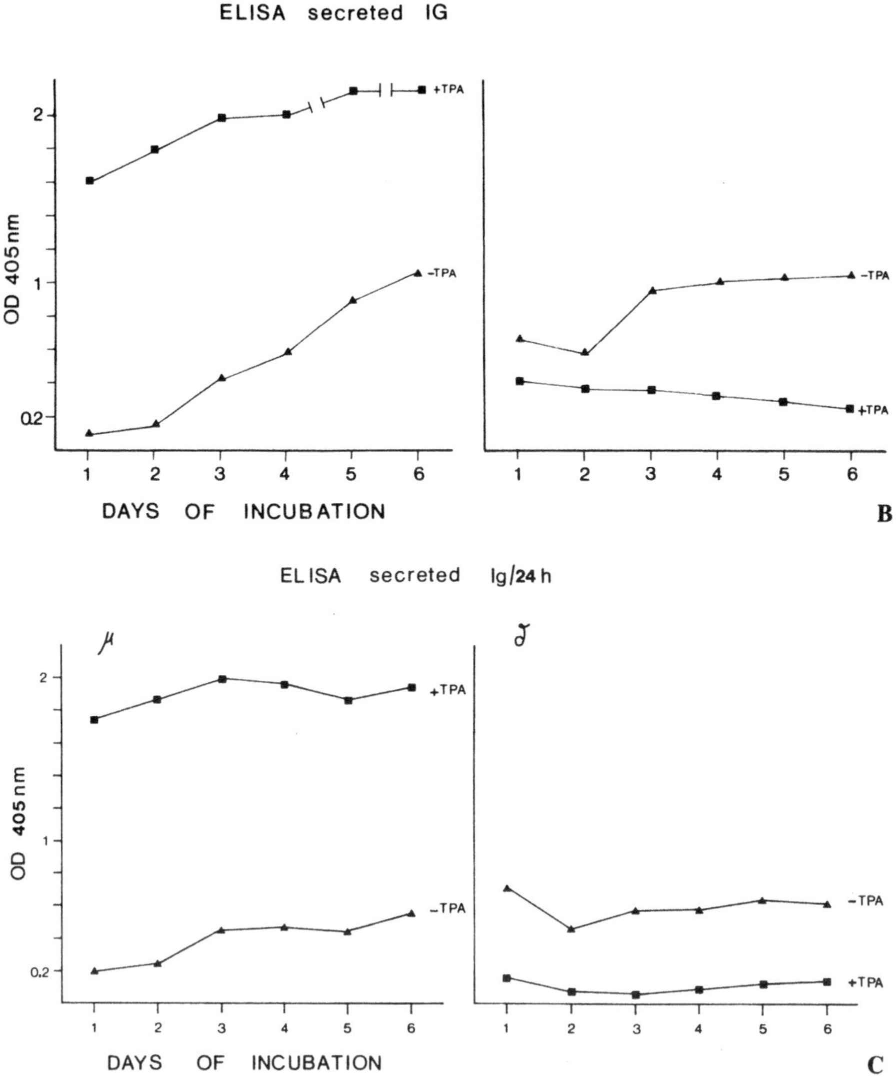

Fig. 46.3. (*Continued*)

tion of DNA synthesis in leukemic cells by TPA was excluded by cell cycle analysis of the respective cells (Fig. 46.5). Over 90% of the analyzed cells consisted of leukemic cells whereas only a small subpopulation of the cells (5% in the case shown) were in the S-phase. It may well be that T cells which could have been present accounted for the TPA-induced DNA proliferation. By means of immunofluorescence staining, 5% of the cells of the presented example were found positive for OKT3.

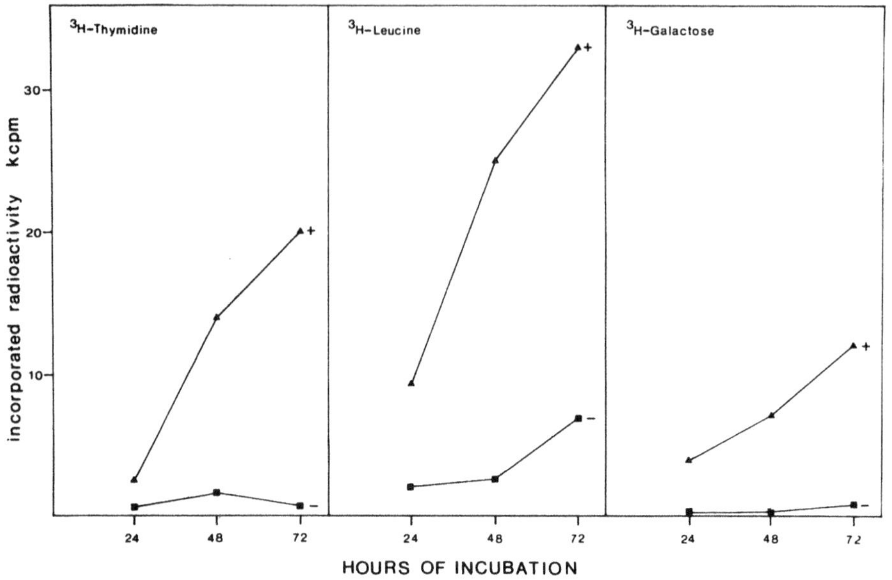

Fig. 46.4. Rates of incorporation of radioactive precursors into TPA-treated B-CLL cells.

Discussion

Four of our monoclonal antibodies against B cell antigens have been shown to be appropriate tools to monitor the TPA-induced differentiation processes of various B type leukemias. These antibodies reacted exclusively with antigens on B cells. Only HD28 showed minor cross-reactions with other lymphoid cells. Thus the observed alterations pointed to a specific effect of TPA on cells of B lineage. Secondly, the monoclonal antibodies recognized cells at different stages of differentiation. The antigens detected by HD6 and HD39 appeared to be restricted to mature and late stages of B cell maturation. HD28 recognized preferentially early stages while HD37 had the broadest reactivity pattern. Detailed descriptions of the monoclonal antibodies are provided in Chapters 3, 7, and 33 of this volume. First experiments with PWM-stimulated normal B cells derived from peripheral blood confirmed the reactivity pattern of the antibodies and substantiated their suitability for *in vitro* differentiation experiments (Table 46.3).

The antibody reaction patterns of TPA-treated leukemias suggest that TPA is able to drive leukemic cells to more mature stages of differentiation. The general direction of alterations in antigen expression during 3 days of TPA treatment is demonstrated in Table 46.4. By reaction kinetics of HD6/HD39 it can be seen that B-CLL cells progress to a more mature

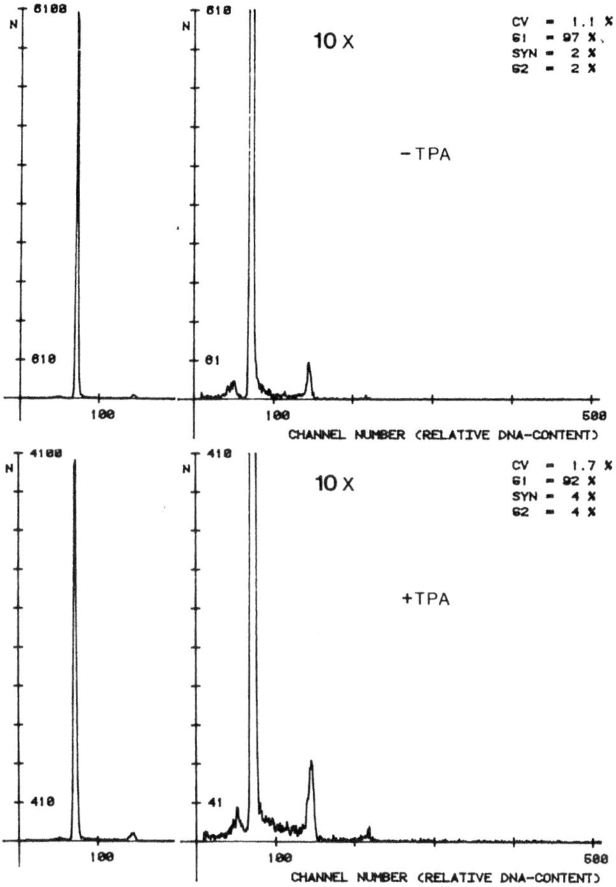

Fig. 46.5. Impulse cytophotometric (ICP) determination of DNA content of TPA-treated B-CLL cells (same cell sample as used for experiments shown in Fig. 46.4). ICP analysis was kindly performed by Mrs. M. Vogt-Schaden.

stage which is represented by B-PLL and HCL cells in terms of antigen composition. In this context it is noteworthy that TPA-treated CLL cells sometimes had a hairy cell-like appearance (cf. Fig. 46.2). The expression of HD6/HD39 antigens on PLL and HCL cells was not altered except in the case of patient Eis (c.f. Table 46.2) whose disease seemed to be at a more progressed stage. It may well be that three-days incubation with TPA is not sufficient to detect alterations in these types of leukemia. B-ALL cells, being arrested at an earlier stage of differentiation than the other leukemias studied, responded to TPA with an increased reaction with HD6/HD39. Monoclonal antibody HD28 showed a reduced reaction with all TPA-stimulated CLL cells; its expression remained, however, unchanged on PLL/HCL cells. It is apparent that the HD28 antigen is not

Table 46.3. Reaction of monoclonal antibodies with pokeweed mitogen (PWM)-stimulated normal B cells separated from peripheral blood.[a]

	Days of culture				
Antibody	0	2	4	6	8
HD6/HD39	+[b]	++	+	−	−
HD28	++	++	++	+	−
HD37	+	+	+	+	+

[a] B cells were separated from peripheral blood consecutively by Ficoll–Hypaque centrifugation, Sephadex G10 filtration, and several E-rosetting procedures. The separated B cells were cultivated in the presence of 1% PWM and irradiated T cells.
[b] Cells were harvested at times indicated and reaction with the monoclonal antibodies was evaluated by indirect immunofluorescence under light microscopy.

restricted to closely defined stages like HD6/HD39. The mechanisms of its reactivity in this kind of *in vitro* stimulation experiments are as yet unknown. HD37 as an internal pan B control antibody remained unchanged in its reactions after TPA treatment. Additionally, we applied the monoclonal antibody B1 described by Nadler *et al.* (10) in our experiments. As already reported (11), the expression of the B1 antigen on CLL cells was reduced after TPA treatment. HLA-DR antigen expression, as detected by a monoclonal antibody against the framework structure of HLA-class II antigens [HD11 (produced by our group], was not altered in our experiments in contrast to the findings of Okamura *et al.* (3). Changes in immunoglobulin expression and secretion observed after TPA stimulation also pointed to a differentiation-like process induced by TPA. Secretion of immunoglobulins reached maximum values already after 24 hr of TPA stimulation. Taking secretion of immunoglobulins as a distinct property of mature plasma cells, the time order of events during TPA stimulation does not seem to be absolutely comparable to that of normal differentiation.

Although it is not clear whether the TPA-induced alterations described reflect a true B cell differentiation process, the changes observed with our panel of monoclonal antibodies occurred in a regular fashion. Hence, *in vitro* activation of normal and malignant cells by TPA or other agents may

Table 46.4. Trends of TPA-induced antigen alterations in leukemia.

Type of leukemia	HD6/HD39	HD28	HD37
CLL	↗	↘	↦
PLL/HCL	↦	↦	↦
ALL	↗	↦	↦

be helpful to characterize the reactivity range of monoclonal antibodies against B cell-specific antigens and also to gain a deeper understanding of the mechanisms regulating the development of leukemia.

Summary

Cell surface changes of peripheral blood lymphocytes from patients with various B-type leukemias after *in vitro* exposure to the phorbol ester TPA were determined by their reaction patterns with monoclonal antibodies raised against B cell leukemias. These monoclonal antibodies recognize B cell differentiation antigens of preferentially late (HD6, HD39) and early stages (HD28). Antibody reactions were measured by indirect immunofluorescence using a cytofluorograph for quantitative evaluation. Cells of the different classes of B cell leukemias reacted uniformly on TPA treatment. During a time course of 3 days all 13 cases of chronic lymphocytic leukemia (B-CLL) examined displayed enhanced expression of the HD6/HD39 antigen and decreased expression of the HD28 antigen; 1 case of acute lymphoblastic leukemia (B-ALL) and 1 case of lymphoid chronic myelogenous leukemia (B-CML) in blast crisis showed only a slight increase in HD6/HD39 expression whereas the relatively high expression of the three antigens in 2 prolymphocytic leukemias (PLL) and 2 hairy cell leukemias (HCL) remained virtually unchanged after TPA addition. For all types of leukemias the strong reaction with a pan B monoclonal antibody, HD37, remained unchanged after TPA stimulation.

In addition to alterations of differentiation antigens, a diminished expression of sIgD, an enhanced secretion of Ig, and an increase of protein and glycoconjugate synthesis were observed in chronic lymphocytic leukemias after TPA treatment. The changes caused by TPA in these parameters suggest that leukemic cells respond to TPA exposure with a differentiation-like process which is correlated in its extent to the stage of developmental arrest of the respective B cell leukemia. Furthermore, the data show that the monoclonal antibodies applied here are useful for the study of B cell differentiation and for the characterization of B cell neoplasms.

References

1. Tötterman, T.H., K. Nilsson, and C. Sundström. 1980. Phorbol ester-induced differentiation of chronic lymphocytic leukemia cells. *Nature* **288:**176.
2. Sugawara, I. 1982. The immunoglobulin production of human peripheral B lymphocytes induced by phorbol myristate acetate. *Cell. Immunol.* **72:**88.
3. Okamura, J., M. Letarte, L.D. Stein, N.H. Sigal, and E.W. Gelfand. 1982. Modulation of chronic lymphocytic leukemia cells by phorbol ester: increase in Ia expression, IgM secretion and MLR stimulatory capacity. *J. Immunol.* **128:**2276.

4. Lindsten, T., C.B. Thompson, F.D. Finkelman, B. Andersson, and I. Scher. 1984. Changes in the expression of B cell surface markers on complement receptor-positive and complement receptor-negative B cells induced by phorbol myristate acetate. *J. Immunol.* **132:**235.
5. Pantazis, P., N. Pavlidis, G.E. Demetrakopoulos, and J.E. Dahlberg. 1982. Morphological changes in cultured human leukemic cells (K 562) treated with the tumor promotor 12-*O*-tetradecanoylphorbol-13-acetate. *Biol. Cell.* **46:**143.
6. LeBien, T.W., F.J. Bollum, W.G. Yasmineh, and J.H. Kersey. 1982. Phorbol ester-induced differentiation of a non-T, non-B leukemic cell line: model for human lymphoid progenitor cell development. *J. Immunol.* **128:**1316.
7. Gordon, J., H. Mellstedt, P. Aman, P. Biberfeld, and G. Klein. 1984. Phenotypic modulation of chronic lymphocytic leukemia cells by phorbol ester: induction of IgM secretion and changes in the expression of B cell-associated surface antigens. *J. Immunol.* **132:**541.
8. Lennert, K. 1978. *Handbuch der speziellen pathologischen Anatomie und Histologie. Malignant lymphomas.* Springer-Verlag, Berlin.
9. Morich, F.-J., F. Momburg, G. Moldenhauer, K.-U. Hartmann, and K.J. Bross. 1983. Immunoperoxidase slide assay (IPSA)—a new screening method for hybridoma supernatants directed against cell surface antigens compared to other binding assays. *Immunobiol.* **164:**192.
10. Nadler, L.M., P. Stashenko, J. Ritz, P. Hardy, J.M. Pesando, and S.F. Schlossman. 1981. A unique cell surface antigen identifying lymphoid malignancies of B cell origin. *J. Clin. Invest.* **67:**134.
11. Stashenko, P., L.M. Nadler, R. Hardy, and S.F. Schlossman. 1981. Expression of cell surface markers after human B lymphocyte activation. *Proc. Natl. Acad. Sci. U.S.A.* **78:**3848.

CHAPTER 47

Functional Studies of p24: Platelet Aggregation Inhibition by Fab Monomers of BA-2

Jo Ellen Brown, James G. White, R.D. Hockett, Jr., Kathleen R. Hagert, and John H. Kersey

Introduction

The cell surface antigen p24, first detected by monoclonal antibody BA-2 (1), has been found to occur in a variety of cells including both normal and leukemic bone marrow lymphohematopoietic progenitor cells (1), leukemia/lymphoma cell lines (2), activated peripheral blood T cells (3,4), myeloid progenitors, and platelets (4,5). Human and non-human renal cells and platelets (5–6,7), smooth muscle and capillaries (5,7), human foreskin fibroblasts (unpublished data), and a variety of epithelial cells (5) also express this antigen. The widespread distribution of this molecule suggests that it may play a vital role in basic cell physiology. To investigate the role of p24 in normal cell functions, we have performed platelet aggregation studies with both intact BA-2 IgG and BA-2 Fab monomers. Similar studies were performed by Boucheix *et al.* (8) using monoclonal antibody ALB_6 which we have known by blocking studies to be directed against the same or adjacent epitope of p24.

Materials and Methods

Monoclonal Antibody Production

BA-2 antibody was produced in mice as previously described (1). Antibody was purified from ultracentrifuged ascites fluid by passage over a Sepharose–protein A column which was washed sequentially with pH 8.0 phosphate buffer and pH 6.0, 5.5, 4.5, and 3.5 citrate buffers as described

(9). The pH 5.0 and 4.5 buffers did not contain azide. BA-2, an IgG_3 antibody, eluted at pH 4.5. Fractions (1.7 ml) were collected to which 0.3-ml 1 *M* Tris base, pH 8.8, was added immediately to raise the fraction pH to 7.3. Protein-containing fractions were pooled. The antibody was then desalted and equilibrated into phosphate-buffered saline (PBS) on a Sephadex G10 column. The pH was adjusted, if necessary, to pH 7.3. ALB_6 was kindly provided by C. Rosenfeld.

W6/32, a monoclonal antibody directed against monomorphic HLA-A and -B determinants (10) was produced in mice and purified in the same manner as described for BA-2.

IgG3 myeloma protein, produced by FLOPC-21 hybridoma, was obtained from Cappel Laboratories (Cooper Biomedical, Inc., Malvern, PA) in purified form.

Fab Monomer Production

Fab monomers of BA-2, W6/32, and IgG3 myeloma protein were produced by a modification of the papain digestion procedure of Porter (11) which was described by Kaye and Janeway (12). Antibody in PBS (0.1–2.0 mg/ml) was diluted 1 : 1 with Sorensen buffer, pH 7.3, with 8 m*M* EDTA. Carboxymethylcellulose beads, coated with papain, were washed two times in Sorensen's buffer to remove borate salts, in which they are stored. Diluted antibody was then added to the beads (0.8 units papain per mg antibody) and β-mercaptoethanol was added to give a final concentration of 10 m*M*. The digestion mixture was rocked on a rocker platform at 37°C for 4–6 hr. Digestion was shown to be complete by SDS–PAGE. Digestion was stopped by centrifuging the mixture at 1000 rpm for 10 min and removing supernatant from beads. β-Mercaptoethanol was removed from the digest and antibody fragments were equilibrated in PBS on a Sephadex G10 column. Fab fragments were separated from Fc fragments by a protein A–Sepharose "mini column." Column volume is 1/2 ml, which can bind 10 mg of IgG. Fab fragments pass through the column when the digest is applied. FC fragments are bound to the protein A and are eluted with pH 3.5 buffer.

Preparation of Column-Washed Platelets

Peripheral blood, collected in heparin, was centrifuged at 120 × *g* for 15–20 min to obtain platelet-rich plasma (PRP). A 50-ml column of Sepharose 2B was washed with 100-ml Tyrode's buffer with 0.1% BSA. PRP was applied to the column, followed by more Tyrode's buffer. When eluate drops became cloudy, washed platelets were collected until drops became clear again. Platelets for iodination and indirect immunofluorescence were prepared this way.

Indirect Immunofluorescence

Column-washed platelets were incubated 30 min at 4°C with BA-2, W6/32 (1 : 100 dilutions of mouse ascites), or a 1 : 100 dilution of IgG3 myeloma protein. After washing twice in PBS-A, platelets were incubated with FITC-conjugated goat anti–mouse antibody for 30 min at 4°C. Platelets were washed three times in PBS-A and examined by fluorescence microscopy.

Radioimmunoprecipitation

Platelets were labeled with ^{125}I by the lactoperoxidase method. Cells were lysed with detergent and lysates were reacted with either BA-2 antibody or IgG3 myeloma protein, followed by incubation with protein A–Sepharose beads. Immunoprecipitates were recovered by centrifugation and immune complexes were dissociated by boiling. Samples were reduced by β-mercaptoethanol and the molecular weight of the precipitated protein was determined by SDS–PAGE followed by autoradiography.

Preparation of Washed Platelets for Aggregation Studies

Whole blood was anti-coagulated with citrate–citric acid–dextrose solution (CCD), pH 6.5, at a ratio of 9 ml blood : 1 ml CCD. PRP was obtained by centrifugation at 200 × *g* for 20 min. PRP was further diluted 1 : 1 with CCD plus adenosine and theophylline, pH 7.3, and centrifuged at 1500 rpm. The supernatant was decanted and the platelet pellet was resuspended in calcium- and magnesium-free Hank's balanced salt solution (HBSS). The resuspended platelets were incubated for 30–60 min at room temperature before use.

Platelet Aggregation

$CaCl_2$ was added back to washed platelets to a final concentration of 1 m*M*. Aliquots of washed platelets were incubated in a Payton aggregometer at 37°C with stirring. Whole antibody at micrograms/ml, Fab fragments at 100 μg/ml, or other aggregating agent was added to platelets and change in light transmission was recorded.

Aggregation Inhibition

$CaCl_2$ was added back to washed platelets to a final concentration of 1 m*M*. Aliquots of platelets were incubated with Fab fragments, at 100 μg/ml and 10 μg/ml, with stirring at 37°C for 2.5 min before one of the following aggregating agents was added: thrombin (0.4 units), collagen (70 μg/ml), ADP (10^{-4} *M*), or calcium ionophore, A23187 (10^{-5} *M*). Changes in light transmissions were recorded.

Results

BA-2, ALB_6 and W6/32 were shown to bind to platelets by indirect immunofluorescence microscopy.

Immunoprecipitation of detergent-lysed platelet membranes was carried out. The molecule precipitated by BA-2 was shown by SDS–PAGE and autoradiography to migrate as a 24-Kd protein indistinguishable from p24 which has been previously observed in leukemia cell membranes (Fig. 47.1).

We studied the effect of antibodies directed against p24 on platelet aggregation. We found that two intact antibodies, BA-2 and ALB_6, induced platelet aggregation; however, a class-matched myeloma protein (FLOPC-21) and a class I HLA antibody (W6/32) did not induce aggregation (Fig. 47.2). Fab monomers of BA-2 did not induce aggregation (Fig. 47.3).

Preincubation of platelets with BA-2 Fab monomers inhibited aggregation induced by thrombin (Fig. 47.4). One hundred μg/ml of Fab was effective in inhibiting aggregation, although 10 μg/ml was not effective. The W6/32 Fab control did not inhibit thrombin-induced aggregation at either concentration (Fig. 47.5). Aggregation by calcium ionophore

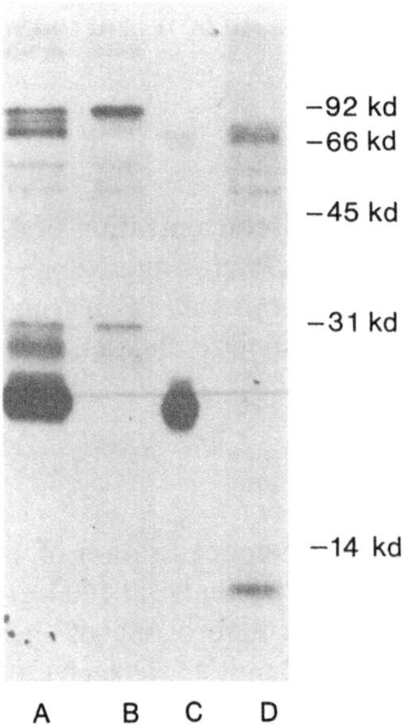

Fig. 47.1. Autoradiograph of p24 radioimmunoprecipitated from platelets. Lanes A and B: Platelet lysate precipitated with BA-2 (A) or control IgG3 myeloma protein (B). Lanes C and D: Nalm-6 (pre–B cell leukemia cell line) lysate precipitated with BA-2 (C) or control IgG3 myeloma protein (D).

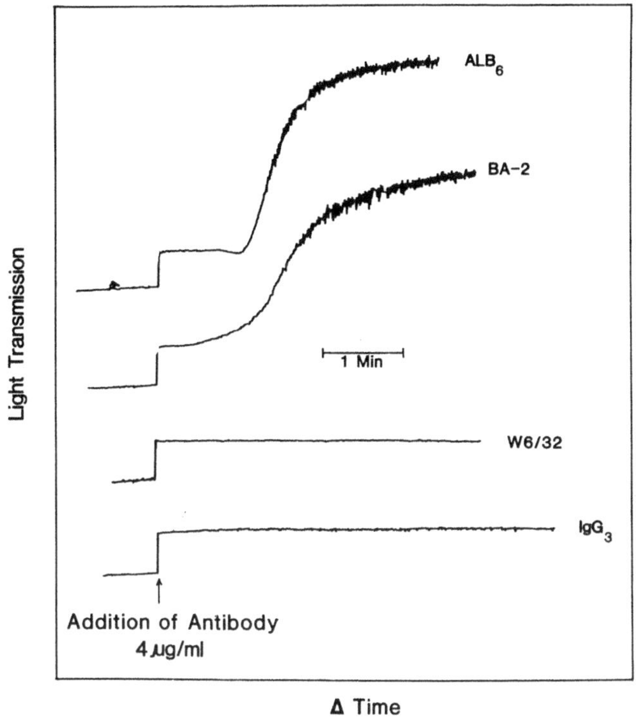

Fig. 47.2. Platelet aggregation using anti-p24 antibodies, ALB_6 and BA-2, and control antibodies W6/32 and IgG3 myeloma protein.

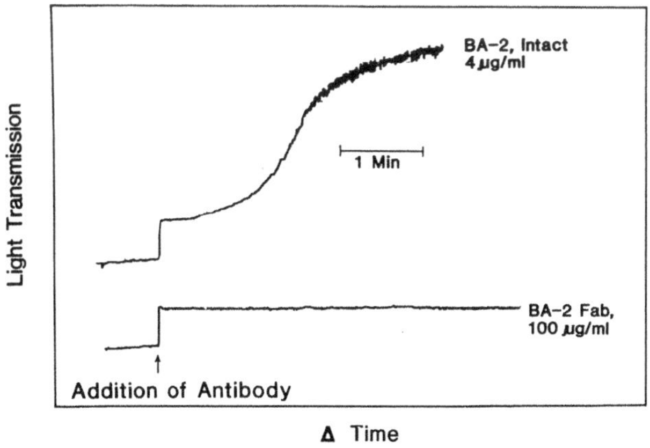

Fig. 47.3. Effect of intact BA-2 and BA-2 Fab monomers on platelets

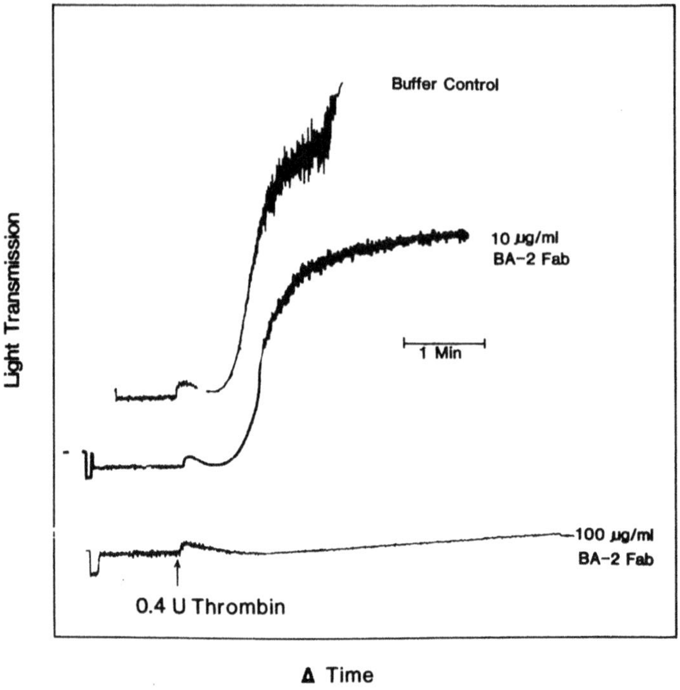

Fig. 47.4. Effect of BA-2 Fab monomers on thrombin-induced platelet aggregation.

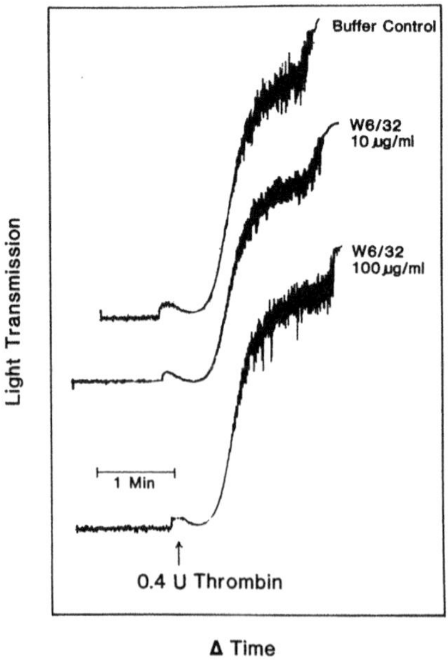

Fig. 47.5. Effect of W6/32 control antibody Fab monomers on thrombin-induced platelet aggregation.

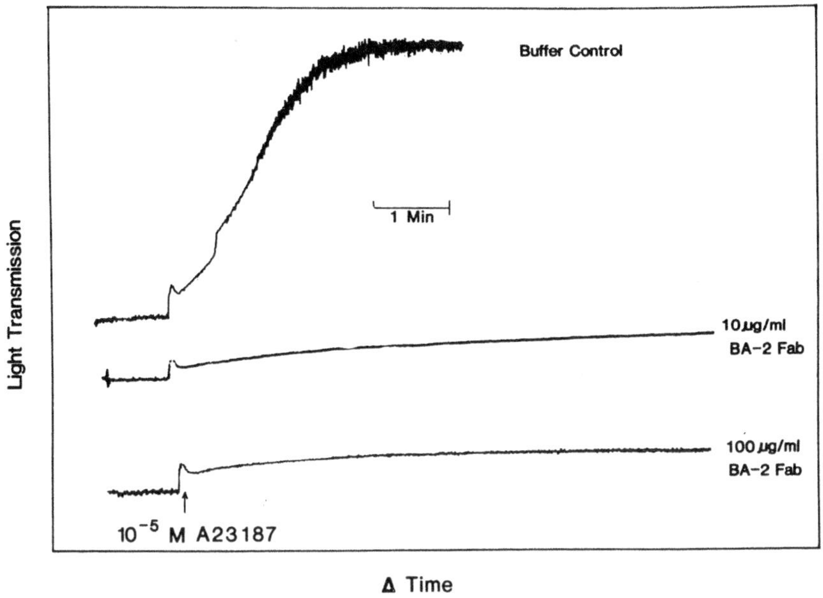

Fig. 47.6. Effect of BA-2 Fab monomers on A23187-induced platelet aggregation.

A23187 was inhibited by BA-2 Fab at both 100-μg/ml and 10-μg/ml concentrations (Fig. 47.6), but not by W6/32 Fab (Fig. 47.7). BA-2 Fab also inhibited aggregation induced by ADP and collagen (results not shown). Aggregation induced by ADP and collagen was also inhibited by control antibody W6/32 Fab.

Discussion

The monoclonal antibody BA-2 binds to the p24 molecule on the surface of a number of cell types including the platelet. Since the platelet can be activated by a number of activating agents, we utilized this system to evaluate the role of p24 in platelet activation.

We first observed that intact BA-2 induced platelet aggregation, whereas control antibodies did not. In order to evaluate the possibility that the BA-2 effects were in part due to cross-linking, we produced Fab monomers of our monoclonal antibodies. In this study of the effects of antibody binding to antigen on platelet membranes without cross-linking, we demonstrated clearly that the BA-2-induced aggregation required cross-linking. Similar results were reported by Boucheix *et al.* using ALB_6, a BA-2-like antibody (8). We also found platelet aggregation inhibition by Fab monomers of BA-2. Our studies demonstrated that Fab monomers of BA-2 inhibit platelet aggregation induction by four aggregating

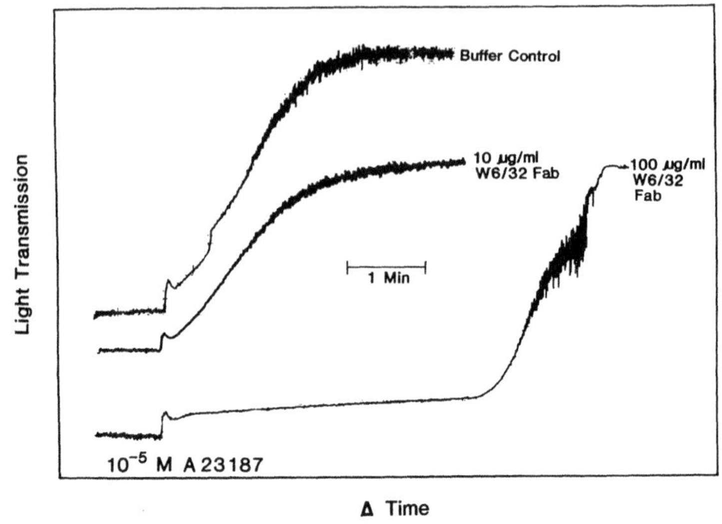

Fig. 47.7. Effect of W6/32 control antibody Fab monomers on A23187-induced platelet aggregation.

agents: thrombin, ADP, collagen, and a calcium ionophore, A23187. Our study included, as a control, Fab monomers of an HLA-class I antibody, W6/32, prepared in the same manner as BA-2 Fab monomers. The control Fab monomers had no effect on aggregation induced by thrombin and A23187, but inhibited aggregation induced by ADP and collagen. A selective inhibition of platelet aggregation had previously been reported for a monoclonal antibody directed against β_2-microglobulin, the light chain of class I HLA antigens (13,14). This antibody, UMR-304, inhibited aggregation by ADP, but not by thrombin (above threshold levels) or A23187.

Our observation that BA-2 Fab monomers, but not W6/32 Fab monomers, inhibit platelet aggregation induced by thrombin and A23187 clearly indicates that the preparation of Fab monomers does not result in nonspecific toxicity. The exact mechanism by which BA-2 Fab monomers inhibit platelet aggregation is unknown. Inhibition of several different agents would seem to indicate that p24, which is blocked by BA-2 Fab monomers, may play an important role in the aggregation cascade, rather than acting as a specific receptor for a particular agent. Evidence for a basic physiologic role is also seen in its ubiquitous nature: it occurs in the membranes of cells with many different functions. Further studies of platelet aggregation and inhibition effects of divalent and monomeric anti-p24 antibodies, as well as studies of other cell types, are necessary to further elucidate the function of p24.

Summary

In the present study we have shown that monoclonal antibody BA-2 binds to a 24,000-dalton molecule on platelet membranes and that this binding induces aggregation of platelets. Univalent antibody, in the form of Fab monomers of the antibody, does not induce aggregation which demonstrates a role for cross-linking in BA-2-induced aggregation. Additionally, Fab monomers of BA-2 inhibit aggregation induced by standard aggregating agents, e.g., ADP, thrombin, collagen, and a calcium ionophore, A23187. The inhibitory effect upon all agents studied suggests that p24 plays an important membrane role in an activation pathway common to many standard aggregating agents.

Acknowledgment. This work was supported in part by grants R01-CA-25097 and HL-11880 from the National Institutes of Health.

References

1. Kersey, J.H., T.W. LeBien, C.S. Abramson, R. Newman, R. Sutherland, and M. Greaves. 1981. p24: A human leukemia-associated and lymph-hemopoietic progenitor cell surface structure identified with monoclonal antibody. *J. Exp. Med.* **153:**726.
2. LeBien, T.W., J.H. Kersey, S. Kakazawa, K. Minato, and J. Minowada. 1982. Analysis of human leukemia/lymphoma cell lines with monoclonal antibodies BA-1, BA-2 and BA-3. *Leuk. Res.* **6:**299.
3. Newman, R.A., D.R. Sutherland, T.W. LeBien, J.H. Kersey, and M.F. Greaves. 1982. Biochemical characterization of leukemia associated antigen p24 defined by the monoclonal antibody BA-2. *Biochim. Biophys. Acta* **701:**318.
4. Hercend, T., L.M. Nadler, J.M. Pesando, E.L. Reinherz, S.F. Schlossman, and J. Ritz. 1981. Expression of a 26,000-dalton glycoprotein on activated human T cells. *Cell. Immunol.* **64:**192.
5. Jones, N.H., M.J. Borowitz, and R.S. Metzgar. 1982. Characterization and distribution of a 24,000 dalton antigen defined by a monoclonal antibody (DU-ALL-1) elicited to common acute lymphoblastic leukemia (c-ALL) cells. *Leuk. Res.* **6:**449.
6. Dowell, B.L., F.L. Tuck, M.J. Borowitz, T.W. LeBien, and R.S. Metzgar. 1984. Phylogenetic distribution of a 24,000 dalton human leukemia-associated antigen on platelets and kidney cells. *Dev. Com. Imm.* **8:**187.
7. Platt, J.L., T.W. LeBien, and A.F. Michael. 1983. Stages of renal ontogenesis identified by monoclonal antibodies reactive with lymphohemopoietic differentiation antigens. *J. Exp. Med.* **157:**155.
8. Boucheix, C., C. Soria, M. Mirshaki, J. Soria, J.-Y. Perrot, M. Fournier, M. Billard, and C. Rosenfeld. 1983. Characteristics of platelet aggregation induced by monoclonal antibody ALB_6 (acute lymphoblastic leukemia antigen p24): inhibition of aggregation by ALB_6 Fab. *FEBS Lett.* **161:**289.

9. Ey, P.L., S.J. Prowse, and C.R. Jenkin. 1978. Isolation of pure IgG1, IgG2a, and IgG2b immunoglobulins from mouse serum using protein A–Sepharose. *Immunochem.* **15:**429.
10. Brodsky, F.M., P. Parham, C.J. Barnstable, M.J. Crumpton, and W.F. Bodner. 1979. Monoclonal antibodies for analysis of the HLA system. *Immunol. Rev.* **47:**3.
11. Porter, R.R. 1959. The hydrolysis of rabbit r-globulin and antibodies with crystalline papain. *Biochem. J.* **73:**119.
12. Kaye, J., and C.A. Janeway. 1984. The Fab fragment of a directly activating monoclonal antibody that precipitates a disulfide-linked heterodimer from a helper T-cell clone blocks activation by either allogeneic Ia or antigen and self-Ia. *J. Exp. Med.* **159:**1397.
13. Vercelotti, G., W. Mullins, R. Curry, G. Gaudernack, C. Moldow, and R.P. Messner. 1982. A unique monoclonal antibody inhibits the second wave of platelet aggregation. *Clin. Immunol. Immunopathol.* **23:**691.
14. Curry, R.A., R.P. Messner, and G.J. Johnson. 1984. Inhibition of platelet aggregation by monoclonal antibody reactive with β_2-microglobulin chain of HLA complex. *Science* **224:**509.

Index